Tuber and Root Crops

Horticulture Science Series - 09

Other Volumes of the Series

HORTICULTURE SCIENCE SERIES

Series Editor : Prof. K.V. Peter

Former Vice Chancellor and Presently Professor of Horticulture
Kerala Agricultural University, Thrissur - 680 656, Kerala

1. Aromatic Plants
   Baby P. Skaria et al.
2. Medicinal Plants
   Kurian Alice et al.
3. Fruit Crops
   Radha T. et al.
4. Vegetable Crops
   Gopalakrishnan, T.R.
5. Spices
   Nybe E.V. et al.
6. Propagation of Horticultural Crops
   Rajan S. et al.
7. Post Harvest Technology of Horticultural Crops
   Sudheer K.P. et al.
8. Commercial Crops Technology
   Kurian Alice et al.
10. Flowers for Trade
    Sheela V.L.

- Management of Horticultural Crops
  Kumar Pradeep et al.
- Ornamental in Garden
  George Sabina
- Coastal Horticulture
  Kanthaswamy, V.
- Basics of Horticulture
  Peter K.V.

Horticulture Science Series Vol. 09

Series Editor:

Prof. K.V. Peter, Former VC and Presently Professor of Horticulture Kerala Agricultural University, Thrissur - 680 656, Kerala

TUBER & ROOT CROPS

Authors

DR. M.S. PALANISWAMI

Project Coordinator (Tuber Crops)
AICRP Tuber Crops, Central Tuber Crops Research Institute
Trivandrum - 695017, Kerala, India

PROF. K.V. PETER

Former VC and *Presently* Professor of Horticulture
Kerala Agricultural University, Thrissur - 680 656, Kerala

New India Publishing Agency

Pitam Pura, New Delhi- 110 088

ISBN 978-81-89422-53-0

Typeset at:

Laxmi Art Creations
Shiva Mkt., Pitam Pura, New Delhi
Cell : 9811482328

Published by:
Sumit Pal Jain *for*

New India Publishing Agency

101, Vikas Surya Plaza, CU Block, L.S.C. Mkt.,
Pitam Pura, New Delhi- 110 088, (INDIA)
Phone: 011-27341717, Fax: 011-27341616
E-mail: newindiapublishingagency@gmail.com
Web: www.bookfactoryindia.com

National Academy of Agricultural Sciences

Prof. M.S. Swaminathan
President

Foreword to the Series

A second green revolution aimed to meet the food requirements of about 1.8 billion people in India by 2010 is on the anvil. It is, however, imperative that the attainment of the food and nutritional security of this vast population should be combined with the preservation of clean environment. Horticulture encompassing fruits, vegetables, ornamentals, plantation crops, spices, tuber crops, medicinal and aromatic plants, besides harvest and post harvest technologies has a special role in providing food and nutritional security, rural poverty alleviation, employment generation as well as improving the quality of life in our country sides.

India, incidentally, is the world's leading producer of fruits and is followed by Brazil; in vegetables it occupies the second place after China. India is also the land of spices producing an array of such crops. Indeed, the varied eco-climatic conditions prevailing in different parts of this country make it possible to grow a disparate range of fruits, vegetables and ornamental plants at any time of the year.

With the opening up of the world markets under the WTO regime, however, international competition for Indian horticultural products has reached unprecedented levels, necessitating strategies for quality control, branding and product development. In particular, production of organic vegetables, fruits, cut flowers, pepper, ginger, tea, coffee, rubber and cardamom has of late become demand-driven.

Among the inputs for higher productivity in horticultural crops, seeds and quality planting materials account for 20 to 30% of the potential productivity. Transgenic crops are a reality now; yet the technology for production of transgenics is a far cry in many situations. Furthermore, water is a limiting factor in several environments. Hence, water saving crops and those with high water use efficiency should be preferred

NASC Complex
DPS Marg, P.O. Pusa
New Delhi - 110 012

Tel. 91-11-25846055
91-11-25846051
91-11-25846052
91-11-25846053
Fax 91-11-25846054
E-mail naas@vsnl.com
Website
//education.vsnl.com/naas
P.O. Box No. 11325

over other crops and management practices. Likewise, the use of biodegradable plastics, biological control of pests and diseases, and new product development are receiving increasing attention, especially in the arena of horticultural crop management. Marketing, village knowledge centres, packaging industry, zero-energy cool chambers and a variety of economically viable technologies, however, remain to be streamlined.

The present 12-volume series on **Horticulture Sciences,** edited by Professor K.V. Peter, Vice-Chancellor, Kerala Agricultural University, is a good compilation of the scientific knowledge and experience gained over the past 30 years in this important sector of Indian economy. Each volume in the series is authored by accomplished researchers such as Dr. T. Radha and Dr Lila Mathew (*Fruit Crops*), Dr. T.R. Gopalakrishnan (*Vegetable Crops*), Dr. P.K. Valsalakumari (*Flowering Trees*), Dr. P.K. Rajeevan (*Garden Plants*), Dr. Alice Kurian and Dr. K.V. Peter (*Commercial Crops Technology*), Dr. Alice Kurian and Dr. M. Asha Sankar (*Medicinal Plants*), Dr. Baby P. Skaria (*Aromatic Plants*), Dr. M. S. Palaniswami (*Tuber and Root Crops*), Dr. E.V. Nybe and Dr N. Mini Raj (*Spices*), Dr. S. Rajan and Dr Baby Lizy Markose (*Propagation of Horticultural Crops*), Dr. K.P. Sudheer and Dr. V. Indira (*Post Harvest Technology of Horticultural Crops*), and Professor K.V. Peter (*Basics of Horticulture*). The New India Publishing Agency, New Delhi deserves special credit for publishing this useful series, especially in view of the fact that the Government of India is currently implementing the National Horticultural Mission.

Prof. M.S. Swaminathan, FNA, FRS
President, National Academy of Agricultural Sciences
and Chairman, National Commission on Farmers
New Delhi 110 012

भाकृ अनुप
ICAR

Indian Council of Agricultural Research
Office of the Secretary, DARE & Director General, ICAR
Krishi Bhavan, New Delhi - 110 001

Foreword

The tropical climatic conditions, especially in the southern, eastern and North eastern states, offer a congenial environment for cultivation of root and tuber crops in our country. These crops are also called as the energy banks of nature and can flourish even in marginal land.

Given the projected importance of tuber and root crops in developing countries, it is essential to treat these crops as an integral component of the overall strategy for sustained food and nutritional security and alleviation of poverty, hunger and malnutrition. Besides being a source of food, these underground Treasures offer tremendous opportunities for entrepreneurial initiatives, employment generation and income from a wide range of value added foods and feeds, bio-fuels, pharmaceuticals, nutraceuticals, industrial starches, biodegradable plastics and many other products.

The publication in its present form is a comprehensive and an updated compilation of all relevant information on various aspects of tuber and root crops with special reference to India. This book broadly touches on history, botany, biodiversity, nutraceutical and pharmaceutical values, improved varieties, production aspects, protection techniques, utilization and post harvest technologies, tuber crops starches, biotechnological approaches and future impetus in research and development of tropical root and tuber crops.

The authors, Dr. MS Palaniswami. Project Coordinator, All India Coordinated Research Project on Tuber Crops, CTCRI, Sreekaryam, Trivandrum and Prof K.V. Peter, Former Vice-Chancellor, Kerala Agricultural University deserve special appreciation for their concerted efforts. It is hoped that this publication will prove to be a valuable reference source to all those interested in tropical root and tuber crops research and development.

(MANGALA RAI)
Secretary, Department of Agricultural
Research & Education and
Director General, Indian Council of Agricultural Research
Dr. Rajendra Prasad Road, Krishi Bhawan
New Delhi 110 001, India

December 1, 2006

Preface

A short fall of 41% in food production is envisaged in India by 2020.Growing tuber and root crops and enhancing their productivity are options to meet the calorie needs of about one fifth of worlds population. Among 30 genera producing edible roots and tubers; cassava, sweet potato, aroids (taro, swamp taro, elephant foot yam and tannia), yams (lesser yam, greater yam and aerial yam) and minor crops like yam bean, country potato (coleus) and arrow root are commonly cultivated. As availability of energy will also be a limiting factor as coal and petroleum based fuels get depleted and are non-renewable. Ethanol from starch from tuber and root crops will meet energy requirement to a considerable extent. Tuber and root crops based starch is the raw material for innumerable products in bakery, confectionary, beverages, drug industry, animals, bird and fish feeds; fermentation industry, textiles, paper and sanitary napkins, pulp industry and even in warfare.

Tuber and root crops are sure bets against famine and cereal food shortages. During 1950s, when India faced cereal shortage, it was tuber crops on which millions depended. There is considerable progress in research and development of tuber crops. The Central Tuber Crops Research Institute was established in July 1963 at Sreekaryam,Trivandrum. An exclusive All India Co-ordinated Research Project on Tuber Crops started functioning during IVth plan. An exclusive Horticultural and Forestry University Dr.Y.S. Parmar University of Horticulture and Forestry, Solan- was started. Out of 40 State Agricultural Universities in India, 35 Universities have seperate Divisions of Horticulture and Agronomy, where root and tuber crops are researched, taught and extended. Council of Scientific and Industrial Research (CSIR) and its constituent institutes like Central Indian Medicinal and Aromatic Plants (CIMAP) Lucknow and Regional Research Laboratory, Trivandrum have research programmes in tuber crops. The ICAR, DBT, DST, DOEF, DOE, ICMR *etc.* fund R and D projects in root and tuber crops.At International level,International Institute of Tropical Agriculture,Ibadan, Nigeria MARDI, Kualalumpur, Malaysia, CIP, Lima, Peru; Embrapa, Brazil and CIAT, Cali Columbia, conduct scientific research on tuber and root crops.

The present book Tuber and Root Crops covers in 17 chapters, 30 plant genera consisting of edible roots and tubers in its totality. Plant biotechnology, tissue cuture, new products development,processing equipments, storage and manifold uses of starch and management technologies are dealt with. Authoritative statistics on root and tuber crops are provided as a ready reckoner to researchers and planners. The book is authored by the leading authority on the subject, Dr.M.S. Palaniswami, Co-ordinator, All India Co-ordinated Research Project on Tuber Crops, ICAR,CTCRI,Trivandrum and edited by Dr. K. V. Peter, Professor of Horticulture and Former Vice-Chancellor, Kerala Agricultural University.

Authors

Contents

1

Introduction

As per the projection of International Food Policy Research Institute, Washington, there is likelihood of shortfall of 41% in food production in India by 2020 and need to produce 300mt of food by 2020. India is the seventh largest country in the world with a population of more than 1.03 billion and with 1.93 population growth rate. To overcome such problems we have to go for horizontal or vertical increase in production. As the cultivable area and water resources are limited to bring under cropping, except by way of increasing the cropping intensity/cropping system, horizontal increase has little scope. While the vertical increase in production has tremendous scope with better value added varieties and farmer friendly sustainable management and production technologies.

Tropical Root and Tuber Crops are known as the energy banks of nature serving either as primary or secondary staple to meet the calorie needs of about one fifth of world's population. These crops have myriad and complex roles to play in the food security and reduction of hunger and elimination of poverty. Under angiosperms there are about 18 families which have more than 30 genera producing edible roots and tubers. Economically important tropical root and tuber producing species commonly used for food, feed, medicine and in industries are enunciated in this compendium. Among more than 25 species of root and tuber crops, cassava, sweet potato, aroids (taro, bunda, swamp taro, elephant foot yam and tannia), yams (lesser yam, greater yam, white yam and aerial yam) and minor tuber crops such as yam bean, country potato (coleus) and arrow root are commonly cultivated (Table 1).

Table 1. Tropical Tuber and Root Crops under cultivation

Crop	Scientific name	Family	Edible part
Cassava	*Manihot esculenta* Crantz	Euphorbiaceae	Root, leaves
Sweet potato	*Ipomoea batatas* (L.) Lam	Convolvulaceae	Root, leaves
Greater yam	*Dioscorea alata* L.	Dioscoreaceae	Tuber
White yam	*Dioscorea rotundata* Poir	Dioscoreaceae	Tuber
Lesser yam	*Dioscorea esculenta* (Lour.) Burls	Dioscoreaceae	Tuber
Aerial yam	*Dioscorea bulbifera* L.	Dioscoreaceae	Tuber, bulbils
Intoxicating yam	*Dioscorea hispida Dennst.*	Dioscoreaceae	Tuber
Yam	*Dioscorea floribunda Mart.*	Dioscoreaceae	Tuber
Yellow yam	*Dioscorea cayenensis Lam.*	Dioscoreaceae	Tuber
Yam	*Dioscorea pentaphylla L.*	Dioscoreaceae	Tuber, bulbils
Cush-cush yam	*Dioscorea trifida L.*	Dioscoreaceae	Tuber
Arvi	*Colocasia esculenta* var. antiquorum (L.) Schott	Araceae	Corm, cormels, leaves, petiole
Bunda	*Colocasia esculenta* var. esculenta (L.) Schott	Araceae	Corm, leaves
Swamp taro	*Colocasia esculenta* var. stoloniferum (L.) Schott	Araceae	Stolon
Tannia	*Xanthosoma sagittifolium* (L.) Schott	Araceae	Corm, Leaves, Petiole
Elephant foot yam	*Amorphophallus paeoniifolius* (Dennst.) Nicolson	Araceae	Corm, pseudo-stem
Giant taro	*Alocasia macrorrhizos* (L.) G.Donf	Araceae	Corm
Giant swamp taro	*Cyrtosperma chamissionis* (Schott) Merr.	Araceae	Corm
Country potato	*Solenostemon rotundifolius* (Poir) J.K. Morton	Labiatae	Tuber
Arrowroot	*Maranta arundinacea* L	Marantaceae	Rhizome
Canna	*Canna edulis* Ker-Gawler	Cannaceae	Rhizome
Winged bean	*Psophocarpus tetragonolobus* (L.) DC.	Leguminoceae	Tuber, young pods
Yam bean	*Pachyrhizus erosus* (L.) Urban *P. ahipa, P. tuberosus Lam.*	Leguminoceae	Root
Vigna	*Vigna capensis* Walp	Leguminoceae	Root
Typhonium	*Typhonium* spp.	Araceae	Corm
Costus	*Costus speciosus* (Koenig) Sm.	Costaceae	Tuber
Shoti	*Curcuma zedoaria* (Berg.) Roscoe.	Zingiberaceae	Rhizome
Jerusalem artichoke	*Helianthus tuberosus L.*	Compositeae	Tuber

Roots and tubers will be many things to many people by 2020. Their adaptation to marginal environments, their contribution to household food security, and their great flexibility in mixed farming systems make them an important component of strategies to help to improve the welfare of the people especially rural and semi urban population. At the same time, they will link small holder farmers with emerging markets, providing a diversified range of high quality, competitive products for food, feed and industry. The starch content, pharmaceutical and nutraceutical value endow these crops with an extra ordinary range of potential and uses.

Among other things, farm households see the value of roots and tubers in their ability to produce higher edible energy production per hectare per day than other commodities. Yams give 182 MJ ha^{-1} day^{-1}, sweet potato give 152, while in cassava it is 121 which is equivalent to Rice (Horton and Fano, 1985). Root and tubers have the capacity to out produce cereals in terms of quantities of edible energy harvested per unit area per unit time. What is more important is that they produce under conditions where other crops may fail.

The projections for roots and tubers for 2020 using the International Food Policy Research Institute's IMPACT (International model for policy analysis of Agricultural Commodities and Trade) model is estimated at 10.5 percent share of the total economic value of all the major food commodities in the global food system, including cereals, soybean, and meat products and it is 18.6 percent share in cereals, roots and tubers and soybean (Scott *et al.,* 2000).

The projected growth rates in output are particularly strong for potato (2.7 percent/year) and yam (2.9 percent/year). Production of cassava and sweet potato will expand at a more modest pace of 1.95 percent and 1.0 percent per year respectively. While these projected growth rates may appear high, they actually represent a considerable slowdown compared to recent rates of expansion for these crops. Nevertheless, future growth rates calculated for cassava, potato and yam exceed those estimated for rice and wheat (Scott *et al.,* 2000).

The versatility of the root and tuber crops will remain an enduring attraction for producers and consumers. Cassava, sweetpotato and other roots and tubers such as canna will be used more and more in processed form for food, feed and starch-derived products. Expansion in non-food, non-feed uses will depend on multidisciplinary research (in breeding, biochemistry, and socioeconomics, for instance) to enhance key characteristics and lower the crops cost as a source of raw material.

Breeding improvements, more technically and economically efficient procedures for producing raw material and finished products and strengthened grower-to-processor linkages will be fundamental in achieving the full growth of these crops. In India with a vast biodiversity value-added germplasm repositories of cassava, aroids, sweet potato and yam as well as collections of several other root and tuber crops, which will provide the genes to the year 2020 and beyond that, can be used to stabilize /increase the yield and improve the quality of these crops. Given the projected importance of roots and tubers in developing countries to 2020 and ahead, it is fundamental to retain these crops as an integral part of a global strategy to increase food production and utilization in Asia, Africa, and Latin America in the decades to come (Scott *et al.,* 2000).

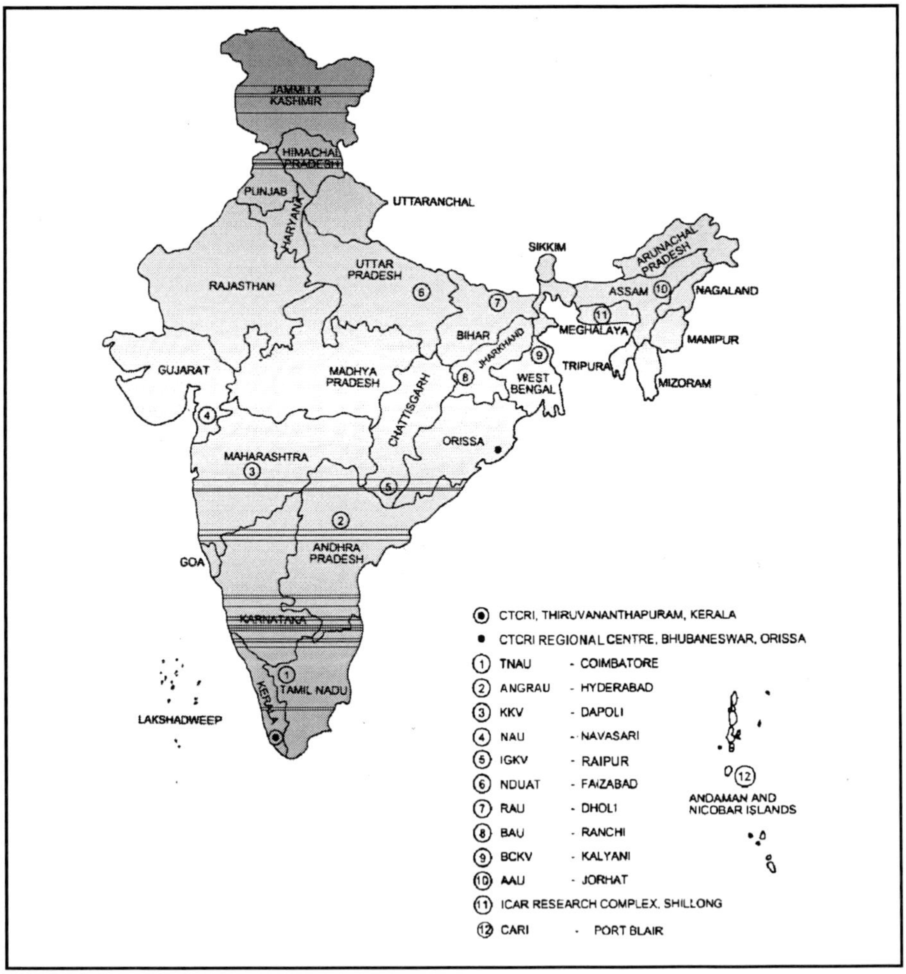

Fig. 1. Major Tuber crops research centers in India

Cassava

Nigeria is the country with maximum cassava cultivation and production and it is closely followed by Brazil and Congo. India ranks 17th in area but 9th in production, but India has the highest productivity in the world. The top 20 cassava growing countries are given in the table-2 with its production and productivity.

In India, cassava is being cultivated on an area of 0.24 m ha, with a production of 6.7 million tonnes. Cassava has been an important crop in South India as a subsidiary food as well as an industrial raw material especially in Kerala, Tamil Nadu and Andhra Pradesh. With the changes in the dietary habits, industrial uses of cassava, particularly in sago, textile starch, gelling agent, pharmaceutical fillers, capsules, biodegradable plastics, alcohol production, feed industries etc., are being exploited. In Thailand, starch is employed to produce monosodium glutamate and plywood.

More than 80% of the area under cassava cultivation is in Kerala and Tamil Nadu. Now it is slowly spreading to Andhra Pradesh, Maharashtra, Meghalaya, Assam, Sikkim, Mizoram, Karnataka and Rajasthan states of India (Table 3, Fig.1). The area of cassava cultivation in India has come down from 0.368 m ha in 1973-74 to 0.207 m ha in 2002-2003. Though there was decline of 0.161 m ha, the total production had more or less maintained status quo with the result the productivity had increased from 17 to 26 t ha^{-1}. During the last 25 years (1980-81 to 2004-2005) though the total area of cultivation in India was reduced by 33% under cassava, the total production increased by 15% and productivity increased by 52% during the period from 1980-2005 (Fig 2) (Anonymous, 2005). Similarly, in Kerala also the cassava cultivation had come down from 0.13 m ha to 0.104 m ha, during the last 10 years. However the production was more or less maintaining same, ranging only 23 to 26 lakhs tonnes (Fig. 3). In Tamil Nadu, 67,951 ha is cultivated mostly concentrated in Salem, Dharmapuri, Perambalur and Kanyakumari districts. The mean productivity of cassava with irrigation in Tamil Nadu is 31.59 t ha^{-1} which is the world's highest (Table 4). In Andhra Pradesh, the cassava cultivation is largely restricted to East Godavari district (20,000 ha) where it is cultivated on a commercial scale as rainfed crop (Table 5).

Table 2. Major Cassava, Sweet potato, Taro and Yam growing countries in the world with its area, production and productivity

Sl. No.	Country	Area (m ha1)	Production (mt)	Productivity (t ha^{-1})
Cassava				
01	Nigeria	4.118	38.2	9.2
02	Brazil	1.933	26.6	13.7
03	Congo	1.845	15.0	8.1
04	Indonesia	1.223	19.5	15.9
05	Mozambique	1.050	6.1	5.8
06	Thailand	0.985	16.9	17.2
07	Ghana	0.783	9.7	12.4
08	Angola	0.748	8.6	11.4
09	Tanzania	0.670	7.0	10.4
10	Uganda	0.407	5.5	13.5
11	Vietnam	0.390	5.7	14.6
12	Madagascar	0.352	2.2	6.2
13	Paraguay	0.310	4.9	15.8
14	Cote D' Ivoire	0.300	1.5	5.0
15	Guinea	0.270	1.3	5.0
16	China	0.250	4.2	16.8
17	India	0.240	6.7	27.9
18	Benin	0.240	3.1	12.9
19	Philippines	0.204	1.6	8.0
20	Cameroon	0.200	1.2	6.0
Sweet Potato				
01	China	5.008	107.2	21.4
02	Uganda	0.602	2.6	4.4
03	Nigeria	0.516	2.5	4.9
04	Tanzania	0.500	1.0	2.0
05	Indonesia	0.176	1.8	10.4
06	Rawanda	0.148	0.9	5.9
07	Angola	0.143	0.6	4.5
08	Burundi	0.125	0.8	6.6
09	Philippines	0.120	0.5	4.5
10	Madagascar	0.105	0.5	5.1
11	Papua New Guinea	0.104	0.5	5.0

12	India	0.100	0.9	9.0
13	Cuba	0.073	0.5	6.7
14	Kenya	0.061	0.6	9.5
15	Japan	0.041	1.1	25.6
16	Egypt	0.0010	0.30	30.0
17	Israel	0.0006	0.02	34.9
Taro				
01	Nigeria	0.735	4.027	5.47
02	Ghana	0.270	1.800	6.67
03	Cote D' Ivoire	0.265	0.370	1.40
04	Cameroon	0.200	1.100	5.50
05	China	0.092	1.639	17.72
06	Papua New Guinea	0.040	0.260	6.50
07	Madagascar	0.030	0.200	6.67
08	Japan	0.017	0.184	10.87
09	Philippines	0.019	0.102	5.37
10	Egypt	0.003	0.100	29.41
Yams				
01	Nigeria	3.106	26.587	8.6
02	Ghana	0.310	3.892	12.5
03	Cote D' Ivoire	0.310	3.000	9.7
04	Benin	0.172	2.257	13.0
05	Ethiopia	0.174	0.310	4.2
06	Sudan	0.057	0.137	2.4
07	Togo	0.055	0.570	10.3
08	Central Africa	0.058	0.350	6.0
09	Japan	0.009	0.175	19.9
10	Philippines	0.006	0.029	4.8
11	Solomon Islands	0.0013	0.003	22.3

Source : FAO Stat Data 2005

Sweet Potato

China is in the forefront with respect to sweet potato cultivation (5.008 m ha) and production (107.2 m t) among the sweet potato growing countries. The major sweet potato producing countries are listed in table-2 with its area, production and productivity. The highest productivity was noted in Israel (34.9 t ha^{-1}) followed by Egypt (30.0 t ha^{-1}) and Japan (25.6 t ha^{-1}). However the total area of cultivation under sweet potato in these countries is rather low (Table 2). China produces noodles, sorbitol and manitol from sweet potato starch. Sweet potato cultivation in India is largely in Orissa, West Bengal, Uttar Pradesh, Assam, Bihar, Jharkhand, Madhya Pradesh, Meghalaya, Maharashtra and Karnataka. About 80% of the total sweet potato growing area is occupied by Orissa, West Bengal and Uttar Pradesh (Table 2). During the last 10 years, the area of cultivation under sweet potato has declined from 1,40,700 ha during 95-96 to 1,00,000 ha during 2004-05. But at the same time the productivity has increased from 8 to 9 t ha^{-1}. In Orissa, Bolangir, Koraput, Baragarh, Ganjam, Dhenkanal, Koenjhar and Sundergadh districts cultivate sweet potato in larger areas as rainfed crop (Table 6). In Uttar Pradesh, Etah, Badaur, Farookabad and Sultanpur are largely involved in sweet potato cultivation. Sweet potato is cultivated in 6900 ha with a productivity of 8 t ha^{-1} in Madhya Pradesh. In Maharashtra, sweet potato is grown in an area of 4700 ha. Its large scale cultivation has been reported to be restricted to Kolhapur division where the crop is taken on an area of 2800ha. In Konkan region of Maharashtra, 200 ha are under sweet potato cultivation with an average yield of 13.5 t ha^{-1}. In Karnataka, sweet potato is largely grown in Belgaum, Dharwad, Bijapur, Uttar Kannada and parts of Raichur. There was a sharp decline in the area under cultivation in the country to half and 40% reduction in total production during 1980-1981 to 2004-2005. However the productivity had increased by 28.6% (Fig. 4).

Andaman and Nicobar Islands popularly known as Emerald Isles have the archipelago of 572 islands with varied topography and vegetation and accounts for exceptionally high endemism. About one thousand hectares are under sweet potato, *Colocasia*, elephant foot yam and yam, mostly as intercrops in coconut and arecanut gardens.

Aroid and Yams

Among the important taro and yam growing countries in the world, the highest productivity in taro was observed in Egypt while in yams, the maximum productivity was recorded in Solomon Islands (Table 2). Cultivation of aroids and yams is spread all over India. In Tamil Nadu, about 900 ha each of taro and elephant foot yam is being cultivated and the productivity is 10 and 30 t ha^{-1} respectively (Table 4).

Taro and elephant foot yam are largely grown in Uttar Pradesh, West Bengal, Bihar, Chhattisgarh, Jharkhand, Maharashtra, Gujarat, Andhra Pradesh and to a certain extent in Tamil Nadu and Kerala. In the Konkan region of Maharashtra, 110 ha of area was under *Xanthosoma* and *Colocasia* cultivation with an average productivity of 5.5 t ha^{-1} during the year 2000.Yams are also cultivated in the Konkan part of Maharashtra on an area of 61 ha.

In Madhya Pradesh, of the 11 agro-climatic zones, tropical root and tuber crops (taro, elephant foot yam and yams) are grown in three agro-climatic zones of the state namely Bundelkhand region, Kymore plateau and Satpura Hills and Northern Hills. Tikangarh, Datia, Chhatarpur and Guna districts are known for tuber crops cultivation. In Orissa, aroids and yams are cultivated in Ganjam, Dhenkanal, Koenjhar, Mayurbhanj and Balasore. Taro cormel preparations are given as offering in the temple of Puri. In Rajasthan, Banswara, Bhilwara, Bundi, Chittor, Dundarpur, Kota and Udaipur are major tuber crops (aroids and other tubers) growing areas on 2381 ha.

Other tuber crops

Country potato is largely cultivated in Tamil Nadu, Kerala and Karnataka. Swamp taro cultivation is mostly found in West Bengal and North Eastern States of India. Yam bean is cultivated in Bihar, Jharkhand, Chhattisgarh, Uttar Pradesh and West Bengal.

More than 840 million people do not have enough food to meet their basic daily energy needs. Far more - an estimated 3 billion - suffer the insidious effects of micronutrient deficiencies leading to hidden hunger. Current efforts to combat micronutrient malnutrition in the developing world focus on providing vitamin and mineral supplements for women and young children. In regions with adequate infrastructure and well-established markets for food processing and delivery, food fortification has greatly improved the micronutrient intake of vulnerable populations, particularly the urban poor. Unfortunately, there are limits to commercial fortification and supplementation. In the developing countries, consumption of commercially fortified foods is rather minimal. Furthermore, the recurrent costs associated with supplementation and commercial fortification are significant. By conservative estimates, providing vitamin A supplements and iron fortified foods to one half of those in need would cost $ 100 million every year in South Asia alone. Biofortification is a new paradigm shift for agriculture and a tool for improved human health and nutrition.

New breed of ultra-nourishing crops capable of alleviating malnutrition in even the most hard-to-reach populations has been explored in tropical root and tuber crops for vitamin A and minerals. These crops would need no commercial fortification and could be grown on family plots throughout the developing world. It is now possible to breed plants for increased vitamin and mineral content, making "biofortified" crops one of the most promising new tools in the fight to end malnutrition and save lives. Such roots and tubers can play a vital role in the mid day meal in developing countries. The introduction of biofortified crops - varieties bred for increased mineral and vitamin content - would complement existing nutrition approaches by offering a sustainable and low cost way to reach people with poor access to formal markets or health care systems. Biofortification can provide ongoing benefits throughout the developing world at a fraction of the recurring cost of either supplementation or post-production fortification. The bio-fortification approach backed by sound science makes sense as part of an integrated food systems approach in reducing micronutrient malnutrition and targets the poor people, uses built-in delivery mechanisms and is scientifically feasible and cost effective, in enabling rural households to improve family health and nutrition in sustainable ways.

Root and tuber crops are also greater sources of industrial raw materials for biodiesel and starch derived materials. Toyota Group Company, Tokyo in their strive for better tomorrow, are engaged in bioplastics development using Carbohydrates obtained from sweet potato cultivated on a larger scale and they indicated that this new plant derived material may save the mother earth.

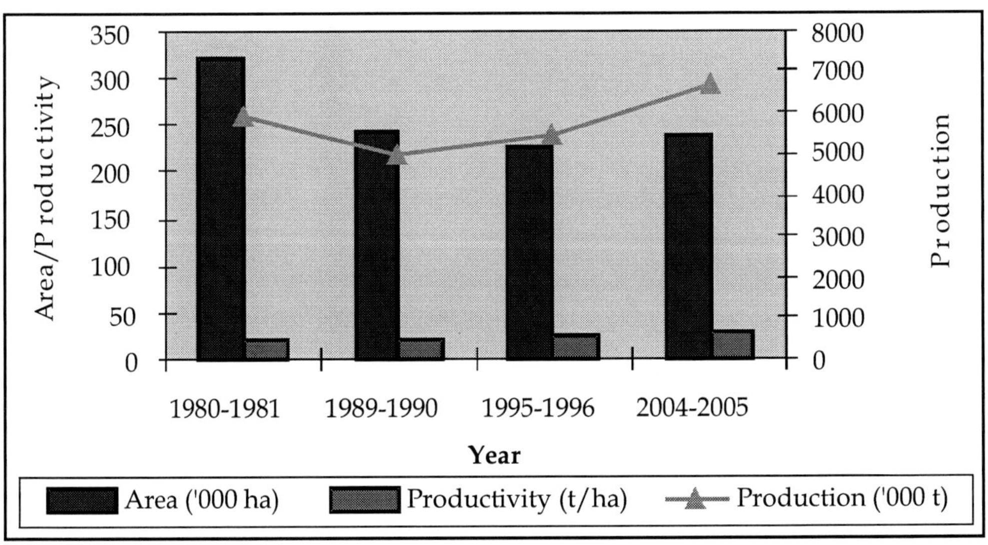

Fig. 2. Status of Cassava in India (1980-2005) *Source :* Anonymous 2005

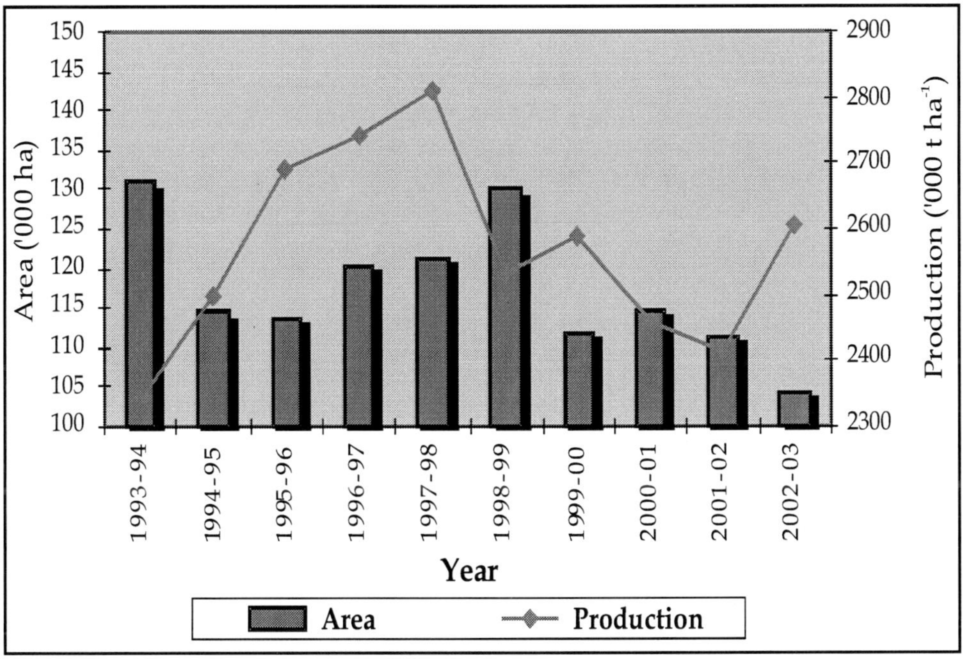

Fig. 3. Area and production of cassava in Kerala. *Source* : Anonymous 2005

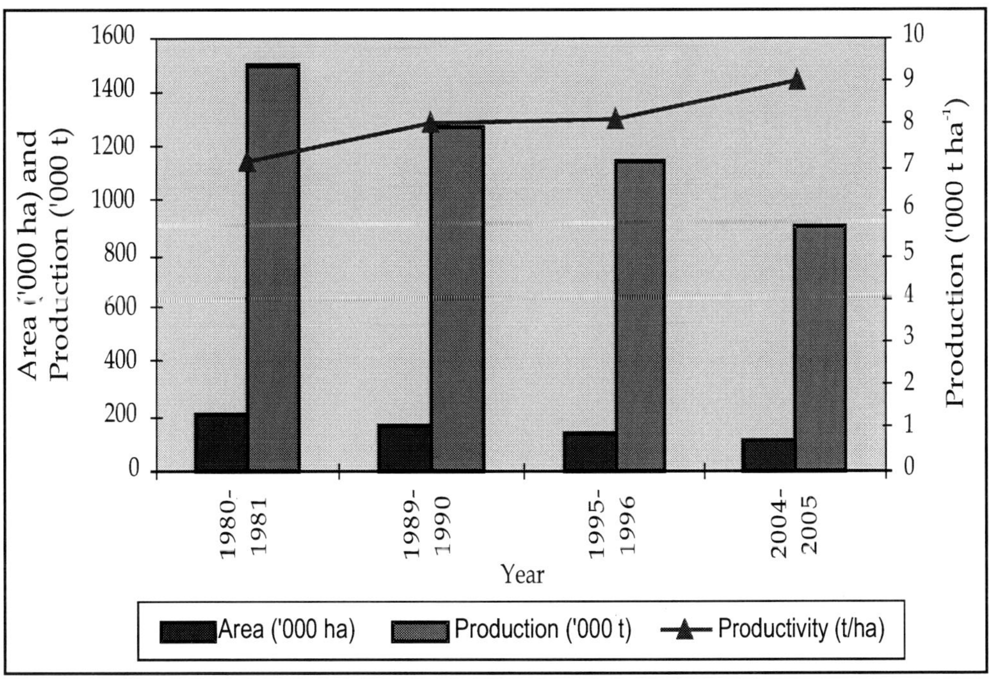

Fig. 4. Status of Sweet potato in India (1980-2005) *Source* : Anonymous 2005

Table 3. Area, Production and Productivity of Sweet potato and Cassava in India (2001-2002)

State	Area ('000 ha)	Production ('000 t)	Productivity (t ha^{-1})
Sweet potato			
Orissa	37.7	272.1	7.22
West Bengal	24.0	172.0	7.15
Uttar Pradesh	22.6	259.8	11.50
Assam	8.8	30.4	3.45
Bihar	5.0	75.6	15.12
Jharkhand	7.6	103.0	13.55
Madhya Pradesh	6.9	42.2	6.12
Meghalaya*	5.2	17.6*	3.40
Maharashtra	4.7	66.7	14.20
Karnataka	3.5	26.4	7.63
Andhra Pradesh	1.3	13.4	10.40
Rajasthan	1.2	2.7	2.30
Tripura	1.1*	9.7*	9.08
Tamil Nadu	0.8	11.1	13.66
Kerala	0.8	9.0	10.98
India	131.9	1130.3	8.57
Cassava			
Kerala	109.3	2471.8	22.61
Tamil Nadu	68.0	2146.5	31.59**
Andhra Pradesh	17.7 (13.1)	353.9 (79.5)	20.00 (6.05)**
Meghalaya	4.0	20.6	5.15
Assam	2.9	13.7	4.72
Nagaland	1.3	16.0	12.31
Karnataka	0.9	7.2	8.20
Sikkim	0.5	1.2	2.40
Mizoram	0.3	1.5	5.00
Rajasthan	0.1	0.2	2.00
India	207.0**	5426.2**	26.21**

*2000 data **2002-03 data

Source Anonymous 2005 : Agriculture, Centre for Monitoring Indian Economy. March 2005 P-283-291.

Table 4. District wise area and production of major tuber crops in Tamil Nadu – 2002-03 (Area in hectare, Production in tonnes and Productivity in t ha^{-1})

Sl. No	District	Cassava			Sweet potato			Elephant foot yam			Taro			Total tuber crops	
		Area	Prodn.	Prodvy	Area	Prodn	Prodvty	Area	Prodn	Prodvty	Area	Prodn	Prodvty	Area	Prodn.
01	Coimbatore	518	15576	30.07	-	-	-	5	150	30.00	-	-	-	523	15726
02	Cuddalore	4923	205397	41.72	69	540	7.83	-	-	-	11	110	10.00	5003	206047
03	Tharmapuri	14712	305811	20.78	27	369	13.67	1	30	30.00	-	-	-	14740	306210
04	Dindigul	205	6476	31.59	64	874	13.66	-	-	-	3	30	10.00	272	7380
05	Erode	3745	76725	20.49	3	41	13.67	375	11250	33.30	-	-	-	4123	88016
06	Kanchipuram	7	221	31.57	47	642	13.60	6	180	30.00	-	-	-	60	1043
07	Kanyakumari	7405	211442	28.55	-	-	-	12	360	30.00	16	160	10.00	74.33	211962
08	Karur	680	21480	31.59	146	2095	14.35	2	60	30.00	-	-	-	828	23635
09	Madurai	98	3096	31.59	63	861	13.67	17	510	30.00	16	160	10.00	194	4627
10	Nagapatanam	160	5055	31.59	15	205	13.67	-	-	-	2	20	10.00	177	5280
11	Namakkal	6827	312351	45.75	10	137	13.70	-	-	-	-	-	-	6837	312488
12	Perambalur	2578	120981	40.93	24	328	13.67	-	-	-	165	1650	10.00	2767	122959
13	Pudukottai	16	505	31.55	3	41	13.67	-	-	-	2	20	10.00	21	566
14	Ramanathapuram	-	-	-	-	-	-	-	-	-	-	-	-	-	-
15	Salem	11828	326310	27.59	1	14	14.00	3	90	30.00	8	80	10.00	11840	326494
16	Sivaganagai	-	-	-	5	68	13.60	-	-	-	2	20	10.00	7	88
17	Thanjavur	40	1264	31.60	2	27	13.50	25	750	30.00	53	530	10.00	120	2571
18	The Nilgiris	102	3222	31.59	3	41	13.67	20	600	30.00	2	20	10.00	127	3883
19	Theni	28	885	31.61	2	27	13.50	3	90	30.00	-	-	-	33	1002
20	Thiruchirapally	5510	261040	47.38	24	328	13.67	-	-	-	4	40	10.00	5538	261408
21	Thirunelveli	170	5370	31.59	20	273	13.65	12	360	30.00	6	60	10.00	208	6063
22	Thiruvallur	128	4043	31.59	9	123	13.67	25	750	30.00	-	-	-	162	4916
23	Thiruvannamalai	605	19112	31.59	12	258	21.50	137	4110	30.00	159	1590	10.00	913	25070
24	Thiruvarur	24	758	31.58	2	27	13.50	3	90	30.00	5	50	10.00	34	925
25	Thoothukudi	4	126	31.50	51	697	13.67	134	4020	30.00	2	20	10.00	191	4863
26	Vellore	98	3096	31.50	4	55	13.75	94	2820	30.00	117	1170	10.00	313	7141
27	Vilipuram	7517	235465	31.32	138	2094	15.17	-	-	-	233	2330	10.00	7888	239889
28	Virudhunagar	23	727	31.61	70	956	13.66	36	1080	30.00	51	510	10.00	180	3273
	Total	**67951**	**2146534**	**31.59**	**814**	**11121**	**13.66**	**910**	**27300**	**30.00**	**857**	**8k570**	**10.00**	**70532**	**2193525**

Source : Department of Horticulture and Plantation, Government of Tamil Nadu.

Table 5. District wise area and production of cassava in Andhra Pradesh

District	Area (ha)	Production (tones)
Srikakulam	345	2249
Vizianagaram	323	3050
Visakhapatnam	629	2048
East Godavari	20144	208036
West Godavari	10	102
Krishna	4	41
Nellore	24	245
Chittore	17	173
Rangareddy	2	20
Adilabad	2	20

Table 6. Area and Production of sweet potato in Orissa (2004-05)

Sl. No.	Name of the District	Area (in ha)	Production (in Tonnes)
1	Balasore	127	1050
2	Bhadrak	107	889
3	Bolangir	3622	30434
4	Sonepur	258	2179
5	Cuttack	636	5228
6	J.S. Pur	191	2480
7	Jaipur	129	1051
8	Kendrapara	233	1895
9	Dhenkanal	3298	27809
10	Angul	1780	14828
11	Ganjam	8534	71963
12	Gajapati	2415	20117
13	Keonjhar	2698	22613
14	Kalahandi	258	2157
15	Khariar	1030	8595
16	Koraput	5459	45421
17	Malkangiri	1399	11942
18	N.W. Pur	129	1065
19	Rayagada	1737	14525
20	Phulbani	1961	16415
21	Boudh	371	3058
22	Mayurbhanj	3082	25050
23	Puri	50	406
24	Khurda	77	636
25	Nayagarh	258	2121
26	Sambalpur	1758	14610
27	Deogarh	627	5322
28	Bargarh	1643	13715
29	Jharsuguda	26	218
30	Sundargarh	3219	26495
	Total:	47112	394287

Source : Department of Agriculture, Government of Orissa

India exports about 90 t of fresh cassava tubers, 928 t of sago and sago starch / meal and 1156 t of cassava as substitutes and about 110 t of fresh sweet potato (*Source:* Various issues of Foreign Trade Statistics of root and tuber crops and its products, Director General of Commercial Intelligence and Statistics, Kolkata, 2004-2005).

The Global Crop Diversity Trust and United Nations with the support of the Bill and Melinda Gates foundation and the Govt. of Norway developed programmes to save genetic resoures of 21 food crops in which root and tuber crops are included (http : // www.nature.com/news).

Consultative Group on International Agricultural Research (CGIAR) in collaboration with, International Potato Centre (CIP), Peru, Lima; Central International Agriculture Tropical (CIAT), Cali, Columbia; International Institute for Tropical Agriculture (IITA), Ibadon, Nigegia, International Plant Genetic Resources Institute (IPGRI, Rome) and many other organization is surging ahead for food and nutrition security and environment safety. CGIAR is giving lot of emphasis on root and tuber crops as future harvest. In India also Indian Council of Agricultural Research through Central Tuber Crops Research Institute, All India Coordinated Research Project on Tuber Crops and many State Agricultural Universities and NGOs the thrust is being given for research and development leading to value added varieties and sustainable production and management technologies and post harvest utilization for end users.

References

Anonymous, 2005. Area, Production and Productivity of Sweet potato and Cassava in India. State wise details. (In:)Centre for Monitoring Indian Economy. pp. 283-291.

FAO, 2005. Area, production and productivity of Root and Tubers FAO. Statistics 2005.

Horton, D and H. Fano. 1985. Potato Atlas. Lima, Peru: International Potato Centre (CIP).

Scott, G.J., Best, R., Rosegrant, M. and Bokanga, M. 2000. Roots and tubers in the global food system. A vision statement to the year 2020. International Potato Centre, Peru, Lima. 111 pp.

❑❑❑

2

History, Origin and Geographical Distribution

Among angiosperms, there are about 18 families which have more than 30 genera which produce tubers. Most genera have many species which are tuberizing, most of which are unknown to man even now. The history, origin and distribution of only a few species which are economically important root and tuber crops and which are commonly used by men for edible as well as medicinal purpose are elucidated in this chapter.

1. Sweet Potato

Exact centre of origin of sweet potato *(Ipomoea batatas)*, and the routes and times for its dispersal to some of its present locations, are still in dispute and are likely to remain so until further archaeological or ethnobotanical evidence is available (O'Brien, 1972).

All archaeological, linguistic and historical evidences so far establish origin of sweet potato as the New World, in either the Central or South American lowlands. Columbus apparently discovered sweet potato in Hispaniola and Cuba during his first voyage in 1492 and, on his return, introduced it into Spain, whence it spread to rest of Europe. There it was first known as *batata* or *padada,* later being called Spanish potato or sweet potato to avoid confusion with Irish potato *(Solanum tuberosum* L.), which reached Europe 80 years later than *I. batatas* (Chris Ballard, *et al.*, 2005).

Further spread of sweet potato in historic times was by two lines of transmission: (a) the *batatas* line, which followed on from Spanish introduction into Europe, continuing after 1500 A.D. by transfer of European-grown clones to Africa, India and the East Indies through Portuguese exploration; and (b) the *kamote* line whereby Mexican clones were carried to Philippines by Spanish trading galleons (Yen, 1982).

Sweet potato crop, which has its origin in tropical Americas, spread to most of the world's tropical, sub-tropical and warmer temperate regions. According to the FAO, sweet potatoes are grown in 111 countries, of which 101 are classified as 'developing nations'. Among the world's root crops, sweet potato ranks second only to potato in economic importance (Horton, 1988).

2. Cassava

Cassava *(Manihot esculenta* Crantz) is a native of tropical America from where it diffused to other continents. But exact data to pinpoint the original habitat of cassava on continent of America are still not available. Until recently North-Eastern Brazil was considered the most likely point of its origin. Theory is in favour of a Brazilian origin. It gave reliance to the fact that the largest number of cultivars and numerous closely related wild species of *Manihot* were found there. But bulk of botanical evidence in support of its origin in Brazil was abandoned later for the reason that wild and herbaceous species of *Manihot,* some of which bear morphological traits similar to cassava cultivars, were found distributed in other areas as well, particularly in Mexico and Central America. Based on this and ethnological considerations, Rogers (1965) suggested that Southern Mexico and portions of Central America, particularly Guatemala and Honduras (Meso-American Complex) are equally probable centres of its origin. Rogers pointed out that numerous wild species are found both in Meso-America and in many regions of South America and that *M. aesculifolia,* the most closely related wild species of *M. esculenta,* is widely distributed in former area. In each area of cultivation, wild species could have hybridized with locally growing cultivars to give rise to a number of new genotypes which could diffuse to other areas as entirely new forms. Though the natural distribution pattern of *M. esculenta* was obscured by transference of varieties during human migration, Rogers and Appan (1973) suggested that it was probably first domesticated in low lands of Southern Mexico, Guatemala and Honduras and was distributed from there to other areas.

It is likely that progenitors of *M. esculenta* were shared both by Meso-America and South America indicating possibility of separate origin and domestication as cultivated plants in two regions. Historical evidences are, however, in favour of its first spread throughout South America from where the distribution to other continents was initiated by Portuguese.

Although original home of cassava is not clearly known, historical and archaeological evidences show that spread of the crop to Africa, India, countries of South East Asia and the Pacific Islands started from Brazil. It is believed that Portuguese introduced the crop into Africa in the 6th century where cultivation gained ground much before Africa came under European political control.

3. Yams

Within tropics, species of *Dioscorea* are to be found almost everywhere where rainfall is sufficient for their growth. They occur in both Old and New Worlds in the wild state, while even Australian continent, whose flora is in many ways unique, has a few indigenous species. Virtually the only tropical areas where no *Dioscorea* are to be found are swamps, near-desert districts where rainfall is less than about 80 cm per annum, and on mountains of altitude sufficiently great for heavy frost to occur.

Vavilov (1950) suggested that *D. alata* and *D. esculenta* are derived from Indian centre of origin in Burma and Assam, while former of these had a subsidiary distribution from Indo-Malayan centre rather farther South-East, from which centre *D. hispida, D. pentaphylla* and *D. bulbifera* also came into cultivation. Two more temperate species of section Enantiophyllum, *D. opposita* and *D. japonica,* were derived from Chinese centre of origin.

All the species referred to are found wild in the areas concerned, with the exception of *D. alata,* which is unknown in wild state anywhere in the world. Even in this case, however, there is little doubt that it originated in cultivation in this area, as the species *D. hamiltonii* and *D. persimilis,* with which *D. alata* undoubtedly has a common ancestry, are found wild in Burma, Assam and neighbouring countries (Burkill, 1935). So far as they go, Vavilov's opinions appear entirely justified (Vavilov, 1950).

Wild forms of *D. cayenensis* are reported (Chevalier, 1946) in West Africa. The most important species of all, *D. rotundata,* is unknown anywhere in wild state, but is definitely African in origin. It does not occur, apart from small numbers in botanical gardens and experimental farms, outside Africa except in Caribbean area where it is an introduction of historical times, and where, in any case, it could not have originated, as section Enantiophyllum to which it belongs has no indigenous representatives in Western hemisphere. Both these species exist in West Africa in an enormous plurality of varieties and forms which is in itself a strong indication of their local origin. Many closely related species are also known in wild state in West Africa, including *D. praehensilis,* a possible ancestor of the cultivated *D. rotundata.* The great antiquity of yam cultivation in West Africa is also indicated by number and importance of cultural and superstitious practices associated with cultivation of crop and by African tradition. Prohibition in certain areas on use of iron tools for digging of yams to be used in New Yam Festivals, strongly suggests that yam cultivation antedates commencement of iron age about two thousand years ago.

Available evidence, however clearly indicates four distinct centres of origin for edible yams. Firstly, a site somewhere in Indo-Chinese peninsula, where *D. alata, D. pentaphylla* and probably *D. esculenta, D. bulbifera* and some minor species were brought into cultivation. Secondly, a centre in Southern China, where the more temperate-adapted species *D. opposite* and *D. japonica* and possibly *D. esculenta* originated. Thirdly, a location on fringe of West African forest belt either within the savannah, or possibly in Dahomey gap, where savannah conditions penetrate south to the sea where originated *D. cayenensis, D. rotundata* and *D. dumetorum.* There may have been a subsidiary centre nearer or in Congo basin. Lastly, *D. trifida* and probably some other less important species of *Dioscorea* have a centre of origin in Caribbean area : again, there may be a subsidiary centre in South America (Coursey, 1967).

4. *Colocasia*

Taro originated in Southeast Asia including India (Watt, 1889; Chang, 1958) and Malaysia (Keleny, 1962). There is considerable evidence that taro and other edible aroids were distributed from east India to Formosa and the Solomon Islands (Spencer, 1966). Taro was probably also in Japan and Okinawa as *Colocasia antiquorum* Schott including varieties *esculenta, globulifera, typical, nymphaeifolia,* and *gigantean* (Kumazawa *et al.,* 1966). Taro seed is dispersed by birds (Panoff, 1972) and palm civets (Hambali, 1979). The latter may have been responsible for distribution of taro to Timor island in Indonesia. Oriental drosophilids, including *Drosophila colocasiae,* which breed in *Alocasia adora* as well as *C. antiquorum (C. esculenta* var. *antiquorum)* in Formosa, Okinawa, Japan, and Java may have been widely responsible for distribution of taro in Western Pacific (Okada, 1975).

Coursey (1977) also maintained that the 2000-3000 years BC date for domestication of taro is too recent and that taro culture probably preceeded that of yams during Mesolithic times which preceded the Neolithic domestication of grains during 7000-10000 years BC. Taro was probably cultivated by littoral or riverine people primarily dependent on fish for food. Cohen (1977) reviewed evidence of rice and perhaps taro and yam domestication earlier than 6000-9000 years BC in Thailand.

There is evidence that Austronesian horticulturists dispersed taro from Lungshanoid and Yuanshan provinces in Southern China (Yen and Wheeler, 1968) about 10000 years BC to Formosa by 9000 years BC, reaching Philippines in 7000-5500 years BC (Shutler and Marck, 1975). From Philippines, Austronesians spread in two directions : (a) east to Halmahera and Sulawesi by 4500 years BC and Papua New Guinea, and (b) west to Indonesia by 4000 years BC.

5. *Xanthosoma*

Xanthosoma is a genus belonging to tribe *Colocasiae* with about 38 species. *X. sagittifolium* otherwise known as tannia is an important edible aroid ranking next to taro. Tannia originated in tropical America and was first brought under cultivation there (Onwueme, 1978). It is known to early Spanish and Portuguese explorers but spread of this species is rather ambiguous and is supposed to have reached Southeast Asia much before reaching Pacific Islands and Africa probably in nineteenth century. The crop was introduced into Ghana from West Indies by missionaries in 1941. In continental U.S.A., cocoyam acreage is very limited - three other species namely *X. caracu, X. violaceum* and *X. atrovirens* are grown in South Florida but to a limited degree. *X. violaceum* has recently been introduced to Philippines for food and ornamental purposes (Ghani, 1984; Pancho, 1984).

6. *Alocasia*

Alocasia macrorrhiza is one of the largest aroids, with a thick stem usually 1-2 m high and 0.3 m in diameter. Giant taro (*Alocasia macrorrhiza* (L.) G. Don.) originated in Sri Lanka, but has become widely distributed in South-East Asia subcontinent, Malaysia, Indonesia and Polynesia, and spread to parts of tropical America.

A genus of herbs with a rootstock bearing a short succulent stem with large variegated leaves, it includes about 65 species, distributed in tropical Asia, of which 12 are found in India.

In 1925, Ridly reported that genus *Alocasia* has 20 species in tropical Asia of which seven are found growing in wild in Malaysia of (Ridley 1925). Among them *A. macrorrhiza* (L) Schott, *A. indica* (Roxb.) Schott, and *A. cucullata* (Schott) are used as food in parts of Asia (Kundu, 1967). In Malaysia the only two semi-cultivated species are *A. macrorrhiza* (Keladi Birah or Sente) and *A. indica* (Keladi Sebaring).

A. indica is much like *A. macrorrhiza* but with darker coloured petioles. It grows wild and is hardly ever cultivated. Stem can be eaten. Origin of *Alocasia indica* is through spontaneous mutation resulting in desynapsis during meiosis (Rajendran *et al.,* 1972).

A. cucullata was introduced from Bengal to Peninular Malaysia as an edible crop but it is better known for its medicinal value among Chinese herbalists (Burkill, 1935).

A. denudata grows wild in damp, shady habitats but not in water-logged conditions. Characteristic grey-black marbled petioles and peduncles of plant are likened to a snake-skin pattern and hence its vernacular name Keladi Ular or snake yam. Standing 69-80 cm tall, it has four to five petioles and peduncles in twos or threes are generally shorter. Green immature inflorescence on peduncle resembles head of a snake form.

7. *Cyrtosperma*

With 12 species, *Cyrtosperma* is a small genus. All species with exception of *Cyrtosperma chamissionis* (now *C. merkusii*) are confined to New Guinea and associated islands (Thompson, 1982), the Philippines, Borneo, Sumatra, Java, and Oceania. *C. chamissionis* or Giant swamp taro is terrestrial, usually grown in atoll pit gardens and in freshwater swamp or marsh area and is the largest of taros. Originated in South-East Asia, perhaps in Indonesia, it spread throughout Malayan Archipelago and Pacific territories. It is a minor crop in Melanesia and parts of Polynesia, but a major food in Micronesia (Plucknett, 1976). Subfossil leaf fragments of giant swamp taro *(C. chamissonis)* were recovered from archaeological contexts dating as early as A.D. 1489 (mean date) on Henderson Island (22° 22' S, 128° 18' W), Pitcairn group—a raised limestone *(makatea)* island isolated at extreme margin of South-Eastern Polynesia and the Indo-West Pacific biotic province (Hather and Weisler, 2000).

8. *Amorphophallus*

Occurrence of wild plant was reported all over the plains of North India, extending to West Bengal and Assam and in Konkan (Maharashtra). Its origin is however prescribed to be in Eastern India. It is widely cultivated throughout Upper Gangetic plains and in peninsular India (CSIR, 1950).

Amorphophallus paeonifolius (Dennst.) Nicolson or elephant foot yam has its centre of origin in India. Many wild spp. were explored from Western Ghats like *A. dubius, A. wightii* and *A. sylvaticus,* Kunth. A mention of *A. paeonifolius* (Dennst) Nicols. var. paeonifolius which is a thorny type is made in *Hortus Malabaricus* (Van Rheedes, 2003). Other species are *A. oncophyllum, A. variabilis, A. rivieri* (Ghosh *et al.*, 1988) which produce small tubers which are not edible. Corms of *A. paeonifolius* vary - One type with conspicuous corms and less or no cormels and the other type with one corm and more cormels.

Old World genus *Amorphophallus* has long been dispersed with migration of sea-faring people of Indo-Malaysia-Pacific region and believed to predate culture of rice (Ghani, 1984). Cultivation of *A. campanulatus* otherwise known as elephant foot yam, is rather limited to Malaysia and drier parts of India and Sri Lanka. It is, however, reported to grow wild in Southeast Asia extending to Java, Philippines and Pacific (Purseglove, 1974). Generally, *Amorphophallus* and *Alocasia* are considered only as famine foods in Pacific Islands (Thaman, 1984).

Genus *Amorphophallus* consists of about 90 species (Willis, 1955) and elephant foot yam is the most important economic species. Other species like *A. blumei* and *A. prainii* are growing wild in lowland secondary forests. *A. rivieri* is found in China and Japan (Ghani, 1984).

9. Coleus

Coleus (*Solenostemon rotundifolius* Poir. Syn. *Coleus parviflorus* Benth or *Plectranthus rotundifolius*) is an important minor tuber crop grown extensively as a vegetable in most of homestead gardens in Kerala and Tamil Nadu (Doraipandian, 1973; Hrishi and Mohankumar, 1976). Genus *Coleus* belonging to family labiatae, contains about 200 species which include ornamental and medicinal plants and plants with edible tubers. They are distributed in tropics, sub-tropics of Asia, Africa and Pacific Islands. Of these, *Coleus parviflorus* Benth, or *Solenostemon* commonly known as Chinese potato, produces edible tubers which can be used as a vegetable. It is a native of tropical Africa.

Coleus amboinicus is a native of Indonesia and *C. blumei* Benth. is also a native of Indonesia (Purseglove, 1974). *Coleus forskohli* is a perennial member of the mint family, first discovered in lower elevations of India. It is now grown around the world as an ornamental plant. Root is used medicinally.

10. *Pachyrhizus*

Pachyrhizus ferrugineus (Piper) Sorensen is the only wild species in the genus which is evergreen (with the exception of wild accessions of *P. tuberosus*) and where parts of plant above ground are perennial (all species have perennial tuberous roots). Both tuberous root(s) and stem(s) are woodier and more lignified than in other species. Root is less tuberous, although plants grown in greenhouses have produced fair-sized tubers (± 0.5 kg).

Species is ecologically associated with evergreen to deciduous rain forest with soils low in available phosphorus, a remarkable characteristic in a tuberous plant (Sorensen, 1990). Habitat of *P. ferrugineus* is often in the vicinity of rivers, along river banks, and its pods possess considerable resistance to humidity, a useful character when considering *P. tuberosus* under Amazonian conditions.

Distribution area is mainly along Atlantic coast of Central America from Mexican state of Vera Cruz to Panama, where it shifts to Colombian department of Choco on Pacific coast. This species may constitute an attractive source of rotenone (an insecticide) and may eventually be involved in breeding of new varieties to introduce to evergreen habit. Of even greater likely interest for breeding purposes is adaptability of *P. ferrugineus* to soil with a low phosphorus content.

P. panamensis is another wild species of *Pachyrhizus*. On the basis of existing records, wild species exhibit a peculiar dis-junct distribution. The best-documented area, covers the seasonally dry regions of central and Pacific Panama extending into coastal deciduous forest in the Santa Marta department of Colombia. Remaining records are from provinces of Guayas and El Oro in western part of central Ecuador (Sorensen, 1988) and (one) from southernmost part of Venezuelan state of Bolivar, 125 km north of Puerto Ayacucho, at foot of Guyana high-lands. It is therefore likely that the species may have originally been distributed all the way from its northernmost current distribution areas in Panama to the relatively dry coastal plains of Ecuadorian province of Guayas.

Wild materials of *Pachyrhizus erosus* (Yam bean) are recorded in Mexican states of Jalisco, Guanajuato, San Louis Potosi, Michoacan, Morelos, Puebla, Guerrero, Oaxaca, Vera Cruz and Chiapas; Central and Western Guatemala; El Salvador; Western Honduras, Western Nicaragua and Northwestern Costa Rica. The species is also cultivated in Mexican states of Nayarit, Tamaulipas as well as in Yucatan and Quintana Roo, where pre-Columbian landraces of *P. erosus* were introduced. In states where a wild distribution is recorded, different cultivars are often found as an escape from cultivation. The plant is widely cultivated in most southern states of Mexico. This situation also applies to El Salvador and Northwestern Honduras, where cultivation of the crop is widely practiced.

A description of present-day yam bean cultivation in Thailand and the local cultivars used is given in Ratanadilok and Thanisawanyangkura (1994). Although the crop is widely grown in all countries in Southeast Asia, this is the only comprehensive source on current cultivation of yam bean in the region. *P. erosus* was introduced to India during 1880 and 1890 (Tiwari *et al.* 1977).

P. ahipa is found in sporadic cultivation in Bolivia, and in the provinces of Jujuy and Salta in Argentina in subtropical Andean valleys along eastern side of the range. The Argentinian genotypes probably originated from seeds introduced from southern part of Bolivia. Bolivian farm labourers working in Argentina recall importing seed material from Bolivian province of Tarija (Parodi, 1935; Burkart, 1952; Sørensen, 1990; Ørting *et al.* 1996). It may be cultivated in Peru in the area surrounding Tarapoto, and is distributed in fertile sub-tropical valleys between altitudes of 1500 and 3000 m.

Geographical origin of species is still obscure. Brücher (1977, 1989) states that the likely location must have been in the 'ceja de montañas' Andean region, although recently claimed to have found wild *P. ahipa* near Sorata, Bolivia, i.e. the locality where the type specimen was collected by G. Mandon (no. 747,1856). Until material of this genotype becomes available for morphological and molecular analyses, phylogeny of species cannot be conclusively determined. The possible locations where a wild progenitor may be found are the Peruvian valleys of Apurimac, Ene and Mantaro (Sorenson, 1996).

P. tuberosus is distributed from southern lowlands of Colombia and Venezuela (the Andean mountain ranges of Sierra de Perijá and Cordillera de Merida), where it is known as 'nupe' through the Amazonian parts of Ecuador as far west as the province of Manabí, and in much of Brazil. In colonial Brazil, *P. tuberosus* was well known in cultivation (Munos Otero, 1945).

11. West Indian Arrowroot

Maranta arundiaceae L. west Indian arrowroot, a herbaceous, tropical perennial plant, with a creeping rhizome, indigenous to tropical America to Northern South America and the Lesser Antilles, was known to Indians during pre-Columbian days. The plant is naturalized in Florida, but it is chiefly cultivated in West Indies (Jamaica and St. Vincent), Australia, Southeast Asia, and South Africa. Because of this, Napoleon supposedly said the real reason for British love of arrowroot was to support their colonies.

12. Canna

Canna genus contains 55 species of herbs widely distributed in tropics and sub-tropics particularly in Western Hemisphere. *Canna edulis* or purple arrowroot is a perennial herb grown for its edible rhizomes and is a native of tropical America.

13. Winged bean

Origin of *Psophocarpus tetragonolobus* (L.) DC is uncertain, but probably in tropical Asia, where it is now mainly cultivated. Burkill (1935) considered that it originated on African side of Indian Ocean, possibly in Madagascar or Mauritius. However, it is not cultivated on African mainland. Although, probably originating in central Africa (Harder and Smartt, 1992) the greatest variation and only natural occurrence of the cultivated winged bean occurs in parts of Papua New Guinea and Indonesia. The plant readily naturalized in tropical and subtropical areas following introduction. In Papua New Guinea, plants are grown for tuberous roots, a revered food of highlands, where varieties are predominantly short-podded and sometimes express purple coloration. In Indonesia, plants are grown mainly for edible pods and seeds and local selection produced varieties that are long podded with lighter seed colours. A few Indonesian varieties do not produce tubers (Harder, 1992). It seems to have been long known to the Melanesians and is fairly extensively cultivated in New Guinea. It was first recorded by Rumpf in the Moluccas in the second half of 17th century, who considered that it was brought there from elsewhere, possibly Java. Later it was introduced to the West Indies.

14. Jerusalem artichoke

Helianthus tuberosus L., is North American in origin and cultivated by Indians in north-eastern United States in pre-Columbian times for its edible tubers. It was introduced into Europe before 1616 and has now been taken to most countries. It is grown in tropics for its tubers.

15. Asparagus

From edge of civilization, poets have sung song of asparagus, artists painted it, craftsmen recreated it, photographers captured it, healers prescribed it, and epicures sighed over it. Asparagus, one of nature's most beautifully designed vegetables, inspired even kings. Greeks and Romans enjoyed asparagus as a food and valued it for supposed medicinal properties as well.

Because asparagus was found growing wild in so many places around the world, there are many opinions as to its actual place of origin. Some scholars place this tasty tuber in Eastern Mediterranean area or in Asia Minor. Others say East Central Europe is the ancestral home of asparagus. Regardless of its origin, asparagus is known cultivated for at least 2,500 years. The earliest recorded growing of asparagus in California was in Northern delta in 1852. Market was strictly local until August 1900. On that date, the first trainload of asparagus was sent east. Shipped from Hickmott Cannery on Bouldin Island, the twenty cartloads began a marketing revolution. With easy and regular availability of canned asparagus, Americans across the country fell under the spell of nature's most perfect vegetable.

Asparagus is native of Eastern Mediterranean lands and Asia Minor. Romans cultivated the plant in 200 BC. It has, over centuries, been taken into gardens from wild and progressively improved by selection. In China, there are several wild species; the tuberous roots of some of which are used as food and in medicine. Asparagus is now an important crop, especially in Taiwan where shoots are canned for export. It is widely grown in East Africa and Malaysia.

Asparagus was found 'wild' in many places that there was much argument as to where it actually originated," says Boswell. Yet he, along with Sturtevant and Vavilov, placed origin of plant in Eastern Mediterranean and Asia Minor. Here it was cultivated for over 2,000 years. Thompson and Kelly noted that, "It was prized as a food by the Greeks and Romans and valued for medicinal properties of all its parts." "Before asparagus was used for food," says Boswell, "it had quite a reputation as a medicine for almost anything from prevention of beef stings to heart trouble, dropsy, and tootache!" Also, "Suetonius, about the beginning of second century, informs us how partial the Emperor Augustus was to asparagus." Boswell noted that the Emperor originated a saying, "Quicker than you can cook asparagus" (Boswell, 1949).

Boswell described its usage and spread as follows :"The Greeks apparently collected asparagus only from wild, since they gave no directions for cultivating it. The Romans, however, as early as 200 B.C. gave detailed gardening instructions that would be considered good today. In Roman times asparagus was not only eaten 'in season' but was dried for later use. It was simply and quickly prepared by boiling the dried shoots. Its introduction into Americas and other lands made no ripple worth noting at the time, but because of its old popularity it was presumably taken to those lands by early voyages (Boswell, 1949; Vavilov, 1950).

Early writers praised taste of wild over cultivated form of asparagus. Pomponius, who lived in second century, noted existence of both forms and said, "The wild asparagus is more pleasant to eat". Palladius, an author of third century, praises sweetness of wild form found growing among rocks. Along with this preference there was a rationale, "Nature has made the asparagus wild so that any one may gather as found," wrote Pliny in the first century. Boswell acknowledged preference of ancients for wild over cultivated form or cultivated form grown from wild seeds or plants, but he says, "such progress in development has been made (by 200 A.D.) that the cultivated forms were consistently as good as the best wild plants" (Vavilov, 1950).

Asparagus was grown in American gardens since the earliest settlements were established. It was not, however, until after 1850 or 1860 that asparagus was planted extensively by commercial growers in America. The average annual acreage increased from 20,000 ha for the period 1918-1927 to a 1959-63 average acreage of 75,000 ha. Today asparagus is one of the two most important perennial vegetable crops grown in USA (http://food.oregonstate.edu/ faq/uffva/asparagus.2.html).

16. *Vigna* spp.

Zombi Pea (*Vigna capensis* Walp. Syn. *Vigna Vexillata* L.)

In India, it is largely grown in north-eastern hill region. It grows in grassy places, thickets, grasslands, bush lands, forests and abandoned cultivated areas from sea level to 2250 m. It is widely distributed from tropics to warm temperature regions, often in humid and eroded areas. It tolerates annual precipitation of 64-429 cm, annual mean temperature of 18.3-27.4°C and pH of 5.0-7.8. *Vigna* spp. a climbing, woody perennial with broad ovate to lanceolate leaflets and pink or reddish purple flowers is a native of tropical Africa. It is a pantropical species distributed nearly throughout tropical parts of India, ascending to an altitude of 2400 m in Himalayas, and also to Malaysia and Australia.

Pillipesera (*Vigna trilobata* (L.) Verdcourt)

This species is found wild or cultivated from Himalaya to Sri Lanka ascending to an altitude of 2100 m in the North-West. It is also grown in China and southern region of India.

Rice bean (*Vigna umbellata* (Thunb.) Ohwi and Ohashi)

Rice bean is a native of South and South East Asia. It occurs wild in the Himalayas and central China to Malaysia, spreading by cultivation to other parts of tropical Asia. It is widely cultivated in India, China, Korea, Japan, Myanmar, Malaysia, Indonesia, Philippines, Java, Fiji, Bangladesh, Srilanka and Nepal (Purseglove, 1974; Rachie and Roberts, 1974). In India, its distribution is mainly confined to the tribal regions of north-eastern hills and the western and eastern Ghats in peninsular region, often in hilly tracts (Arora *et al.*, 1980). In the north-eastern region, rice bean is grown predominantly under shifting cultivation, rainfed mixed farming system, or as a kitchen garden and backyard crop particularly in Assam, Meghalaya, Manipur, Mizoram, Arunachal Pradesh, Nagaland, north Bengal and Sikkim. It is also grown on a limited scale in eastern peninsular tract covering Chhotanagpur region, Bihar and parts of Orissa, in the western peninsular tract particularly southern hills and sub-temperate hilly regions in Himachal Pradesh and Uttar Pradesh. The distribution pattern indicates its diverse adaptation ranging from humid sub-tropical to warm and temperate climate. The cultivated form seem to have originated from the wild populations occurring in the Indian sub-continent (Chandel and Pant, 1982). It is an excellent fodder as the crop remains green for a longer period.

17. Hyacinth bean *(Lablab purpureus* (L.)

Hyacinth bean is believed to be native to India or Southeast Asia. It is regarded as an Asiatic species possibly due to its wide cultivation in whole of the tropical Asia (Purseglove, 1974). However, there are reports (Zeven and Zhukovsky, 1975) which indicate its origin in the African Centre of diversity. It has spread over the entire tropics and is now cultivated in warmer regions throughout the world. In India, it is largely grown as a pulse crop in Tamil Nadu, Karnataka and Madhya Pradesh. However, viny varieties are grown for vegetable purpose all over the country.

18. *Curcuma*

Curcuma angustifolia Roxb.

This species occurs wild in India in hilly tracts of Central Provinces, Bengal, Bombay, Madras and lower Himalayan ranges. Tubers or rhizomes which contain starch, is used as a substitute for arrowroot powder from *Maranta arundinacea* Linn. (Kundu, 1967). The plant grows wild in many places.

Curcuma zedoaria Rosc. (Palua)

This species grows wild in Eastern Himalayas and in moist deciduous forests of Coastal tracts of South India. It is cultivated in many parts of Srilanka and China for production of 'shoti starch'. It grows as a crop after summer fallows (CSIR, 1950; IBPGR, 1981).

19. *Tacca*

Tacca leontopetaloides (L) Kuntze (East Indian Arrowroot)

Indigenous to eastern tropics, it is found wild or cultivated on smallholdings through-out Asia, tropical Africa, the Pacific Islands and Queensland. There appears to be two distinct types growing in the Pacific Islands, one produces large number of potato-sized tubers and other produces a single large tuber. Massal and Barrau (1956) considered that this plant probably originated in South-East Asia and was introduced to South Sea Islands in very early times. Its importance declined, and at one time, it was a staple food in parts of Polynesia. It is also cultivated in Thailand. In Africa, tubers from wild plants are used as food in a few areas (Purseglove, 1972).

20. Safed Musli

Naturally occurs in forests of Gujarat, Madhya Pradesh and Maharastra and is listed under endangered species of India. There are around 256 varieties of *Chlorophytum* in the world. In India, we have around 17 of them, of which, *C. borivilianum* has good market demand. *(www.IndianGyan.com).*

21. *Costus specious*

The plant *Costus speciosus* (Koen) Sm., belonging to the family Zingiberaceae is among five species of stout perennial herbs indigenous to Indo-Malayan Region. Hooker (1984) described two varieties of *Costus speciosus* viz., var. *nipalensis* Rosc., found in Nepal and Arunachal Pradesh and var. *argyrophyllus,* having wide distribution in India. Distribution extends from almost sea level to about 1500 m elevations in Arunachal Pradesh and Meghalaya. Most of the areas where *C. speciosus* was encountered under natural conditions have humid tropical or subtropical climate. It is completely absent in arid and semi-arid areas of Punjab, Haryana, Rajasthan, Gujarat and parts of South India.

Plant occurs in scattered pockets in Reasi, Udhampur and Billawar districts of Jammu & Kashmir. Plant has rare occurrence in major parts of Central India. It is only in the South-Eastern Chattisgarh and Bastar regions of Madhya Pradesh where it is gregarious especially under sal forests or around cultivated fields. The diosgenin content is 2.5%. *C. speciosus* is quite common in coastal areas extending from Bombay to Trivandurm. It frequents perennially in moist localities including palm grooves, orchards and rice fields.

The plant also occurs as undergrowth of evergreen forests in Kerala (Sarin *et al.,* 1982).

22. *Gloriosa superba*

Glory lily, a native of tropical Africa is now found growing naturally throughout much of tropical Asia including India, Sri Lanka, Malaysia, Burma (Jayaweera, 1982). *G. superba* is also planted outdoors in Southern United States. In cool temperate countries, it is treated as a greenhouse or conservatory plant.

23. *Ullucus tuberosus*

Ullucu is grown in scattered pockets from Venezuela to Northern Argentina, usually between 30,000 and 4,000 m above sea level. In Peru, the crop is grown in lower valleys, between, 1,000 and 2,000 meters. Plant prefers soils that are rich in organic matter. The period between planting and harvest varies from 140 to 200 days, depending on variety, latitude, and altitude. Ullucu is the most important and commercially viable of most of the Andean roots and tubers.

24. *Artemisia dracunculus*

The crop originated in Central Asia, probably Siberia. It is not known when and by whom the aromatic varieties were first bred, nor when the plant was introduced to Europe. A herb *dragantea* is mentioned in the *Capitulare de villis* of Charlemagne, but its identity is not unambiguously clear.

In middle ages, tarragon was known as *tragonia* and *tarchon,* which is an Arabic loan; in Modern Arabic, the name is *tarkhun.* Origin of Arabic name is not clear, but it might be a loan from Old Greek, perhaps akin to *drakon* "dragon, snake". Apparently, the plant was linked to dragons because of serpent-shaped rhizome, and there was a wide-spread belief that tarragon could not only ward off serpents and dragons, but could also heal snake bites.

25. *Pueraria (Kudzu)*

Kudzu appeared in Japanese literature in poems dating back to 600 AD (Shurtleff and Aoyagi, 1977). Reference, apart from taxonomical ones that treat the formidable synonymy produced since 1753, are usually on economic botany. The appreciable folk taxonomy indicated that kudzu is well in Asia and often used.

Pueraria is a genus distributed in South and Southeast Asia, China and Japan, and parts of Oceania. Tropical kudzu was spread by human intervention as a green manure in plantation crops. Kudzu was introduced from China or Japan to the highlands of New Guinea and New Caledonia much before yams, taro and sweet potatoes. Introduced into USA in 1876 as an ornamental, it became a pasture crop around 1910, and in 1935 its main use was for erosion control. In 1950 its coverage was 3 million ha, but due to the uncontrollable growth covering forests, smothering trees and electricity poles, the enthusiasm for kudzu declined. Around 1970, the area was reduced to 30,000 ha, after the removal in 1954 from the list of acceptable cover crops. Eradication was needed, and in 1981 kudzu was about to be declared a pest affecting 22,000 ha of commercial forests in Georgia alone. In Switzerland, Zanzibar, and in Argentina, kudzu did not develop so aggressively. In South Africa, it is a new plant invader (Maesen, 1985, 1992).

26. *Lathyrus tuberosus*

Lathyrus tuberosus, also known as the Tuberous Pea, Earthnut Pea, or Aardaker, is a small, climbing perennial plant, native in moist temperate parts of Europe and western Asia.

27. *Arracacia xanthorrhiza*

The arracacha is a garden root vegetable originally from the Andes, somewhat intermediate between the carrot and celery. Its starchy taproot is a popular food item in South America, especially in Brazil where it is a major commercial crop.

The name *arracacha* (or *racacha*) was borrowed into Spanish from Quechua, and is used in the Andean region. The plant is also called *apio criollo* ("Creole celery") in Venezuela, *zanahoria blanca* ("white carrot") in Ecuador, *virraca* in Peru, and *mandioquinha* ("little cassava") or *batata-baroa* in Brazil. It is sometimes called white carrot in English, but that name properly belongs to white varieties of the common carrot. It was imported into Brazil in the 19th century and has been grown commercially since the 1960s. Brazilian crop improvement programs have developed varieties that grow in seven months. The harvest season in the Southern Hemisphere spans from January to September. The roots must be picked promptly lest they become woody. They have a short shelf life and must reach consumers within a week of harvest. The plant grows west of the Andes at altitudes varying from 200 m to 3600 m, but optimally between 1800 and 2500 m. It is frequently grown with other crops such as maize, beans, and coffee (http://en.wikipedia.org).

28. *Arthropodium*

Arthropodium is a genus of herbaceous perennial plants native to the Southern Hemisphere. *A. cirratum* is native to New Zealand, where it may once have been farmed. It is used for medicine as well as food, and has symbolic importance in traditional Mâori culture.

29. *Tropaeolum tuberosum* Ruíz and Pavón

The mashua or añu is a perennial plant grown in the Andes for its edible tuber. It is a major food source there. The plant is well adapted to high-altitude subsistence agriculture. It grows vigorously even in marginal soils and in the presence of weeds, and gives high yields; the tubers comprise as much as 75 percent of the mature plants by dry weight (40 percent is typical for cereals), and 70 tons per hectare have been produced under research conditions.

The anti-aphrodisiac property of the tuber has been recorded by the Spanish chronicler Cobo that mashua was fed to their armies by the Inca Emperors, "that they should forget their wives".

30. *Eleocharis dulcis* (Burm. f.) Trin. ex Hensch.

The Chinese water chestnut is native to China and is widely cultivated in flooded paddy fields in southern China and parts of the Philippines.

31. *Mirabilis expansa*

Mirabilis expansa (mauka or chago) is cultivated as a root vegetable in the Andes, at cold, windy altitudes above 2700 meters. *Mirabilis expansa* was an important root crop to the Inca empire and was considered a "lost" crop until being rediscovered in the 1960's and 1970's in three separate distant locations in Peru, Ecuador and Bolivia. There is a possibility that the plant's continued use and survival in three separate locations was due to the Inca policy of transplanting valuable food crops and communities throughout the empire (Davidson, 1999).

32. *Cyperus esculentus*

Cyperus esculentus (Chufa Sedge, Yellow Nutsedge, Tigernut Sedge, Earth almond) is a species of sedge native to warm temperate to subtropical regions of the Northern Hemisphere. The tubers were originally cultivated by Ancient Egypt's populations at the Nile Valley; their cultivation was subsequently extended throughout other areas with temperate climate and fertile soil. Presently, they are mainly cultivated - at least for extended and common commercial purposes - in Spain, where they were introduced by Arabs, almost exclusively in the Valencia region (http://en.wikipedia.org).

It is extremely difficult to remove permanently when it is considered to be an intrusive weed in lawns and gardens. This is due to the plant having a stratified and layered root system, with tubers and roots being interconnected to each other to a depth of 50 cm or more.The tubers are connected by fragile roots that are extremely prone to snapping when pulled on, making the plant extremely difficult to remove with its entire root system intact, and the plant can quickly regenerate if even a single tuber is left in place.

33. *Psoralea esculenta*

Psoralea esculenta is an herbaceous perennial plant *native to* prairies and dry woodlands of central North America. Abundant, palatable, and nutritious, the root was once a wild-gathered staple of native Americans and early European explorers. Its characteristics make it an obvious candidate for possible domestication.

34. *Smallanthus sonchifolius* (Poeppig and Endlicher) H. Robinson *syn.Polymnia sonchifolia* Poeppig and Endlicher

The Yacón is a perennial plant grown in the Andes for its crisp, sweet-tasting tuberous root. The texture and flavour have been described as a cross between a fresh apple and watermelon which is why it is sometimes referred to as the apple of the earth .After the first few frosts the tops will die and the plants are ready for harvest. It is generally best to leave some in the ground for propagating the following spring. Alternatively, the propagating roots can be kept in the refrigerator or buried away from frost until spring. While usable-sized roots develop fairly early, they taste much sweeter after some frost.

Yacón can easily be grown in home gardens in climates with only gentle frosts. It grows well in southern Australia, Tasmania and New Zealand, where the climate is mild and the growing season long. It has recently been introduced to the Philippines, and is now widely available in markets.

35. *Oxalis tuberosa* Savign.

The oca or oka is a perennial plant grown in the central and southern Andes for its starchy edible tuber, used as a root vegetable. Its leaves and young shoots can be eaten as a green vegetable as well. Introduced to Europe in 1830 as a competitor to the potato and to New Zealand as early as 1860, it has become popular in that country under the name New Zealand yam and is now a common table vegetable. The oca is one of the important staple crops of the Andean highlands, second only to the potato due to its easy propagation, and tolerance for poor soil, high altitude and harsh climates.

36. *Tragopogon porrifolius*

Purple Salsify is native to Mediterranean regions of Europe but introduced into Britain and northern Europe, North America, and southern Africa; in the United States it is now found growing wild in almost every state, including Hawaii, except in the extreme south-east.

The root, and sometimes the young shoots, of Purple Salsify are used as a vegetable, and historically the plant was cultivated for that purpose; it is mentioned by classical authors such as Pliny the Elder. However in modern times it has tended to be replaced by Spanish Salsify or Black Salsify as a cultivated crop. The root is noted for its taste of oysters, from which the plant derives its alternative name of Oyster Plant.

References

Arora, R.K., Chandel, K.P.S., Pant, K.C and Joshi, B.S. 1980. Rice bean – a tribal pulse of eastern India. *Econ. Bot.* 34 (3): 260-263.

Boswell. 1949. Our Vegetable Travelers. *National Geographic.* Aug. pp. 179.

Brücher, H. 1977. Tropische Nutzpflanzen: Ursprung, Evolution and Domestikation. Springer-Verlag, Berlin.

Brücher, H. 1989. Useful Plants of Neotropical Origin and Their Wild Relatives. Springer-Verlag, Berlin.

Burkart, A. 1952. (2nd ed.) Las Leguminosas Argentinas Silvestres y Cultivadas. Acme Agency, Buenos Aires.

Burkill, I.H. 1935. A dictionary of the economic products of the Malay Peninsula. Min. of Agric. and Coop, Kuala Lumpur, Malaysia. I, 755 pp.

Chandel, K.P.S and Pant, K.C. 1982. Genetic Resources of *Vigna* species in India: their distribution, diversity and utilization in Crop Improvement. *Ann. Agric. Res.* 3: 19-34.

Chang, T.K. 1958. Dispersal of taro in Asia. *Ann. Assn. Am. Geogrs.* 48, 255-6.

Chevalier, A. 1946. 'Nouvelles recherches sur les ignames cultivées', *Revue int. Bot. appl. Agric. Trop.,* 26 : (279-80) 26-31

Chris Ballard., Paula Brown, R., Michael Bourke. and Tracy Harwood. 2005. The sweet potato in Oceania – a reappraisal. Ethnology monographs 19, Oceania Monograph 56. The University of Sydney.

Cohen, M.N. 1977. The food crisis in prehistory: over-population and the origins of agriculture. Yale Univ. Press New Haven and London.

Coursey, D.G. 1967. Yams. London : Longman.

Coursey, D.G. 1977. The comparative ethnobotany of African and Asian yam cultures. (*In:*) Leakey, C.L.A. (ed) *Proc. 3rd Symp. Inter. Soc. Trop. Root Crops.* IITA, Ibadan, Nigeria. pp. 164-169.

CSIR 1950. The Wealth of India II. P. 58, 400, 405-406.

Davidson, Alan. 1999. Oxford Companion to Food. "Mauka". pp. 484-485 ISBN 0-19-211579-0

Doraipandian, A. 1973. 'Sirukizhangu'. A valuable tuber crop. *Indian Hort.*, 18 : 14-15

Ghani, F.D. 1984. The potential of aroids in Malaysia. (*In:*) Edible Aroids. (Ed. S. Chandra), Clarendon Press, Oxford, pp. 10-16.

Ghosh, S.P., Ramanujan, T., Jos, J.S., Moorthy, S.N. and Nair, R.G. 1988. Tuber Crops. Oxford IBH Publishing Co Pvt. Ltd., New Delhi

Hambali, G.G. 1979. The dispersal of taro by common palm civets. *Proc. 5th Symp. Inter. Soc. Trop. Root Crops,* Philippines.

Harder, D.K and Smartt, J. 1992. Further evidence on the origin of the cultivated winged bean, *Psophocarpus tetragonolobus* (L.) DC. Chromosome numbers and the presence of a host-specific fungus. *Econ. Bot.* 46, 2 -187-191.

Harder, D.K. 1992. Temporal Mineral allocation in plants of the cultivated winged bean, *Psophocarpus tetragonolobus* (L.) DC. :Nitrogen or Calcium. *(In:) Proceedings of the first International Symposium of tuberous legume,* Guadeloupe FWI, 21-24 April 1992.

Hather, J.G. and Weisler, M.I. 2000. Prehistoric Giant Swamp Taro *(Cryptosperma chamissionis)* from Henderson Island, Southeast Polynesia. Pacific-Science. 54(2): 149-156.

Hooker, J.D. 1984. The Flora of British India Vol. VI. London: L. Reeve and Co. pp. 249-250.

Horton, D. 1988. Underground crops. Long-term trends in production of roots and tubers. Winrock International Institute for Agricultural Development, Morrilton, AR

Hrishi, N. and Mohankumar, C.R. 1976. Coleus for homestead gardens. *Indian Farming,* 26: 33-35.

http://en.wikipedia.org

http//food.oregonstate.edu/faq/uffva/asparagus2.html.

IBPGR. 1981. Root and Tuber Crops. IBPGR, Rome. pp. 23, 49, 81, 87 and 107.

Jayaweera, D.M.A. 1982. Medicinal plants used in Ceylon. Colombo, National Science Council of Sri Lanka (part 3).

Keleny, G.P. 1962. The origin and introduction of the basic food crop of New Guinea People. *Papua New Guinea Agric. J.* 15: pp 7-13.

Kumazawa, S., Niuchi, K., and Honda, F. 1966. Classification of the taro varieties in Japan. *J. Hort. Association, Japan.* 25: 1-10.

Kundu, B.C. 1967. Some edible rhizomatous and tuberous crops of India. *Proc. 1st Symp. Inter. Soc. Trop. Root Crops,* Trinidad, 124-30.

Maesen, L.J.G. van der. 1985. Revision of the genus *Pueraria* DC. With some notes on Teyleria Backer (Leguminosae).- *Agric. Univ. Wageningen papers* 85,1 : 1-132.

Maesen, L.J.G. van der. 1992. *Pueraria,* the Kudzu and its relatives, an update on the taxonomy. *(In:) The proceedings of the first International Symposium on tuberous legumes.* Guadeloupe, F.W.I., 21-24 April, 1992. pp:55-86.

Massal, E. and Barrau, J. 1956. Food Plants of the South Sea Islands. Noumea : South Pacific Commission, Tech. Paper No. 94.

Munos, Otero, S. 1945. "Nupe, a promising crop for Venezuela." 3rd Conf. *Interamer. Agric., Caracas* 38:5-34. [In Spanish, no English summary].

O'Brien, P.J. 1972. The sweet potato; its origin and dispersal. *Am. Anthropologist.* 74:342-365.

Okada, T. 1975. The Oriental Drosophilids breeding in flowers. *Kentyu* (Tokyo), 43(3), 356-63.

Onwueme, I.C. 1978. The tropical tuber crops. John Wiley and Sons, New York, 234 pp.

Ørting, B., Gruneberg, W. J and Sorensen, M. 1996. Ahipa (*Pachyrhizus ahipa* (Wedd.) Parodi) in Bolivia. Genetic Resources and Crop Evolution.

Pancho, J.V. 1984. Folk uses of cultivated aroids in the Philippines. (*In:*) Edible Aroids (Ed. S. Chandra), Clarendon Press, Oxford, pp. 3-9.

Panoff, F. 1972. Maenge taro and cordyline element of a Melenesian Key. *J. Poly. Soc.* 81(3): 375-388.

Parodi, R.L. 1935. Relaciones de la argricultura prehispanica con la agricultura Argentina actual. *Anales Aca. Nac. Agr. Vet. Buenos Aires* 1:115-167.

Plucknett, D.L. 1976. Edible aroids. (*In:*) Simmonds, N.W. (ed.), Evolution of crop plants. Longman, London and NY, pp. 10-12.

Purseglove, J.W. 1972. Tropical Crops : Monocotyledons. 2 Vols. London : Longman

Purseglove, J.W. 1974. Tropical Crops : Dicotyledons. Vols. 1 and 2 combined London : Longman.

Rachie, K.O. and Roberts, L.M. 1974. Grain legumes of the lowland tropics. *Ad. Argon.* 26: 1-132.

Rajendran, P. G., Ravindran, P. S. and Jos, J. S. 1972. Desynapsis in *Alocasia indica* Schott. *Genetica.* 43 (2) : 231 - 235.

Ratanadilok, N and S. Thanisawanyangkura. 1994. Yam Bean (*Pachyrhizus erosus* (L.) Urban) status and its cultivation in Thailand. (*In:*) *Proceedings of the First International Symposium on Tuberous Legumes;* Guadeloupe, FWI, 21-24 April 1992. (M. Sørensen, ed.) Jordbrugsforlaget, København. pp. 305-314.

Ridley, H.N. 1925. The flora of the Malay Peninsula 5, L. Reeves, Kent.

Rogers, D.J. 1965. Some botanical and ethinological considerations of *M. esculenta. Econ. Bot* (USA) 19: 369-377.

Rogers, D.J. and Appan. S.G. 1973. Manihot and Manhotoides Euphorbiaceae – a computer aided study, organization of Flora Neotropica, Monograph No. 13, 278 pp.

Sarin, Y.K., Singh Ajit., Bedi, K.L., Kapur, S.K., Kapahi, B.K., Atal, C.K. 1982. Costus specious Rhizome as commercial source of Diosgenia *(In:)* Cultivation and Utilization of Medicinal plants (eds.) Atal, C.K and Kapur, B.M. RRL, CSIR, Jammu-Tawi.

Shurtleff, W and Aoyagi, N.A. 1977. The book of kudzu, A culinary and heading guide. Autumn Press.102 pp.

Shutler, R. Jr. and Marck, J.C. 1975. On the dispersal of the Austronesian horti-culturists. *Arch. and Phy. Anthrop. In Oceania* 10(2), 83-100.

Sorenson, M. 1988. A taxonomic revision of the genus *Pachyirhizus* Rich. exDc. Nom. Nord. *J. Bot.* 8(2): 167-192.

Sorenson, M. 1990. Observation on distribution, ecology and cultivation of the tubers bearing legume genus *Pachyirhizus* Rich. exDc. (Fabaceae : Phaseoleae), Wageningen paper 90-3: 1-38.

Sorenson, Marten. 1996. Yam bean (*Pachyrhizus* DC). Promoting the conservation and use of underutilized and neglected crops. 2. Institute of Plant Genetics and Crop Plant Research, Gatersleben/International Plant Genetic Resources Institute, Rome. 141 pp.

Spencer, J.E. 1966. Shifting cultivation in southeast Asia. U. Cal. Publ. Geogr. V.19, Figs. 4 and 5.

Thaman, R.R. 1984. Intensification of edible aroid cultivation in Pacific Islands. (*In:*) Edible Aroids (Chandra, S., Ed.) Clarendon Press, Oxford. pp. 102-122.

Thompson, S.A., 1982. *Cyrtosperma chamissonis* (Araceae): ecology, distribution and economic importance in the South Pacific. (now *C. merkusii*). *J. Agric. Trad. Bot. Appl.* 29(2): 185 - 203

Tiwari, R., Praksah, R. and Haque, F. 1977. Effect of gamma radiation on germination and survival in Yam Bean. *Tropical Grain Legume Bulletin* 10:6-7.

Van Rheedes. 2003. Hortus Malabaricus (Reprinted) Vol. 11. 2003. University of Kerala. 245 pp.

Vavilov, N.I. 1950. The origin, variation, immunity and breeding of cultivated plants. English edition. Chronica Botanica. Waltham ; Mass, USA

Watt, G. 1889. Dictionary of the economic plants of India, Supt. Government Printing Calcutta, 2:509-513.

Willis, J.C. 1955. A dictionary of the Flowering plants and Ferns. Univ. Press, Cambridge.

www.Indian Gyan.com.

www.uni-graz.at/~katzer/engl/Arte_dra.html.

Yen, D.E and Wheeler, J.M. 1968. Introduction of taro into the Pacific: the indications of the chromosome numbers. *Ethnology* 7(3), 259-267.

Yen, D.E. 1982. Sweet potato in historical perspective. *(In:)* Sweet potato (Eds. R.L. Villareal and T.D. Griggs), AVRDC, Taiwan, pp. 17-30.

Zevan, A. and Zhukovsky, P.M. 1975. Dictionary of Cultivated Plants and their Centers of Diversity: Excluding Ornamentals, Forest Trees and Lower Plants. Wageningen, Centre for Agricultural Publishing and Documentation (PUDOC).

❑❑❑

3

Botany, Physiology and Biodiversity

Plant biodiversity, in particular, is arguably the single greatest resource that humankind has garnered from nature during our long cultural development. Nowhere is the value of biodiversity more evident than in our food supply. Roughly one third of all plant species have edible fruits, tubers, nuts, seeds, leaves, roots, or stems. The World Health Organization estimates that 3.5 billion people in developing countries rely on plant-based medicine for their primary health care, in which tuber crops also play a major role (Tuxill, 1999). There are 25 hot spots of biodiversity of which Western Ghats and Eastern Himalayan are the most important ones in India. About 7% of world flora and 6.5% of world fauna are in India. Indo-Burma region is one of the eight hottest hot spots of the world, and is the centre of origin of tuber crops like *Dioscorea, Colocasia, Amorphophallus* and *Alocasia.*

Tropical tuber crops form an important group of food crops for millions of people (about one fifth of world population), especially to those living in the tropics and sub-tropical zone. Tuber crops are the third important food crops after cereals and grain legumes (Chadha and Nayar, 1994). The carbohydrates synthesised in the leaves are stored in the form of starch in the underground tubers and roots, which not only form food but also a source of raw material for several industries. Among the many tuber crops available only a few have been domesticated and widely cultivated for edible and industrial purposes. The major tuber crops are cassava, sweet potato, Asiatic yam, African yam, lesser yam, elephant foot yam, taro/bunda, tannia and country potato which are widely cultivated and used for edible purpose and industrial uses. In addition there are minor tuber crops like arrow root, winged bean, yam bean, *Canna,* safed musli, Jerusalem artichoke etc.

1. Sweet Potato (*Ipomoea batatas.* L.)

Family : Convolvulaceae

A herbaceous, perennial vine cultivated as an annual. There are several hundred cultivars of sweet potatoes and great variation is found in the form and growth habit. Forms that have twining and trailing long stems 0.9-4.5 m of slender to moderate thickness and moderately to widely spaced leaves are the most prevalent but types with short thick stems, short internodes and semi-erect to erect growth habits also occur. Leaves are highly variable in colour and degree of purple pigmentation, spirally arranged along the stems in a pattern of 2/5 phyllotaxy, they may be shouldered, toothed, entire, deeply cleft or variously lobed and the petioles can also vary in length, thickness and degrees of erectness. The flowers occur solitary or cymose and vary in colour from white to purple. The fruit is a glabrous or hirsute dehiscent capsule 5-8 mm, containing 2-4 angular, brownish-black seeds with a very hard testa. The root tubers are formed by thickening of parts of the adventitious roots close to the subterranean part of the stem or at the nodes which rest on the soil.

Species diversity of *Ipomoea*

Species	Remarks
Ipomoea mauritiana Jacq.	Medicinal plant common on the coastal plain of Malabar (Hort. Malabaricus 11:101-102, t.49-1692)
I. alba L.	Found in most region of coastal Malabar (Hort. Malabaricus 11: 101-102, t.50 -1692)
I. aquatica Forssk.	Medicinal – found in stagnant fresh water ponds throughout Malabar
I. campanulata	Uplands in moist shaded places in Malabar – antidote for snake poison. (Hort. Malabaricus 11: 115, t.56 -1692)
I. marginata (Desr.) Verdc	Leaflets are crushed and used as shampoo and occurs in the uplands in Malabar. (Hort. Malabaricus 11: 109-110, t.53 -1692)
I. pescaprae (L.)	Common in the beaches of Malabar (Hort. Malabaricus 11: 117, t.57-1692)
I. paniculata R. Br.	This glabrecent twining shrub is seen throughout India preferring moist areas (Deepa *et al.*, 1999).
I. asarifolia (Desr.) Roem and Schutt	Found in the uplands throughout Malabar.
I. pes-tigridis (L.)	Occurs throughout Malabar (Hort. Malabaricus 11: 121, t.59 -1692)
I. quamoclit (L.)	Cultivated as a garden plant in Malabar, with beautiful red flowers.

Though the centre of origin is South America, wild relatives were found in Malabar coast. Van Rheedes (2003) reported a number of *Ipomoea* spp. like *I. mauritiana, I. alba, I. aquatica, I. marginata* and *I. campanulata* etc., most of which are medicinal, from the Malabar coast, in his *Hortus Malabaricus.*

Though the genus *Ipomoea* has received intense cytogenetical attention, *I. batatas* is rather neglected probably due to its high chromosome number and small size of chromosomes. Sweet potato has 90 somatic chromosomes and considered a hexaploid.

Physiology

Sweet potato is well adapted to a wide range of environmental conditions and yield over 20 t ha^{-1} in four to five months duration showing its superiority over cereals in solar energy utilisation capacity. As the tuber forms the major proportion of total dry weight of plant, productivity of sweet potato crop is largely governed by the process of tuberisation and photosynthetic efficiency of the leaf canopy in support of the storage root sink. Both the processes are being controlled by environmental factors (Ghosh *et al.,* 1988).

The process of tuberisation in sweet potato consists of two distinct phases, namely tuber initiation, differentiation and tuber development or tuber bulking. In the root system of sweet potato, certain roots exhibit lateral swelling which are called tuberous roots. Wilson (1973, 1982) differentiated the root system of sweet potato into string roots, pencil roots and tuber bearing roots. Togari (1950) reported that young roots of sweet potato turned into either tuberous roots by increased meristematic activity of the cambium to produce starch storing parenchyma or pencil roots by more lignification of the stele cells.

Studies by Kays *et al.,* (1982) using labelled ^{14}C found that the labelled assimilates accumulated along the vascular cambium of the storage root in 24 hours after exposing the leaves to $^{14}CO_2$. This demonstrated clearly that the cambium is the zone of maximum meristematic activity of the storage root in sweet potato. The process of tuber initiation in sweet potato would normally be completed in eight weeks after planting. Using tracer technique, it was established that early deposition of carbohydrates was concentrated at distal end of tuberous roots and bulking proceeded longitudinally towards the proximal end (Chua and Kays, 1982).

Tuber development in sweet potato is largely governed by soil-environmental interaction and is non-inductive in nature. Like cassava, the tuber bulking in sweet potato also follows the phenomenon of balanced partitioning which normally commences by 45 to 60 days after planting. Tuber development is associated with the rate of cell division and cell expansion and the density of starch in the cells is associated with synthesis and accumulation of starch granules. The number and size of starch granules per cell increased from the twentieth to the sixtieth day after planting and thereafter it remained constant. Therefore, further increase in tuber size is associated with increase in cell number (Togari, 1950).

Both water logging due to floods and water stress conditions due to drought are the major physical stress conditions found to limit the productivity of sweet potato. Sweet potato when grown in association with annuals or perennials under multiple cropping systems, often records a reduction in tuber yield due to accompanying shade effect. Sweet potato can tolerate drought better than flooding (Sung, 1981). Moisture stress imposed during the early phase of growth inhibited tuber formation while stress imposed during tuber bulking phase did not affect productivity. This was probably due to the fact that LAI of sweet potato is very high (CTCRI, 1984). Storage root formation or development is affected under inundation. Inhibition on the storage root development in sweet potato under flooded environment is attributed to high levels of ethylene production under such conditions (Paterson *et al.,* 1979).

Shaded environment affected the yield of sweet potato. Reduction in flowering, increase in chlorophyll, and developing climbing habit are a few of the external manifestations of sweet potato plants under shaded environment. Zara *et al.* (1982) reported 32 to 92% yield reduction among the cultivars grown under shade of coconut trees in the Philippines, while under artificial condition, Martin (1983) observed no significant reduction in tuber yield in 20% shade, whereas 80% shade resulted in as high as 94% reduction in yield.

2. Cassava (*Manihot esculenta* Crantz)

Family : Euphorbiaceae

A perennial shrub, with enlarged tuberous roots; there are over 100 cultivars, whose form varies greatly, ranging in height from about 0.9-4.6 m or more; with variously branched stems, green, white, pale or dark grey in colour. The leaves are large and palmate, with 5 to 9 or more lobes. The inflorescence is a panicle and the flowers, which are produced on peduncles within the axis of a branch, are monoecious. The fruit is a six-winged capsule and the seeds are about 1 cm in length. The plant has fibrous root system, but some of the roots develop round the base of the plant into root tubers by a process of secondary thickening.

Sometimes bitter and sweet cassavas are referred to as separate species: the former being *M. esculenta* and the latter *M. palmate* (syn. *M. dulcis)* but this is not justified as the toxicity of a cultivar varies according to location; although for practical purposes it is usual to divide the cultivars into sweet and bitter types and sometimes into short and long season cultivars.

Though the genus *Manihot* comprised of 98 species, only 15 species were cytologically studied and the somatic chromosome number was invariably reported to be 2n=36 (Jos and Sreekumari, 1994).

Table 2. Species diversity of *Manihot*

Species	**Remarks**
Manihot esculenta	Native of Brazil, cultivated for edible tubers
M. glaziovii Muell. Arg.	Ceara rubber - yield rubber of good quality but latex coagulated immediately on exposure (Purseglove, 1974)
M. dichotome Ule	From Brazil
M. piauhyense Ule	From Brazil tried for rubber production
M. anomala	Tuber bearing and high leaf retention
M. caerulescens	Resistance to drought and excessive soil toxicity.
M.tristis	Wild species used in breeding programme for improving the protein content.
M. grahami	Having some tolerance to seasonal low temperature.
M. anisophylla	Having some tolerance to seasonal low temperature.
M. brachyloba	A Mexican species found in Brazil.

Physiology

Cassava productivity in terms of calories/unit land in unit time appears to be significantly higher than that of other staple food crops (De Vries *et al.*,1967). Coursey and Haynes (1970) reported that cassava can produce 250 x 10^3 calories/ha/day as compared to 176 x 10^3 and 110 x 10^3 calories/ha/day produced by rice and wheat respectively indicating its superiority over major cereals in terms of biological value (Ghosh *el.al.*, 1988).

Cassava requires moist soil for bud sprouting and establishment. The initial phase of growth of the axillary buds depends on reserves of stem cuttings. The axillary buds begin to sprout during second week of planting with formation of callus tissue which give rise to root initials. Normally two to three axillary buds elongate and produce leaves which are small with three to five lobes. Besides soil moisture, temperature also plays a vital role on establishment of cassava plant. Though the crop tolerates a hot climate, it cannot establish below the daily mean temperature of 18 to 20^{0}C, considered as critical for cassava (Jones, 1959).

The mechanism of tuber differentiation has also been studied in detail (Doku, 1969; Indira and Sinha, 1970; Paliawal and Kavathekar, 1971; Williams, 1972). The tuber differentiation is associated with initiation of secondary growth of the roots due to cambial activity followed by starch deposition. The activity of root cambium causes stele to enlarge, filled with xylem parenchymatous cells, which store bulk of starch grains. Viswanathan and Krishnan (1965) studied various enzymes associated with starch synthesis in leaves and tubers of cassava. It is reported that starch synthetate and UDPG-Starch glucosyl transferase, the important enzymes of starch synthesis, are found in tubers and not in the leaves.

Photoperiod, temperature, light intensity, soil and water availability are the most important factors governing the onset of tuber differentiation and rate of tuber bulking. The thick roots (>1 cm diameter) are classified as tuberous roots in cassava which can be noticed between the 45th and the 90th day after planting depending upon cultivars. Differences in the time of initiation of tuberous roots noticed among the cultivars may be attributed to the differences in the critical day length requirement for tuberisation (Kopetz and Steineck, 1954); the inherent genetical variation controlling the endogenous hormonal-balance to initiate tuberisation; and the availability of surplus photosynthates to meet demand of activity of root sink.

Bolhius (1966) studied tuberisation using six cassava cultivars under four different photoperiods (10 hours, 12 hours, 14 hours and 16 hours) by supplementing artificial light over and above 10 hours natural light in Netherlands for a duration of seven months. The mean tuber yields under a long day were much lower (0.1-0.2 kg/plant) when compared to 0.6 and 0.7 kg/plant recorded under 10 and 12 hours, found optimum for tuberisation. It was found that a short day promoted tuber growth and reduced shoot growth, while a long day promoted shoot growth and inhibited tuber growth (Lowe *et al.*, 1976; Keating *et al.*, 1985) suggesting an antagonistic relationship between shoot and storage root growth in response to a change in day length.

The root anatomical studies for starch deposition showed that the process of tuberisation in all cultivars was delayed significantly under shade by about three weeks when compared to open environment and there was also significant reduction in number of tuberous root formed per plant under shade

(Ramanujam *et al.*, 1984a). Enyi (1973) estimated optimum LAI for maximum crop growth rate in Sierra Leone which lies between 4.0 and 6.5 depending upon the cultivars. In cassava, accumulation of dry matter in the shoot increased during the early phase of growth till commencement of tuber bulking. The partitioning of dry matter (DM) for growth of shoot and tuber was linear and there was a high positive correlation between the total biomass and DM accumulation in shoot and tuber (Ramanujam and Laskhmi, 1984).

The direct relationship between total root dry weight and maximum photosynthetic rate (Williams, 1972) indicated that root demand for assimilate increases photosynthetic activity. It is shown that tuber bulking in cassava is largely influenced by LAI. However, an increase in LAI above 3.5 (Enyi,1972) and 4.0 (Cock *et al.*,1979) lead to a decrease in tuber bulking rate.

Moisture stress and salinity are the two major environmental stress conditions which limit crop production. Cassava has the tendency to adapt better, under drought situation than under salt accumulation. Cassava can come up well even in low rainfall regions and the plants can withstand prolonged dry periods. Response to moisture stress is mainly by reducing transpiration, thereby conserving soil-plant water status. Under extreme condition, the plants shed their leaves and remain biologically active in contrast to many other crops and put new canopies when the stress condition is removed. Degree of leaf shedding and productivity under limited water supply vary among the cultivars.

Even though a steep fall in LAI below 0.5 has been noticed under laterite soils during prolonged moisture stress, the lateral buds in cassava remain active even during severe drought. Close pruning of stem at this period induced sprouting of lateral buds forming a new canopy at expense of storage roots. Subsequently, the new canopy thus established could support the storage root growth resulting in higher yield (Ghosh *et al.*, 1988)

Cassava is highly sensitive to salt injury. Decay of the cortical parenchyma and formation of tyloses in the vessels are the major anatomical changes noticed in the roots in relation to the toxic effect of salinity (Indira, 1978). Cassava is a sun loving crop and its yield potential could be realised to a maximum only under full sunlight with optimum leaf canopy (Ramanujam *et al.*, 1984a). Cassava accumulates cyanogenic glucoside as a secondary metabolite in all parts of the plant which ultimately limit utilization of crop. Linamarin and lotaustralin are the two cyanogenic glucosides (CNG) found in cassava (Butler, 1965) and the metabolic pathways for synthesis and degradation of these two glucosides in cassava have been elucidated in detail (Nartey, 1973). Significant positive correlations were observed between the HCN content of the leaves and rind of the tuber (Moh, 1976) and flesh of the tuber (Ramanujam *et al.*, 1984b). Bala Nambisan and Sundaresan (1985) worked out various processing techniques for reducing CNG level in tubers and found that crushing followed by sun-drying was the most effective method for elimination of CNG (95%) from cassava tubers.

3. Yams

Family : Dioscoreaceae

Yams are a group of plants belonging to the genus *Dioscorea* which consists of about 600 species, under monocotyledons. They produce underground as well as aerial tubers. Yams are dioecious, with male and female flower borne on separate plants but are essentially vegetatively propagated (CSIR, 1952).

Indo-Burma region is one of the centres of origin of yams (Coursey,1976). Wild relatives of yams are seen in Western Ghats, especially in Southern parts of Western Ghats (Agasthyamalai ranges) in almost all the reserve forests (Ramanathan, 1978). *D. wallichii, D. pentaphylla, D. oppositifolia, D. tomentosa, D. spicata, D. intermedia, D. hispida, D. bulbifera,* and *D. alata* were distributed throughout this range (Prain and Burkill, 1938).

Chromosome counts are made on a considerable number of both wild species and cultivated forms of *Dioscorea*. In case of the most primitive section, Stenophora, the counts are usually 2n=20, although *D. deltoidea* has 2n=40. Among the more typical sections, which include the food yams, 2n=40 is also the most usual, but higher degrees of polyploidy are not uncommon. These forms are, however generally based on multiples of ten (Smith, 1937; Simmonds, 1954; Sharma and De, 1956; Raghavan, 1958; Ramachandran, 1962), the highest number recorded being over 120 (Miege, 1954). As yams are propagated by vegetative means some variation from the characteristic chromosome number occur in some of the cultivated forms as in *D. alata* (2n=38, 52, 55, 66, 81) (Smith, 1937; Sharma and De, 1956), *D. bulbifera* (2n=36), *D. cayenensis* (2n=54), *D. dumetorum* (2n=36,45,55) (Miege, 1954).

Greater Yam (*Dioscorea alata* L.)

A large climber, which can reach 15 m in height, with quandrangular winged stems, twinning to the right. Leaves opposite, variable in size and shape, but essentially ovate, with a deep basal sinus. The male flowers are borne on panicles, sometime 30 cm long; the female flowers are on small axillary spike; most cultivars only rarely produce fertile seed and some are completely sterile. Bulbils are sometimes formed in leaf axils, but not so freely as with certain other species. The tubers are usually single and show a great deal of variation in size, shape and colour. The flesh of some cultivars can be pink or even deep reddish-purple.

Table 3. Species diversity of *Dioscorea*

Species	Remarks
D. dumetorum (Kunth) Pax	Bitter yam, trifoliate leaf and stem pubescent. Tubers large and coarse tubers are bitter and sometimes poisonous. Native : Tropical Africa
D. rotundata (Poir)	White yam, stem twines to the right, wingless and circular in cross section. Tuber cylindrical, smooth skinned, white firm fleshed. Native : West Africa
D. alata L.	Greater yam or water yam, winged stem, stellate in cross section twining to the right, tuber shape extremely variable but mostly cylindrical, tuber flesh is white or purple and watery in texture. Native : South East Asia (Burma)
D. cayenensis Lam.	(Yellow yam) The tuber flesh is yellowish in colour due to the presence of carotenoids, stem twine to left, possess thorns. Native : West Africa
D. esculenta (Lour) Burk	(Lesser yam) Stem twines to the left, stem thin often pubescent and cylindrical, tubers small, characteristically borne in clusters, tubers are white fleshed and sweet in taste when cooked. Native : Indo China
D. bulbifera L.	Aerial yam, characterised by production of large number of bulbils (aerial tubers) at axil of a leaf. Tubers are generally single, round and unbranched and covered with roots. Grows widely in the forest areas in Kerala. Native: Asia and Africa
D. trifida L. (Cush-cush yam)	Vines twine to left, produce small tubers. The stem is winged rectangular in cross section. The tuber flesh is white, yellow, pink or purple and highly palatable. Native : North of South America
D. opposita Thunb. (*D. batatus* syn.)	(Cinnamon yam/Chinese yam) Tolerate colder climate cylindrical and twines to the right, tuber is thin and very long and grows downward to the soil. Bulbils occur in leaf axils. It is the hardiest of yam. Native : China

D. japonica	As food and medicine. Native : Japan
D. hispida	Found growing wildly in Ponmudi forest areas in Kerala. Tubers are single, spherical, irregularly branched and covered with roots. The tubers irritate the throat and are not edible.
D. mexicana	Used for extraction of pharmaceuticals (diosgenin) on commercial basis
D. composite	Used for extraction of pharmaceuticals (diosgenin) on commercial basis
D. floribunda	Used for extraction of pharmaceuticals (diosgenin) on commercial basis
D. wallichii Hook f.	Is a wild species growing in South Wyanad and Car Nicobar Islands. This plant produces many tubers in a cluster. The tubers are long and narrow and grow parallel to the ground. Tuber flesh is fibrous and cream in colour.
D. pentaphylla	Is a wild species producing elongated bulbils in abundance on the vines. Underground tuber is elongated and narrow. The flesh is fibrous, but smooth at the bottom and this portion is eaten by the hill tribes.
D. tomentosa	Is a wild species occurring in CarNicobar Islands,which does not produce bulbils. Underground tubers are one to many, 1 to 2 metres long and narrow with fibrous flesh. Trace of sapogenin content is reported in this species (Barua *et al.*,1956).
D. oppositifolia	Is a wild species of *Dioscorea* which is eaten by hill tribes.
D.cirrhosa Lour	Is indigenous to Southern China and has spread to Indochinese peninsula.It is remarkable for its high content of tannin(6.4%) in its tubers.

Bitter Yam (*Dioscorea dumetorum* Kunth.)

Many forms, the vine twines clockwise, unlike most other yams of economic importance. The leaf is trifoliate and tomentose and the stems robust with spikes, the flowers are small, globose, dioecious and fertile. Bulbils are only rarely formed; the tubers are usually medium sized and sometimes fused together to form a cluster. The wild forms are very poisonous, but the cultivated forms are usually non-toxic. *D. dumetorum* is very closely related to the Asiatic species *D. hispida.*

Chinese Yam (*Dioscorea opposita* Thunb.)

The vine stems are round, more or less spineless and climb to a height of about 3 m, twining to the right. The acuminate leaves are arranged oppositely. Bulbils are formed in the leaf axils. The flowers are cinnamon-scented, sessile in 1 to 2 simple raceme-lime spikes from the axils.

Intoxicating Yam (*Dioscorea hispida* Dennst.)

D. hispida is a climber with a prickly stem twining to the left, with trifoliate hairy leaves. The flowers are small, pale yellow and borne on compound inflorescences, which are usually large. The fruits are about 5 cm long and divided into 3 thin lobes, which are more than twice as long as wide. The species does not produce bulbils. The tubers are borne near the surface and are often lobed, sometimes elongated.

Lesser Yam (*Dioscorea esculenta* Lour. Burk.)

A climber, usually grows to a height of 1.8-3 m. The cylindrical stems twine in a clockwise direction and are usually covered with spines. Leaves are small, alternate, smooth textured, bulbils are absent. Flowers are rare in most cultivars, but when they occur they are larger than in many other species. The tubers are small and resemble sweet potatoes, usually 5-20 in a plant and are borne near the soil surface. Two distinct types of *D. esculenta* are recognized *fasiculata* and *spinosa,* depending upon the spination of the roots.

Potato Yam (*Dioscorea bulbifera* L.)

D. bulbifera is found in wild state, throughout Africa and Asia and has spread to South and Central America and Oceania. The Asiatic forms are generally superior to the African.

Stems of the vines are cylindrical, normally spineless and can attain a height of 6 m or more. The leaves are large, simple, ovate and may be alternate or opposite. The flowers are larger than in most species of *Dioscorea,* with spreading perianths. The tuber is usually small, hard, bitter and unpalatable and in some cultivars almost entirely absent. Edible bulbils or aerial tubers are produced in leaf axils of the vine.

White Yam (*Dioscorea rotundata* Poir.)

There are many cultivars, showing great variation in form and palatability of the tubers. Stem of the vine is cylindrical and in some forms covered with spines, especially near the ground, but other forms are almost spineless. The leaves are simple, usually opposite and normally a dark glossy-green. Under favourable conditions the vine can reach a height of 9-12 m. The tubers are normally cylindrical with rounded or pointed ends although larger specimen often occur, sometimes with much distorted shapes.

Yellow Yam (*Dioscorea cayenensis* Lam.)

Many cultivars show considerable variation in form. The stem is cylindrical and covered with spines, especially near the ground, the degree of coverage varying greatly. The leaves may be opposite or alternate, simple, pointed, light green in colour, with a leathery texture. The flowers are born singly or in pairs. The tubers vary in size and form but are usually light yellow in colour.

Dioscorea abyssinica Hochst.

A native of Ethiopia and widespread in the savannah regions of Africa. It is cultivated to a limited extent as a food crop, especially in Uganda (Burkill, 1939)

Dioscorea belophylla Voigt. (syn. *D. glabra* var. belophylla)

Common at 4000-5000 ft. (CSIR, 1952). A species demanding less rain than most of the others; met with along the Himalaya towards the west as far as the Kashmir valley and found also on the Salt Range; eastwards, as far as is known, its distribution terminates in Sikkim; southwards it occurs on hill crests as far as Nilgiri Range and it has once been recorded from the Khasia Hills. It is common within the area of Upper Gangetic Plain, especially in Dehra Dun and on the Siwalik range also in N. Oudh, Bandelkhand and in the Mirzapur districts (Vern toree Turhur) (Duthie, 1973). The tubers are eaten after repeated boiling, washing and baking. It is found among the fissures of the rocks and is reported as one of the excellent yams by the tribals of Jharkhand, Orissa and West Bengal.

D. glabra Roxb

A climber with stem twining to the right, armed with prickles toward the base, unarmed and smooth above. Leaves elliptic-ovate or ovate. Bulbils absent. Tuber 1,2 or many long stalked and buried deep in the ground, skin earth coloured with a few rootlets; flesh white and edible. The species occur in Assam, Bengal, Bihar, Orissa and in Andaman and Nicobar Islands, and also in Central and E. Himalayas, in Nepal, Darjeeling, Aber hills and Khasia hills.

D.oppositifolia Linn

A climber with glabrous or finely pubescent stem twining to the right. Leaves alternate, sometimes opposite. Bulbils absent. Tuber usually single, descending deep with the soils, cylindrical with a few rootlets; skin reddish; flesh white; soft and edible. It is a native of India and occurs throughout the hills of Deccan at elevation of 2000-4000 ft. (Duthie, 1973).

D. pubera

Discontinuously distributed, largest area across India about the tropic from Indore to upper Burma, in South West India and in Java. It has been found in the state of Travancore (Prain and Burkill, 1938). In the hills of the Ganges valley it is eaten. Further to the South it is called Pani alu or Kanta alu. Stem at the base upto 8 mm in diameter never prickly but with elongated brownish green warts close to the ground, covered with white hairs and in age faintly lined with brown. Main flowers in spikes on special leafless branches which are 16-18 cm long and density pubescent; female flowers in simple or branched spikes. Tubers one or two, penetrating into soil very deeply, often to 2 m with a diameter of 3-8 cm with tawny - orange skin sparingly beset with small rootlets with lemon yellow flesh which is firm at maturity. It is also found in Car Nicobar Islands. In Assam, it is called 'Haldia alu' means the yam with colour of turmeric (Prain and Burkill, 1938).

D.vexans

It is found in Car Nicobar Islands among the tribals.

D.deltoidea Wall.

An alpine rhizomatous species, occurring in the Himalaya, Yunnan and Tonking at altitudes of upto 3,500 metres. The rhizomes contain a toxic sapogenin, and preparations of these are used by some peoples of the area for killing lice in hair or clothing and as a vermifuge and for other medicinal purposes. It is a good source of diosgenin 2.3-3.7% (Coursey, 1967).

Many different forms occur naturally and some of these are described as separate species by certain authors, eg *D. crispate* Roxb., *D. heterophylla,* Roxb., *D. papilaris* Blanco, *D. pulchella* Roxb., *D. sativa* Thunb., *D. tamnifolia* Salisbury, *D. tunga* Hamilton, *D. latifolia* Benth. and *D. anthropophagum* Chev. *D. vexans* is found in Car Nicobar Islands among tribals.

Physiology

Yam tubers exhibit a definite period of dormancy shortly after the harvest and duration varied from two to three months depending upon the species. The high level of tuber dormancy was related to low glutathione content (Campbell *et al.*, 1962b). The dormancy level of freshly harvested tuber is usually high and the time taken for field emergence is prolonged when such materials are used as planting setts. Parsam (1977) reported that storage at low temperature increased the dormancy period, while storage at high relative humidity decreased the rest period. The tuber dormancy of yam was explained to the presence of polyphenolic compounds known as *batatasins* (Hashimoto *et al.*, 1972) and abscissic acid (Hasegawa and Hashimoto, 1973). Bulbils also exhibited a definite period of dormancy. Attempts were made to break dormancy of yam tubers using various chemicals. Dormancy period of yam tubers can be broken by treating tubers with 2-Chloro ethanol (Campbell *et al.*, 1962a) or by gamma radiation (Adesuyi, 1973 and 1976; Martin *et al.*, 1974).

Tuberisation is a special morphogenetic process which requires an inductive environment. Photoperiod, temperatures and edaphic factors usually contribute a major role in tuber differentiation and its subsequent development. Allard (1945) reported that a day length of above 12.5 hours increased growth of vine while a day length of less than 12 hours promoted production of aerial and the subterranean tubers of *Dioscorea batatas* and *D.alata*. Positive influence of short days on tuberisation was demonstrated for yams *(Dioscorea* spp.) by Njoku (1963). Moisture stress delays considerably the onset of tuberisation in yams (Onwueme, 1975).

Four distinct phases of growth are recognized for yams after the period of dormancy (Sobule, 1972; Onwueme, 1978). The first phase lasts up to sixth week during which period, elongation of root and vine and root development can be noticed. During phase I, depending upon the size of seed sett, the vine attains considerable length with rudimentary leaves called 'cataphylls' (Oyolu 1982), which are absent in plants propagated through seeds (Okezie, *et al.* 1980). The second phase coincides with growth of primary shoot and root, starting from sixth week up to tenth week after sprout emergence. Extensive leaf growth and reduction in root growth are the characteristic features associated with phase II. The phase III coincides with rapid tuber bulking with little or no further increase in vine growth, which commences from tenth week after emergence. The dry matter accumulation in the tuber was slow during early part of phase III and attained maximum rate during twentieth to twenty-eighth week after planting and then slowed down considerably towards maturity (Phase IV). There is usually a loss of tuber dry matter during maturity period (Sobule, 1972).

Positive correlation between seed yam size and yield was demonstrated by many workers. Enyi (1973) obtained significant increase in tuber yield of lesser yam from 15 to 20.5 t ha^{-1} (for small setts) and from 23.3 to 25.8 t ha^{-1} (for larger setts) and suggested that the increase in tuber yield with increase in seed sett could be due to associated increase in tuber number, leaf area duration and bulking rate.

4. Taro (*Colocasia esculenta* (L.), Schott.)

Family : Araceae

A perennial, cormous plant, normally 0.4-2.3 m tall, with large heart-shaped leaves variable in size and colour, arising from an underground, farinaceous corm, surrounded by a number of secondary corms or cormels, which also vary greatly in size and shape, and are also farinaceous. The inflorescences, usually 2-5 together in the leaf axils on long peduncles, 15-30 cm long flowering consecutively. Female flowers are uncommon, but when they occur, consist of a few ellipsoid or ovoid ovaries; males have 2-3 sessile anthers. Berries are thin-walled green and only rarely contain a few albuminous seeds.

There are about 1,000 recognized cultivars, but these fall mainly into two groups; the *eddoe* type of taro, which has a relatively small corm surrounded by large well developed cormels and the *dasheen,* which has a large central corm and fewer, smaller, more compactly clustered cormels. The two types of *C. esculenta* are frequently referred to as separate species in the literature: *C. antiquorum* and *C. esculenta,* but it is more generally accepted that the taros are a polymorphic species *C. esculenta* and under this classification the *eddoe* is *C. esculenta* var. *antiquorum* syn. *C. esculenta* var. *globulifera* and the *dasheen* is *C. esculenta* var. *esculenta.*

In *C. esculenta* var. *stoloniferum,* stolons are very much distinct and are the edible portion.

Linnaeus was the first to describe taro in 1752 as belonging to two species *Arum colocasia* and *Arum esculentum* (Hill, 1939). Later, Schott in 1832 established the genus *Colocasia,* which is supposed to have derived from *Colcas,* an Egyptian word indicating taro. He renamed the two Linnaeus' species as *C. antiquorum* and *C. esculenta* but in 1856, he reconsidered the position and used only one name *C. antiquorum.* Although in the meantime a number of other edible aroids were put together (Hill, 1939). *C. esculenta* is a highly polymorphic species where the variability is wide that the morphologists ascribed species status to some varieties.

When all the respects of morphology in relation to taxonomy of species are taken into consideration, Haudricourt (1941) and Purseglove (1972) concluded that it would be more proper to consider *C. esculenta* (L.) Schott as a single polymorphic species with several botanical and cultivated varieties. It was pointed out that as per the accepted nomenclature system for the species, *C. esculenta* took precedence over *C. antiquorum* (Hill,1939) and if one polymorphic species is to be recognized, it should be *C.esculenta* and that *C. antiquorum* would then become a variety. However, Plucknett (1983) further differentiated the two types. (1) *C. esculenta* var. esculenta (Dasheen or bunda or Kanda) having large main corm with a few cormels, can be grown under a wide range of moisture conditions from flooded as in Hawaii and parts of Pacific Islands to rainfed upland conditions. Generally this group has 28 chromosomes. It may be propagated by the leaf bearing tips of mature corm or by small cormels. In Orissa, bunda is known as Sankh Saru and Mankachu in Assam and West Bengal. (2) *C. esculenta* var. antiquorum (sometimes called as *C. esculenta* var. globulifera) (eddoe or taro or arvi) having small or medium sized main corm that often may be inedible because of high acridity and has a large number of side cormels (about 5 to 20 and may be as many as 40 or more) which are edible. Some cormels may have some degree of dormacy. It is raised as an irrigated or rainfed upland crop. Generally this group of plants shows 42 chromosomes and is supposed to have developed in Japan or China. The plant is generally propagated through cormels.

Another type *C. esculenta* var. stoloniferum is found in parts of West Bengal and Assam which produce stolons which may run some distance from the main corm and develop into new plants at the tips. These are found and cultivated on swamp lands and stolons are the edible portion.

Colocasia gigantea Hook. f. is reported to occur in Burma and Java. The leaves and petiole are cooked and eaten. It is said to contain hydrocyanic acid (CSIR, 1950).

Colocasia virosa occurs wild in North Eastern parts of India especially Assam and is reported to be poisonous (CSIR, 1950). It is a threatened species (Narzary, 2002).

A new species of *Colocasia* Schott, *C. lihengiae* restricted to the rainforest of Xishuangbanna in Southern Yunnan, China is reported. There are diagnostic characters that distinguish the new species from the related species,

C. fallax Schott. Chromosome number of $2n$=28 was observed in plants of *C. lihengiae* cultivated at the Kunming Botanical Garden (Chun Lin and KeMing, 2001).

5. Tannia (*Xanthosoma* spp. (L.) Schott.)

Family : Araceae

There is considerable confusion between the genus *Xanthosoma* and that of *Colocasia.* Tannias are often confused with taros, since superficially they appear to be rather large coarse taros. Tannia plants reach a height of 1.8-2.1 m and have a short-stem and large stalked, sagittate or hastate leaves, with sharp pointed tips, deep, wide basal lobes, a prominent marginal vein, light green in colour and 0.45-0.9 m in length when fully developed. The inflorescence is borne below the leaves, but some cultivars rarely flower and seeds are produced very infrequently. A corm is produced at the base of plant and this bears 10 or more lateral cormels, usually 10-25 cm long.

Xanthosoma is cultivated on a small scale in our country mostly in homesteads or as an intercrop in banana and other plantation crops. However, the crop is getting popularity in Maharashtra, Gujarat and other parts of India for making roll of leaves (patra). It is locally called Vadiche aaloo in Marathi, Rikawatch in Hindi and Manakachu in Bengala and Palchaembu in Malayalam.

Many different cultivars of tannia are grown but it is usual to refer to most of the edible ones as *Xanthosoma sagittifolium.* Other recognized species include *X. atrovirens,* a yellow fleshed tannia, which constitutes one of the staple starchy foodstuffs of Dominica and is of importance in Puerto Rico, *X. caracu,* a white fleshed species grown in Puerto Rico for its edible tubers and *X. violaceum,* the blue tannia, which has violet coloured leaves and petioles and large pinky tubers, is common in Phillippines and cultivated for its edible tubers. The foliage is also used as a vegetable. *X. brasiliense* is another edible species grown not for tuber but for its leaves to be used as salad.

The somatic chromosome number of tannia is 26. The haploid chromosome complement at pachytene could be resolved into 12 types based on the morphology and staining pattern (Magoon *et al.,* 1970).

Another species *X. nigrum* is distinguished by purple violet petiole, bluish pruinose young leaves and basal nerves with purple tinge whereas *X. sagittifolium* is distinguished by green petiole, green young leaves and pale basal nerves.

6. Giant Taro (*Alocasia macrorrhiza* L. Schott.)

Family : Araceae

Tall succulent herbaceous plant, usually at least 1.8 m in height, but can reach 4.5 m, with large elongated stems. The leaves are about 0.9 m long, generally arrow-shaped, with rather shallow and rounded lobes. The leaves point upwards forming a straight line with the main axis of the petiole, unlike *Colocasia* and *Xanthosoma* spp., which point down to form an acute or right angle, with the axis of petiole. The leaves have a conspicuous mid-rib and are green in colour, except the variegated forms which can be blotched or mottled with white. The spathe has a glaucous, yellowish-green blade. Ovules are one to a few in each ovary cell in sub basal placenta. Leaves nearly peltate. Caudex well developed. A number of species of *Alocasia* were identified from different parts of North East India. *A. macrorrhiza* forms underground cormels.

Table 5. The species diversity of *Alocasia*

Species	Remarks
Alocasia cucullata (Lour) G. Don	Chinese Ape, Chinese Taro-Leaves and tubers used as vegetable
A. lowii Hook.	Leaves olive green and purple. Cultivated as ornamental
A. sanderiana (Schott) G. Don.	These species have been recorded as being irritant (Burkill, 1935).
A. denudata Engl.	This species has been recorded as being irritant (Burkill, 1935). The species has been used as an irritant in dart poison and is also a contact poison since if the powdered plant is mixed with bamboo hairs and rubbed over the body, severe itching and sometimes death may occur.
A. indica Schott	Indomalayan Alocasia - This species has been recorded as being irritant. The rhizome may be used as a rubefacient.
A. macrorrhizos G. Don. (syns. *Arum macrorrhizum* L., *Alocasia macrorrhiza* Schott, *Colocasia macrorrhiza* Schott)	Giant Alocasia, Giant Elephant's Ears, Giant Taro, Taro, Cunjevoi, Cunjerboy, Ape, Spoon Lily, Australian Cabbage. The whole plant is irritant; the sap of the leaves and petioles will blister the lips and is powerfully and sometimes destructively irritant to the eyes. The sliced tubers and leaves may be used as a rubefacient, and the acrid juice has been used to combat the caustic effect of the sap of *Melanorrhoea laccifera* Pierre

A. portei Becc. & Engl.	Reported as having irritant juice
A. metallica Hook.	Has rich bronze and purple leaves, cultivated as ornamental
A. acuminata Schott.	Cultivated species
A. montana (Roxb.) Schott.	The roots of this plant is reported to be used for poisoning tigers.
A. fornicata (Roxb.) Schott.	Medicinal

A. cucullata is an ornamental small herb with lateral nerves, excepting the upper most a few, congested at base in a nearly palmate arrangement. *A. fallax, A. macrorrhiza* and *A. fornicata* are rather big herbs with lateral nerves almost equidistant, pinnately arranged. Chromosome number of most species is 2n = 28.

7. Giant Swamp Taro (*Cyrtosperma chamissonis* (Schott,) Merr.)

Family : Araceae

A giant herbaceous perennial, usually 1.8-2.4 m high, even more, with 6-8 large arrow-shaped leaves about 1.5 m in length, carried on long thick petioles in a spiral arrangement with a relatively short leaf sheath. The leaf blades have deep and sharply pointed basal lobes. The petiole has a diameter of about 10 cm and in some cultivars the lower part is covered with spines. There is considerable variation among cultivars in the leaf shapes and colour, and in colour and degree of spineness of the petioles. The young purplish spadix is tubular and covered with a bract gradually tapering towards the top. The fruit is a berry and seeds of cultivated forms are frequently fertile. The corm develops by thickening at base of stem and almost cylindrical in shape and externally resembles a banana sucker. The chromosome number of *Cyrtosperma* is 2n=26 (Mangenot and Mangenot, 1962).

Species that are potentially excellent horticultural subjects, both for their long lasting inflorescence and spectacular foliage, are *C. cuspidispathum, C. macrotum, C. carii*, and *C. beccarianum* (www.aroid.org/genera/ cyrtosperma/ Cyrtospermalist.html).

8. Elephant foot yam (*Amorphophallus* spp.)

Family : Araceae

A robust herbaceous plant, with an erect solitary stem usually 3.2-8.5 ft (1-2.5 m) in height and bearing at top a single tripartite leaf each part of which is divided into numerous segments. Towards the end of the plant's cycle (usually 4-6 yrs.) a large terminal inflorescence is produced, consisting of a short stalk and spathe and a spadix, which emits a malodorous smell, reminiscent of rotten meat. The corms are large globulose depressed tubers, usually dull yellow or brownish-yellow in colour and these produce 5-10 corms or cormels at the end of each growing season.

The species diversity of *Amorphophallus*

Species	**Remarks**
Amorphophallus paeonifolius (Dennst.) Nicolson	Cultivated species.
A. prainii Hook. f.	This species irritates the oral cavity when eaten (Burkill, 1935).
A. rivieri Durand (syn. *A. konjac* K. Koch) Devil's Tongue, Leopard Palm, Umbrella Arum, Snake Pal	The sap of the flower produces marked itching.
A. sylvaticus (Roxb) Kunth. Wild Suran	Found in deccan peninsula and suspected to be poisonous to human beings and livestock. The fruit causes a lasting burning of the tongue and lips together with salivation and numbness. In Indian indigenous medicine, the crushed seeds are applied externally to bruises and also into cavities in teeth as an analgesic and in the treatment of glandular enlargement.
A. variabilis Blume	The tubers of this species are inedible except in Java where they are eaten after suitable preparation (Burkill, 1935).
A. commutatus (Schott) Engl.	The appendages eaten as vegetable but the flowers emit a good deal of stench
A. dubius Blume. Purple stalked dragon	The tuber is administered in piles and apoplexy

A. paeonifolius and *A. variabilis* have chromosome number 2n=28, where as *A. blumei* possessed chromosome number 2n=39.

Many wild spp. were explored from Western Ghats like *A. dubius, A. wightii* and *A. sylvaticus,* Kunth. A mention of *A. paeonifolius* (Dennst) Nicols. var. paeonifolius which is a thorny type is made in *Hortus Malabaricus* (Van Rheedes, 2003). Other species are *A. oncophyllum, A. variabilis, A. rivieri* (Ghosh *et al.,* 1988) which produce small tubers which are not edible. The corms of the *A. paeonifolius* vary - One type with conspicuous corms and less or no cormels and the other type with one corm and more cormels.

Physiology of Aroids

Colocasia and *Xanthosoma* are together called cocoyams. Corms are the storage organs of cocoyams which represent morphologically the main stem of plant and rings of leaf scars are prominently noticed on surface. Lateral buds are located just above the leaf scars which give rise to secondary corms, which in turn produce the tertiary corms. Both secondary and tertiary corms are called cormels which develop into suckers. Some cultivars produce stolons which may run some distance from the main corm and develop into new plants at the tips. Growth of roots and leaves begin simultaneoulsy when the apical portion of corm with petiole base is used as planting material. However, root growth commences much earlier than that of leaf growth, when corms devoid of petiole base are used as planting material.

It is demonstrated that plant hormones like cytokinins are controlling the process of initiation and development of underground storage organs (Palmer and Smith, 1969). The presence of cytokinin like activity and growth inhibitory activity has been detected in the extracts of cormels (Radosevich and Maldonado, 1984). These endogenous substances may contribute to the process of cormel intitiation and its subsequent development. Tsukamoto and Inaba (1961) observed positive influence of short days on tuberisation in cocoyam (*Colocasia esculenta* var. antiquorum Schott).

Three phase of growth in taro are recognized (Reynolds and Netram, 1973; Sivan, 1979). The Phase I coincides with regeneration of mother corm mainly characterised by emergence of roots and leaves extending up to eight weeks after planting. The phase II represents the grand growth period characterized by rapid vegetative growth commencing from second to sixth month after planting (MAP), while the phase III coincides with crop maturity and duration of which varied with cultivation and growing conditions.

Tannia is more robust in growth and taller than taro, perennial and grows up to a height of 60 mm or more. The growth pattern of *Xanthosoma* was almost similar to that of *Colocasia* as described earlier. Sprouting of seed corm in *Xanthosoma* starts within two weeks after planting, lasting up to four weeks. Although lateral buds are present at the axil of leaf scars of main corm, only some of the buds develop into suckers in tannia. Root initials are noticed within a week after planting and root growth increased rapidly up to four to five months stage and then declined with commencement of corm bulking.

Cocoyams and elephant foot yam are perennial species but most often cultivated as annuals. In Cocoyams there is no clear-cut symptom for maturity while in elephant foot yam the psuedostems wilt and perish when the corms are mature.The leaves of cocoyams become yellowish towards the end of a growth cycle and the ratio in the weight of top to corm decreased towards harvest which can be considered as a useful index for maturity. Besides, the quality of tuber (corms, cormels) with particular reference to DM and starch content can also be considered to predict the optimum time of harvest. It has been observed that prolonged dry condition accelerated maturity while interruption of the long dry spell with late rains promoted regeneration of the sucker using the stored food materials resulting in a drastic reduction in the cormel quality. Specific gravity of the corm may also be considered as a useful parameter to determine the maturity of cocoyams.

Taro varieties differ significantly in their maturity period which range from six to 15 months according to the cropping system being followed. The distribution pattern of rainfall largely determines duration of cocoyam in upland while in lowlands, the type of irrigation systems/ practices influences the crop duration. Usually upland taro matures earlier (six to nine months) than the lowland taro (12 to 15 months). It is reported that under high fertility wetland conditions, corm bulking commences very late (3 to 5 MAP) while under upland conditions bulking commences early (2 to 3 MAP) and the crop duration in the latter is shortened due to prevailing dry conditions. The specific gravity of 1.0 to 1.1 for main corm and 1.1 to 1.58 for sucker corms were optimum to determine the stage of harvest of wetland taro (Ezumah and Plucknett, 1973).

Leaves of cocoyam are being used as vegetable in Africa, Caribbean, the Philippines, Pacific Islands etc. and hence defoliation is often practiced. Studies showed that light defoliation did not affect yield of *Colocasia* but removal of the youngest two or three leaves during rapid bulking, reduced corm yield significantly (Jackson *et al.,* 1980). Karikari (1973) also found significant reduction in the yield of *Xanthosoma* when defoliation was made during the first 23 weeks after planting or when more than four leaves per plant were removed after the twenty-third week. Wherever leaf removal is in practice, plant types with larger leaf area and higher rate of leaf production have to be identified to compensate defoliation without being detrimental to yield.

Cocoyams are propagated vegetatively as flowering is considered a rare phenomenon and even if flowered, seed set is not noticed in many cultivars. In Solomon Islands, a large number of cultivars of taro produce flowers and viable seeds (Jackson *et al.,* 1977). Strauss *et al.* (1979) reported that some cultivars do not flower under field condition but flowering occur spontaneously when grown in a green house. From the breeding point of view, both flowering and seed set are essential. Shaw (1975) reported production of viable seeds through sexual cross in taro and Volin and Zettler (1976) in tannia. Flowering can be induced in field-grown taro by application of gibberellic acid. In tannia, foliar application of GA_3 (50 and 250 ppm), one month after planting, promoted flowering (McDavid and Alamu, 1976).

9. Hausa Potato (*Solenostemon rotundifolius* (Poir.), J.K. Morton)

syn. *Plectranthus rotundifolius* (Poir.) Spreng. *(Coleus parviflorus)*

Family : Labiatae

A small herbaceous annual, 15-30 cm high, prostrate or ascending, with a succulent stem and somewhat thick leaves having an aromatic smell resembling that of mint. Flowers small, pale violet in colour produced on an elongated terminal raceme. Small, dark brown, tubers and produced in clusters at base of stem. The chromosome number is 2n=64, 84.

Coleus (*Solenostemon rotundifolius* (Poir.), J.K. Morton) is also known as Chinese potato or country potato or coleus potato in English and Koorka in Hindi. It is cultivated in our country on a small scale in the South including Kerala, Tamil Nadu and Karnataka and in North Eastern states, Madhya Pradesh and tribal areas all over India.

Coleus amboinicus Lour. commonly known as wild oregano, is rather large succulent herb with aromatic leaves and small, pale purple flowers commonly cultivated in gardens throughout India. It is a native of West Indies.

C. forskohlii Briq. syn. *C. barbatus* is a perennial member of the mint family having medicinal properties. It is a branched aromatic herb 1-2 ft. height with a thick root stalk, distributed in subtropical Himalayas of Kumaon and Nepal ascending to 8000 ft. and in the Deccan Peninsula, Gujarat and Bihar. It is common on dry, barren hills and is cultivated in Bombay and other regions for roots and pickled and eaten (CSIR, 1950).

Coleus vetteveroides is a succulent herb, 45-53 cm height with a procumbent stem and thick, purplish, pubescent leaves, cultivated in Tamil Nadu. The plant bears long (35-50 cm) fibrous roots which are straw coloured and strongly aromatic when fresh (CSIR, 1950).

10. Yam Bean (*Pachyrhizus erosus* (L.))

Family : Leguminoceae

A coarse hairy climbing or trailing vine with large fleshy tuberous roots, the stems of which can reach about 5.5 m in length and are woody at the base, somewhat angular and hairy. The leaves are alternate trifoliate, with large toothed leaflets. The flowers are violet or white and are borne on fascicled pedicels in racemes. The seed pods are usually 7.5-15 cm long, almost smooth at maturity and containing 8-10 yellow, brown, or red, square more or less flattened seeds. There are usually several tubers, which can be simple or compound and of various shapes. The underground part of the plant is perennial, but the aerial portions are annual.

The genus consists of six species of perennial herbs with trifoliate leaves and tuberous roots. *P. erosus* and *P. tuberosus* are cultivated for their edible tubers. *P. ahipa* (Wedd.) Perodi, a non-climbing species, is cultivated for its small fusiform tubers (Purseglove, 1974).

Yam bean also called potato bean in English and Mishrikand in Hindi is known as Kesaru in Bihar. It is called Sank alu in West Bengal, Assam and Orissa. The tuber is rich in ascorbic acid and has high sugar content. In many developed countries the tubers are processed, canned and many sweet preparations are made.

The diploid chromosome complement of yam bean is 2n=22 (Vimala, 1994).

11. Arrowroot (*Maranta arundinacea* L.)

Family : Marantaceae

An erect herbaceous dichotomously branched perennial, 60-180 cm high, with large, fleshy, cylindrical, obvoid subterranean rhizomes, large lanceolate leaves and white flowers arranged in twin clusters, which very rarely produce red seeds.

Maranta arundinacea L is commonly known as West Indian Arrowroot. It is primarily grown for its quality starch which is valued as a food stuff, particularly for infants and invalids and has a number of industrial uses in making various bakery products, special glue and paste, as a base for face powder, as ice cream stabilizer and in carbonless paper used for computer printouts. In India it is grown in North Eastern States, West Bengal, Assam and Kerala in homesteads.

Two main cultivars are recognized in St Vincent, 'Creole' which has long thin rhizomes, which spread more widely and penetrate more deeply into the soil and 'Banana', which has shorter thicker less fibrous rhizomes produced near soil surface. The latter is more easily adapted to mechanical harvesting. The chromosome number is 2n=18 (Sato, 1960), 48 (Simmonds, 1954).

The genus consists of about 25 spp. of perennial herbs in tropical America. *M. arundinacea* is the arrowroot. *M. bicolor* Ker. and *M. leuconeura* Morr. are cultivated as ornamentals for their attractive variegated leaves

(Purseglove,1972). *M. leuconeura* or Prayer Plant is a native of Brazil. The leaves fold upwards at night, hence the common name.

12. Queensland Arrowroot (*Canna edulis* Ker-Gawler)

Family : Cannaceae

A perennial, herbaceous, monocotyledon, rather variable in many characteristics such as colour of foliage, height, size, shape and composition of the rhizomes. The stems are usually purple, normally 0.9-1.8 m in height, but can reach 3.0 m or even higher and are fleshy and put up in clumps. The large, broad, pointed leaves are entire, normally 30 cm long and about 12.5 cm wide with a marked, thick, mid-rib; they are often suffuse purple beneath. The unisexual flowers have orange-red petals about 5 cm long and three petal-like staminodia. The fruit is a 3-celled capsule with round black seeds. The rhizomes have fleshy segments resembling corms and are borne in clumps, which can reach 60 cm in length. In the Andes two clones are recognized: 'Verdes' with dirty-white corms and bright green foliage and 'Morados' with corms covered with violet coloured scales. *Canna edulis* has a chromosome number of 2n=18 (Simmonds, 1954), 27 (Venkatasubban, 1946).

13. Winged bean

Family : Leguminoceae

The Winged bean (*Psophocarpus tetragonolobus* (L.) DC.) also known as the Goa bean, is a tropical legume plant native to Papua New Guinea. It grows abundantly in hot, humid equatorial countries, from the Philippines and Indonesia to India, Burma and Sri Lanka. It does well in humid tropics with high rainfall.

The winged bean plant grows as a vine with climbing stems and leaves, 3-4 m in height. It is herbaceous perennial, but can be grown as an annual. It is generally taller and more massive than the Common bean. The fibrous roots are numerous with the main laterals running horizontally near the soil surface; after a few months they usually become thickened and tuberous, near the base of plant, though this does not occur in all strains. The roots are normally heavily nodulated. The stem is moderately thick, slightly ridged and grooved, and can reach 3-3.6 m in height, if given support. The leaves are trifoliate, on long, stiff petioles; the leaflets are ovate, 7.5-15 cm long with the terminal one usually longer than the laterals and attached to the petiole by a marked pulvinus. The inflorescence is borne on an axillary raceme, up to 15 cm in length, with 2-10 flowers, which may be blue, white or lilac. They contain 5-20 seeds, which can vary in colour from white, through varying shades or yellow and brown to black, and may be mottled. The bean pod is typically 15-22 cm long and has four wings with frilly edges running lengthwise. The skin is waxy and the flesh partially translucent in the young pods. When the pod is fully ripe, it turns an ash-brown color and splits open to release the seeds. The large flower is pale blue. The beans themselves are similar to soybeans in both use and nutritional content (being 29.8% to 39% protein).

Another species *P. palustris* Desv. grows wild throughout tropical Africa.

The plant is one of the best nitrogen fixers with nodulation accomplished by the soil bacterium *Rhizobium*. Because of its ability to fix nitrogen from the atmosphere, the plant requires very little or no fertilizers. The chromosome count is 2n=26.

14. Jerusalem Artichoke (*Helianthus tuberosus* L.)

Family : Compositae

An erect, hardy tuberous perennial, but normally cultivated as an annual. It attains a height of 0.6-1.2 m in the tropics and 1.8-3.0 m in more temperate areas. The leaves are approximately 10-20 cm long, opposite below, alternate above; finely pubescent beneath. The number of flowers varies greatly according to climate and cultivar, but the flower heads normally have a diameter of 6-7.5 cm with yellow florets. The plant produces a number of small tubers, which resemble potatoes, but have much larger eyes and are often knobbly; there is a wide range of forms, one which was developed in France has long fusiform tubers of good quality and flavour and is often designated as *H. tuberosus* var. *fusiformis.* The chromosome number is n=34

15. Asparagus

Family : Liliaceae/Asparagaceae

Asparagus rubicundus

A fibrous-rooted, hairless shrub with smooth, dark stems that can grow to 1.5 m in height, especially in shade. Sharp, slightly curving spines up to 6 mm long occur on all but the flowering branches. The 6 mm long flowers, usually solitary in the axils of the 3-20 mm long needle-like leaves, give rise to red berries.

The chromosome number of asparagus is 2n=2x=20.

16. Vigna species

Family : Fabaceae alt. Leguminosae

Zombi Pea (*Vigna capensis* Walp. Syn. *Vigna vexillata* L.)

Zombi pea *(Vigna vexillata* (L) A. Rich), also known as Aka sasage, Pois zombie, Pois poison, is a perennial climbing or trailing herb, 0.6-3.0 m long; stems sparsely or densely covered with brown hairs; leaves compound; stipulate, alternate, trifoliate and pinnate with bisexual flowers, zygomorphic with papilonaceous corolla consisting of 5 petals, comprising one standard, 2 wings and 2 keels, 10 stamens, 9 fused to form a staminal column and one free and a carpel with the style borne laterally on the ovary, flowers greenish yellow tinged with purple; pod sessile, linear, straight, 7.4 cm long, 10-11 seeded, glabrescent or pubescent and covered with brown bristly hairs; seed oblong-reniform, buff to black, unspotted or brown or dark red with black specks; root tuberous, spindle shaped and edible and only one tuber is borne on each plant. However, 2 tubers or branched tubers are also observed in some plants. The tubers are 5.0-19.0 g, having an easily peelable skin. The seeds can be used as a nutritious vegetable. Vigna species are diploids, with a basic chromosome number of x = 11. The species has 2n=22 chromosomes.

Pillipesera (*Vigna trilobata* (L.) Verdcourt)

Vigna trilobata (L) Verdcourt syn. *Dolichos trilobus* (Linn.) sensu Ait & Arut, is an annual or perennial plant with stems slender, trailing, 30-60 cm long, glabrous or with a few decidous hairs; leaflets more or less deeply 3 lobed, glabrous or with a few short hairs; flowers in close deltoid heads; pods 2.5-5.0 cm long, sub-cylindrical, glabrous, recurved; seeds 6-12, long cylindrical to oblong, truncate at each end; seed coat colour mottled brown or black.

Rice bean (*Vigna umbellata* (Thunb.) Ohwi and Ohashi)

Rice bean *(Vigna umbellata* (Thunb.) Ohwi and Ohashi, syn. *Phaseolus calcaratus* Roxb.), also known as climbing mountain bean, mambi bean, oriential bean and haricot bean, is a highly branched annual with erect or sub-erect stem tending to be viny and tendillar, axillary raceme inflorescence, bright yellow flowers occurring in clusters, pods slender and somewhat curved, seeds oblong with seed coat smooth, dark vine, green yellow, brown, black, speckled or mottle. Wild forms are perennial, very viny, thin stemmed with tuberous root system. Rice bean is largely a self-pollinated diploid (2n=22) but there is some evidence of natural cross pollination (Sastrapradja and Sutarno, 1977).

17. Hyacinth bean *(Lablab purpureus* (L.)

Hyacinth bean *(Lablab purpureus* (L.) Sweet syn. *Dolichos lablab* L., *Dolichos purpureus* L., *Lablab vulgaris* Savi), also known as Lablab bean, Field bean, Indian bean, Dolichos bean, Egyptian bean, is a perennial species (2 or more years) but cultivated as an annual or biennial, with bushy erect and viny, extremely variable forms; stems twining, hairy or glabrous, usually 2-3 m, but often upto 10 m long, other forms dwarf and bushy; leaves pinnately trifoliate; flowers purple, pink or white, 2-4 at each node in an elongated raceme upto 2.5 cm long; pods flat or inflated, 5-20 cm long, pubescent or smooth, papery, straight or curved, green or purple; seeds 3-6, black or white perennial forms produce an edible root tuber also (Smartt, 1990). Hyacinth bean is a self-fertilized crop with chrosome number 2n=22, 24.

18. Shoti (*Curcuma zedoaria* (Berg.) Roscoe.)

Family : Zingiberaceae

Robust perennial tillering plant, with erect stems sprouting from the tops of the lateral rhizomes, reaches about 0.45 cm in height and bears green leaves with brownish-purple veins. The subterranean rhizome is fleshy, creeping branched and gives rise to tuberous roots. Ramachandran (1961) reported a chromosome number of 2n=63 and Chakravorti (1948) as 2n=63,64.

19. East Indian Arrowroot (*Tacca leontopetaloides* (L) Kuntze)

Family : Taccaceae

The genus *tacca* includes about 30 species of perennial herbs with tuberous or creeping rhizome. A perennial herb, 60-90 cm tall with a tuberous, starchy rootstock and a petiole 60-90 cm high, with broad, deeply lobed, divided single leaf, about 30 cm across. The single leaf is erect, long-stalked, glabrous, 3-partite, each segment being further divided and pinnati-partite, with the lower segments separated and the upper ones connected. The inflorescence, borne on a long, rather slender peduncle, has a terminal umbel of green flowers, subtended by about 6 outer involucral bracts, which are 3-4 cm long, spatulate, acuminate, strongly nerved outside, and a number of long threadlike inner bracts, which are purpish in colour. The outer flowers of the umbel mature first ; flowers regular, pedicelled ; perianth short, 6-lobed ; stamens 6 inserted on tube; ovary inferior, 1-locular, with short style and curving, somewhat petaloid stigmas. The fruit is ovoid, smooth, yellowish berry, six-ribbed, about 3.5 cm in length with persistent perianth. There appears to be two distinct types growing in the Pacific Islands, one produces large number of potato-sized tubers and the other produces a single large tuber. Seeds numerous, pale brown, ribbed. Baldwin and Spece (1951) reported a chromosome number of 2n=30 for tacca.

20. Safed musli

Family : Liliaceae

Safed Musli belongs to family, Liliaceae. It was originally grown in thick forest in natural form, and is a traditional medicinal plant. Safed Musli is a herb with sub-erect lanceolate leaves and tuberous root system. It can grow upto a maximum height of 45 cm. Tubers can grow upto a depth of 30". Safed Musli is a tiny annual herb that grows well in tropical and sub-tropical climates with altitudes upto 1500 meters. Safed Musli has its origin in the Indian Subcontinent. Safed Musli is an annual herb with tubers, crown, leaf and flowers as useful parts. It has a chromosome number of 2n=4x=28.

21. *Costus speciosus* (Koenig) Sm.

Family : Costaceae

It is readily recognized in the field by its broadly lanceolate, dorsally silky leaves, borne spirally on stout, erect or sometimes ascending stems. The flowers are white, with a large incurved lip. These are clustered in terminal globose heads characterized with large shiny brown or red bracts.

Both diploid (2x = 18) (Sato, 1960) and tetraploid forms occur in nature. The tetraploid plants are generally characterized with an ascending habit, while diploids have tall erect canes densely crowded with leaves. Triploids (3x = 27) (Simmonds, 1954) have also been reported in literature.

C. speciosus perennates through underground rhizome, which lies a few cm below soil surfaces. The rhizome is differentiated into nodes and internodes. The vegetative activity above the ground lasts for about seven to eight months. Stems sprout during April, flowering commences during July and continues till end of September, the fruits ripen during middle of November, after which the leaves are shed and majority of canes start drying up. The underground portions remain dormant from December to March or even April.

22. *Gloriosa superba*

Family : Liliaceae/ Colchicaceae

Gloriosa superba is a semi-woody herbaceous, branching climber, reaching approximately 5 metres in height. 1 to 4 stems arise from a single V-shaped fleshy cylindrical tuber. The leaves are stalkless, lance-shaped, alternate or opposite or in whorls of up to 3; leaf size, 5 to 15 cm long by 4 to 5 cm wide with parallel veins and tips ending in spiral tendril which are used for climbing. Large, showy, long- stalked flowers are made up of 6 long reflexed petals usually with wavy margins. Flower size, 6 to 10 cm long by 1 to 2.5 cm wide. Flower colour, usually very bright ranging from red with yellow margins to very pale yellow forms with a mauve or purple stripe; pale white forms also occur. Many additional colour forms have arisen through cultivation. The fruit is oblong, 6 to 12 cm by 2 to 2.5 cm and contains about 20 globose red seeds in each valve (Huxley, 1992; Neuwinger, 1994; Burkill, 1995).

The plant grows in sunny positions in free-draining soil; it is very tolerant of nutrient-poor soils. In warm tropical countries, it occurs in thickets, bushland, forest edges and cultivated areas up to a height of 2530 metres above sea level. It is widely grown as an ornamental in cool temperate countries under glass or in conservatories (Neuwinger, 1994). It is native to tropical Asia and Africa and it is commonly known as Glory lily or Malabar lily. Different chromosome numbers are reported for *Gloriosa superba* - 2n=22 (Miller, 1930), (Tjio, 1949), 88 (La Cour, 1951), 90 (Sharma and Sharma, 1961).

23. Ullucus tuberosus

Family : Basellaceae

Ullucus tuberosus Caldas, Ullucu, is endemic in the Andes and is an important food crop at high altitudes from Bolivia and central Peru to Ecuador and Colombia. The plant is highly resistant to frost and gives good yields of edible tubers. A twining herb with alternate, cordate leaves and axillary inflorescences. The tubers may be either small and round or elongated and curved; they are usually pale magenta or yellow in colour.

Ulluco tubers have many shapes, colors and sizes in the Andes. There are two types of plants: a sprawling plant with red-tinted stalks and leaves, and an upright plant, smaller in size, with large, intensely green leaves. The chromosome number is 2n=24.

24. *Artemisia dracunculus*

Family : Asteraceae

It grows to a height of about 60 cm and has long, narrow leaves, which, unlike other members of its genus, are undivided. It blossoms in August, the small flowers, in round heads, being yellow mingled with black, and rarely fully open. Inflorescence is an ample panicle; involucre 2-4 mm. high, imbricate and dry; corollas all tubular, yellowish, the outer pistillate and fertile, the inner sterile with abortive ovary; The roots are long and fibrous, spreading by runners. The plant is nearly odourless to strongly odorous perennial herb, usually glabrous, from a stout rhizome. Leaves are linear or broadly linear, 3-8 cm. long and 2-10 mm. wide, entire, or a few cleft, the lower generally deciduous. Fruits are achenes which are glabrous. The chromosome number of *Artemisia dracunculus* is 2n=18, 36.

25. *Typhonium* species

Family : Araceae

The genus Typhonium includes perennial tuberous herbs distributed throughout India. *Typhonium trilobatum* Schott. is found distributed in the north eastern region of India. The plants are erect with trilobed leaves grown to a height of 50-75 cm. The globose tubers are acrid and used as a vegetable (CSIR, 1976). This is a popular tuber crop in Tamil Nadu, known for its culinary and medicinal value.

Tubers are nearly round and white having fibres issuing from the apex. Leaves are radical, stalked, deeply 3-lobed, lobes ovate, painted, petioles erect, round, tapering, striated, scape axillary and very short. Spathe shorter than the petioles. Spadix of the base surrounded with ovaries.

In *Typhonium flagelliforme,* the tubers are globose and leaves are hastate and sagittately cordate in shape. *T. divaricatum* is a perennial herb with sagitate leaves and sub-globose small tubers. The tuber yield varied from 25-50 g/ plant.

The diploid chromosome number of *Typhonium* spp. is 2n=2x=10.

26. Kudzu (*Pueraria*)

Family : Fabaceae

Pueraria species are strong climbers, rarely shrubs. The leaves have three leaflets, as usual for Phaseoleae. The stipules are attached either at the base or below the middle, the basal lobe is single, twin-lobed or fringed. The inflorescence is crowded, difficult to interpret from herbarium, with either 3, seldom 2, or else 4 to 7 flowers per brachyblast, the more or less thickened node. The flowers are subsessile or short-pedicellate, bracts are readily caduceus, whereas the bracteoles near the calyx are quite persistent. The upper calyx boles are connate or nearly so. The vexillum has a clawed base, with or without auricles, is inflexed and callosities are present or absent. Flower colour varies from white to blue or purple, sometimes with a green or yellow patch on the vexillum. Pods are commonly flat, hairy or glabrous, sometimes septate, pale brown to black. The pods bear up to 20 flattened-oblong seeds that look smooth, but are minutely potted when viewed under strong magnification.

Chromosome counts in *Pueraria* give both 2n=22 (and n=11) or 2n=24 for *P. lobata and P. phaseoloides, P. collettii* has 2n=20 or 2n=44 chromosomes. The counts other than 2n=22 seem less likely, as the Glycinicae usually has basic chromosome numbers of 10 and 11 (Maesen, 1985; Lackey, 1977).

27. Tuberous Pea *(Lathyrus tuberosus)*

Family : Fabaceae

Lathyrus tuberosus, also known as the, Earthnut Pea, or Aardaker, is a small, climbing perennial plant, native to moist temperate parts of Europe and western Asia. The plant is a trailer or weak climber, supported by tendrils, growing to 1.2 m tall. The leaves are pinnate, with two leaflets and a branched twining tendril at the apex of the petiole. Its flowers are hermaphroditic, pollinated by bees. The plant can also spread vegetatively from the root system (Bhag Mal, 1994).

It is occasionally grown as a root vegetable for its edible tuber. The sweet, starchy tubers are edible, cooked or raw. Although palatable and nutritious, the crop is hampered by low productivity.

The chromosome number of *Lathyrus tuberosus* is 2n=14 (Darlington and Wylie, 1955; Fedorov, 1974).

28. Arracacha (*Arracacia xanthorrhiza* Bancroft)

Family : Apiaceae

The arracacha is a garden root vegetable originally from the Andes, somewhat intermediate between the carrot and celery. Its starchy taproot is a popular food item in South America, especially in Brazil where it is a major commercial crop.

The leaves are similar to parsley, and vary from dark green to purple. The roots resemble fat short carrots, with lustrous off-white skin. The interior may be white, yellow, or purple. The most important part is the starchy root. It cannot be eaten raw, but when cooked it develops a distinctive flavor and aroma that have been described as «a delicate blend of celery, cabbage and roast chestnuts» (http://en.wikipedia.org).

The diploid chromosome number of *Arracacia* is 2n=44 (Fedorov,1974).

29. *Arthropodium*

Family : Liliaceae

Arthropodium is a genus of herbaceous perennial plants native to the Southern Hemisphere. It is sometimes placed in the family Asphodelaceae or

Lomandraceae. The rhizomes of some species can be eaten as root vegetables, including *A. cirratum, A. milleflorum, A. minus,* and *A. strictum. A. cirratum* is native to New Zealand, where it may once have been farmed. It is used for medicine as well as food, and has symbolic importance in traditional Mâori culture.

The different species show a chromosome number of 2n=22,44,36 (Darlington and Wylie, 1955).

30. Mashua *(Tropaeolum tuberosum* Ruíz and Pavón*)*

Family : Tropaeolaceae

The mashua is a perennial plant grown in the Andes for its edible tuber. It is a major food source there. The tuber is rather peppery in flavor when raw, but this quality disappears when cooked. It is related to nasturtiums.

The plant is well adapted to high-altitude subsistence agriculture. It grows vigorously even in marginal soils and in the presence of weeds, and gives high yields. Tubers comprise as much as 75 percent of the mature plants by dry weight (40 percent is typical for cereals), and upto 70 tons per hectare tubers could be produced under research conditions.

Its extraordinary resistance to insect, nematode, and bacterial attack is attributed to high levels of isothiocyanates. In Colombia, it is planted as a companion crop to repel pests in potato fields.

The chromosome number is 2n=42 (Darlington and Wylie,1955).

31. Chinese water chestnut (*Eleocharis dulcis* (Burm. f.) Trin. ex Hensch.)

Family : Cyperaceae

The *Eleocharis dulcis* (synonyms *E. equisetina, E. indica, E. plantaginea, E. plantaginoides, E. tuberosa, E. tumida*), more often called simply the water chestnut, but not to be confused with the unrelated water caltrop which also goes by that name, is a grass-like sedge grown for its edible corms. It has tube-shaped, leafless green stems that grow to about 1.5 metres high.

The basic chromosome number of the genus is x=5,8,9 (Darlington and Wylie, 1955).

32. Mauka (*Mirabilis expansa*)

Family : Nyctaginaceae

Mirabilis expansa (chago) is cultivated as a root vegetable in the Andes, at cold, windy altitudes above 2700 meters. The above-ground portion dies back with frost, but the root is quite hardy. It grows to a height of 1 meter, and bears edible tuberous roots that can reach the size of a man's forearm.*Mirabilis jalapa* ,the four o' clock plant is also a perennial plant with tuberous roots but cultivated for ornamental purpose.

The chromosome number of *M.jalapa* and *M. longiflora* is 2n=58 (Darlington and Wylie, 1955).

33. Chufa Sedge (*Cyperus esculentus)*

Family : Cyperaceae

Cyperus esculentus (Yellow Nutsedge, Tigernut Sedge, Earthalmond) is a species of sedge native to warm temperate to subtropical regions of the Northern Hemisphere. It is an annual or perennial plant, growing to 90 cm tall, with solitary stems growing from a tuber. The stems are triangular in section, and bear slender leaves 3-10 mm wide. The flowers of the plant are distinctive, with a cluster of flat oval seeds surrounded by four hanging leaves positioned 90 degrees from each other. The plant foliage is very tough and fibrous, and is often mistaken for a grass.

There are several varieties:*Cyperus esculentus* var. *esculentus* from Mediterranean region east to India., *Cyperus esculentus* var. *hermannii* from Florida, *Cyperus esculentus* var. *leptostachyus* and *Cyperus esculentus* var. *macrostachyus* from United States and *Cyperus esculentus* var. *sativa* from Asia. The tubers are edible, with a slightly sweet, nutty flavour, compared to the more bitter tasting tuber of the related *Cyperus rotundus* (Purple Nutsedge). They are quite hard and are generally soaked in water before they can be eaten. They have various uses; in particular, they are used in Spain to make horchata. They are sometimes known by their Spanish name, "chufa" (http://en.wikipedia.org).

The chromosome number of *Cyperus esculentus* is 2n=108 (Darlington and Wylie,1955).

34. Indian Breadroot (*Psoralea esculenta)*

Family : Fabaceae

Psoralea esculenta is an herbaceous perennial plant native to prairies and dry woodlands of central North America, which bears a starchy tuberous root edible as a root vegetable. It is also known as Indian Potato, Pomme Blanche, Teepsenee.

Several densely-haired stems emerge from the ground and reach up to 30 cm, bearing palmately compound leaves divided into five leaflets. During summer it produces abundant blue or purple flowers in terminal clusters 5 to 10 cm long, leading to flattened, slender-tipped pods. The plant grows from one or more sturdy brown roots which form rounded tuberous bodies about 7 to 10 cm below the surface, each 4 to 10 cm long. These can be eaten raw, dried, or cooked. The raw root is moderately sweet and tastes like the turnip. The dried root can be ground into a flour. Abundant, palatable, and nutritious, this species was once a wild-gathered staple of Native Americans and early European explorers. Its characteristics make it an obvious candidate for possible domestication (http://en.wikipedia.org).

The chromosome number of the species is 2n=22(Fedorov,1974).

35. Yacon (*Smallanthus sonchifolius* Poeppig and Endlicher H. Robinson) *syn.Polymnia sonchifolia* Poeppig and Endlicher

Family : Asteraceae

The Yacón is a perennial plant grown in the Andes for its crisp, sweet-tasting tuberous root. The texture and flavour have been described as a cross between a fresh apple and watermelon which is why it is sometimes referred to as the apple of the earth. The root is composed mostly of water and fructo-oligosaccharides. It has recently been introduced into farmer's markets and natural food stores in the US.

Although sometimes confused with jicama, yacón is actually a close relative of the sunflower and Jerusalem artichoke. The plants produce two types of roots: propagation roots and storage roots. Propagation roots grow just under the soil surface and produce new growing points that will become next year's aerial parts. These roots resemble Jerusalem artichokes. Storage roots are large and edible.

Yacón plants can grow to over 2 meters in height and produce small, yellow inconspicuous flowers at the end of the growing season. Unlike many other root vegetables domesticated the yacón is not photoperiod sensitive, and can produce a commercial yield in the tropics.

The chromosome number of *S. sonchifolius* is 2n=60 (Fedorov,1974).

36. Oca (*Oxalis tuberosa*)

Family : Oxalidaceae

The oca or oka is a perennial plant grown in the central and southern Andes for its starchy edible tuber, used as a root vegetable. Its leaves and young shoots can be eaten as a green vegetable as well. Introduced to Europe in 1830 as a competitor to the potato and to New Zealand as early as 1860, it has become popular in that country under the name New Zealand yam and is now a common table vegetable.

The oca is one of the important staple crops of the Andean highlands, second only to the potato due to its easy propagation, and tolerance for poor soil, high altitude and harsh climates.

The flavor is slightly tangy, and texture ranges from crunchy (like a carrot) when undercooked, to starchy or mealy when fully cooked. Though the original Andean varieties are widely variable in color from purple to yellow, the standard NZ variety is a fleshy pink.

Ocas need a long growing season, and are day length dependent, forming tubers when the day length shortens in the fall. In areas with harsh winter climates, the cold weather that accompanies shorter days may kill the plant before tubers have a chance to form. Likewise in tropical areas where the days are uniformly longer, the oca will not set a crop successfully, since the days are never short enough.

The species showed variable chromosome numbers of 2n=63-70,66 (Darlington and Wylie,1955).

37. Chinese artichoke (*Stachys affinis* Bunge)

Family : Lamiaceae

Stachys affinis (knotroot, artichoke betony, or crosne), is an herbaceous perennial plant of the family Lamiaceae. Although it can be grown as a root vegetable, it is a rare sight in the garden. From a cultivation standpoint this is rather odd - the plant is easy to grow, requiring neither staking nor earthing-up. The reason that it is so unpopular is the nature of the tubers - small, convoluted and indented, so that it is the cook rather than the gardener or the family who finds this vegetable frustrating. The thin skin is of whitish-brown or ivory-white. The flesh underneath, under proper cultivation, is white and tender. It is in season generally commencing with October.

The diploid chromosome number of the above species reported by different workers varied between 66-70 (Fedorov, 1974).

38. Purple Salsify (*Tragopogon porrifolius*)

Family: Asteraceae

Purple Salsify is one of the most widely known species of the salsify genus. It is also known as Oyster Plant, Vegetable Oyster, or Common Salsify, or simply as Goatsbeard or Salsify - though as these last names are also applied to other members of the genus, or to the genus as a whole, they are better avoided.

Purple Salsify is a common wildflower, native to Mediterranean regions of Europe but introduced elsewhere, for example, into Britain and northern Europe, North America, and southern Africa; in the United States it is now found growing wild in almost every state, including Hawaii, except in the extreme south-east.

The plant grows to around 60cm in height. As in other goatsbeards, its stem is largely unbranched, and the leaves are somewhat grasslike. In Britain it flowers from June to September, but in warmer areas such as California it can be found in bloom from April. The flower head is about 5cm across, and each is surrounded by green bracts which are longer than the petals (technically, the ligules of the ray flowers). The flowers are hermaphrodite, and pollination is by insects.

The root, and sometimes the young shoots, of Purple Salsify are used as a vegetable. Cultivated varieties include 'White French', 'Mammoth Sandwich Island' and 'Improved Mammoth Sandwich Island'; they are generally characterised by larger or better-shaped roots. The root is noted for tasting of oysters, from which the plant derives its alternative name of Oyster Plant; young roots can be grated for use in salads, but older roots are better cooked, and they are usually used in soups or stews.

The chromosome number is 2n=12 (Darlington and Wylie,1955).

39. Valerian (*Valeriana officinalis L. & Maillefer*)

Family: Valerianaceae

Valerian or All heal is a hardy flowering plant native to Europe and parts of Asia. Valerian has uses in herbal medicine. The roots contain pharmaceutical principles. In North India V. jatamansi is grown for its roots. V.officinalis showed variable chromosome numbers of 2n = 14, 28, 56 (Darlington and Wylie, 1955) where as V. jatamansi showed 2n = 32.

40. Utleria salicifolia Bedd. ex benth. & Hook. f.

Family : Periplocaceae

This is commonly found in the Western Ghats in Anamallai Mts., at an attitude of 3000-4000 ft.

It is a glabrous tree with stout branches, leaves, narrow, subcrenulate crowded at the end of branches. Flowers minute, towards the ends of the long branches of terminal peduncled dichotomous cymes. Tubers medicinal.

References

Adesuyi, S.A. 1973. Advances in yam storage research in Nigeria. 3rd International Symposium on Tropical Root crops, IITA, Ibadan, Nigeria. 13 pp.

Adesuyi, S.A. 1976. The use of gamma radiation for control of sprouting in yam *(Dioscorea rotundata)* during storage. Niger. *J. Plant Protection* 2: 34-39.

Allard, H.A. 1945. Some behaviours of yam (*Dioscorea*). *Castanea* 10: 8-13.

Bala Nambisan and Sundaresan, S. 1985. Effect of processing on the cyanoglucoside content of cassava. *J. Sci. Food Agric.* 36: 1197-1203.

Baldwin, J.T. and Spece, B.M. 1951. *Tacca involucrata* its chromosomes. *Bull. Torrey. Bot. Club.* 78,1:70-72.

Barua, A.K., Chakravarti, D. and Chakravarti, R.N. 1956. 'Steroidal sapogenins from Indian *Dioscorea* plants, II', *J. Indian chem. Soc.*, 33, 799-803.

Bhag Mal. 1994. Underutilized grain legumes and pseudocereals – their potentials in Asia. Regional Office for Asia and the Pacific. FAO, Bangkok, Thailand. 162 pp.

Bolhius, G.G. 1966. Influence of length of the illumination period on root formation in cassava. *Neth. J. Agric. Sci.* 14: 251-254.

Burkill, H.M. 1995. The useful plants of West Tropical Africa, 2nd ed., vol. 3. Royal Botanic Gardens, Kew.

Burkill, I.H. 1935. A dictionary of the economic products of the Malay Peninsula. Crown Agents, London.

Burkill, I.H. 1939. Notes on the genus *Dioscorea* in the Belgian Congo. *Bull. Jord. Bot., Etat, Brux; 15* (4) : 345-392.

Butler, G.W. 1965. The distribution of the cyanoglucosides, linamarin and lotaustralin in higher plants. *Phytochemistry* 4:127-131.

Campbell, J.S. Chukueke, V.O., Teriba, F.A. and Ho-A-Shu, H.V.S. 1962a. Some physiological experiments with the white Lisbon yam (*Dioscorea alata* L.) 1. The breakage of the rest period in tubers by chemical means. *Emp. J. Expl. Agric.* 30: 108-114.

Campbell, J.S. Chukueke, V.O., Teriba, F.A. and Ho-A-Shu, H.V.S. 1962b. Some physiological experiments with the white Lisbon yam (*Dioscorea alata* L.) 2. Growth period and out of season production. *Emp. J. Expl. Agric.* 30: 232-238.

Chadha, K.L. and Nayar, G.G. 1994. History of Tuber Crops Research in India. *(In:) Advances in Horticulture vol. 8-Tuber Crops.* (Eds.) Chadha, K.L. and Nayar, G.G. Malhotra Publishing House, New Delhi. pp. 1-16

Chakravorti, A.K. 1948. Multiplication of chromosome numbers in relation to speciation in *Zingiberaceae. Sci. and Culture.* 14,4: 137-140.

Chua, L.K and Kays, S.J. 1982. Assimilation patterns of ^{14}C photosynthate in developing sweet potato storage roots. *J. Amer. Soc. Hort. Sci.* 107: 866-871.

Chun-Lin Long and Ke-Ming Liu. 2001. *Colocasia lihengiae* (Araceae: Colocasieae), a new species from Yunnan, China. *Bot. Bull. Acad. Sin.* 42: 313-317.

Cock, J.H., Joseph, K.J. and Ramli, K. 1979. Influence of soil applied micronutrients on cassava *(Manihot esculenta)* in Malaysian tropical oligotrophic peat. *Expl. Agric.* 14:105-111.

Coursey, D.G. 1976. The origins and domesticaiton of yams in Africa *(In:)* Harlan, J.R (ed.) Origin of African Plant Domestication, Mouton, La Hague. pp. 383-408.

Coursey, D.G. 1967. Yams. Longman, London 230 pp.

Coursey, D.G and Haynes, P.H. 1970. Root Crops and their potential as food in the tropics. *World Crops.* 22-261-265.

CSIR. 1950. Wealth of India. Vol. II, CSIR, New Delhi.

CSIR. 1952. Wealth of India, Raw Materials. Vol. III. CSIR, New Delhi.

CSIR. 1976. Wealth of India, Raw Materials. CSIR, New Delhi. Vol. X: 401-402.

CTCRI, 1984. Annual Report. Central Tuber Crops Research Institute, Trivandrum.

Darlington, C.D and Wylie,A.P.1955.Chromosome Atlas of Flowering plants .Second edition. 519 pp.

Deepa, G., Indira, P.V., Sasidharan Pillai, C.R., Kurian Jose and Fatima Beevi. 1999. Tubers – A short cut to the health of mothers and children – A review based on Ayurveda (In:) Tropical Tuber Crops in Food Security and Nutrition. Oxford IBH. New Delhi. pp. 15-19.

De Vries, C.A, Ferweda, I.D and Flach, M. 1967. Choice of food crops in relation to actual and potential production in the tropics. *Neth. J. Agric. Sci.* 15: 241-248.

Doku, K.V. 1969. Cassava in Ghana. Ghana Univ. Press.

Duthie, J.F. 1973. Flora of the Upper Genetic Plain. Vol, II-III Part I-II. M/s. Bishan Sing Mahendra Pal Singh, Dehradun, M/s. Periodical Experts, Delhi. pp. 255.

Enyi, B.A.C. 1972. Effect of shoot number and time of planting on growth, development and yield of cassava (*Manihot esculenta* Crantz). *J. Hort. Sci.* 47:457-466.

Enyi, B.A.C. 1973. Growth development and yield of some tropical root crops. (*In:*) *Proc. 3rd Symp. Inter. Soc. Trop. Root Crops*, Ibadan, Nigeria, pp. 87-103.

Ezumah, H.C and Plucknett, D.L. 1973. Response of taro, *C. esculenta* (L) Schott. to water management, plot prepartion and population. *(In:) Proc.* 3rd *Symp. Inter. Soc. Trop. Root. Crops*, Ibadan, Nigeria. pp. 362-268.

Fedorov. 1974. Chromosome numbers of flowering plants.Otto Koelyz Science Publishers,West Germany.926p

Ghosh, S.P., Ramanujan, T., Jos, J.S, Moorthy, S.N. and Nair, R.G.1988.Tuber Crops. Oxford IBH Publishing Co Pvt Ltd., New Delhi.

Hasegawa, K. and Hashimoto, T. 1973. Quantitative changes in batata sins and abscissic acid in relation to the development of dormancy in yam bulbils. *Plant cell Physiol.* 14: 369-377.

Hashimoto, T., Hasegawa, K. and Kawarada, A. 1972. Batatasins : New dormancy inducing substances of yam bulbils. *Planta.* 108:369-374.

Haudricourt, A. 1941. Le Colocasiees Alimentaires (Taros et Yantias). *Rev. Bot. Appl. Agric. Trop.* 21, 40-65.

Hill, A.F. 1939. The Nomenclature of the Taro and its varieties. Harvard Univ. Bot. Mas-Leaflets, 7, 113-22.

Huxley A.1992. Dictionary of gardening. The Royal Horticultural Society. Vol 2. London, MacMillan Press.

Indira, P. 1978. Salinity effects on plant growth and Tuberization in cassava. *J. Root Crops.* 4 : 19-23.

Indira, P. and Sinha, S.K. 1970. Studies on the initiation and development of tubers in *M. esculenta. Indian J. Plant Physiol.* 13: 24-39.

Jackson, G.V.H, Ball, E.A. and Arditti, J. 1977. Seed germination and seedling proliferation of *taro. J. Hort. Sci.* 52: 169-171.

Jackson, G.V.H, Collifer, D. E and New Hook, F.J. 1980. Studies on taro leaf blight fungus *Phytophthora colocasiae* in Solomon Island. Control by fungicides and spacing. *Ann. Appl. Bot.* 96(1):1-10.

Jones, W.O. 1959. Manioc in Africa. Stanford. Univ. Pres. Stanford, California, pp. 315.

Jos, J. S and Sreekumari, M.T. 1994. Genetic Resources in cassava. *(In:) Advances in Horticulture. Vol. 8-Tuber Crops.* (Eds.) Chadha, K.L. and Nayar, G.G. Malhotra Publishing House, New Delhi. India. pp. 17-33

Karikari, S.K. 1973. Preliminary evaluation of 14 Puerto Rican and 6 Ghananian varieties of cocoyam (*Colocasia* and *Xanthosoma*) under Ghanian condition. (*In:*) *Proc. V. Tuber. Symp. Trop. Root Crops.* Philippines. pp. 615-627.

Kays, S. J., Majnuson, C. E and Fares. Y. 1982. Assimilation patterns of carbon in developing sweet potatoes using ^{11}C and ^{14}C. (*In:*) Sweet Potato (Eds. R.L. Vilareal and T.D. Griggs), AVRDC, Taiwan. pp. 95-118.

Keating, B.A., Wilson, G.L. and Evenson, J.P. 1985. Effect of photoperiod on growth and development of cassava (*Manihot esculenta* Crantz). *Aust. J. Plant Physiol.* 12: 621-630.

Kopetz, L.M. and Steineck, O. 1954. Photoperiodische Untersuchungen an Karotofelsamlingen. Zucher 24: 69-77.

La Cour, L.F. 1951. Annual Report. John Innes Hort. Institute. (1950). UK. 41:23.

Lackey, J.A. 1977. A synopsis of the *Phasedeae* (Leguminosea, Papilionoideae) - Ph.D. dissertation, Dept. Bot. Pl. Path. Iowa State Univ. Ames. USA. 293 pp.

Lowe, S.B., Mohan. J.D. and Hunt, L.A. 1976. The effect of day length on shoot growth and formation of root tubers in young plants of Cassava *(Manihot esculenta* Crantz). *Plant Sci. Letter* 6:57-62.

Maesen, L.J.G. van der, 1985. Revision of the genus *Pueraria* DC. with some notes on Teyleria Backer (Leguminosae). *Agric. Univ. Wageningen Papers* 85, 1: 1-132.

Magoon, M.L., Krishna, R. and Vijaya Bai, K. 1970. Cytological studies in *Xanthosoma sagittifolium. Nucleus* 13:142-148.

Mangenot, S., and Mangenot, G. 1962. Enquete surles nambres chromosomiques dam une collection d'especes tropicales. *Rev. Cyt. Et Biol. Veg.* 25,3-4:411-447.

Martin, F.W. 1983. Variation of sweet potatoes with respect to the effects of water logging. *Trop. Agric.* (Trinidad) 60 :117-121.

Martin, F.W, Koo, F.K.S and Cuevas, J. 1974. Stimulation of yam *(Dioscorea)* tuber growth by gamma irradiation. *J. Am. Soc. Hort. Sci.* 99-282-284.

McDavid, C.R and Alamu, S. 1976. Promotion of flowering in tannia (*Xanthosoma sagittifolium)* by gibberellic acid. *Trop. Agric.* (Trinidad) 53: 373-374.

Meige, J. 1954. Nombres chromosomiques et repartition geographique de quelques plants tropicales et equatorials. *Rev. Cytol. Biol.veg.* 15(4), 312-48.

Miller, E.W. 1930. A preliminary note on the cytology of the *Melanthioideae* section of the Liliaceae. *Pric. Univ. Burham Philos. Soc.*, 8:267-271.

Moh, C.C. 1976. Correlation between hydrocyanic acid level in leaf and root of cassava (*Mahihot esculenta* Crantz). *Turrialba* 26: 132-133.

Nartey, F. 1973. Biosynthesis of cyanogenic glucoside in cassava (Manihot spp). (*In:*) Chronic cassava Toxicity. B.L.Nestel and R. Mac Intyro (Eds.) IDRC Canada Publication No.010, 73-87.

Narzary, B.D. 2002. Final report of ICAR Ad-Hoc Scheme on Collection, evaluation and taxonomical classification of germplasm of *Colocasia, Alocasia* and *Xanthosoma* of

North-East India. Department of Horticulture, Assam Agricultural Universtiy, Assam. 154 pp.

Neuwinger, H.D.1994. African ethnobotany. Poisons and drugs. Chemistry, pharmacology, toxicology. English translation by A Porter. Weinheim, Chapman & Hall.

Njoku, E. 1963. The propagation of yams (*Dioscorea* spp.) by vine cuttings *J. West African Sci. Assoc.* 8: 29-32.

Okezie, C.E., Okankwo, S.N.C and Noveke, F.I. 1980. Growth pattern and growth analysis of the white guinea yam raised from seed. (*In:*)Trop. Root. Crops : Research Strategies for the 1980's (Eds. E.R. Terry, K.A. Odure, and F. Caveness), IDRC 163 e pp. 180-188.

Onwueme, I.C. 1975. Tuber formation in yams. (*Dioscorea* spp.). Effect of moisture stress; contribution of the parent sett. *J. Agric. Sci. Camb.* 85:267-269.

Onwueme, I.C. 1978. The tropical crops . John Wiley and Sons. England 234p.

Oyolu, C. 1982. Inherent constraints to high productivity and the production cost in yam (*Dioscorea* spp.) with special reference to *Dioscorea rotundata* Poir. (In:) Yams. (Eds.) J. Miege and S.N. Lyonga. Clarendon Press, Oxford. pp. 147-160.

Paliawal G.S. and Kavathekar, A.K. 1971. Anatomy of vegetative food storage organs. *Acta Agronomica : Academias Scientiarum, Hungarical* 20: 261-270.

Palmer, C.E. and Smith, O.E. 1969. Cytokinin and tuber initiation in potato tuber *Solanum tuberosum* L. *Nature* 221:79-80.

Parsam, H. C. 1977. Sprouting and apical dominance of yam tubers. *Tropical Science* 19: 29-39.

Paterson, D.R., Earhart, D.R. and Fugue, M.C. 1979. Effect of flooding level on storage root formation, ethylene production and growth of sweet potato. *Hort. Sci.* 77: 452-457.

Plucknett, D. L. 1983. Taxonomy of the genus *Colocasia (In:)* Taro, (Ed.) J.K. Wang. Univ. Hawaii Press. Honalulu, pp. 14-33.

Prain, D and Burkill, I.H. 1938. An account of the genus *Dioscorea* in the East. Part III. *Ann. R. Bot. Gdn.* Calcutta. 14: 211-528.

Purseglove, J.W. 1972. Tropical Crops : Monocotyledons. 2 Vols. London : Longman.

Purseglove, J.W. 1974. Tropical Crops : Dicotyledons. Vols 1 and 2 combined. London : Longman.

Radosevich, M and Maldonado, A.C. 1984. Cytokinin activity in storage organs of tanier. *J. Agric. Univ. Peurto Rica.* 68:281-287.

Raghavan, R. S. 1958. 'A chromosome survey of Indian Dioscoreas'. *Proc. Indian Acad. Sci. Sec.* B., 48(1), 59-63.

Ramachandran, K. 1961. Chromosome numbers in the genus *Curcuma* Linn. *Current. Sci.* 30, 5:194-196.

Ramachandran, K. 1962. Studies on cytology and sex determination of Dioscorcaceae'. *J. Indian Bot. Soc.*, 41(1), 93-97.

Ramanathan, S. 1978. Management plan for the Kalakkad wild life sanctuary, Govt of Tamil Nadu, Forest dept., Madras.

Ramanujam, T. and Lakshmi, K.R. 1984. The pattern of dry matter production and partitioning in cassava. Indian *J. Plant Physiol.* 27: 138-144.

Ramanujam, T., Nair, G.M. and Indira P. 1984a. Growth and development of cassava (*Manihot esculenta* Crantz) genotypes under shade in a coconut garden. *Turrialba* 34: 267-274.

Ramanujam, T., Indira, P. and Rajendran, P.G. 1984b. Distribution of HCN in cassava and its relationship with anthocyanin pigments. *South Indian Hort.* 32:132-137.

Reynolds, S.G and Netram, P. 1973. The effect of rainfall and time of planting on yield of taro (*Colocasia esculenta* (L) Schott) in Western Samoa. (*In:*) *Proc. 3rd Symp. Inter. Soc. Trop. Root Crops.* IITA, Nigeria pp. 374-376.

Sastrapradja, S. and Sutarno, H. 1977. *Vigna umbellata* (L.) DC. in Indonesia. *Annales Bogoriensis* VI (3): 155-167.

Sato, D. 1960. The karyotype analysis in *Zingiberales* with special reference to the prokaryotype and stable karyotype. *Sci. Papers. Coll. Genet. Education Univ. Tokyo, Biol.* 10, 2:225-243.

Sharma, A.K. and De, D. W. 1965. Polyploidy in *Dioscorea*. *Genetica.* 28 (1): 112-20.

Sharma, A.K and Sharma, A. 1961. An investigation of the cytology of some species of Liliaceae. *Genetica Iberica.* 13, 1-2 : 25-42.

Shaw, D.E. 1975. Illustration notes on flowering, flowers, seeds and germination in taro. (*Colocasia esculenta*). *PNG. Dept. Agric. Stock Fish Res. Bull.* 13: 39-59.

Simmonds, N.W. 1954. Chromosome behaviour of some tropical plants. *Heredity* 8(1), 139-145.

Sivan, P. 1979. Growth and development of taro (*Colocasia esculenta*) under dryland conditions in Fiji. *Proc. 5th Symp. Inter. Soc. Trop. Root Crops*, Philippines, pp.637-646.

Smartt. J. 1990. Grain Legumes:Evolution and Genetic Resources. Cambridge, U.K. 379 pp.

Smith, B.W. 1937. Notes on the cytology and distribution of the Dioscoreaceae. *Bull. Torrey Bot.club.* 64(3), 189-197.

Sobule, R.A. 1972. Studies on white yam *Dioscorea rotundata* (Poir) III Foliar analysis for nitrogen nutrition. *Expl. Agric.* 1: 271-279.

Strauss, M.S., Michan, J.D. and Arditt, J. 1979. Seed storage germination and seedling proliferation in taro. *Ann. Bot.* 43: 603-612.

Sung, F.J.M. 1981. The effect of leaf water status on stomatal activity, transpiration and nitrate reductase of sweet potato. *Agrl. Water Management.* 4: 465-470.

Tjio, J.J. 1949. The somatic chromosome of some tropical plants. *Hereditas.* 34, 1-2:135-146.

Togari, Y. 1950. A study on the tuberous root formation of sweet potato. *Nat. Agric. Expt. Station. Bull.* Tokyo, Japan. 68: 1-96.

Tsukamoto, Y and Inaba, K. 1961. The effect of day length upon the cormel formation in taro (*Colocasia antiquorum*). *Mem. Res. Inst. Food Sci. Kyoto Univ.* Japan. 23:15-22.

Tuxill, J. 1999. Appreciating the benefits of plant biodiversity. (*In :*) "State of the World, 1999". Worldwatch Institute, 1776 Massachusetts Avenue, N.W., Washington, D.C. USA.

Van Rheede. 2003. Hortus Malabaricus (Reprinted Eng. Ed) Vol. 11. 2003. University of Kerala. 245 pp.

Venkatasubban, K.R. 1946. A preliminary survey of chromosome numbers in Scitamineae of Bentham and Hooker. *Proc. Indian Acad. Sci.* Sect. B. 23(6):281-300.

Vimala, B. 1994. Genetic Resources and Varietal Improvement in Minor Tuber Crops. . (*In:*) Advances in Horticulture vol. 8-Tuber Crops. (Eds.) Chadha, K.L. and Nayar, G.G. Malhotra Publishing House, New Delhi-110064. India. pp. 139-153

Viswanathan, P.N. and Krishnan, P.S. 1965. Metabolic activity of starch granules from the tapioca plant. Part I. UDPG-Starch glucosyle transferase. *Indian J. Biochem.* 2:16-21.

Volin, R.B and Zettler, F.W. 1976. Seed propagation of Cocoyam, *Xanthosoma caracu* Koch and Bouche. *Hort. Sci.* 11: 459-460.

Williams, C.N. 1972. Growth and productivity of tapioca (*Manihot utilissima*) III Crop ratio, spacing and yield. *Expl. Agric.* 8:15-23.

Wilson, L.A. 1973. Effect of different levels of nitrate-nitrogen supply on early tuber growth of two sweet potato cultivars. *Trop. Agric.* (Trinidad) 50:53-54.

Wilson, L.A. 1982. Tuberization in sweet potato (*Ipomoea batatus* (L) Lam). (*In:*) Sweet potato (Eds. R.L. Villareal and T.D. Griggs) AVRDC, Taiwan. pp. 79-94.

Zara, D.L., Cuevas, S.E. and Carlos, J. T. Tr: 1982. Performance of sweet potato grown under coconuts. (*In:*) Sweet potato (Eds. R.L. Villareal and T.D. Griggs, AVRDC, Taiwan pp. 233-242.

http://en.wikipedia.org

www.aroid.org/genera/cyrtosperma/cyrtospermalist.html.

❑❑❑

4

Biochemical Constituents and Pharmaceutical and Nutraceutical Values

The dry matter production is maximum in cassava (3.0 m t ha^{-1}) and is followed by yams (2.4 m t ha^{-1}) and sweet potato (2.1 m t ha^{-1}) and are higher than rice, wheat, maize and banana (1.3 to 1.9 m t ha^{-1}). The energy production is maximum in yams (182 M J ha^{-1} day^{-1}) and it is followed by sweet potato (152 M J ha^{-1} day^{-1}) and cassava (121 M J ha^{-1} day^{-1}) (Horton and Fano, 1985).

1. Sweet Potato

Both sweet potato roots and tops are nutritious foods which could be used to advantage in combating nutritional deficiencies in parts of the developing world if means could be found to overcome resistance to their increased consumption. The roots are used primarly as human food and are eaten boiled, steamed, fried or baked. They are not only a moderately good energy source, but also contain significant quantities of the water-soluble vitamins *viz.*, ascorbic acid and thiamin, besides supplying part of the daily requirement for riboflavin and niacin. Their contents of pyridoxine, folic acid and pantothenic acid may be relatively high. Raw leaves and tender tips are also excellent sources of ascorbic acid and some of the B-vitamins, especially riboflavin which is deficient in many Asian diets. However, high percentages of water-soluble vitamins are lost on cooking (Woolfe, 1992).

Biochemical composition of sweet potato tubers varies widely according to cultivar, climatic conditions, degree of maturity and the duration of storage after harvesting. The usual range of values for the edible portion is: energy

490 kJ 100 g^{-1}; water 65-81 per cent; protein 0.95-2.4 per cent; fat 0.4-6.4 per cent; carbohydrate 25-32 per cent; fibre 0.9 per cent; ash 0.9-1.4 per cent; calcium 30-34 mg 100 g^{-1}; iron 0.8-1 mg 100 g^{-1}; magnesium 24 mg 100 g^{-1}; phosphorus 49 mg 100 g^{-1}; potassium 373 mg 100 g^{-1}; sodium 13 mg 100 g^{-1}; carotene trace 12 mg 100 g^{-1}; thiamine 0.1 mg 100 g^{-1}; riboflavin 0.05 - 0.06 mg 100 g^{-1}; niacin 0.6-0.9 mg 100 g^{-1}; ascorbic acid 23-25 mg 100 g^{-1} (Kay, 1987; Bradbury and Holloway, 1988; Wheatley *et al.*, 1995).

The deep yellow- or orange-fleshed roots and the leaves and tips are rich sources of beta-carotene and other biologically active carotenoids and would be valuable as one means in a multiple approach to prevention of vitamin A deficiency. HPLC analysis showed that orange fleshed varieties contained beta carotene as principal carotenoid with less amount of xanthophylls while in cream and pale yellow varieties the reverse was true (Lila Babu, 2006). Baked products like golden bread buns biofortified with beta carotenoid rich sweet potato to add value to rural diets and increase profits of rural bakers in Central Mozambique have been developed (Low and Jaarsveld, 2006). Sweet potato leaves are rich sources of provitamin A carotenoids, chiefly beta carotene, the lowest value being 600 μg per 100 g (fwb) (Haytowitz and Mathews, 1984). Thus sweet potato has the unique distinction of providing sufficient calories alongwith carotenoids and other bioactive principles. It is possible, that cultivars high in carotenoid could be encouraged for planting in home gardens for both roots and leaves. Furthermore ways could be found to incorporate the roots for example in a dry form as flour in other foods, something which has already been investigated in Taiwan (Woolfe, 1992).

The purple-coloured sweet potatoes containing anthocyanin is used for commercial production of food colourant. In Japan ,approximately 160 tons of food colourant is produced annually and its sales reach about 9 million US $ with three high anthocyanin cultivars 'Ayamurasaki'released in 1995, 'Murasakimasari' released in 2001 and 'Akemurasaki' released in 2005 (Yoshinaga *et al.*,2006).

Sweet potato roots are also sources of some minerals and trace elements. They are a good source of P and though not having outstanding contents of Fe and Ca, they can make modest contributions to the recommended daily intakes of these minerals in a quantity of as little as 100 g, which also provide of tuber part of the daily allowance of Mg, Cu and Mn.

Sweet potato leaves are potentially much richer source of Fe and Ca than the roots. But the Fe which they contain may be very poorly absorbed, and their oxalate content is such as to reduce availability of Ca. Phytate levels in roots and leaves, oxalate levels in leaves and resulting availabilities of minerals should be investigated with a view to selection of cultivars containing the minimum possible concentrations of these anti-nutrients. The role of naturally occurring ascorbic acid in both roots and leaves in promoting absorption of iron could be clarified. Methods of preparation should be devised to increase mineral availabilities. A potentially very important contribution which sweet potato leaves can make to nutrition lies in their relatively high level of riboflavin which is generally deficient in Asian rice-based diets (Villareal *et al.,* 1985). They are also moderately good sources of other B-vitamin namely thiamin, niacin and pyridoxine.

Both parts of the sweet potato plant used for food are moderately good sources of dietary fibre, being made up of soluble and insoluble fibre, which could promote certain physiological effects such as faecal transit times and reduction of blood cholesterol levels. The total dietary fibre of raw sweet potato samples from Solomon Island and Papua, New Guinea ranged from 1.20% to 2.62% (fwb) (Bradbury *et al.,* 1985). Other medical applications, particularly the use of roots in oral rehydration therapy, could be studied further.

The leaves are used to treat diabetics, hookworm, haemorrhage and abscesses (http://en. wikipedia.org). Despite the name "sweet", it is actually a good food for diabetics as preliminary studies on animals have revealed that it helps to stabilize blood sugar levels and lowered insulin resistance (Kusano and Abe, 2000; Suzuki *et al.,* 2002; Ylonen *et al.,* 2003).

Ipomoea paniculata R. Br. is glabrecent twining shrub seen throughout India preferring moist areas and are highly medicinal. It cures liver and spleen enlargement as cholagogue, cures spermatarrhoea and menorrhagia. It is an ingredient in snakebite compositions and recommended for urinary disorders. Powder of tuber along with milk is given as a general tonic to improve the health of mother and child as it increases breast milk. It is useful in treating sterility and allaying immature abortion. It increases voice, strength and immunity and it is vitalizing, diuretic and is used in burning conditions. Most of the wild Ipomoea spp. are medicinal (Deepa *et al.,* 1999).

2. Cassava

Cassava tubers are utilized primarily as human food, and can be eaten boiled, fried, baked or steamed. They are used in a variety of food products like sago, noodles, chips and in flakes and puddings. Leaves and tubers also form an important component of animal feed and are extensively used in feed for cattle, poultry and swine. A third major use of cassava is in industry where it is used for production of starch, liquid glucose, dextrins and high fructose syrup.

The biochemical constituents in cassava tuber on fresh weight basis were : dry matter 40%, starch 32.5%, total sugar 0.5%, protein 0.8%, fiber 0.5%, fat 0.2%, ash 1%, calcium 0.05%, phosphorous 0.04%, vitamin C 0.025%, cyanoglucoside 25-400 μg cyanide g^{-1}. The biochemical constituents in cassava leaf on fresh weight basis were 22% dry matter, 8% starch, 8% protein, 2% fibre, 1.7% fat, 1.5% ash, 0.125% calcium, 0.08% phosphorous, 0.25% vitamin C and cyanoglucoside 350-600 μg cyanide g^{-1} (Kay, 1987; Bradbury and Holloway, 1988). Some genotypes are rich in β carotene. The rind of the tuber and seed extract were found to have biopesticidal properties especially for nematodes and insect pests.

3. Yams

Dioscorea species like *D. alata, D. bulbifera. D. rotundata* and *D. esculenta* have been cultivated for a long time for their edible tubes. Other than a few edible species, the tubers of many species are highly toxic. The toxicity of yams is due to the presence of one of the three main components like alkaloids like dioscorin, tannins and saponins. Saponins on hydrolysis give sapogenins which are of much interest and importance in the field of medicine. All the important yam sapogenins are steroidal in character. Three sapogenins that have been isolated from yam species are diosgenin, botogenin and kryptogenin. Interest in these steroidal sapogenins has developed rapidly because of their close relationship to various natural products that perform vital functions in the human body (Applezweig, 1962). These vital products include hormones elaborated by the sex glands and by the adrenal cortex. After much research, it could be fairly claimed that yam sapogenins offered great promise as practicable raw materials for steroid drug production and of these diosgenin is the most favoured one. With the help of certain microorganism diosgenin can be oxidized for the production of corticosteroid type drugs (Coursey, 1967). Corticosteroids are used as anti-inflammatory agents, as metabolic stimulants and as general stress reaction protectives. Yam sapogenins forms the basis of modern oral contraceptives (Applezweig, 1962). More than 50 species of *Dioscorea* contain diosgenin. Most of the world production of diosgenin today is met from Central American species mainly *D. floribunda* Mart and Gal and *D. composite* Hemsl, both growing wild in their native home. In India *D. deltoidea* and *D. prazeri* occurring wild in North-west and North-east Himalayas respectively contain good percentages of diosgenin (Barua *et al.,* 1954, 1956).

Phytoeostrogens were reported in *Dioscorea* spp. These have many antitoxidant properties such as slowing the ageing process and improving overall health especially in women.

Cultivation of indigenous as well as exotic species of *Dioscorea* was tried in late sixties in different parts of the country. At the same time a Project for the improvement of *Dioscorea* sp. was started with the joint collaboration of Ford Foundation, ICAR and CSIR (Asolkar and Chadha, 1979). For the past ten years, efforts have been directed towards the development of one or more *Dioscorea* species, rich in diosgenin, as cultivated crop in order to ensure regular supplies of the raw material to the industry. Among cultivated species *D. floribunda* has so far given the best results. Experimental cultivation of *D. deltoidea* was also tried at various places in India but was not successful because of its slow growing nature. *D. composite* on the other hand is quite vigorous, hardy and promising source of diosgenin but the area under this crop has not expanded due to difficulties in its propagation. At present no viable scheme for its cultivation exists, barring a few firms and institutions namely Cinchona Deptt., Ootacamund, Medicinal Plants Deptt., West Bengal, Wyeth, Searle and Cipla, etc who are raising the crop through seeds in few hectares of land.

Recently, *Dioscorea* tubers were determined to exert immune response-modifying effects. These effects include immunomodulation, anti-inflammatory effects, anti-tumor effects and anti-osteoporotic effects. In addition, the anti-oxidative effects of dioscorin, a storage protein harbored by *Dioscorea* tubers, have been also extensively investigated. It has been demonstrated that the consumption of *Dioscorea* tubers may ameliorate some metabolic abnormalities, including hyperglycemia, gut function and lipid metabolism (Jeon *et al.*, 2006).

Many members of the *Dioscorea* are used in traditional medicine in Africa (Watt and Breyer-Brandwijk, 1962), among the Chinese (Burkill, 1935) and other Asiatic peoples. It is not unlikely that the saponins or other steroids contained in the tubers of many species may actually have some useful medicinal effects. The use by the Zulu of extracts of *D. sylvatica* for the treatment of uterine and mamilliary disorders in cattle is particularly interesting in the context. Many of the uses, however, probably depend more upon superstition and the principles of sympathetic magic, than on any detectable physiological effect. It is interesting to note that the rhizomatous species of temperate North America, *D. glauca* and *D. villosa* and possibly some other, were used by early European settlers in the treatment of internal disorders, from which use their vernacular name in the United States, 'Colic Root', is derived (Bartlett, 1910).

In many of the remoter parts of West Africa, poisonous varieties, particularly *D. dumetorum* and *D. bulbifera*, are sometimes deliberately planted in fields of edible yams, in order to discourage thieves who, not knowing where they are planted, might gather them in mistake for the edible varieties (Dalziel, 1937).

The antiscorbutic properties of yams used as famine foods have been reported by Corkill (1948). Some Asiatic species, notably *D. cirrhosa,* contain enough tannin to be used by some of the fishing people of Indo-China for the preservation of fishing line and nets. The dried stems of some Asiatic and African yams are used as cordage by the inhabitants of the districts in which they grow (Burkill, 1935 ; Watt and Breyer-Brandwijk, 1962).

The small rhizomes of some of the section Stenophylla, notably *D. deltoidea,* have a high content of saponins, and are made into decoctions by many of the people of South-East Asia for use as shampoo, the saponins being sufficiently toxic to destroy head lice. The tubers of *D. poscatorum* can be used to prepare an insecticidal powder similar in its effects to derris dust. This species is also used to destroy rice parasites in paddy fields in Malaysia.

The studies on the effect of ethanol extract of dried Chinese yam (*Dioscorea batatas* or *D. opposita)* flour containing dioscin on gastrointestinal function in rat model (Jeon *et al.,*2006) indicated that Chinese yam extract may prove helpful as a digestion-aiding agent for patients suffering from hyperglycemia or hyperlipidemia.

Yams are essentially carbohydrate foods with relatively high protein and ascorbic acid contents. Typical figures for the composition of the edible portion of fresh tuber are: energy 439 kJ 100 g^{-1}; water 72.4 per cent; protein 2.4 per cent; fat 0.2 per cent; carbohydrate 42.1 per cent; fibre 0.6 per cent; calcium 22 mg 100 g^{-1}; iron 0.8 mg 100 g^{-1}; thiamine 0.09 mg 100 g^{-1}; riboflavin 0.03 mg 100 g^{-1}; niacin 0.5 mg 100 g^{-1}; ascorbic acid 10 mg 100 g^{-1}. Yams contain large amounts of ascorbic acid. In general avitominosis C (scurvy) is rare in yam growing countries.

4. *Colocasia*

The starch grains in *Colocasia* tubers are among the smallest in food plants and the tuber is reported to be more easily digested than other starchy foods and a good source of calcium and phosphorus. Taro lactin and taro malt prepared from the flour are said to form good infant and invalid foods. The presence of a sapotoxin in the tubers has been reported.

In Hawaii, and parts of Polynesia, the tubers are grated and the resultant slurry allowed to ferment to produce 'poi', a steamed pudding made from grated taro and coconut which is very popular in Hawaii. The peeled tubers, after pre-cooking and drying can be used to produce a flour, similar to potato flour, which is used in the preparation of soups, biscuits, bread, beverages, infant foods and puddings. In recent years there has been interest in the use of taro as a speciality food in the prevention of allergic disease in potentially allergic infants and as a cereal substitute in coeliac diseases.

Colocasia esculenta (L.) Schott. is one of the important food plants introduced into the Philippines during the early prehistoric times. It is commonly called gabi in most dialects. One cultivar, called binata, is used as medicine by the Pinatubo Negritos of the Philippines. A boiled mixture of the leaves and the corms of this cultivar are eaten by women experiencing a difficult childbirth. In some remote areas in Mindanao and the Visayas, when women have dysmenorrhoea they are made to sit on gabi leaves. Why they sit on rather than ingest the leaves is difficult to explain but it seems the ecbolic properties of the plant are recognized.

Kirtikar and Basu (1918), Degener (1930), and Chopra (1923) report that the juice of the petioles is used as a styptic or astringent and may be used to arrest arterial haemorrhage. It is sometimes used in ear-ache and otorrhoea and also as an external stimulant and rubefacient. Kirtikar and Basu (1918) also assert that the juice of the corm of this species is used in cases of alopecia. It is used as a laxative in cases of haemorrhoids and congestion of the portal system, and is also an antidote to the stings of wasps and other insects.

Nadkarni (1927) further states that the heated tubers are locally applied on painful parts in rheumatism. The ash of the tuber mixed with honey is a cure for apthae in mouth. Honigberger (1852) records that in the Punjab and in Kashmir the roots are used in catarrh and colic. Degener (1930) states that in Hawaii the raw juice of gabi mixed with sugar is taken orally to reduce fever.

Taro corms and leaves and petioles are accredited to have medicinal properties. In Brazil, crushed corm is considered to be an effective abortifacient. People sick from morphia nourish exclusively on cooked tubers to ease pain. It is claimed to reduce tuberculous ulcers. It is also used for fungal abscess in animals. Tubers are also considered anthelmintic.

In Hawaii, taro juice is used for styptics and poultices. Mixed with sugarcane juice, it is used as a medicine for pulmonary congestion. A remedy recommended for fungal infection called thrush is a mixture of taro and burnt coconut meal.

It has been reported that *Colocasia* mucilage produced significant hypoglycemic and lipid lowering effect when administered to rats, suggesting that these polysaccharides could have immense health value (Bala Nambisan, 2006).

Petioles of *Colocasia esculenta* have been used for wounds and snake bites (Burkill and Haniff, 1930) in Malaysia.

Corms and cormels are rich in starch ; the flesh is mealy to smooth and usually has a somewhat nutty flavour. The composition of the edible portion of the corms has been given as : energy 373-406 kJ 100 g^{-1}; dry matter 24-26%; starch 15.50-18%; Total sugar 1.75-1.90%; water 73-78 per cent; protein 1.4-3.20 per cent; fat 0.1-1.5 percent; carbohydrate 19-21 per cent; fibre 0.4-2.9 per cent; ash 0.6-1.3 per cent; calcium 32-44 mg 100 g^{-1}; iron 0.8-5.27 mg 100 g^{-1}; phosphorous 64-135 mg 100 g^{-1}; manganese 0.26-0.19 mg 100 g^{-1}; potassium 514-575 mg 100 g^{-1}; sodium 7-9 mg 100 g^{-1}; carotene trace - 67 IU 100 g^{-1}; thiamine 0.09-0.18 mg 100 g^{-1}; riboflavin 0.03 - 0.04 mg 100 g^{-1}; niacin 0.4-0.9 mg 100 g^{-1}; ascorbic acid 0-10 mg 100 g^{-1} (Kay, 1987).

5. *Xanthosoma*

Xanthosoma plant decoction is reported to be used to reduce the body temperature of a feverish patient (Burkill and Haniff, 1930) while the burnt petiole in a poultice is used for elephants sprained muscle (Maxwell, 1960). The inflorescence and sap juice are used to heal wounds and as an antidote for stings of wasps and other insects (Pancho, 1984). Special herbal tea using *Xanthosoma* leaves has been made against diarrhea.

Coursey (1968) gives the approximate composition of *Xanthosoma* tubers as : water 70-77 per cent ; protein 1.3-3.7 per cent; fat 0.2-0.4 per cent; carbohydrate 17-26 per cent ; fibre 0.6-1.9 per cent; ash 0.6-1.3 per cent.

There is considerable variation in the composition of tannias and starch contents ranging from 17 to 34.5 per cent have been reported. Average approximate composition of the edible portion has been quoted as : energy 556 kJ 100 g^{-1}; water 70-77 per cent; protein 1.3-3.7 per cent; fat 0.2-0.4 per cent; carbohydrate 17-26 per cent; fibre 0.6-1.9 per cent; ash 0.6-1.3 per cent; calcium 20 mg 100 g^{-1}; iron 1 mg 100 g^{-1}; thiamine 1.1 mg 100 g^{-1}; riboflavin 0.03 mg 100 g^{-1}; niacin 0.0005 mg 100 g^{-1}; ascorbic acid 6-10 mg 100 g^{-1} (Kay, 1987).

Xanthosoma nigrum cormels are very rich in starch and are said to be more nutritious than potatoes. It is grown on a small scale as a vegetable in South India and also as an ornamental part. The plant yields a milk like fluid used as a blood coagulant. The herb is also credited with anthelmintic properties (CSIR, 1950, 1956).

6. *Alocasia*

The rootstock of *A. macrorrhiza* is pungent, fragment, cooling, useful in inflammation and diseases of the abdomen and spleen. After boiling, it is given as a mild laxative and is useful as diuretic and in piles. The flour from the direct rootstock is boiled along with rice flour until all the water has evaporated and is given for anasarca. The leaves are styptic and astringent. The juice of the petiole is dropped into the ears of children for otorrhora. Application of acrid juice to the stings of the giant nettle (*Laportea gigan* Wedd.) gives instant relief. In Java, the chopped leaves and roots are reported to be used as rubifacient. The leaf is used against tumours. The petioles are used for toothache and their juice for coughs. The roots of *A. mantane,* is used for poisoning tigers and *A. fornicata* is used medicinally. *A. cucullata* was introduced from Bengal to Peninsular Malaysia as an edible crop but is better known for its medicinal value among Chinese herbalists (Burkill, 1935).

The food value of the edible portion of the raw tubers of giant taro has been reported as : energy 293-599 kJ 100 g^{-1}; water 63-81 per cent; crude protein 0.6-3.3 per cent; fat 0.1-0.2 per cent; carbohydrate 17-27 per cent; ash 1.1-1.3 per cent; calcium 46-153 mg 100 g^{-1}; iron 0.5-1 mg 100 g^{-1}; phosphorous 45-72 mg 100 g^{-1}; niacin 0.4 mg 100 g^{-1}; riboflavin 0.02-0.03 mg 100 g^{-1}; thiamine 0.09-0.1 mg 100 g^{-1}; ascorbic acid trace. Much of the calcium is in calcium oxalate crystals (Kay, 1987).

7. *Cyrtosperma*

According to Guerrero (1921), the spadix of *Cyrtosperma chammissonis* in decoction is used as an emmenagogoe and ecbolic.

Analysis of tubers grown in South Pacific have been given as : energy 598 kJ 100 g^{-1}; water 60-70 per cent; protein 0.5-1.4 per cent; fat 0.1-0.5 percent; carbohydrate 28-36 per cent; fibre 1-1.6 per cent; ash 1-1.9 per cent; calcium 301-598 mg 100 g^{-1}; iron 0.9-1.4 mg 100 g^{-1}; phosphorous 28-79 mg 100 g^{-1}; thiamine 0.03-0.06 mg 100 g^{-1}; riboflavin 0.08 - 0.11 mg 100 g^{-1}; niacin 0.6-1.1 mg 100 g^{-1}; ascorbic acid trace 1 mg 100 g^{-1} (Kay, 1987).

8. *Amorphophallus* spp.

Usually eaten as a vegetable, boiled or baked. In Japan, a type of vermicelli is prepared after prolonged boiling of the corms. Young petioles are boiled and eaten as vegetable, similar to asparagus. Older and tougher stems find use in livestock feeding.

A. paeoniifolius is considered as best among tubers and is highly nutritious. Even the stem and flowers are used as food (Raghu *et al.*, 1999). The plant is used in indigenous medicine, it is used as a single drug and in combination form in many diseases. It is used in different pharmaceutical preparations mentioned in Ayurveda, in the treatment of piles, gastric problems, stomach pains and liver diseases. The corms are administered internally particularly in the treatment of dysentery, piles and haemorrhoids. If the tuber alone is consumed continually for a month using butter milk as an adjunct, even highly complicated piles will be enrooted. The corms are aperient, carminative and expectorant and increase appetite and taste. They are applied externally as an irritant to treat acute rheumatism. In filarial swelling and rheumatoid swellings, the ground tuber can be externally used after washing it in hot water. In wart like big growth and in certain type of papules, the tuber is used externally after boiling and grinding with jaggery (Raghu *et al.*,1999).

The active principle of the stem is reported to inhibit the growth of bovine strain *Mycobacterium tuberculosis*. Methanolic extract of the corms showed significant effect on the uterus of guinea pig. The fermented juice of the petioles is use to cure diarrhea. The seeds are also applied externally as an irritant in treatment of rheumatic swellings. According to Guerrero (1921) the corms are employed in anti-rheumatic poultices as rubefacient.

It is used in Indian Alchemy in the common purification of copper (Cu) and as a catalyst in the melting of lead (Raghu *et al.*, 1999).

In the Philippines, the corms are boiled and fed to pigs. In Indonesia, a sp. *A. oncophylla* is used to produce flour for industrial purpose and in Japan, *A. rivieri* is used for mannan extraction. Small sized corms and one year old corms of *A. rivieri* are considered delicacies.

Chopra (1923) and Nadkarni (1927) reported that in India, the corm is stomachial and a tonic. It is used to treat haemorrhoids and as a restorative in complaints of dyspepsia and debility. It is also effective against boils, and ophthalmia and as an emmenagogue. The petiole and tuber juice extract is supposed to be a cure for hypertension.

A.dubius tuber is administered in piles and epilepsy.

A.sylvaticus is supposed to be poisonous to human beings and livestock. The fruits and seeds are crushed into a paste and used for toothache and bruises. They are also used in the treatment of glandular enlargements.

The composition of the edible portion of the corms of *A. paeoniifolius* has been reported as : energy 330 kJ 100 g^{-1} (approx); water 72-79 per cent; protein

1.7-5.1 per cent; fat 0.2-0.4 percent; carbohydrate 18-24 per cent; fibre 0.6-0.8 per cent; ash 0.7-1.3 per cent; calcium 50-56 mg 100 g^{-1}; iron 0.6-1.4 mg 100 g^{-1}; phosphorous 20-53 mg 100 g^{-1}; vitamin A 434 IU 100 g^{-1}; thiamine 0.04-0.06 mg 100 g^{-1}; riboflavin 0.05-0.08 mg 100 g^{-1}; niacin 0.7-0.75 mg 100 g^{-1}; ascorbic acid trace -3 mg 100 g^{-1}. Most of the carbohydrate is starch (75-80 per cent); the starch granules vary in shape and size (about 5.5-19 microns).

9. Coleus

The tubers of *Solenostemon rotundifolius* (Poir.). J.K. Morton, syn, *Coleus parviflorus* are used as vegetable having special flavour and taste. In India, tubers are used as potato in curry preparation. It is preferred for its particular aromatic flavour and sweetness. It can be baked like potato or made into chips.

The Saponins (2-20%) and Alkaloids (15-25%) present in the plant are the primary source of its significant medicinal properties. In Africa, the Hausa potato is sometimes used in the treatment of dysentery and in the treatment of certain eye disorders.

Coleus forskohli is a perennial member of the mint, or Lamiaceae, family that was first discovered in the lower elevations of India. It is now grown around the world as an ornamental plant. The root is used for a myriad of medicinal purposes. As recorded in ancient Sanskrit texts, coleus has been used as a medical herb to treat heart and lung diseases, intestinal spasms, insomnia, and convulsions. Forskolin is a chemical found in the coleus herb that activates the enzyme adenylate cyclase. Adenylate cyclase compound initiates a myriad of critical events and processes within all the cells of the human body and are responsible for executing a number of important hormonal processes. Stimulation that is caused by forskolin supposedly leads to blood vessel dilation, inhibition of allergic reactions, and possibly an increased secretion of thyroid hormone. Forskolin has other reported uses as well, including inhibition of the pro-inflammatory substance known as platelet-activating factor (PAF) and inhibition of the spread of cancer cells.

Some studies have indicated that forskolin may help to reduce blood pressure and improved heart function in people with cardiomyopathy. However, there is no clear evidence as to whether or not oral supplementation with coleus extracts would be an efficient way of taking this herb. One study did find evidence that inhaling forskolin could decrease lung spasms for people with asthma.

The leaves of *Coleus amboinicus* Lour. is used in the local folk medicine for whooping cough and anti-inflammatory treatment (Rejesus, *et al.,*1999). It is reported to be useful for urinary diseases. The juice of the leaves mixed with sugar is a powerful aromatic carminative. It is employed in dyspepsia, although it is said to have intoxicating properties. A decoction of the leaves is given for chronic coughs and asthma. The leaves have a pleasant aromatic odour and pungent taste and are used for flavouring meat and salad. It is a good substitute for borage *(Borago officinalis* Linn.) for flavouring wines and beer. The aromatic properties are attributed to a volatile oil.

The fresh fragrant roots of *Coleus vettiveroides* K.C. Jacob., are used for the decoration of temple images and also for dressing hair (Gopala Chetty, 1939, CSIR, 1950). The decoction of the leaves of *C. blumei* Benth., is said to be given internally for dyspepsia (Burkill, 1935).

The composition of the edible portion of the tubers has been quoted as: water 75 per cent; protein 1.4 per cent; fat 0.5 percent; carbohydrate 21 per cent; fibre 0.7 per cent; ash 1 per cent; calcium 17 mg 100 g^{-1}; iron 6 mg 100 g^{-1}; thiamine 0.05 mg 100 g^{-1}; riboflavin 0.02 mg 100 g^{-1}; niacin 1 mg 100 g^{-1}; ascorbic acid 1 mg 100 g^{-1}. The principal amino acids in the protein are arginine, aspartic and glutamic acids (Kay, 1987).

10. Yam bean (*Pachyrhizus* spp.)

P. tuberosus is cultivated for its edible tuberous roots which contain 1.5%-2.9% protein and 20% starch (Alvarenga, 1987). Although seeds produced by this species are rich in proteins and lipids, the high content of isoflavanoids make them inevitable as a food (Alvarenga and Valio, 1989). Rotenone and pachyrhizin are found in *Pachyrhizus*. Pachyrhizin is present in high amounts in the seeds of *P. erosus* (L.) Urban and *P. tuberosus* (Magalhaes *et al.*, 1988).

Rotenone is important as a natural insecticide controlling insects which attract mainly the leaves of the plants (Villar and Valio, 1994).

In Brazil, the cooling effect produced by consumption of the tubers is believed to check various kinds of fever. The starch content is commonly used in the preparation of lemonades to treat infections of the bladder and various ailments of the urinary system. In doses of 3-4 teaspoons per day, mixed with water, sugar and lime, it is also used to treat haemorrhoids. The juice obtained from the tubers is considered an effective diuretic and is commonly used in the treatment of nephritis. An aqueous-alcoholic extract of fresh tubers administered in dozes of 0.05-0.25 g is also used a diuretic (Peckolt, 1922). It has been reported to improve lactation in breastfeeding mothers and have a curative effect on digestive ailments in children. The latter corroborates the finding of the digestibility studies of *P. erosus* conducted by Tadera *et al.* (1984), who found that yam bean starch is highly digestible and is therefore suitable for use in diets to treat a range of disorders.

Bertoni (1913) reported that the Indian at the Parana river (Paraguay) used an extract of *P. tuberosus* leaves as an insecticide and seeds are used as a decoction or powder to control vermin. Fresh pods/legumes ground and mixed with lard may be used as an ointment to cure itching and mange (Perry and Metzger, 1980).

Escobar (1943) reports that in the state of Veracruz, the seeds of *Pachyrhizus erosus* are commonly used in the treatment of skin diseases. In the pre-Columbian period, the tuber was used to cure fevers, peeling skin and flux, and crushed tubers to cure itching or mange. The old Mexican pharmacopoeia mentions a tincture made from the seeds, used to treat pruritus and mange, as well as head lice, cattle louse (Huart, 1902). A 1/10 concentration of ground seed to wettable sulphur was applied to treat three species of cattle louse with the following results: 100% control in *Bovicola bovis* L. and *Linignathus vituli* L., and 90% control in *Solenopotes capillatus* Enderlein (Matthysse and Schwardt, 1943). Yang and Tang (1988) describe the controlling effect of insecticides derived from yam bean seed on the rice leafhoppers *Heliothis assulta* and *Aulacophora femoralis-chinensis.* The rotenone and rotenoids present in mature seeds may be used as a piscicide (Greshof ,1900; Anonymous, 1889; Sillevoldt, 1899). There are reports of the seeds having being utilized as a piscicide in Indonesia (Meijer, 1946).

The seed oil can be used as a purgative Martínez (1936). In Cuba, the tubers are considered to have an analeptic effect when consumed; and the use of the flour obtained from them is recommended in the treatment of dysentery and haemorrhoids (Roig y Mesa, 1988). In China, 250-5000 g of peeled fresh tubers are consumed each morning and evening to treat fever with thirst; 500 g of the juice of fresh crushed and squeezed tuber is sipped slowly three times daily to treat inflammation of the throat or tonsils; 250 g of fresh peeled tuber is eaten three times a day as a laxative, and fresh tuber juice mixed with sugar is considered a remedy for hangovers (Yin-fang and Cheng-jun, 1987). Pételot (1952) reports that pulverized seeds are used in an ointment to cure skin diseases and boils in Java, and that the ingestion of half a seed is thought to have a laxative effect. Pételot (1952) further reports that the seeds are consumed as vermifuge in North Vietnam. *P. erosus* is fairly common in Taiwan, where it is valued as a refreshing food.

There are no reports of the use of *P. ahipa* for insecticidal purposes or ethnobotanical uses. But during the interviews conducted in Bolivia by Ørting *et al.* (1996), it was repeatedly stated that the consumption of *P. ahipa* tubers has a cleansing effect upon the body, is beneficial to the lungs and cures coughs, etc. A recent field report from Bolivia did not confirm the report by Cárdenas (1969) that *P. ahipa* tubers are considered to be cure for gout (Ørting *et al.,* 1996).

The tuberous roots of *P. erosus* contain both starch and sugar and are a moderately good source of ascorbic acid. Average figures for the edible portion have been published as: energy 186-264 kJ 100 g^{-1}; water 82.4-87.8 per cent; protein 1.5-2.4 per cent; fat 0.09-1.3 percent; carbohydrate 10.6-14.9 per cent; fibre 0.6-0.7 per cent; ash 0.5 per cent; calcium 16-18 mg 100 g^{-1}; iron 0.8-1.1 mg 100 g^{-1}; thiamine 0.05-0.1 mg 100 g^{-1}; riboflavin 0.02 - 0.03 mg 100 g^{-1}; niacin 0.2-0.3 mg 100 g^{-1}; ascorbic acid 14-21 mg 100 g^{-1}.

11. Arrowroot

Maranta arundiaceae L. (West Indian Arrowroot)

Arrow root is primarily grown for its quality starch which is valued as a food stuff. The rhizomes taste like corn when boiled. Arrow-root is used as an article of diet in the form of biscuits, puddings, jellies, cakes, etc., and also with beef tea, milk or veal broth, or boiled with a little flavoring added, as an easily digestible food for children and people with dietary restrictions. Arrowroot starch posses demulcent properties and is used in the treatment of intestinal disorders. It is also employed in the preparation of barium meals and in the manufacture of tablets. It is valued as a food stuff for infants and invalids. The fibrous material which remain after extraction of starch is used as a cattle feed or manure (CSIR, 1950, 1956).

Typical analysis of 'Creole' rhizomes are: water 69.1 per cent; protein 1 per cent; fat 0.1 per cent, starch 21.7 per cent; fibre 1.3 per cent; ash 1.4 per cent. Typical analysis of 'Banana' rhizomes are : water 72 per cent; protein 2.2 per cent; fat 0.1 per cent; starch 19.4 per cent; fibre 0.6 per cent; ash 1.3 per cent (Kay, 1987).

12. Canna

Canna edulis (Queensland Arrowroot)

It is a rhizomatous perennial herb, a native of tropical America, cultivated for the tuberous edible rhizomes from which a kind of arrowroot starch is produced (IBPGR, 1981). The tuber and tops of plant are used as live stock feed and starch is easy to digest and hence used as a food for children and invalids. The young rhizomes are eaten as a vegetable. The cooked tubers are delicious whereas the young shoots and petioles are used as fodder.

Analysis of the rhizomes has been given as: water 67-72 per cent; protein 1-1.7 per cent; fat 0.1 percent; carbohydrate 24-30 per cent; fibre 0.6 per cent; ash 1.4 per cent; calcium 18 mg 100 g^{-1}; phosphorous 63 mg/100 g; ascorbic acid 7 mg 100 g^{-1}.

13. Winged bean

Psophocarpus tetragonolobus (L.) DC

As a seed it can be roasted or boiled after removal of the sometime resistant seed coats (seed containing 29-38% protein and 28-30% carbohydrate on a DW basis). As a sprouted seedlings or at any vegetative stage, leaves and shoot tips (5-10% protein DW) can be eaten after steaming or boiling. Pods (1.9%-2.9% protein DW) or immature and mature seeds can be consumed after roasting, boiling or steaming. The flowers (5.6% protein DW) may also be eaten as a vegetable. Also at the end of the growing season, the tubers can be dug from the ground, cooked and eaten as a source of high quality carbohydrate (27% DW) and protein (12-15% DW). It has been suggested as a green manure, cover crop, fodder crop and a restorative fallow crop because its exceptional capacity for forming very large and numerous root nodule (Purseglove, 1974; Masefield, 1961).

The water content of the fresh tubers is 52-68 per cent. The composition of the dry matter has been given as: protein 13-20 per cent; fat 0.6-1.4 per cent; carbohydrate 63-77 per cent; fibre 1.5-21 per cent; ash 1.7-3.9 per cent; calcium 40 mg 100 g^{-1}; iron 3 mg 100 g^{-1}; phosphorous 64mg 100 g^{-1} (Kay, 1987).

14. *Helianthus tuberosus* L.

The tubers may be eaten raw or boiled. They are also pickled made into chips or ground into flour. In food value they are considered equal to potatoes. Jerusalem artichoke has aroused much interest as a commercial source of levulose used as a sweetening agent for diabetics. Fresh tubers are sliced and juice is acidified and heated to hydrolase inulin and inulides. Neutralisation with lime yields calcium leuvulate which is separated by precipitation and filtration. Carbonation of calcium levulose yields a syrup containing levulose from which sugar is crystallized. Fructose syrups suitable for use as sweetening agents may be prepared. It is also used for the preparation of industrial alcohol by fermentation and beer like beverages.

Tubers are used as feed and a rich palatable feed of good digestibility comparable to sugar beet in feeding value is obtained by ensilage. The average nutritive value of the tuber is 1.2 per cent digestible protein, 15.9 per cent total digestible nutrients and 12.3 nutritive ratio. Green tips and stems of young plants are used as feed for cattle, by ensilage. The yield of green fodder is 12.0-22.0 t ha^{-1} but the yield of tubers is greatly reduced and nutritive value of tops is 27.2 dry matter, protein 1.4, fat 0.3, fiber 4.9, mineral matter 2.4, calcium 0.44, phosphorus 0.03, potassium 0.37, digestible protein 0.8 and total digestible nutrients 18.1 per cent, and nutritive ratio 21.6. The stalks may be treated by soda chlorine process to give a pulp which is suitable for the manufacture of certain type of papers (Vimala, 1994).

Analysis show a wide range of variation: published average figures for the edible portion are: energy 332.3 kJ 100 g^{-1}; water 78.9 per cent; protein 2.44 per cent; fat 0.41 percent; carbohydrate 15.8 per cent; fibre 0.7 per cent; ash 1.74 per cent; potassium 478 mg 100 g^{-1}; calcium 10 mg 100 g^{-1}; iron 3.7 mg 100 g^{-1}; phosphorous 78 mg 100 g^{-1}; thiamine 0.2 mg 100 g^{-1}; riboflavin 0.16 mg 100 g^{-1}; niacin 1.3 mg 100 g^{-1}; ascorbic acid 4 mg 100 g^{-1} (Kay, 1987).

15. *Asparagus*

A. officinalis is known from olden times as food and medicine (Deepa *et al.*, 1999). The young shoots of several species are edible and can be eaten like the commercial asparagus. This vegetable is of great importance in the diet because of its valuable salts, vitamins A and C and minerals potassium and phosphorous and for its large amount of cellulose contents. Beautiful and delicious, asparagus is also low in calories. The asparagus is a perennial plant, with underground fleshy root-stock. The young shoots are eaten when 6 to 15 cm high.

The asparagus has multipurpose therapeutic properties. It is an alkaline foodstuff. It is an excellent body cleanser and its use will prevent or combat blood acidity. It has a nitrogenous content called 'asparagin' which has diuretic properties. It is said that the liberal use of the young shoots acts as a rejuvenator. However, the use of asparagus should be avoided in goitre, liver and kidney diseases and in case of excess mucus, especially catarrhal mucus.'

The roots are more diuretic than the shoots. Modern medicine uses the root of the plant as a diuretic. They are recommended in dropsy. They have an insipid taste without odour and are administered in the form of a syrup and is a powerful cardiac sedative. The asparagus is an excellent food for strengthening the heart. A food medicine for weak or enlarged hearts is prepared by mixing the freshly expressed juice of this vegetable with honey and taking a teaspoonful three times daily. Patients with heart disease will also benefit by eating the cooked vegetable provided the cooking is done conservatively, preferably in steamed form.

An infusion of the roots is used against jaundice. The roots have been used as an aperitif and as remedy for schistosomiasis. They exhibit inhibitory activity against tuberculosis and other pathogenic organisms *in vitro*. The asparagus species are also used for bladder and kidney ailments and lung conditions. It should, however, be avoided in advanced stages of kidney diseases.

Eleven substances were isolated from the methanolic extract of the roots. The asparagusic acid extracted from the roots is reported to be toxic to several plant parasitic nematodes. The plant is a demulcent, diuretic, laxative. It is reported to be irritant to skin and cause dermatitis. A tincture of the whole plant is used for urinary infections and rheumatism. The asparagus is a valuable food medicine in rheumatism on account of its potash content. A steamed infusion prepared from this vegetable is considered highly beneficial in the treatment of rheumatism and chronic gout. It should be given in doses of 30 to 60 ml.

Phytoestrogens were reported in asparagus species. Its use slows the ageing process. The dried roots of asparagus are used in Unani medicine as an aphrodisiac. It is available in the market as *Safed musli*. Roots boiled in milk are used as demulcent or soothing medicine. Its regular use is valuable in impotency. A decoction of the fresh or dry fruits is reputed as a contraceptive in Europe. The plant is poisonous to cattle.

The roots of *A.racemosus* Willd. are prescribed in Chamba to cure diarrhea and dysentery. They are used as a substitute of *A. officinalis*. The root bark is a reputed aphrodisiac. The roots contain a steroidal saponin which inhibits the growth of several pathogenic organisms *in vitro*.

The roots of *A. filicinus* are considered tonic ,astringent ,vermifuge, and taenifuge. In Indo-China,they are given as a powerful diuretic in cholera and rheumatism. The roots exhibited inhibitory activity against *Rhizopus arrhizus* Fiscer and the plant extract against *Entamoeba hystolytica in vitro*. The fibre free root powder gave satisfactory results in bulk sizing and finishing in the textile industry. It has also been found to be suitable as a base in Calico-printing.

The root of *A. gonocladus* is considered nourishing and is reputed as an aphrodisiac. Boiled with oil it is applied to cutaneous diseases. Internally the root is given in gonorrhea.

Asparagus aethiopicus is a variable plant that may be a woody climber 7 m tall. Spines are present on all nodes, except sometimes on the flowering stems. The flowers are sweet-scented and give rise to red berries.

16. Vigna species

Zombi Pea (*Vigna capensis* Walp. Syn. *Vigna vexillata* L.)

It has a tuberous, fusiform rootstock which is reported to be eaten and considered superior to sweet potato in flavour and nutrients. The seeds are also said to be eaten. The plant can also be grown for ornament. Edible wild one is important in drought and extracts for medicine to cure foot-and-mouth disease in livestock (Bhag Mal, 1994).

Zombi pea is of great potential as source of seeds and tubers. The tubers contains about 15 per cent protein and the seed contains 20.7 per cent protein at 110 days growth stage and the tuber protein yield is much higher than sweet potato (Lal *et al.*, 1988). Its tuberous roots are eaten like sweet potatoes. It is widely cultivated as green pasture, cover and soil-improvement crop as well as for soil erosion control.

Pillipesera (*Vigna trilobata* (L.) Verdcourt

The grains are used as pulse, generally by the poor people. The root is reported to be a source of a kind of arrowroot. The plants are valued as green fodder which is succulent and nutritious and is relished by cattle. The fodder contains (air dry basis) 11.4 per cent protein, 1.3 per cent fat, 41.4 per cent nitrogen free extract, 22.1 per cent fibre, 13.4 per cent ash, 2.69 calcium, 0.45 per cent phosphorus. The plants can also be converted into hay which can be fed to animals along with rice straw. It can also be used as a cover crop and a green manure crop (Bhag Mal, 1994).

Rice bean (*Vigna umbellata* (Thunb.) Ohwi and Ohashi)

Rice bean is a promising multipurpose crop with a good potential to be used as food, fodder, green manure and a cover crop. Wild form possess tuberous root system. The dried seeds are usually eaten boiled, in soups, or as pulse with or without rice. Young immature pods and leaves are used as vegetable; sprouts are also consumed occasionally. Immature pods and bean sprouts are highly recommended in the nutritional diet of the Philippines. Whole plant is used as forage for livestock. It provides forage at a time when other sources of fodder are scarce. The foliage, green pods, immature seeds and flowers are readily eaten by animals. It is also grown as a green manure and an excellent cover crop. In China, Japan and Korea, the beans are generally ground into flour and meal; beans are often boiled alone or with onion and carrot and eaten as a dish or as soup.

The nutritional quality or rice bean is excellent due to high amount of protein and appreciable quantities of limiting amino acids, tryptophan and methionine which rank it as one of the best among pulses (NAS, 1979).

17. Hyacinth bean *(Lablab purpureus* (L.)

Hyacinth bean is useful in several ways. It is largely grown for use as human food. Young pods are cooked and eaten as vegetable or in salads and as pickles. The leaves can be pickled and used as a salad or spinach. Mature seed can also be used but should be properly cooked before consumption. Perennial form produce an edible root tuber also (Smartt, 1990). In Egypt, pounded seeds are substituted for broad beans in the fried bean cake 'tanniah'. Plants are used as forage for livestock. It is grown for grazing and is also useful as a green manure and a cover crop manure and for soil improvement.

It has a good medicinal value. Seeds are alexeritic, antispasmodic, aphrodisiac, febrifugal and stomachic; also used for menopause. The leaves are used for the treatment of colic, gonorrhea, and leucorrhea and also poulticed onto snakebites and eczema. Juice from pods is applied to ears and throat for curing inflammation. Forage contains 14.2 per cent crude protein, 3.5 per cent fat, 28.1 per cent fibre, 39.4 per cent carbohydrates, 14.8 per cent ash, 1.98 per cent Ca and 0.26 per cent P (Bhag Mal, 1994).

18. Curcuma

The tubers or rhizomes of *Curcuma angustifolia* Roxb. contain starch which is used as a substitute for arrowroot powder from *Maranta arundinaceae* Linn. The tubers are washed, pulped and the starch separated from fibre and other impurities by repeated washing and straining through cloth. It is sun-dried and ground into flour. The product resembles arrowroot starch and is easily digestible (Kundu, 1967). It is used in the preparation of milk puddings and is recommended for children and invalids.

An analysis of the rhizomes has been given as water 69-70 per cent; starch 12-13 per cent; fibre 18-19 per cent. The starch grains have an average size of 1.6-4.2 microns and are elongated or ovoid in shape, closely resembling those of arrowroot. An analysis of a commercial sample of Indian shoti starch gave : water 13.1 per cent; starch 82.6 per cent; ash 1.01 per cent. The starch has 31.3 per cent amylase content. Shoti starch is readily hydrolysed by acids and possesses a high viscosity similar to potato starch.

The rhizomes of *Curcuma zedoaria* Rosc. are large, fleshy, branched and the inner part of which is pale yellowish brown. The young rhizomes can be eaten fresh or after being cooked as vegetable. They are also cut into thin transverse sections and dried. The sliced and dried rhizomes are used for medicines and cosmetics. They have a pungent and somewhat bitter taste. The yield of tubers vary from 7-12 t ha^{-1} (CSIR, 1950). They are rich in starch, which is extracted in India to give shoti starch (Purseglove, 1972). This is valued in diet, particularly for infants and invalids.

'Shoti starch' (flour) is produced in Bengal on a cottage industry basis. The starch gives highly viscous paste with water. Steam distillation of the rhizome yields 1-2 per cent light yellow oil. The rhizomes possess antibacterial, antiulcer properties and is a gastrointestinal stimulant. It is a constituent of a wide variety of ayurvedic preparations like Dasamularishtam, Valiya Rasnadi Kashayam etc. It is extensively used in traditional Chinese medicine to treat various cancers. The methanol extract of *Curcuma zedoaria* rhizome was found to be cytotoxic against Daltons Lymphoma Ascites (DLA) and Ehrlich Ascites Carcinoma (EAC) cells at a concentration of 100 $\mu g\ ml^{-1}$ using Trypan Blue exclusion technique and against Human cervical cancer cells (HeLa) at a concentration of 250 $\mu g\ ml^{-1}$ using MTT assay. The inhibitory concentration required for 50% cytotoxicity (IC_{50}) was also determined. These findings suggest the potential of the plant to be exploited further to develop as an antitumour drug (Laskhmi *et al.*, 2006).

A decoction of the rhizome administered along with long pepper, cinnamon and honey is beneficial for cold. It has a local effect on the digestive organs similar to but milder than ginger and has been occasionally employed as a gastrointestinal stimulant in flatulent colic. Zedoary is used in the manufacture of liquors, stomach essences and bitters and in the perfumery and cosmetics. The leaves are used for flavouring high and other foods in Java (Purseglove, 1972).

19. *Tacca leontopetaloides* (L) Kuntze (East Indian Arrowroot)

The tubers were once a staple foodstuff in Polynesia and were used as a source of starch very similar to arrowroot, often in the treatment of dysentery and for feeding infants.

Sometimes the plant is utilized as a vegetable similar to Irish potatoes, but is said to be rather indigestible; it is regarded as a famine food in parts of W. Africa. The tubers contain about 2.2 per cent of a bitter extract and a bitter principles taccalin has been isolated from the dried tubers.

The tubers, which are very bitter and inedible when raw, require suitable preparation before eating. The tubers are peeled and rasped. The meal is washed in several changes of water, after which it is strained through a coarse cloth. The starch, which passes through with the water, is allowed to settle, and then washed several times to remove the bitter principle before being dried. The product was known as 'Tahiti' or 'Fiji arrowroot', and was exported in small quantities at one time. The starch is cooked and eaten in some Pacific islands. In Tahiti (in Cook Is.) it is used for making poi, a traditional food, in which the starch is mixed with the pulp of bananas, papayas and other fruits, flavoured with lemon and vanilla, wrapped in green leaves, and cooked in a native oven. In Zanzibar the starch is used for making a local Turkish delight. The tubers, after suitable preparation, are eaten in northern Nigeria (Purseglove, 1972). The tubers of *T. pinnatifida* Forst & Forst. f. are used in the treatment of piles. Rubefacient, a bitter extract is prepared by washing the grated tubers in running water, and that is given in diarrhea and dysentery (CSIR, 1976).

An analysis of African tubers on a dry weight basis has been given as: protein 5.1 per cent; carbohydrate 89.4 per cent; ash 3.2 per cent; calcium 0.27 per cent; phosphorous 0.2 per cent. The principal amino acids present in the protein are arginine, glutamic and aspartic acids, leucine, lysine and valine.

20. Safed Musli

The tuberous roots are mainly used in Ayurvedic medicines. Roots are used for the preparation of nutritive tonic. These roots contain spermatogenetic, spermatorrhoea and chronic leucorrhoea factors due to some chemical content.

21. *Costus speciosus* (Koenig)Sm.

The rhizome is edible and is used after cooking. It is mucilaginous, astringent and has no aroma. It is rich in starch and fibre. In some parts of India it is cooked in syrup and preserved. The rhizome is credited with purgative and tonic properties (Burkill, 1935). It is used in UP as a tonic and anthelmintic.

Dasgupta and Pandey (1970) have reported 2.12% of diosgenin out of a total of 3.86% of total sapogenins present in the glycosidic form in the alcoholic extract. They have further reported diosgenin and tigogenin to be present in free form in addition to five sapogenins, of which they could isolate only β-sitosterol and diosgenin.

22. *Gloriosa superba*

Gloriosa superba seeds and tubers have been reported to be rich in colchicine, and worth exploring for the manufacture of colchicines. It has been observed that seeds are richer in colchicine than that of tubers. The tubers of *G. superba* are available in plenty in Tamil Nadu (CSIR, 1950, 1956).

Different parts of the plant have a wide variety of uses especially within traditional medicine practiced in tropical Africa and Asia. The tuber is used traditionally for the treatment of bruises and sprains, colic and chronic ulcers, haemorrhoids, cancer, impotence, nocturnal seminal emissions, leprosy and also for inducing labour pains and abortion. Because of its similar pharmacological action, the plant is sometimes used as an adulterant of aconite (*Aconitum* sp.). The juice of the leaves is used to kill head lice and also as an ingredient in arrow poisons. The flowers are used in religious ceremonies. The tuber has commonly been used as a suicidal agent among women in rural areas and it has also been used for homicide. The tuber also claims antidotal properties to snake-bite and in India it is commonly placed on window sills to deter snakes. Many cultures believe the species to have various magical properties (Watt and Breyer-Brandwijk, 1962; Neuwinger, 1994; Burkill, 1995).

23. *Ullucus tuberosus*

Ulluco has high water content (more than 75%), about 1% protein, depending on the variety, and 10-14% carbohydrates. Its mineral content is low, though it is rich in vitamin C. Cooked tubers are used in several dishes. The most popular in Peru is the "olluquito con charqui." Ulluco is also added to soups and stews. The boiled leaves, rich in calcium and carotene, are used as a vegetable. Some women in the Andes believe that eating ulluco will ease childbirth. The tubers are prepared and eaten like potatoes ; they may be dried and stored.

An analysis of the edible portion of the tubers has been published as: energy 214 kJ 100 g^{-1} water 85.9 per cent; protein 1 per cent; fat 0 percent; carbohydrate 12.5 per cent; fibre 0.6 per cent; ash 0.6 per cent; calcium 3 mg 100 g^{-1}; iron 0.8 mg 100 g^{-1}; phosphorous 35 mg 100 g^{-1}; vitamin A 0 mg 100 g^{-1}; thiamine 0.04 mg 100 g^{-1}; riboflavin 0.02 mg 100 g^{-1}; niacin 0.3 mg 100 g^{-1}; ascorbic acid 23 mg 100 g^{-1} (Kay,1987).

24. *Artemisia dracunculus*

It is highly cordial and friend to the head, heart and liver. In Continental cookery its use is advised to temper the coolness of other herbs in salads. The leaves, which have a fragrant smell in addition to their aromatic taste, make an excellent pickle and vinegar. Fresh Tarragon possesses an essential volatile oil, chemically identical with that of Anise, which becomes lost in the dried herb. Tarragon vinegar is the only correct flavouring for Sauce Tantare, but must never be put into soups, as the taste is too strong and pungent. French cooks usually mix their mustard with Tarragon vinegar. Russian Tarragon is eaten in Persia to induce appetite. The root of Tarragon was formerly used to cure toothache.

25. *Typhonium species*

The fresh tuber is very acrid and a powerful stimulant and is employed as poultice in scirrhous tumours. But the acrid principle is very volatile and the tubers become innocuous on heating or drying. The tubers are reported to relax the bowels and provide relief in haemorrhoids. They are eaten with bananas to cure the stomach complaints (CSIR, 1956). The tubers of *T. divarticatum* Decne posses rubefacient properties and in diarrhea.

The analysis of edible portion of tuber revealed dry matter: 33%, starch : 22.50%, sugar : 0.71%, lipid : 0.92%; protein 1.4, fibre 1%, mineral 1.6%, calcium 35%, phosphorous 20%, iron 1.3, carotene 78mg.

26. *Pueraria*

Pueraria (candollei var.) *mirifica* roots produce the rejuvenating miroestrol, an estrogen, but toxic side effects of the refined product limited its use in the western world. Folk medicine in Thailand still uses it. The isoflavones Puerarin and *Puerarin -?-* xylasides are found in roots of *P. lobata*, as is *D*- pinitol, a cyclitol, and saponins and tannins. *Pueraria tuberosa* contains inulin allied saccharine matter, an easily oxidizable resin and a resin acid, the property as a fish poison in some accessions has not been elucidated. *P. peduncularis* roots have insecticidal properties and also used as fish poison (Maesen, 1994).

27. *Lathyrus tuberosus*

It is occasionally grown as a root vegetable for its edible tuber. The sweet, starchy tubers are edible, cooked or raw (Bhag Mal, 1994).

28. *Arracacia xanthorrhiza*

The most important part is the starchy root. It cannot be eaten raw, but when cooked it develops a distinctive flavor and aroma that have been described as "a delicate blend of celery, cabbage and roast chestnuts" (http://en.wikipedia.org).

The boiled root has about the same uses as boiled potatoes, including side dishes, purées, dumplings and gnocchi, pastries, etc., with the advantage of its flavor and (depending on the variety) its intense color. In the Andes region it is made into fried chips, biscuits, and coarse flour. Because it is highly digestible (due to the small size of its starch grains), purées and soups made from it are considered excellent for babies and children. The young stems can be eaten cooked or in salads, and the leaves can be fed to livestock.

Fresh arracachas keep in the refrigerator for 2 to 3 weeks. 100 grams of arracacha provide about 100 calories (26 g of dry matter, 23 g being carbohydrate, and less than 1 g of protein). The plant is rich in calcium (four times as much as potatoes).

The yellow cultivar contains substantial amounts of carotenoid pigments, precursors to vitamin A, to the point that excessive consumption of arracachas may cause yellowing of the skin (a condition that is not considered to be harmful) (http://en.wikipedia.org).

29. *Arthropodium*

It is used for medicine as well as food, and has symbolic importance in traditional Mâori culture.

30.*Tropaeolum tuberosum* Ruíz and Pavón

Its extraordinary resistance to insect, nematode, and bacterial pests is attributed to high levels of isothiocyanates. In Colombia, it is planted as a companion crop to repel pests in potato fields. It has been used to treat nephropathy and as a diuretic.Popularization of mashua may be limited by its strong flavor, and its reputation as an anti-aphrodisiac (http://en.wikipedia. org).

31.*Eleocharis dulcis* (Burm. f.) Trin. ex Hensch.

The small, rounded corms have a crispy white flesh and can be eaten raw, slightly boiled, grilled, pickled, or tinned. They are a popular ingredient in Western-style Chinese dishes. In China, they are most often eaten raw, sometimes sweetened. They can also be ground into a flour which is used especially to make fried cakes called matigao. They are unusual among vegetables for remaining crisp even after being cooked or canned. They do this because their cell walls are cross-linked and strengthened by certain phenolic compounds, a property shared by other vegetables that remain crisp in this manner, including the tiger nut and lotus root. (http://en.wikipedia.org).

If eaten uncooked, the surface of the plants can transmit Fasciolopsiasis. The corms are rich in carbohydrates (about 90% by DW), especially starch (about 60% by DW), and are also a good source of dietary fibre, riboflavin, vitamin B6, potassium, copper, and manganese.

32. *Mirabilis expansa*

Mirabilis expansa (mauka or chago) is cultivated as a root vegetable in the Andes, bears edible tuberous roots that can reach the size of a man's forearm, with a dry weight composition of about 7% protein and 87% carbohydrate. Yield can reach 50,000 kg ha^{-1} given two years maturation time. Great interest in this root crop has been generated by its ability to be grown in conditions that do not favor other root crops.

The roots of some forms if eaten directly can irritate the mucus membranes, and should be sun-dried and boiled before eating to eliminate the irritating substance. Bolivian forms are more often irritating than Ecuadoran forms. The cooking water of the *M. expansa* makes a satisfying sweet drink while leaves may also be eaten as a leaf vegetable or used raw in salads. Once the root has been exposed to the sun, the astringent, bitter taste is replaced with sweetness. The boiled roots mixed with honey and toasted grain is one of the traditional preparations. Ecuadorians have both sweet and salty preparations (http://en.wikipedia.org).

33. *Cyperus*

Cyperus esculentus (tigernut sedge) tubers are edible, with a slightly sweet, nutty flavour, compared to the more bitter tasting tuber of the related *Cyperus rotundus* (Purple Nutsedge). They are quite hard and are generally soaked in water before they can be eaten. They have various uses; in particular, they are used in Spain to make horchata.

Tigernuts have excellent nutritional qualities with a fat composition similar to olives and a rich mineral content, especially phosphorus and potassium. Tiger nuts are also gluten and cholesterol free, and have a very low sodium content. The oil of the tuber was found to contain 18% saturated (palmitic acid and stearic acid) and 82% unsaturated (oleic acid and linoleic acid) fatty acids.

According to the *Consejo Regulador de Chufa de Valencia* (Regulating Council for Valencia's Tigernuts), the nutritional composition 100 ml^{-1} of a classical Horchata de Chufas, or *Orxata de Xufes* in Valencian language, is as follows: energy content around 66 kcal, proteins around 0.5 g, carbohydrates over 10 g with starch at least 1.9 g, fats at least 2 g. Since the tubers contain 20-36% oil, *C. esculentus* has been suggested as potential oil crop for the production of biodiesel (http://en.wikipedia.org).

34. *Psoralea esculenta*

The raw root of Indian bread root or Indian potato is moderately sweet and tastes like the turnip. The dried root can be ground into a flour.

35.*Smallanthus sonchifolius* (Poeppig and Endlicher) *H. Robinson* syn.*Polymnia sonchifolia* Poeppig and Endlicher

Storage roots of yacon are large and edible. The root of yacón is composed mostly of water and fructo-oligosaccharides. It has recently been introduced into farmer's markets and natural food stores in the US. These edible roots contain inulin, an indigestible sugar, which means that although they have a sweet flavour, the roots contain no calories.

The leaves of the yacón contain quantities of protocatechuic, chlorogenic, caffeic and ferulic acids, which gives tea made from the leaves prebiotic and antioxidant properties. As a result, some researchers have explored the use of yacón tea for treating diabetes and for treating diseases caused by radicals, e. g., arteriosclerosis (Valentova *et al.,* 2003).

36.*Oxalis tuberosa* Savign.

The oca is one of the important staple crops of the Andean highlands, second only to the potato. The flavor is slightly tangy, and texture ranges from crunchy (like a carrot) when undercooked, to starchy or mealy when fully cooked. Though the original Andean varieties are widely variable in color from purple to yellow, the standard New Zealand variety is a fleshy pink.

Ocas are fairly high in oxalates, concentrated in the skin, and traditional Andean preparation methods were geared towards reducing the oxalate level of the harvested vegetable. This is done by exposure to sunlight which increases the glucose content and sweet taste of the oca. Recent oca cultivars have a lower oxalate content, and have also been selected for more flexibility in day lengths.

The oca can be prepared like most root vegetables - boiled, baked or fried. In the Andes it is part of stews and soups; served like potatoes or can be as served as a sweet. Oca is eaten raw in Mexico with salt, lemon and hot pepper (Davidson, 1999).

37.*Stachys affinis* Bunge

The flavor of the tubers of Chinese artichoke is delicate and delicious - they can be treated as Jerusalem artichokes in cooking. It is used as a vegetable, in salad compositions, but more so as a garnish.

Nutritive value of tuber and root crops
(Nutrients per 100 g edible portion FWB)

Food	Food energy	Moisture	Protein	Tot. CHO and fibre	Calcium	Phosph-orous	Iron	Pota-ssium	Caro-tene equiv.	Thia-mine	Ribo-flavin	Niacin	Ascorbic acid	Folic acid
	kJ	%	G	g	mg	Mg	Mg	mg	µg	mg	mg	mg	mg	µg
Sweet potatoes, kumala *(Ipomoea batatus)*														
White	452	72.8	1.0	25.1	21	50	0.9	210	35	0.14	0.05	0.7	21	25 to 52.0
Yellow	481	70.7	1.2	27.1	36	56	0.9	304	1680	0.12	0.05	0.6	30	-
Cassava *(Manihot esculenta)*	565	65.5	1.0	32.4	26	32	0.9	394	0	0.05	0.04	0.6	34	24.2
Via mila (*Alocasia macrorrhiza)*	255	84.0	0.6	14.8	30	50	1.0		0	0.05			5	-
Suran (*Amorpho-phallus campanulatus)*	339	78.5	2.0	18.4	38	38	2.4	416	0	0.06	0.02	1.7	6	-
Taro, dalo *(Colocasia esculenta)*	393	75.4	2.2	21.0	34	62	1.2	448	tr.	0.12	0.04	1.0	8	-
Giant taro, via, babai*(Cyrtosperma chamissonis)*	548	-	0.9	31.0	334	56	1.2	-	0	0.05	0.07	0.88	-	-
Yam, uvi *(Dioscorea alata)*	452	71.8	2.0	25.1	22	39	1.0	294	0	0.10	0.04	0.07	-	-
Taro leaves	255	81.4	4.0	11.9	162	69	1.0	963	5535	0.13	0.34	1.5	63	163.0
Cassava leaves	-	80.5	6.8	9.6	206	81.8	2.0	-	10000 IU	0.16	0.30	1.8	265	-
Sweet potato leaves	-	-	4-6	-	183	264	2.4	3018	3440	0.16	0.35	0.9	55	88.4

Source : Chandra, 1984; William, 1959; Woolfe, 1992

In China, the Chinese artichoke is used primarily for pickling. Its tuber is a part of Osechi cooked for celebrating Japanese New Year. Dyed red by leaves of red shiso after pickling, it is called Chorogi. In French cuisine, its cooked tuber is often served along sidedishes named *japonaise* or Japanese-styled (http://en.wikipedia.org).

38. *Tragopogon porrifolius*

The root, and sometimes the young shoots, of Purple Salsify are used as a vegetable, and historically the plant was cultivated for that purpose. The root is noted for tasting of oysters; young roots can be grated for use in salads, but older roots are better cooked, and they are usually used in soups or stews. A latex derived from the root can be used as a chewing gum. The flowering shoots can be used like asparagus, either raw or cooked, and the flowers can be added to salad, while the sprouted seeds can be used in salads or sandwiches.

The plant has also been used in herbalism, also since classical times (it is mentioned by Dioscorides) and is claimed to have beneficial effects on the liver and gall bladder. The root is regarded as a diuretic (http://en.wikipedia.org).

39. *Valeriana officinalis*

The roots have potential bioactive compounds useful in treating different ailments. It is used in treating insomnia, epilepsy, gastro intestinal pains and spastic colitis. It contains valerianine, valerine, valeric acid and other alkaloids.

40. *Utleria salicifolia*

The rhizome has antiulcer property and is commonly in vogue among tribals in Western Ghats.

Nutritional composition of a 100 g edible portion of various foods

Constituents	Cassava	Sweet potato	Yam	Aroids
Dry matter (% FW)	30-40	19-35	21-40	22-27
Starch (% FW)	27-36	18-28	18-25	19-21
Total sugars (% FW)	0.5-2.5	1.5-5.0	0.5-1.0	2.0
Protein (% FW)	0.5-2.0	1.0-2.5	2.5	1.5-3.0
Fiber (% FW)	1.0	1.0	0.6	0.5-3.0
Lipid (g)	0.1	0.4	0.2	0.2
Ash (mg)	0.6	1.0	1.0	1.2
Vitamin A (mg / 100 g FW)	17	900	117	0-42
Vitamin C (mg / 100 g FW)	50	35	24	9
Starch grain size (micron)	5-50	2-42	1-70	1-12
Amylose (% total starch)	15-29	8-32	10-30	3-45

Source : Bradbury and Holloway (1988); Wheatley et al., (1995); Horton, 1988

Constituents	Cassava	Sweet potato	Yam	Aroids
Food energy (k cal/100 g)	124	114	101	98
Ca (mg /100g)	Na	32	20	28
P (mg /100g)	Na	47	69	61
Fe (mg /100g)	Na	0.7	0.6	1.0
Na (mg /100g)	Na	10	-	7
K (mg /100g)	Na	243	600	514
Thiamine (mg /100g)	Na	0.09	0.10	0.13
Riboflavin (mg /100g)	Na	0.06	0.04	0.04
Niacin (mg /100g)	-	0.6	0.5	1.1

Horton, 1988

References

Alvarenga, A.A. 1987. Estudo de algum aspects do desenvdvimento no feijao jacatupe – *Pachyrhizus tuberosus* (Lam.) Spreng. Ph.D Thesis – Universidade Estadual de Campinas.

Alvarenga, A.A and Valio, I.F.M. 1989. Influence of temperature and photoperiod on flowering and tuberous root formation of *Pachyrhizus tuberosus*. *Ann. Bot.*, 64: 411-414.

Anonymous. 1889. Yam Bean. *Bull. Misc. Inf.* 86 : 62-63. Royal Botanical Gardens, Kew, U.K.

Applezweig, N. 1962. Steroid drugs. McGraw-Hill, New York.

Asolkar, L.V and Chadha, Y.R. 1979. Diosgenin and other steroid drug precursors Publications and Information Directorate, CSIR, New Delhi.

Bala Nambisan, 2006. Composition and properties of mucilages from *Dioscorea, Colocasia* and *Xanthosoma*. Paper presented in the 14th Triennial Symposium of the International Society for Tropical Root Crops from Nov 20-26, 2006, Thiruvananthapuram, Kerala, India.

Bartlett, H.H. 1910. The source of the drug Dioscorea with a consideration of the Dioscorea found in the United States. Bureau of Plant Introduction, U.S.D.A. Bull. No. 189.

Barua, A.K., Chakravarti, D. and Chakravarti, R.N. 1954. 'Steroidal sapogenins from Indian *Dioscorea* plants-I, *J. Indian Chem. Soc.*, 31, 173-178.

Barua, A.K., Chakravarti, D. and Chakravarti, R.N. 1956. Steroidal sapogenins from Indian *Dioscorea* plants - II. *J. Indian Chem.* Soc. 33: 799-803.

Bertoni, M.S. 1913. Catálogo de las especies óvariedades de plantas cultivadas de la Estación Agronomica de Puerto Bertoni, á fines de 1912; con la indicación de aquellas cuyas semillas ó renuevos están generalmente disponibles para el canje, la distribución ó la venta. *Agron., Bol. Est. Agr. Pto. Bertoni* (Paraguay) 5. 7:255-282.

Bhag Mal. 1994. Underutilized grain legumes and pseudocereals – their potentials in Asia. Regional Office for Asia and the Pacific. FAO, Bangkok, Thailand. 162 pp.

Bradbury, J. H., Beatty, R.E., Bradshaw, K., Hammer, B., Holloway, W.D., Jealous, W., Lau, J., Nguyen, T and Singh, U. 1985. Chemistry and Nutritive value of tropical root crops in the South Pacific. *Proc. Nutr. Soc. Austr.* 10:185-8.

Bradbury, J.H. and Holloway, W.D. 1988. Chemistry of tropical root crops: Significance for nutrition and Agriculture in the Pacific. Australian Centre for International Agricultural Research (ACIAR). Monograph series number 6. Canbera, Australia: ACIAR.

Burkill, H.M. 1995. The useful plants of West Tropical Africa, 2nd ed., vol. 3. Royal Botanic Gardens, Kew.

Burkill, I.H. 1935. A Dictionary of the Economic Products of the Malay Peninsula. Crown Agents, London.

Burkill, I.H and Haniff. 1930. Malay Village Medicine. *Gard. Bull. Str.* Settlements Part 2:265 p.

Cárdenas, M. 1969. Manual de Plantas Economicas de Bolivia. pp. 1-421. Imprenta Ichtus, Cochabamba, Bolivia.

Chandra, S. 1984. Edible Aroids. Clarendon Press, Oxford, UK. pp. 220-221.

Chopra, R.N. 1923. Indigenous drugs of India. Calcutta, 655 pp.

Corkill, N.L. 1948. 'The poisonous wild cluster yam, *Dioscorea dumetorum* Pax. as a famine food in the Anglo-Egyptian Sudan', *Ann. Trop. Med. Parasit.* 42 (3/4): 278-87.

Coursey, D.G. 1967. Yams. Longman, London. 230 pp.

Coursey, D.G. 1968. The Edible Aroids. *World Crops,* 20 (4): 25-30.

CSIR 1950. The Wealth of India (Raw Materials) (CSIR, New Delhi), Vol. 2, p. 58, 309, 400, 405-406.

CSIR, 1956. Wealth of India (Raw Materials) (CSIR, New Delhi), Vol. 4, 141 pp.

CSIR, 1976. The Wealth of India- Raw materials. CSIR, New Delhi. 9:257-258.

Dalziel, J.M. 1937. Useful Plants of West Africa. London: Crown Agents.

Das Gupta, B. and Pandey, V.B., 1970. Diosgenin in *Costus speciosus. Experientia.* 26:19-21.

Davidson, Alan. 1999. Oxford Companion to Food. "Oca", p. 547 (ISBN 0-19-211579-0).

Deepa, G., Indira, P.V., Sasidharan Pillai, C.R., Kurain Jose and Fathima Beevi, 1999. Tubers - A short cut to the health of mothers and children - A review based on Ayurveda *(In:)* Tropical Tuber crops in Food Security and Nutrition. Oxford & IBH Publishing Co. Pvt. Ltd. New Delhi. pp. 15-19.

Degener, O. 1930. Ferns and flowering plants of Hawaiian National Park. Honolulu, 312 pp.

Escobar, R.1943. Jicama. pp. 442-444 *in* Enciclopedia Agrícola y de Conocimientos Afines..., Vol. 2: Escuela Particular de Agricultura. Cd. Juárez, Chihuahua, Mexico.

Gopala Chetty. 1939. *Coleus vettiveroides. Madras agric. J.*, 1939, 27, 369.

Greshof, M. 1900. Beschrijving der giftige en bedwelmende planten biji de vischvangst in gebruik. Meded. 'S Lands Plantentuin 29(2):57, 174. Landsdrukkeriji, Batavia.

Guerrero, L.M. 1921. Medicinal uses of Philippine plants. *Phil. Bur. Forestry Bull.* 23(3), 149-246.

Haytowitz, D.B and Mathews, R.H. 1984. Composition of foods: Vegetables and vegetable products, Human nutritious information series, USDA Agric. Handbook No.8-11, Washington D.C.

Honigberger, J.M. 1852. Thirty-five years in the East. London, Vol. 2, 448 pp.

http//www.indiangyan.com/index.htm.

Horton, D. 1988. Undergound crops. Long-term trends in production of roots and tubers. Morrilton, A.R. USA: Winrock International Institute for Agricultural Development.

Horton, D and Fano, H. 1985. Potato Atlas. Lima, Peru: International Potato Center (CIP).

Huart, A. 1902. La jicama, su classificación, su cultivo, sus usos. *Bol. Soc. Agríc. Mex.* 26: 555-558.

IBPGR 1981. Root and Tuber Crops. IBPGR, Rome. pp. 23, 49, 81, 87 and 107.

Jeon, J.R., Lee, J.S., Lee, C.H., Kim, J.Y., Kim, S.D. and Nam, D.H. 2006. Effect of ethanol extract of dried Chinese yam *(Dioscorea batatas)* flour containing dioscin on gastrointestinal function in rat model. *Arch Pharm Res.* 29 (5) : 348-353.

Kay, DE. (revised by Gooding, E.G.B).1987. Crop and Product Digest No.2 Root Crops, Second edition. London : Tropical Development and Research Institute, XV and 380 pp.

Kirtikar, K.R. and Basu, B.D. 1918. Indian medicinal plants. Parts I and II, Allahabad, 1419 pp.

Kundu, B.C. 1967. Some edible rhizomatous and tuberous crops of India. *Proc. Int. Symp. Trop. Root Crop.*, pp. 124 30.

Kusano, S. and Abe, H. 2000. Antidiabetic activity of white skinned sweet potato *(Ipomoea batatus* L.) in obese zucker fatty rats. *Biol. Pham. Bull.* 23(1) : 23-26 PMID: 13880.

Lakshmi, S., Remani, P., Dhanya, G.S and Padmaja, G. 2006. Cytotoxic properties of *Curcuma zedoaria.* Paper presented in the 14th Triennial Symposium of the International Society for Tropical Root Crops from Nov 20-26, 2006, Thiruvananthapuram, Kerala, India.

Lal, S., Tripathi, D.P. and Asthana, A.N. 1988. Minor pulse crops of potential value. Proceed. Natl. Seminar on Underutilized and Underexploited Plants. NBPGR, New Delhi. 7-9 June, 1988.

Lila Babu. 2006. Carotene content in sweet potato varieties and its retention after processing. Paper presented in the 14th Triennial Symposium of the International

Society for Tropical Root Crops from Nov 20-26, 2006, Thiruvananthapuram, Kerala, India.

Low, Jan and Jaarsveld, Paul van. 2006. The potential of golden bread buns biofortified with beta-carotene-rich sweet potato to add value to rural diets and increase profits of rural bakers in Central Mozambique. Paper presented in the 14th Triennial Symposium of the International Society for Tropical Root Crops from Nov 20-26, 2006, Thiruvananthapuram, Kerala, India.

Maesen, L.J.G. van der. 1994. *Pueraria,* the kudzu and its relatives an update of the taxonomy. *(In:) Proceedings of the first International Symposium on tuberous legumes,* Guadeloupe, FWI., 21-24, April, 1992. pp. 55-86.

Magalhaes, A.F., Magel hae, G.E.G., Valio, I.F.M and Sales, B.H.L.N. 1988. Ocorrencia de neotenoma em *Pachyrhizus tuberosus.* 40a. Reuniao Annal da Sociedade brasileira paro o Progrenso da Ciencia. pp 552-553.

Martinez, M. 1936. Plantas utiles de Mexico. 2nd ed. Ediciones Botas, Mexico. pp. 244-247.

Masefield, G.B. 1961. Root nodulation and agricultural potential of the leguminous genus *Psophocarpus. Trop. Agriculture, Trin.,* 38: 225-229.

Matthysse, J.G. and Schwardt, H.H. 1943. Substitutes for Rotenone in Cattle Louse control. *J. Econ. Entomol* 34(5):718-720.

Maxwell, G. 1960. Mantra Gajah. *J. Roy. As. Soc. Str. Br.* 45, 35.

Meijer, T.M. 1946. The insecticidal constituents of *Pachyrrhizus erosus* Urban. I.-Recueil des travaux chimiques des Pays-Bas. 65:835-842.

Nadkarni, K.M. 1927. Indian Materia Medica. Bombay. 1142 pp.

NAS. 1979. Rice bean. *(In:)* Tropical Legumes. Resources for the future. National Academy of Sciences, Washington DC. USA, 331 pp.

Neuwinger, H.D. 1994. African ethnobotany. Poisons and drugs.Chemistry, pharmacology, toxicology. English translation by A Porter. Weinheim, Chapman & Hall.

Ørting, B., Gruneberg, W. J and Sorensen, M. 1996. Ahipa (*Pachyrhizus ahipa* (Wedd.) Parodi) in Bolivia. Genetic Resources and Crop Evolution.

Pancho, J.V. 1984. Folk uses of cultivated aroids in the Philippines. *(In:)* Edible Aroids (Ed. S. Chandra), Clarendon Press, Oxford, pp. 3-9.

Peckolt, G. 1922. Jacutupe. Chácaras and Quintais 25 (3): 187-189 and 25(4):275-276.

Perry, L.M. and Metzger, J. 1980. Medicinal plants of East and Southeast Asia: Attributed properties and uses. The MIT Press, assachusetts. pp. 1-620 (p.221).

Petelot, A. 1952. Les plantes medicinales du Cambodge, du laos et du Viet-nam. Arch. Recherches Agron. Cambodge, Laos, Vietna, 14: 1-408 p. 267.

Purseglove, J.W. 1972. Tropical Crops : Monocotyledon. London : Longman.

Purseglove, J.W. 1974. Tropical Crops : Dicotyledons. Vols 1 and 2 combined London : Longman.

Raghu, A., Deepa, V.C. and Sundaran, K. 1999. A study on Soorana *(Amorphophallus paeonifolius)* The King of Tubers. *(In:)* Tropical Tuber crops in Food Security and Nutrition. Oxford & IBH Publishing co. Pvt. Ltd. New Delhi. pp.10-14.

Rejesus, B.M., Punzalan, E.G., Vasquez, E.A., Kraus and Philippine.,W. 1999. Biological activities of the compound from *Coleus amboinicus. J. Crop. Sci.* (Philippines).24(1):81.

Roig y Mesa, J.T. 1988. Plantas medicinales, aromaticas o venenosas de Cuba. (1st ed. 1945) Editorial cientifico-Tecnica, La Habana.

Sillevoldt, H. E. Th. Van. 1899. Ueber das Derrid und Pachyrhizid, ein Beitrag zur Kenntnis der Indischen Fischgifte. Arch. Pharm. (Berlin), 237(8): 595-616.

Smartt. J. 1990. Grain Legumes: Evolution and Genetic Resources. Cambridge, U.K. 379p.

Suzuki, K., Ito, Y., Nakamura, S. *et al.*, 2002. Relationship between serum carotenoids and hyper glycemia; a population–based cross-sectional study. *J. Epidemiol.* 12(5):357-366.

Tadera, K., Tanguchi, T., Teremoto, M., Arima, M., Yagi, F., Kobayashi, A., Nagahama, T. and Ishihata, K. 1984. Protein and Starch in Tubers of Winged Bean, *Psophocarpus tetragonolobus* (L.) DC., and Yam Bean, *Pachyrrhizus erosus* (L.). *Urban. Mem. Fac. Agric.*, Kagoshima Univ. 20:73-81.

Valentova, K., Cvak, L., Muck, A., lrichova, J., Simanek, V. 2003. "Antioxidant activity of extracts from the leaves of Smallanthus sonchifolius". *Eur J Nutr.* 42 (1): 61-62.

Villar, M.L.D. and Valio, I.F.M. 1994. Rotenone and Pachyrhizin in *P. tuberosus.* 'Effect of endogenous rotenone and Pachyrhizin an early growth of *Pachyrhizus tuberosus* (Lam.) Spreng. *In.* Proceedings of the first International Symposium of tuberous legume, Guadeloupe FWI, 21-24 April 1992.

Villareal, R.L., Tsou, S.C.S., Lo, H.F and Chiu, S.C. 1985. Sweet potato vine tips as vegetables. *(In:)* Bouwkamp, J.C. (ed.), Sweet potato products: a natural resource for the tropics. CRC press, Inc., Boca Raton, F.L., pp. 175-183

Vimala, B. 1994. Genetic Resources and Varietal Improvement in Minor Tuber Crops. *(In:)* Advance in Horticulture vol. 8-Tuber Crops. (Eds.) Chadha, K.L. and Nayar, G.G. Malhotra Publishing House, New Delhi-110064. India. pp. 139-153.

Watt, J.M. and Breyer-Brandwijk, M.G. 1962 The medicinal and poisonous plants of southern and eastern Africa. Edinburgh, E. & S. Livingstone.

Wheatley, C., Scott, G.J., Best, R. and. Wiersema. S. 1995. Adding value to root and tuber crops. A manual on product development. Cali, Columbia: Centro Internacional de Agricultura Tropical (CIAT).

William O Jones. 1959. Manioc in Africa. Stanford University Press, California. pp. 10-11.

Woolfe, J. A. 1992. Sweet potato - an untapped food resource. Cambridge University Press, Cambridge. 643 pp.

Yang, R.Z and Tang, C.S. 1988. Plants used for Pest Control in China: A literature Review. Econ. Bot. 42(3):376-206.

Yin-fang, D and Cheng-jun, L. 1987. Fruit as medicine. Yam Bean. The Rams Skull Press, Kuranda, Australia. 74 pp.

Ylonen, K., Alfthan, G., Groop, L. *et al.*, 2003. Dietary intakes and plasma concentrations of carotenoids and tocopherols in relation to glucose metabolism in subjects at high risk of type 2 diabetics : the Botnia Dietary Study. *Am. J. Clin Nutr.* 77(6): 1434-1441.

Yoshinaga,M., Kai,Y., Katayama, K., and Sakai,T. 2006. Improvement in anthocyanin content of purple fleshed sweet potatoes in Japan. Paper presented in the 14th Triennial Symposium of the International Society for Tropical Root Crops from Nov 20-26, 2006, Thiruvananthapuram, Kerala, India

http://en.wikipedia.org

www.fao.org/Wairdocs/ILRI/x5461E/x5461e18.htm.

❑❑❑

5

Anti-Nutritional Factors

A broad class of polypeptides and proteins occur in plants and other life forms, which inhibit the action of proteolytic enzymes and are known therefore as proteinase inhibitors and this includes those which specifically inhibit the important digestive enzyme trypsin. Others inhibit a related enzyme chymotrypsin. The subject of proteinase inhibitors was reviewed by Ryan (1981). The occurrence of these inhibitors has nutritional implications for human and animal feeding as they are found in many important food plants eaten both raw and cooked. The first known plant proteinase inhibitor was the trypsin inhibitor crystallized from soybeans (Kunitz, 1945). Other proteinase inhibitors have since been described in a variety of plants, especially the Graminae, Leguminosae and Solanaceae families. The exact physiological function of proteinase inhibitors in plants is thought to be in plant protection by inhibiting digestive enzymes of invading insect pests or pathogens (Ryan, 1981).

Free oxalic acid, soluble potassium, sodium and ammonium oxalates and insoluble calcium oxalate are widely distributed in plants. High oxalate levels can cause acute poisoning, resulting in hypocalcaemia, or chronic poisoning in which calcium oxalate is deposited as crystals in the kidneys causing renal damage. The small quantities of soluble oxalates in the sweet potato samples were similar to those in other root crops examined, but the content of calcium oxalate was ten times less than that of either giant swamp taro or elephant foot yam.

1. Sweet Potato

The only organic acid to be investigated in sweet potato tips, because of its implications for their nutritional value, is oxalic acid. The oxalate content of tips of 10 cultivars grown in the same location in Taiwan differed from one another, ranging from (fwb) 280 mg 100 g^{-1} to 450 mg 100 g^{-1} (mean 370 mg 100 g^{-1}) (Villareal *et al.,* 1979). Three cultivars, representing broad, medium and fine leaf types contained 490, 480 and 340 mg oxalate/100 g, respectively.

On a fresh-cooked basis, sweet potato leaves and other tropical leaf vegetables (Evenson and Standal, 1984) contain much higher levels of oxalic acid than do common temperate climate vegetables, with the exception of spinach. The fact that some developing-country diets are deficient in calcium and other minerals, bound and rendered unavailable to the body by oxalic acid, makes selection of low oxalate types an important goal for researchers seeking to popularize sweet potato leaves.

The first non-leguminous plant reported to contain a trypsin inhibitor was sweet potato (Sohonie and Bhandarkar, 1954). The isolated impure inhibitor fraction is very thermolabile. Since this initial report, several researchers investigated the nature of sweet potato trypsin inhibitor. The strong inhibition of trypsin which was demonstrated to occur *in vitro* with sweet potato trypsin inhibitor could indicate an interference with protein digestion *in vivo* thus having nutritional implications in humans, especially those whose protein intake is already marginal. Although no harmful effects were reported in communities which eat raw roots, for example as a between-meal snack, the presence of trypsin inhibitors could decrease protein digestion and utilization.

Heating is the usual method identified for destroying sweet potato trypsin inhibitors, but it is desirable to expend as little energy as possible in sweet potato for animal feed. For both animal and human feeding, excessive use of heat reduces availability of amino acids, notably lysine. Generally speaking, normal methods of cooking for human consumption effectively destroy all or most of the trypsin inhibitor activity (TIA).

It is quite generally recognized that consumption of sweet potatoes causes flatulence (Palmer, 1982). The occurrence of flatulence is suggested as a possible factor contributing to low acceptability of sweet potato (Tsou and Villareal, 1982). The residual flatulence after cooking could be associated with a fraction of the starch remaining resistant to digestive enzymes and hence entering the colon; in addition, substances in the dietary fibre could also be responsible for flatus gases.

From time to time, outbreaks of poisoning of cattle were reported due to incorporation of mouldy sweet potatoes in their feed. The metabolites ipomeamarone and ipomeamaranol were isolated from mouldy tubers and was toxic to liver and other organs, and the latter to cause lung oedema. A disturbing feature is that these metabolites occur in tubers which show only slight blemishes, insufficient to arouse suspicion that they are unsound (Kay, 1987).

2. Cassava

Cassava, whose economic importance is constantly increasing, is an essential foodstuff of tropical populations. Its high carbohydrate content and the relative ease with which it can be grown are its two major advantages. However, because it contains the cyanogenic glucoside, linamarin, cassava must be detoxified before it is eaten. Otherwise, the cyanogenic glucoside, chemically similar to the amino acids, is capable of liberating large quantities of cyanide by hydrolysis and may lead to fatal poisoning and other side effects. Populations consuming cassava employ different detoxification procedures. Cassava plant including the storage root contains linamarin. This cyanogenic glucoside is readily hydrolysed by an enzyme linamarase (a b-glucosidase) to release hydrogen cyanide. While the hydrogen cyanide is readily removed during processing of cassava, the presence of residual linamarin and its breakdown product (acetone cyanohydrin) in cassava-based food products has been a cause for concern because of their possible effects on health.

It has been shown in humans that a substantial part of the ingested linamarin will be absorbed and excreted intact in the urine. Its toxic role remains speculative but one is certain that the cyanide liberated from linamarin is the primary cause of toxicity in cassava. When linamarin comes into contact with its hydrolytic enzyme, linamarase, the molecule is split into glucose and its aglycone, acetone cyanohydrin. The latter can be further degraded by another enzyme or spontaneously under alkaline conditions to form hydrogen cyanide and glucose. Thus if the residual linamarin and its breakdown products are not removed during food processing, they may be retained in the foodstuff. It is believed that in humans, linamarin can be broken down by linamarase found in the bacteria that reside in the intestinal track resulting in release of hydrogen cyanide. Fortunately, humans can readily neutralize about 10 mg of cyanide by a reversible reaction with methemoglobin fraction in the red blood cells. Rodanese can further convert majority of the cyanide to less toxic thiocyanate, which is then excreted in the urine.

Perhaps, a step forward in promoting safe cassava food is to give consumers relevant information on the food they purchase. Reporting the level of residual cyanogens in the food products should be a normal activity of the cassava food industries. As such, developing simple, rapid and inexpensive methods of analysis will be relevant to the food industries and research organizations dealing with cassava.

Levels of cyanoglucoside in cassava leaf and tuber varies with age and climate. Dry period causes more cyanoglucosides in tubers (Balanambisan and Sundaresan, 1990).

The epidemiological, clinical and biological studies carried out in Kivu and Ubangi in Africa indicated that daily ingestion of cassava roots lead to modification in thiocyanate levels and abnormalities in thyroid iodine metabolism. These modifications were identical, qualitatively and quantitatively, to those induced by 1-2 mg thiocyanate ingested daily. This antithyroid effect of cassava is directly linked to endogenous thiocyanate production following cassava ingestion (Ermans, 1980).

The abnormalities in thyroid metabolism induced by cassava or thiocyanate ingestion were characterized by decreased thyroid iodine reserves and the resulting alterations in intrathyroidal hormone production (increased MIT/DIT ratio and decreased T_4 synthesis), in peripheral distribution of thyroid hormones (reduction of $PB^{127}I$), and in thyroid size (goitrogenic action). The abnormalities are identical to those caused by an increase in iodine deficiency. The studies of Bourdoux *et al.* (1980) indicated that the intake of cassava for 3 days produced a significant increase in serum and urinary thiocyanate concentrations which decreased clearly on complete cessation of cassava intake. Thus cassava was found responsible for abnormal thiocyanate load in this population. Studies indicated that iodine deficiency together with ingestion of goiterogenic substances like cassava produces not only goitre and abnormal thyroid metabolism but also severe abnormalities in pregnant women and their offspring especially during fetal and early life.

It is, therefore, important to determine iodine supply and goitre prevalence in all regions of high cassava intake. In addition, the necessity of reducing toxicity of cassava preparations much be emphasized. The populations at risk should be given demonstration on how to prepare cassava in a way that ensures complete liberation of hydrocyanic acid, and agricultural researchers should develop varieties of cassava that have minimal concentrations of linamarin. Also, further research is needed to establish more precisely the thiocyanate/iodine ratio at which prophylactic measures are essential (Ermans, 1980).

Chronic cassava ingestion has been implicated in the pathogenesis of tropical pancreatitis. Cassava fed animals showed pancreatic acinar atrophy, fatty infiltration, ductal/ductular dilation with mucinous metaplasia of the ductal epithelium and eosinophilic material in the lumen. Thickening of blood vessels were marked in low-protein group. No abnormality was noted in the other groups (Shenoy *et al,* 1996). Sandhyamani (1996) established the role of high carbohydrate, Cassava starch-based diets or corn-starch-substituted diets, with low-protein intake in the development of degenerative diseases.

3. Yams

Many yam cultivars contain certain substances which affect the culinary properties of tubers. In addition, several *Dioscorea* spp. contain alkaloids, tannin and sapogenins for which various uses are found. The toxic alkaloid dioscorine is present in the edible species *D. hispida,* diosgenin in *D. mexicana* and others (Kay, 1987) and dihydrodioscorine in the African *D. dumetorum.* Dioscorine and dihydrodioscorine are nerve poisons, and the species that contain them are therefore used as sources of poisons for hunting, fishing, and pharmaceutical purposes. They are soluble in water, so that tubers containing high amounts of alkaloids can be detoxified by prolonged soaking in water (Onwueme, 1978). The alkaloid dioscorine, $C_{13}H_{19}O_2N$, occurs in tubers of *D. hispida* (referred to the earlier literature as *D. hirsute* Blume). Like dioscorine, dihydrodioscorine is toxic, acting as a convulsant poison although the toxicity appears to be rather lower than in the other substance.

Gilbert and Gillman, 1963 indicated that consumption of large quantities of *D. cayenensis* might be harmful.

Presence of tannins in large quantities was recorded in *D. cirrhosa* (Burkill, 1935) and in what is described as *D. atropurpurea* Roxb., *i.e.,* a dark red-fleshed form of *D. alata* and in *D. rhipogonoides* Oliv. (Endres *et al.,*1962), the quantities present in the tubers being 6, 38 and 13% (dry weight) respectively. It is possible, however, that the acrid taste of many wild varieties of yam, which are occasionally used for food, is due to presence of tannins.

Another class of compound found in the tubers of a number of yam species comprises the saponins. Saponins- which, as a class, are quite widely distributed in the plant kingdom – are so called because of their characteristic property of forming a soapy lather in water. They are generally inactive when eaten, either because of hydrolysis in the digestive system or possibly through non-absorption from the intestine. Nevertheless, saponins have a bitter taste and yams containing them are usually regarded as inedible species (Coursey, 1967).

4. *Colocasia*

There are some undesirable properties with taro (Plowman, 1969).

Most cultivars, particularly in *dasheens,* contain oxalic acid (0.1-0.4% fresh weight) mainly in the form of 'raphides'. These are minute bundles of crystals of calcium oxalate (0.1-0.4% fresh weight), which account for irritating effect of the corm. This effect is removed during boiling (Kay, 1987). Small quantities of prussic acid are also present in the taro corm. The amount is much less than in cassava, and is not nutritionally hazardous (Onwueme, 1978). The calcium oxalate content in tubers of acrid taro cultivars was in the range of 0.58-0.84%, while non-acrid cultivars has 0.07-0.15%.

The levels of oxalate accumulation in taro were found to be much higher in leaf tissues compared to that of cormels. The leaf showed around 6% total oxalate, while the cormels had only less than 1% oxalates. No significant differences in the oxalate content were observed between different cultivars of taro. However, the oxalate content of cormels showed large variations between cultivars.

5. *Xanthosoma*

Like taro, the corms of tannia contain raphides, and should not be fed to humans or livestock without cooking. (Onwueme, 1978).

6. *Alocasia*

Glycoside is not present in corms or stems of several cultivars of *A. macrorrhiza* but the young leaves contain up to 0.018% (Kay, 1987).

7. *Amorphophallus*

A. paeoniifolius and *A. konjac* contain calcium oxalate crystals; the wild forms of *A. paeoniifolius* contain more, than the cultivated and are strongly acrid (Kay, 1987). Major differences between acrid and non-acrid cultivars were observed with respect to the distribution of soluble oxalate and calcium oxalate. Nearly 90% of the total oxalate was present in the form of calcium oxalate in acrid cultivars, while calcium oxalate constituted only 50-60% of the total oxalate pool in non-acrid cultivars.

A higher level of protease activity has been found to be associated with the acridity of taro and *Amorphophallus* tubers. All the acrid cultivars examined had significantly higher protease activity compared to that of non-acrid cultivars, indicating the involvement of protease in acridity.

8. Yam bean (*Pachyrhizus* spp.)

An insecticidal compound called rotenone ($C_{23}H_{22}O_6$) is found in mature seeds but not in toxic amounts in the tuber or in any other part of the plant. The insecticidal or fungicidal properties present in seeds, pods, stems and leaves are extensively studied (Magalhaes *et.al.*, 1988; Jimenez,1994). The identification and quantification of rotenone and pachyrhizin in *P. tuberosus* were recently carried out in Brazil. The seeds are toxic and are studied as a possible commercial source of a vegetable insecticide, since they contain 0.12-0.43% of rotenone, pachyrrhizone and pachyrrhizonic acid. The toxic principles of seeds can be eliminated by boiling them (Kay, 1987).

Mature seeds contain high amount of food-quality vegetable oil (approximately 30%) in *P. erosus* (Jimenez.1994), 11-15% yellow-green oil (Bouillenne *et.al.*, 1935), but too much rotenone to be edible (approximately 0.5% pure rotenone and 0.5% rotenoids and saponins). These studies agree that if the insecticidal/anti-nutritional compounds are removed, the oil has a composition comparable to that of groundnut and cottonseed oil: palmitic acid 25.7%, stearic acid 5.7%, oleic acid 33.4% and linoleic acid 34.2% (Broadbent and Shone, 1963). This active principle was identified by Brazilian Scientists working on the Amazonian yam bean and was confirmed more recently by studies on the rotenone content of different plants and species carried out at the University of Costa Rica (Jimenez. 1994).

Duke (1981) reports presence of adenine and choline in tuberous roots and content of adenine, choline and the anti-nutritional compounds rotenone and saponin in mature seeds. Although the seeds are rich in both proteins and lipids/oils, presence of high levels of isoflavonoids (rotenoids and pachyrrizide) make them unsuitable for human consumption (Alvarenga and Valio 1989; Scramin, 1994).

The young seed pods of *P. erosus* are sometimes eaten as a cooked vegetable, similar to French beans, but cannot so be used as seeds develop. There are many reports on the insecticidal and piscicidal effect of rotenone and rotenoids isolated from seeds. In Indonesia the seeds are being used as a piscicide (Meiger, 1946).

Rotenone/rotenoid content also depends on breeding objectives, *i.e.*, whether the seeds are to be used for commercial extraction or for increased pest resistance (which require increased levels), or if the mature seeds are to be used as food/feed (in which case rotenone/rotenoid levels should be decreased to reduce their anti-nutritional effect. In Vietnam, flowers, young leaves and pods are used as a vegetable (Gilbert, 1917).

Hernandez (1790) and subsequently Urbina (1906) report that if the tubers are consumed in excessive amounts, they cause flatulence. Martinez (1936, 1956) and Anonymous (1976) stated that consumption of the fresh tubers caused indigestion in young children. Fine (1991) reported a case of hypersensitive reaction to ingestion of *P. erosus*.

9. Winged bean

The root tubers have high levels of trypsin inhibitory activity (13 500 – 30 100 IU/mg fresh weight); this is inactivated by cooking (Kay, 1987).

10. *Helianthus tuberosus* L.

The carbohydrate consists almost entirely of inulin and inulides with small amounts of starch, fructose and glucose: occasionally the starch may reach 30% of the carbohydrate, but is usually considerably less. As inulin is not digestible by human beings, utilisable energy intake is much lower and allows tubers to be used as a low energy food, which still satisfied volume intake and hunger. Also, they can sometimes cause digestive disturbance (Kay, 1987).

11. Hyacinth bean *(Lablab purpureus* (L.)

Seeds contain trypsin and chymotrypsin inhibitors, may contain cyanide and stigmasterol and is said to be rich source of catechol oxidase (Bhag Mal,1994).

12. *Curcuma zedoaria*

A sample of essential oil extracted from rhizomes was found to have following characteristics: SG (30°C) 0.9724 and N_D (30°C) 1.5002, and to contain d-α-pinene 1.5%; d-camphene 3.5% ; cineol 9.6%; α–camphor 4.2%; d-borneol 1.5%; sesquiterpenes 10% and sesquiterpene alcohols 48%. Phytotoxic compounds were also isolated, and three antibiotic agents were identified, the most abundant being ethyl *p*-methoxycinnamate. An unidentified toxin has been found in impure starch. In the usual process for preparing starch, very thorough washing is employed, until the product is white: in this treatment most of the protein (residual content about 10 g kg^{-1}) and most of the toxin is removed. Less thoroughly prepared starch, containing 155 g protein kg^{-1}, and meal (containing 320 g protein kg^{-1}) was toxic to rats and chicks. It is not known whether toxin is chemically associated with the protein or whether it is separately removed along with protein during starch purification process (Kay, 1987).

References

Alvarenga, A.A. and Valio, I.F.M. 1989. Influence of temperature and photoperiod on flowering and tuberous root formation of *Pachyrrhizus tuberosus. Ann. Bot.* 63(44):411-414.

Anonymous. 1976. La jicama y su cultivo. *Gaceta Agricola* (Mex.) 20, 528:6,15.

Bala Nambisan and Sudaresan, S. 1990. Cyanogluclosides in cassava tubers and leaves at different growth stages. *J. Root Crops* Spl. Issue. 1990. pp 261-263.

Bhag Mal. 1994. Underutilized grain legumes and pseudocereals - their potentials in Asia. Regional Office for Asia and the Pacific. FAO, Bangkok, Thailand. 162 pp.

Bouillenne, R., Bouillenne, M. and Koolhaas, R. 1935. Note sur *Pachyrrhizus angulatus* Rich. *Congres Internationale de Technique et Chimique des Industries Agricoles* 4(2):333-337.

Bourdoux, P., Delange, F., Gerard, M., Mafuta, M., Hanson, A. and Ermans, A.M. 1980. Antithyroid Action of Cassava in Humans. *(In:)* Role of cassava in the etiology and endemic goitre and cretinism (Eds.) Ermans, Mbulamoko, N.M, Delange, F., Ahluwalia, R. Ottawa, Ont., IDRC. 182 pp.

Broadbent, J. and Shone, G. 1963. The composition of *Pachyrrhizus erosus* (Yam bean) seed oil. *J. Sci. Food Agric.* 14 (7): 524-527.

Burkill, I.H. 1935. A dictionary of the economic products of the Malay Peninsula. Crown Agents, London.

Coursey, D.G. 1967. Yams. Longmans, London. 230 pp.

Duke, J.A. 1981. Handbook of Legumes of World Economic Importance. Plenum Press, London.

Endres, H., Howes, F.N. and Von Regel, C. 1962. Gerbstoffe-Tanning materials J. Von Wiesner, *Die Rohstoffe des Pflanzenreichs,* Vol. I, 5th Edn, cramer, Weinheim.

Ermans, A.M. 1980. General Conclusions. *(In:)* Role of cassava in the etiology of endemic goitre and cretinism (Eds.) Ermans, Mbulamoko, N.M, Delange, F., Ahluwalia, R. Ottawa, Ont., IDRC, 1980. 182 pp.

Evenson, S.K. and Standal, B.R. 1984. Use of tropical vegetables to improve diets on the Pacific region. HITAHR Res. Ser. 028, University of Hawaii.

Fine, A.J. 1991. Hypersensitivity reaction to jicama (*Pachyrhizus,* yam bean). *Ann. Allergy* 66:173-174.

Gilbert, H. 1917. Note sur le Dolic bulbeux. *Bull. Econ. Indochine* (Hanoi-Haiphong) 20(123):140-141.

Gilbert, C. and Gillaman, J. 1963.'Yams and liver necrosis'. *Nature. Lond.,* 198 (4876) :196.

Hernandez, F. 1790. Opera, cum edita, tum inedita, ad autographi fidem et integritatum expressa, impensa et jussu regio: historia Novae Hispania, vols. 1-3. Matriti. Ex typographia Ibarrae Heredum.

Jimenez, B, G.G. 1994. Extraccion de rotenone a partir de las semillas de *Pachyrhizus erosus* (jicama). M.Sc. Thesis. Escuela de Ingenieria Quimica, Univ. Costa Rica. 106 pp.

Kay, D.E. (revised by Gooding, E.G.B) 1987. Crop and Product Digest No.2 Root Crops, Second edition. London : Tropical Development and Research Institute, XV and 380 pp.

Kunitz, M. 1945. Crystallization of a trypsin inhibitors from the soyabean. *Science.* 101 (2635): 668-669.

Magalhaes, A.F., Magalhaes, E.G., Valio, I.F.M. and Sales, B.H.L.N. 1988. Ocorrencia de neotenona em *Pachyrhizus tuberosus. (In:)* 40a Reuniao Anual da Sociedade Brasileira para o Progresso da Ciencia. pp. 552-553.

Martinez, M. 1936. Plantas utiles de Mexico. 2nd ed. Ediciones Botas, Mexico. pp. 244-247.

Martinez, M. 1956. Nombres vulgares y cientificos de Plantas del Estado de Mexico. Gobierno del Estado de Mexico, Direccion de Agricult. Y Ganad., Toluca, Mexico.

Meijer, T.M. 1946. The insecticidal constituents of *Pachyrrhizus erosus* Urban. I.-Recueil des travaux chimiques des Pays-Bas. 65:835-842.

Onwueme, I.C. 1978. The tropical crops . John Wiley and Sons. England 234 pp.

Palmer, J.K. 1982. Carbohydrate in sweet potato. *(In:)* Villareal, R.L. and Griggs, T.D. (eds.), Sweet potato. *Proceedings of the First International Symposium.* AVRDC, Shanhua, Taiwan. pp. 137-138.

Plowman, T. 1969. Folk uses of New World aroids. *Economic Botany* 23(2), 97-122.

Ryan, C. A. 1981. Proteinase inhibitors. *(In:)* Stumpf, P.K. and Cann, E.E. (eds.), The bio chemistry of plants. Vol. 6. Proteins and nucleic acids. Academic Press, Inc., New York. pp. 351-370.

Sandhyamani, S. 1996. Low-protein corn-starch based diets exclude role of toxic factors from tapioca in induction of mucoid vasculopathy and associated disorders. *(In:)* Tropical Tuber Crops - Problems, Prospects and Future Strategies (Eds.) Kurup, G.T., Palaniswami, M.S., Potty, V.P., Padmaja, G., Kabeerathumma, S., Santha V. Pillai. Oxford and IBH Publishing Co. Pvt. Ltd. New Delhi, Calcutta. 597 pp.

Scramin, S. 1994. Estudio fitoquimico do genero *Pachyrhizus* Rich. Ex DC. PhD thesis, Universidad Estadual de Campinas, Campinas, Brazil.

Shenoy, K.T., Leena, K.B. and Nair, R.B. 1996. Pancreatic changes (Enzymes and Histological) in an experimental model fed with cassava-based diet. *(In:)* Tropical Tuber Crops - Problems, Prospects and Future Strategies (Eds.) Kurup, G.T., Palaniswami, M.S., Potty, V.P., Padmaja, G., Kabeerathumma, S., Santha V. Pillai. Oxford and IBH Publishing Co. Pvt. Ltd. New Delhi, Calcutta. 597 pp.

Sohonie, K. and Bhandarkar, A.P. 1954. Trypsin inhibitors in Indian Food stuffs. Part 1. Inhibitors in vegetables. *J. Sci. Ind. Res.* 13B: 500-503.

Tsou, S.C.S and Villareal, R.L. 1982. Resistance to eating sweet potato. *(In:)* Villareal, R.L. and Griggs, T. D. (eds.) Sweet potato. *Proceedings of the First International Symposium.* AVRDC, Shanhua, Taiwan, pp. 39,42,43.

Urbina, M. 1906. Raices comestibles entre los antiguos Mexicanos. (Xicama. Pp. 123-125). *Anales Mus. Nac. Mexico* 3 (ser.2):117-190.

Villareal, R.L., Lin, S.K. and Lai. S.H. 1979. Variations in the yielding ability of sweet potato under drought stress and minimum input conditions. *Hort. Sci.* 14:31-32.

❑❑❑

6

Improved Varieties in India

In enhancing the food production there appears to be two solutions, *viz.*, horizontal or vertical increase in production. In the vertical expansion of production varieties are important. High yielding, high starch and value added varieties suitable to each region/ cropping system play vital role in enhancing vertical production. Therefore varieties especially value added consumer preferred ecologically adaptable ones hold the key and is very much true in tuber and root crops which have a significant role in supplementing the food and nutritional requirement in rural and semi urban population of India.

Improved varieties/selections in respect of important tuber and root crops identified in different parts of the country are enumerated with summarized salient characters, yield and suitable region for cultivation.

1. Sweet potato

Sl. No	Variety	Breeding method / Pedigree	Originating source	Recommended region	Year of release	Duration (days)	Average yield (t ha^{-1})	Salient features
1	H-41	Hybrid, Norin x indigenous cultivar	CTCRI, Trivandrum	Tamil Nadu, Karnataka Kerala	1971	120-130	20	Semi-spreading, leaf unifoliate with entire margin, profusely flowering, tuber skin reddish purple with white rind and flesh. Excellent cooking quality, cooks easily, tastes sweet and low in fibre content

2	H-42	Hybrid, Vella damp x Triumph	CTCRI, Trivandrum	Tamil Nadu, Karnataka, Kerala	1971	120-130	22	Semi-spreading plant type, leaf unifoliate with toothed margin, tubers fusiform, tuber skin purplish red and creamy flesh. Cooks easily, tastes sweet and almost free from fibre
3	VL Sakar-kand-6	Clonal selection, Selection of USA Clone 'B'-219	ARS, Almora, UP	Hilly areas of Uttar Pradesh	1974	135-140	20	Late maturing with spreading vine having light green foliage, tubers elongated, bent on edges and purple in colour with light yellowish flesh. Tuber rich in carotene.
4	CO.1	Clonal selection (IB.3)	TNAU, Coimbatore, Tamil Nadu	Tamil Nadu	1976	135	28	Moderately vigorous vine and is less spreading, leaves medium lobed and dark green, tuber skin light pink with white flesh. Starch 24%, soft texture on cooking with higher sugar content
5	CO.2	Clonal selection (IB.81)	TNAU, Coimbatore, Tamil Nadu	Tamil Nadu	1976	110-120	32	Compact plant type, shorter vine and less spreading, leaves 10 lobed and emerging leaves pink, tuber skin light pink with white flesh. Starch 29.5%, carotenoids 3.2 mg/100 g, good consumer acceptability.
6	Dergaon Red	Local selection	AAU, Jorhat	Assam	1976	150	20	Red skinned white fleshed variety. High yielder with better quality tubers.
7	Rajendra Sakar-kand-5	Hybrid, Hybrid-5 x Cross 4	TCA, Dholi, Bihar	Plains of Bihar	1980	105-120	20	Extremely spreading, leaves green and six lobed, number of tubers per plant 4-6, tubers cylindrical with long neck, white tuber skin and greenish white flesh. Can tolerate water logging for a week and tolerant to sweet potato weevil, Cercospora and suitable for two tier cropping as it deep rooted, suitable for fodder.
8	Deragaon White	Local selection	AAU, Jorhat	Assam	1980	130	20	White skin.
9	CO.3	Clonal selection (IB.2837)	TNAU, Coimbatore, Tamil Nadu	Tamil Nadu	1982	110-115	32	A medium duration crop with bushy habit, emerging leaves pink and developed ones dark green, tuber skin light red with dark orange flesh. Rich in

contd...

								carotene, starch 30%, better consumer preference and less root weevil incidence.
10	Varsha (H-268)	Double cross hybrid, (Acc.No.39 x Acc.No.3) x Acc. No.1871 x Acc. No.1103)	CTCRI, Trivandrum	Konkan region of Maharashtra	1983	120	17	Semi-spreading with dark green vines, tubers fusiform with reddish purple skin and light yellow flesh. Drought tolerant.
11	Sree Nandini (76-OP-217)	Seedlings selection, Selection from OPS progeny of S-32	CTCRI, Trivandrum	Kerala	1987	100-105	20	Spreading with light green leaves, number of tubers per plant 3-4, tubers fusiform with creamy skin and white flesh.Drought tolerant, suitable for paddy fallows as a catch crop.
12	Sree Vardhini (76-OP-219)	Seedlings selection from OPS progeny of S-13	CTCRI, Trivandrum	Kerala	1987	100-105	20	Semi-spreading having light brown emerging leaves, fusiform tuber with purple skin and yellow flesh. Rich in carotene, early maturing dual purpose variety, tubers and foliage good for human consumption and as animal feed.
13	Samrat (S-30)	Clonal selection from germplasm	ANGRAU, Hyderabad, AP	All states	1987	110-120	20	Vines medium spreading, petioles violet, tubers globular with light pink skin and white flesh. Medium duration.
14	Kalmegh	Local selection	AAU, Jorhat	Assam	1988	105	30	High yielder.
15	Sonipat-2	Selection (identified)	ICAR, Research Complex for NEH Region, Meghalaya	Meghalaya	-	120	36	Leaves are whole and tubers are pink skinned. Tuber length various from 12-15 cm and tuber diameter 7-8 cm.
16	Rajendra Sakar kand-43	Seedlings selection from germplasm collection H-3	TCA, Dholi, Bihar	Assam, Karnataka, Andhra Pradesh, Maharashtra, Bihar	1990	110-120	20	Medium spreading plant type, tubers 5-6 per plant, tuber skin brownish with white flesh. Moderately tolerant to weevil and Cercospora leaf spot, suitable for late harvesting which will fetch better price.
17	Rajendra Sakar kand-35	Seedlings selections from germplasm	TCA, Dholi, Bihar	Assam, Bihar, Karnataka, West Bengal	1990	105-110	25	Medium spreading having cylindrical tubers, skin dull white with white flesh. Moderately tolerant to weevil and Cercospora disease, suitable for February planting in plains of Bihar, less mortality of

contd...

								vines even when planted in spring season.
18	Kanjan gad (ACV 1)	Clonal selection	College of Agriculture, Vellayani, KAU	Kerala	1992	115	12	Spreading type, with duration of 115 days. Green vines, dark brown young leaves, medium sized spindle shaped, and purple coloured tubers with yellow flesh. Purple tubers with yellow flesh; low weevil incidence and tolerant to partial shade.
19	Rajendra Sakar kand–47	Seedlings selections from seedling progeny X-47	TCA, Dholi, Bihar	Bihar, West Bengal, Uttar Pradesh	1993	120-130	30	Spreading type with green leaves & petiole, emerging leaves purplish white, tubers cylindrical, deep bulking, purple skin with white flesh. Tolerant to sweet potato weevil and Cercospora leaf blight, suitable for late harvesting.
20	Kiran	Seedlings selection from OP seeds	ANGRAU, Hyderabad, AP	Andhra Pradesh	1994	110-130	22	Medium spreading, palmate leaves with 4 to shallow lobes, leaves light green, tubers necked fusiform and tuber skin red with orange flesh. Early bulking.
21	Sree Bhadra (S-1010)	Seedlings selection of Nigeria	CTCRI, Trivandrum	Kerala, Maharashtra, Bihar, MP Meghalaya	1996	90	20	Semi-spreading, greenish brown vine and petiole, dark brown emerging leaf, broad cordate leaves, tuber shape spherical, light pink tuber skin and creamy flesh. Early maturing, no major pests and disease except sweet potato weevil and excellent cooking quality.
22	Sree Rethna (X-108-2)	Hybrid, S-187 x Sree Vardhini	CTCRI, Trivandrum	Kerala	1996	90-105	20	Spreading, light greenish brown vine and petiole, brown emerging leaves, light green shouldered leaves, spherical round tubers, purple tuber skin and orange flesh, starch content 22-23%. Excellent cooking quality.
23	Gouri (H- 85-16)	Hybrid, H-29 x H-42	RC of CTCRI, Bhubaneswar	Orissa	1998	100-120	19	Semi erect, simple leaves with 3-5 lobes, profuse flowering, red skinned, deep orange flesh tubers, shape of tuber is obovate. High carotene content, can tolerate mid season stress, suitable for both kharif and rabi season.

contd...

24	Sankar (85-70)	Hybrid (H-219 x S-73	RC of CTCRI, Bhubaneswar	Orissa	1998	120	14	Red skinned tubers with yellow flesh with excellent cooking quality.
25	Jawahar Sakar kand-145	Clonal selection	JNKV, MP	Madhya Pradesh	1998	130	30	High carotenoid (16mg/ 100mg), deep orange fleshed
26	Vikram (C-I-1853-4)	Selection from AVRDC, Taiwan	UAS, Dharwad	Northern Karnataka	1998	100-110	22	High yielding short duration variety
27	RNSP-1 (S-30/21)	OPS selection	ANGRAU, Rajendranagar Andhra Pradesh	AP	1999	110-120	22	
28	CO-CIP.1	Selection from CIP New Delhi	TNAU, Coimbatore	Tamil Nadu	1999	95-100	31	Shallow lobed leaves, Purple pigmentation at nodes, tubers round elliptic or long elliptic with attractive pink skin and pale yellow flesh.
29	Konkan Ashwini	Selection made from collection of Varor village Tal. Palghar, Dist. thane	CES, Wakawali Konkan region of Maharashtra		2000	105	19	Short duration plant having purple colour elliptical long tubers. Viny plants having light green leaves with purple tinge at nodes on stem and alternate simple leaves. High yielding, short duration, elliptical long tubers with dark purple colour.
30	Indira Madhur (IB-90-15-9)	Selection from CIP collections	IGAU, Raipur, Chhattisgarh	Chattisgarh	2001	120	22	Medium duration Highin Casotene
31	Rajendra Sakar kand-92	Seedling Selection	RAU, Dholi	Bihar	2001	75	25	Spreading, green leaves 6-7 cylindrical tuber. Red skin and white flesh, suitable for flood prone area of North Bihar.
32	Sree Arun (RS III-3)	Clonal selection	CTCRI, Trivandrum	Kerala	2002	90-100	25	Tubers are fusiform short in shape with pink skin and cream flesh
33	Sree Varun (56-2)	Clonal selection	CTCRI, Trivandrum	Kerala	2002	90-100	25	Tubers are fusiform short in shape with cream skin and Flesh colour
34	Sree Kanaka (80/168)	Hybrid S-187x H-633	CTCRI, Trivandrum	Kerala	2004	85	10-15	Cream -skinned and deep orange fleshed tubers. Short duration. High b carotene
35	Bidhan Jagannath (90/101)	Clonal selection	BCKV, Kalyani, West Bengal	West Bengal	2004	110-120	21-24	High in carotenoid content.

contd...

36	Kalinga (90/704)	Clonal selection	RC, CTCRI, Bhubaneswar	Jharkand, Orissa, Bihar	2004	105-110	28	Tubers purple with cream flesh. Cooking quality excellent. Starch content high 28%.
37	Narendra Shakar kand-9/ Narendra Malti (NDSP-9)	Clonal selection	NDUAT, Faizabad, UP	UP	2004	120	26	Mid season variety, white skinned and white fleshed tubers with pleasant taste.
38	Narendra Shakar kand-10 (NDSP-10)	Clonal selection	NDUAT, Faizabad, UP	UP	2004	120	22	Mid season variety, red skinned, white fleshed tubers. Can be grown as irrigated as well as rainfed.
39	Gautam (Pol. 21-1)	Polycross from Dhenkanal local	RC, CTCRI, Bhubaneswar	Orissa	2005	105-110	30	White skinned tubers with cream flesh. Cooking quality very good.
40	Sourin (Pol.4-9)	Poly cross from Acc. No. 1162	RC, CTCRI, Bhubaneswar	Orissa	2005	105-110		Red stem tubers with cream flesh.
41	Kishan (Pol.13-4)	Polycross from Acc. No. 1016.	RC, CTCRI, Bhubaneswar	Orissa	2005	110-120		Reddish purple with cream flesh.
42	Indira Nandhini IGSP-17	Clonal selection	IGAU, Raipur	Chhattis garh	2006	110-120	28	Medium duration
43	RNSP-3	Clonal selection	ANGRAU, Rajendranagar AP	AP	2006	120	15	Suitable for kharif and rabi season
44	IGSP-4 Indira Naveen	Clonal selection	IGAU, Raipur	Konkan Region (Maharasthra) Raipur (Chhatisgarh)	2006	90-100	25	Short duration

2. Cassava

Sl. No	Variety	Breeding method / Pedigree	Originating source	Recomm-ended region	Year of release	Dura-tion (days)	Average yield (t ha^{-1})	Salient features
1	H-97	Hybrid, Manjavella x Acc. No. 300	CTCRI, Trivandrum	Tamil Nadu, Kerala, Karnataka, AP	1971	280-300	25	Erect branching, dark grey stem, broad leaves, green petiole, cylindrical tubers, light brown tuber skin, white flesh, starch content 27-31%. Tolerant to cassava mosaic disease, performs well under drought conditions and field resistant to spider mite and scale insect.

contd...

2	H-165	Hybrid, Chadaya mangalam vella X Kalikalan	CTCRI, Trivandrum	Tamil Nadu, Kerala, Karnatakas AP	1971	240-270	33	Medium tall, erect branching, light grey stem, light brown emerging leaves, broad mature drooping lower leaves light green petiole closely arranged fusiform tubers, golden brown tuber skin, white flesh, starch content 23-25%. Yields good quality starch, easy harvestability, early bulking and suitable for raising as a rotation crop after paddy.
3	H-226	Hybrid, M4 x Ethakka-karuppan	CTCRI, Trivandrum	Tamil Nadu, Kerala, Karnataka, AP	1971	270-300	30	Medium tall, erect branching, grey stem, green emerging leaves, broad leaves, green petiole, cylindrical tubers, tuber skin creamy with purplish patches, white flesh, starch content 28-30%. Preferred by industries, good cooking quality.
4	CO.1	Clonal selection (ME.7)	TNAU, Coimbatore, Tamil Nadu	Tamil Nadu	1977	260-275	30	Plants medium tall (1.5 to 2.5 m), late branching, young stems green, emerging shoot pink, tuber skin whitish brown, creamy rind and white flesh, starch content 35%. Tolerant to CMD and scale insects.
5	Sree Sahya (H-2304)	Multiple Cross Hybrid, Acc.No.468, 174, 3027, 1310, 82, 3939, 3588 and M4	CTCRI, Trivandrum	Assam, AP Karnataka, Kerala, Maharashtra, North Eastern Region, Tamil Nadu	1977	300	35	Tall erect branching dark, grey stem, stipules spiny and prominent, broad leaves, purple petiole, cylindrical, necked medium long tubers, light brown tuber skin, white flesh, starch content 29-31%. Tolerant to Cercospora, mites and scale insects, hardly and highly resistant to drought.
6	Sree Visakham (H-1687)	Hybrid, (Acc.No.1501 x -2312 (exotic))	CTCRI, Trivandrum	Assam, AP, Karnataka, Kerala, Maharashtra, North Eastern Region, Tamil Nadu	1977	270-300	35	Tall, erect branching, dark green stem, light brown emerging leaves, broad leaves, dark green petiole, fusiform compact tubers, brown tuber skin, light yellow flesh, starch content 25-27%. High carotene content, field resistant to Cercospora, less susceptible to spider mite and scale insect.

contd...

7	MVD-1	Selection	Tamil Nadu State Department of Horticulture	Tamil Nadu	1983	270	34	Non-branching type with tubers having 35% starch content. Field tolerance to ICMV. Cassava mosaic absent, tolerant to brown leaf spot and low spider mite incidence.
8	CO.2	Clonal selection	TNAU, Coimbatore	Tamil Nadu, Kerala	1984	260-275	35	Highly branching plant type, tuber skin is brown with creamy white flesh. Good cooking quality, low HCN content, suitable for consumption and starch industry, low incidence of tuber rot
9	Sree Prakash (S-856)	Clonal selection, Indegenous germplasm	CTCRI, Trivandrum	Kerala, Tamil Nadu AP, Assam	1987	210-240	30	Erect, non-branching, high leaf retention, reddish brown stem, dark green leaves, cylindrical tubers, shallow bulking, brown tuber skin and white flesh. Highly tolerant to Cercospora leaf spot, suitable for low land cultivation as a rotation crop in a paddy based cropping system. Early maturing.
10	CO.3	Clonal selection, Clonal selection (ME.120) IITA, Ibadan, Nigeria	TNAU, Coimbatore, Tamil Nadu	Tamil Nadu	1993	240-270	36	Branching type, tuber skin dark brown with white flesh good cooking quality, starch content 35.6%. High yielding, field tolerant to CMD, low incidence of tuber rot.
11	Nidhi	Clonal Selection (Koombu vella)	RRS, Kayamkulam, KAU	Onattukara (Coastal sandy tract) of Kerala	1993	180	25.1	Tolerant to cassava mosaic and moisture stress. Short duration. Low HCN. Good cooking quality starch 27%
12	H-119	Hybrid	TNAU, Coimbatore, Tamil Nadu	Tamil Nadu, Karnataka Maharashtra	1995	210-225	30	Non-branching, High tuber starch (35.2%). High yielding short duration crop and good cooking quality.
13	Sree Harsha	Triploid (OP-4 x H 2304)	CtCRI, Trivandrum	Kerala	1996	300	35-40	white fleshed tubers with high starch content (38-41%) and good cooking quality. Drought tolerant, high starch content acceptable for industrial use.
14	Kalpaka (KMC 1)	Clonal Selection (Ramanthala)	RARS, Kumarakom, KAU	Kuttanad, Kerala	1996	180	42	Tubers with high starch content (31.4%); protein content is 10.09 mg per 100 g tubers. Short duration (6 months), non branching variety. Tubers 35-41 cm

contd...

								long. Tolerant to brown leaf spot, cassava mosaic and low spider mite incidence.
15	CO (Tp)-4	Clonal selection from seedling progenies of hybrid SM-1679	TNAU, Coimbatore, Tamil Nadu	Tamil Nadu	2002	300	50	Tubers are long cylindrical with brown skin and white flesh. Starch content 40%. Suitable for both rainfed and irrigated conditions.
16	Sree Vijaya (CI-731)	Selection	CTCRI, Trivandrum	Bihar Assam Kerala	2006 2006 1998	180-210	18 23-25	Erect branching variety with 27-30 % starch content 40-60ppm HCN and excellent cooking quality and the tuber flesh colour is light yellow. Early maturing variety
17	Sree Rekha (TCH-1)	Hybrid TMS 63198 x Sree Visakam	CTCRI, Trivandrum	Kerala	2000	240-300	45-48	Cream fleshed tubers with 28-30% starch content and excellent cooking quality. Suitable for cultivation under upland and lowland conditions.
18	Sree Prabha (TCH-2)	Hybrid TMS 63173 x H 1687	CTCRI, Trivandrum	Kerala	2000	240-300	40-45	Light yellow fleshed tubers with 26-29% starch and excellent cooking quality. Suitable for cultivation under upland and lowland conditions.
19	Sree Jaya (CI-649)	Clonal selection	CTCRI, Trivandrum	Andhra Pradesh Kerala	2006 1998	180-210	23	white fleshed tuber with 24-27% starch, 40-50% cyanogen but excellent cooking quality. Early maturing variety.
20	Vellayani Hraswa	Clonal Selection	College of Agriculture, Vellayani, KAU	Kerala	2002	155-180	44	Short duration variety with creamy white tuber flesh. Dwarf plants with branching habit and retaining maximum number of leaves till harvest. Very good cooking quality.
21	Sree Athulya (4-2)	Triploid OP-4 X H 1687	CTCRI, Thiruvan anthapuram	Kerala	—	300	38	Higher extractable starch (30%). Good in the industrial belt.
22	Sree Apoorva (5-3)	Triploid Ambakkadan (2x) X H 2307	CTCRI, Thiruvan anthapuram	Kerala	—	300	38	Higher extractable starch (30%). Good in the industrial belt
23	Sree Padma nabha (MNga-1)	Exotic germplasm selection (TMS 30001) from IITA, Nygeria	CTCRI Trivandrum	Tamil Nadu	2007	270-300	38	Resistant to Mosaic disease. Cupping of leaves under drought.

contd...

a. Lesser Yam

Sl. No	Variety	Breeding method / Pedigree	Originating source	Recommended region	Year of release	Duration (days)	Average yield (t ha^{-1})	Salient features
1	Sree Latha (DE-11)	Clonal selection, Selection from indigenous germplasm	CTCRI, Trivandrum	Kerala, Bihar, AP, Maharashtra, Assam, West Bengal	1983	220-240	25	Vine dark brown, twinning to left, spiny and medium hairy petiole, oblong to fusiform tuber with grayish brown skin and creamy white flesh. Field tolerant to mosaic disease, high yield and wide adaptability.
2	Konkan Kanchan	Clonal selection from local	KKV, Dapoli, Maharashtra	Konkan region of Maharashtra	1984	210	18	High yielding vine having light green coloured simple alternate leaves, spines at the base of petiole. Stem is thin, cylindrical and green in colour. Tubers are elongated, almost cylindrical with round ends. High yielding, equal size tubers, white and sweet in taste.
3	Moa aloo (Local)	Local selection	AAU, Jorhat	Assam	1988	240	12	High yielder. Field tolerant to mosaic, good tuber shape and excellent culinary quality.
4	Lotani	Local selection	AAU, Jorhat	Assam	1988	240	12	High yielder. Field tolerant to mosaic, good tuber shape and excellent culinary quality.
5	Sree Kala (DE-55)	Clonal selection, Selection from exotic germplasm	CTCRI, Trivandrum	Kerala	1993	240-270	20	Vine twining to right, spiny medium hairy, tuber shape oval smooth with brown skin and white flesh. Field tolerant to mosaic, good tuber shape and excellent culinary quality.

b. Greater yam

Sl. No	Variety	Breeding method / Pedigree	Originating source	Recommended region	Year of release	Duration (days)	Average yield (t ha^{-1})	Salient features
1	Sree Roopa (DA-80)	Clonal selection, Selection from indigenous germplasm	CTCRI, Trivandrum	Kerala	1987	280-300	25	Stem green with purple ridges, green emerging leaves, digitate and smooth tuber, black tuber skin and white flesh. Good cooking quality.
2	Sree Keerthi (DA-60)	Clonal selection, Selection from indigenous germplasm	CTCRI, Trivandrum	Kerala	1987	280-300	25	Stem and petiole green, light brown emerging leaf, conical, rough tuber with brown skin and white flesh

contd...

3	Konwari aloo	Clonal selection	AAU, Jorhat	Assam	1988	280-300	18-25	High yielding with better quality tuber.
4	CO.1	Clonal selection	TNAU, Coimbatore	Tamil Nadu	1991	255	32	Tubers are big white flesh, rich in carbohydrate (28%) and protein (2.5%)
5	Indu (KM-DA 1)	Selection TCR-5 IC-44209	RARS, KAU	Kuttanad, Kerala	1998	240	24	Shade tolerant variety with 8 months duration. Trailing type plants with shallow tuberization. Digitate tubers have brownish black skin and white flesh. Good as intercrop in coconut plantation.
6	Sree Shilpa (Da-197)	Hybrid	CTCRI, Trivandrum	Kerala	1998	240	28	Stem showing basal branching with green emerging leaves having purple petiole. Tubers swollen, oval and smooth, tubers skin colour is black with yellow cortex and white flesh. Medium size oval tubers with easy harvestability.
7	Konkan Ghor kand	Selection made from collection of Konsai village Tal. Chiplun	CES, Wakawali	Konkan region of Maharashtra	2002	180	16	It is high yielding and erect growing vine produces aerial tubers having purplish green colour stem and simple leaves.
8	Sree Karthika (Da-199)	Selection	CTCRI, Trivandrum	Kerala	2004	270-280	25-30	Good table variety
9	Orissa elite (Hinjilicut Local)	Selection	RC of CTCRI, Bhubaneswar	Orissa	2004	270	20-24	Local selection with good culinary quality

c. African Yam

Sl. No	Variety	Breeding method / Pedigree	Originating source	Recommended region	Year of release	Duration (days)	Average yield (t ha^{-1})	Salient features
1	Sree Subhra (I-146)	Seedlings selection Exotic variety Iwo (Nigeria),	CTCRI, Trivandrum	Kerala	1987	300	35	Green petiole, vine twining right, spiny stem, green emerging leaves, cylindrical partly hairy tubers, brown tuber skin with white flesh, starch content 21-22%. Drought tolerant, good flavour.
2	Sree Priya (U-195 (2))	Seedlings selection, Seedlings selection	CTCRI, Trivandrum	Kerala	1987	300	35	Green petiole and vine twining to right and spiny, green emerging leaves, cylindrical smooth tubers

contd...

		from the African variety "Umidika"						with brown skin and white flesh. Drought tolerant, good flavour, suitable for intercropping in mature coconut gardens and with banana.
3	Sree Dhanya (I-184)	Seedlings selection	CTCRI, Trivandrum	Kerala	1993	270	20	Green petiole, dwarf bushy variety, 30-50 cm height with green emerging leaves. Male plant. Cylindrical and partly hairy tubers with brown skin and white flesh and good cooking quality. First dwarf bushy variety that do not require staking.

3. *Colocasia* - Taro / Arvi

Sl. No	Variety	Breeding method / Pedigree	Originating source	Recomm-ended region	Year of release	Dura-tion (days)	Average yield (t ha^{-1})	Salient features
1	Satam ukhi (Kovvur)	Clonal selection, Kovvur	ANGRAU, Hyderabad, Andhra Pradesh	AP, Maharashtra, Kerala, Karnataka, West Bengal, Bihar	1985	180	15	Plant height 50-55 cm, medium spreading, 6-7 tillers, dark green sagitate leaf, light purple petiole. Field tolerant to blight
2	Sree Pallavi (C-266)	Clonal selection, Meghalaya	CTCRI, Trivandrum	Kerala	1987	205-210	16	Erect tall droopy leaves with glaucous green upper surface, light green lower surface, big conical corm with club shaped cormels. Field tolerant to leaf blight and dasheen mosaic and a tall variety with large number of small sized tubers.
3	Sree Rashmi (C-149)	Clonal selection, Selection from indigenous germplasm	CTCRI, Trivandrum	Kerala	1987	205-210	18	Erect tall with droopy leaves having entire purple margin, green at upper surface and light green at lower surface, corm is cylindrical and medium sized conical cormels. This variety has acrid free leaves, corms and cormels, gives economy yield under low input level, field tolerant to dasheen mosaic.
4	ML-1	Local selection	ICAR, Research Complex for NEH Region,	Meghalaya	1988	160	27	Plant height around 75 cm with light green leaf and purple green petiole. Corm is medium sized and more number of small to

contd...

			Meghalaya					medium sized cormel. Comparatively tolerant to leaf blight.
5	ML-2	Local selection	ICAR, Research Complex for NEH Region, Meghalaya	Meghalaya	1988	160	22	Plant height around 60 cm. Cormels are small to medium sized and 7-8 per plant. Moderately resistant to leaf blight.
6	ML-9	Local selection	ICAR, Research Complex for NEH Region, Meghalaya	Meghalaya	1988	160	22	Plant height around 60 cm with light green leaves and petioles. It has medium sized corm and small to medium sized cormels. Comparatively tolerant to leaf blight.
7	CO.1	Clonal Selection	TNAU, Coimbatore	Tamil Nadu	1991	240	20	Erect plant type with less acrid tuber and good keeping quality. Tuber starch content 22.5% and protein 2.4%.
8	Bidhan Chaitanya (BCC-1)	Clonal selection	BCKV, Kalyani, West Bengal	West Bengal	1997	180-240	15	Plant height 75–80 cm, with entire inclined waxy leaf with blund tip peltate shape and yellowish greeny colour, branched round to oval cormels. Tubers non-irritant with excellent cooking quality and very mealy and non aromatic.
9	Bidhan Joydeb (BCC-2)	Clonal selection	BCKV, Kalyani, West Bengal	West Bengal	1997	150-210	12	Medium spreading with a plant height of 60-70 cm with entire inclined cup shaped lamina having peltate shape and green in colour. Branched eliptical cormels. Tubers are non-irritant with excellent cooking quality very mealy and non aromatic.
10	White Gouria	Clonal selection	RAU, Dholi	Bihar	1997	180	18	Early maturing variety, erect tall with droopy leaves. Moderately susceptible to leaf blight.
11	Sahasra-mukhi	Local selection	RAU, Dholi	Bihar	2000	120-140	12-15	Cormels are more with good culinary qualities.
12	Bhavpuri	Local selection	ANGRAU, Rajendra nagar	Andhra Pradesh	-	180-200	13 Rainfed 28	Irrigated Rich in culinary qualities.
13	Halu Kesu	Selection	ARS, Sirsi	Karnataka	-	-	-	High yielding
14	Mukta keshi	Clonal selection	RC, CTCRI, Bhubaneswar	Orissa	2002	150-180	20	Good cooking quality. Resistant to leaf blight

contd...

15	Indira Arvi-1 (BK Col-2)	Clonal selection	IGAU, Jadgalpur	Chhattisgarh, Assam	2003	210-240	22	Erect plant type with emerging leaves like greeny pigmented margin. Higher number of main marketable cormels (9-10). Cormels are non acrid with very good cooking quality. Tolerant to water logging condition. Corm shape round and cormel shape club with blunted. Suitable for cultivation in kharif season.
16	Ahina Katchu	Selection	AAU, Jorhat, Assam	Assam	2004	175	17	High yielder better quality tuber.
17	Sree Kiran (H-13)	Hybrid (C-303 x C-383)	CTCRI, Trivandrum Kerala	Kerala	2004	190-210	28	Good cooking quality
18	NDC-1	Clonal selection	NDUAT, Faizabad	Uttar Pradesh	-	170-180	12-15	Mid season variety with less acrid corms, which can be grown in both seasons-summer and rainy.
19	NDC-2	Clonal selection	NDUAT, Faizabad	Uttar Pradesh	-	180-190	12-15	Mid season variety with less acrid corm can be grown in both seasons.
20	Pani Saru-1	Clonal selection Kantilo local	RC, CTCRI, Bhubaneswar	Orissa	2005	180-200	16	Good cooking quality with longer shelf life.
21	Pani Saru-2	Clonal selection Begunia	RC, CTCRI, Bhubaneswar	Orissa	2005	180-200	16	Good cooking quality with longer shelf life.
22	Kadma/ Arvi local	Clonal selection	BAU, Ranchi, Jharkhand	Jharkhand	2006	150	17	Erect, with emerging leaf colour light green, moderate flowering type. Resistant to leaf blight, viral infestation and pests, 18 cormels per plant. Tubers almost non irritant with excellent cooking quality.
23	PKS-1	Clonal selection	NDUAT, Faizabad, Uttar Pradesh	Faizabad	2006	150	20-31	Good for chips making.
24	RNCA-1	Clonal selection	ANGRAU, Rajendra-nagar, AP	Hyderabad	2006	180	14-18	Medium duration
25	Sonajuli	Clonal selection	CTCRI, RC, Bhubaneswar	Orissa	2006	180	12	Tolerant to blight, drought and salinity stress
26	Jhankri	Clonal selection	CTCRI, RC Bhubaneswar	Orissa	2006	180	12	Tolerant to blight, drought and salinity stress

contd...

Colocasia - Bunda

Sl. No	Variety	Breeding method / Pedigree	Originating source	Recommended region	Year of release	Duration (days)	Average yield (t ha-1)	Salient features
27	NDB-21	Clonal Selection	NDUAT, Faizabad	Uttar Pradesh	2006	155	45-50	Early maturing. Corms cook easily
28	NDB-03	Clonal Selection	NDUAT, Faizabad	Uttar Pradesh	2006	121	35-40	Corms cook fast.
29	Narendra Bunda -1 (NDB-1)	Clonal Selection	NDUAT, Faizabad	Uttar Pradesh	2001	180-195	35-40	Plant height 75-125 cm with erect growth habit with clock wise leaf arrangement and erect lamina orientation. Vine colour green with purple streaks on upper and lower portion. Cylindrical corm and eliptical cormels. Resistant to lodging. Corm and cormels are suitable for table purpose as it has very less acridity soft and easy to cook.

4. *Xanthosoma*

Sl. No	Variety	Breeding method / Pedigree	Originating source	Recomm-ended region	Year of release	Dura-tion (days)	Average yield (t ha^{-1})	Salient features
1	Konkan Harit parni	Selection made form Sakhloli village, Tal. Dapoli.	CES, Wakawali	Konkan region of Maharashtra	2000	180	2.5-5 cormels 3 green leaves	Erect growing having dark green leaves with purple tinch. Corm small size and brown coloured. High yielding, more green leaves and corms.

5. *Amorphophallus*

Sl. No	Variety	Breeding method / Pedigree	Originating source	Recomm-ended region	Year of release	Dura-tion (days)	Average yield (t ha^{-1})	Salient features
1	Gajendra (Kovvur)	Clonal selection	ANGRAU, Hyderabad, AP	AP Bihar, West Bengal, Assam	1991	180-210	50	Compact plant with medium canopy spread. Good culinary quality, low acridity.
2	Sree Padma (AM-15)	Clonal selection from Wyanad, Kerala	CTCRI, Trivandrum	Kerala	1998	240-270	42	Plant height 78 cm with a mean canopy spread of 123 cm radius. Fairly tolerant to mosaic disease and colar rot. Non-acrid corm with good cooking quality.
3	Bidhan Kusum	Clonal selection	BCKV, Kalyani	West Bengal	1999	240	50	Non-irritant variety. Smooth corm with excellent cooking quality.
4	NDA-9	Clonal selection	NDUAT, Faizabad	Uttar Pradesh	2006	180-210	65	Early variety. Corms non-acrid with good cooking quality.

contd...

Sl. No	Variety	Breeding method / Pedigree	Originating source	Recommended region	Year of release	Duration (days)	Average yield (t ha^{-1})	Salient features
5	NDA-5	Clonal selection	NDUAT, Faizabad	Uttar Pradesh	2006	190-220	60	Early maturing, Corms non-acrid with good cooking quality with good taste.
6	Sree Athira (AM.H-2)	Hybrid Selection from Am 15 x Am 45	CTCRI, Trivandrum	Kerala	2006	270-300	40	Tuber skin brown with light pink flesh.

6. *Solenostemon rotundifolius* (Coleus)

Sl. No	Variety	Breeding method / Pedigree	Originating source	Recomm-ended region	Year of release	Dura-tion (days)	Average yield (t ha^{-1})	Salient features
1	CO.1	Clonal Selection	TNAU, Coimbatore	Tamil Nadu	1991	180-195	26	Cooked tubers are tasty. Characteristic aroma, oblong shaped tubers and early maturity.
2	Sree Dhara (CP-58)	Clonal selection	CTCRI, Trivandrum	Kerala	1993	150	25	Charactersitic aroma, oblong shaped tubers and early maturity. Tuber flesh cream
3	Nidhi (CP-79)	Clonal Selection IC 85708	RARS, Pattambi, KAU	Central zone of Kerala	2000	120-130	28	Good cooking quality, early maturity. Charactersitic aroma, oblong shaped large tubers.

7. Yam bean

Sl. No	Variety	Breeding method / Pedigree	Originating source	Recomm-ended region	Year of release	Dura-tion (days)	Average yield (t ha^{-1})	Salient features
1	Rajendra Mishri kand-I (RM-I)	Seedling selection from germplasm	TCA, Dholi, Bihar	North Bihar, West Bengal	1999	110-140	35	Round conical type, tuber skin colour brownish white, shallow tuberization, 2-3 tubers per plant. Good quality, resistant to major pests and diseases, suitable for intercropping with Kharif maize and arhar.

References

Palaniswami, M.S. and Shirly Raichal Anil. 2006. Region specific technologies for tropical root and tuber crops in India. All India Research Project on Tuber Crops. CTCRI, Thiruvananthapuram.67 pp.

Rajan, S and Prameela, K.P. 2004. Crop varieties released from KAU. Kerala Agricultural University, Directorate of Research, Vellanikkara. 256 pp.

❑❑❑

7

Agronomy and Production Technologies

Application of manures and adoption of proper cultivation practices have profound influence in yield and quality of tubers. There crops also need macro nutrients (N,P,K which are primary nutrients and Ca, Mg and S which are secondary nutrients) and micro nutrients (Fe, Mn, Zn, Cu, Mo, B) for their maximum yield potenial.

The package of practices vary with crop and again with soil, climate, edaphic and topographic factors, especially with reference to tropical tuber and root crops. The yield can be maximized through good agronomical practice and by adoption of proven recommended production technologies. Soils, climate, topographic and edaphic factors suitable for each crop alongwith salient cultivation practices and production technologies are enunciated for important tuber and root crops in various parts of India.

1. Sweet potato

Soil and Climate

Sweet potato perform well in well-drained, fertile, sandy loam soils having high organic matter. Compact and heavy soil reduces the size of tubers due to improper aeration whereas high sandy soil results in long cylindrical pencil like tubers. The optimum pH range of the soil should be 5.2 to 6.7. Sweet potato is sensitive to alkaline and saline condition. Water loggings not good for its cultivation.

Sweet potato vines require long warm growing season. A temperature range of 20-27°C is best suited to vine growth, and for tuberization the ideal temperature is 20-25°C. The areas with an average day temperature of more than 24°C and an average annual rainfall of 750 mm or more are suitable for its cultivation. No tuber production takes place at very high temperature (35°C-40°C) and temperature below 15°C, and plants are damaged by temperature below 10°C. Short day with low light intensity promote root development while excess of rainfall and long photoperiod encourages vine growth and reduces tuber yield.

Propagation and Planting material

Vine cuttings, obtained from harvested plants or from nursery, is the propagule. The vines obtained from nursery are found to be healthy and vigorous and give maximum tuber yield while the vines obtained from freshly harvested mature crop, are commonly used in areas where two crops are raised.

Nursery Preparation

a) Primary nursery

The nursery is prepared about 2 months before planting in main field. About 100m^2 primary nursery area and 100 kg of medium sized weevil free seed tubers, weighing 125-150 g each, are enough for planting one hectare of land. The tubers are planted 25 cm apart on ridges formed at a spacing of 60 cm. Urea may be applied 15 days after planting @ 1.5 kg urea/100m^2 to ensure quick growth of vines and the nursery may be irrigated on every alternate day for the first 10 days and thrice a week thereafter. The vines are clipped off to a length of 20-30 cm after 40-45 days of planting, for replanting in the secondary nursery.

b) Secondary nursery

The vines obtained from the primary nursery are further multiplied in secondary nursery of area 500 m^2, which produces enough planting material to plant one hectare of land. About 500 kg FYM or compost may be applied at the time of preparation of the nursery. The vines, obtained from primary nursery are planted at a spacing of 20 cm on ridges formed 60 cm apart. Urea @ 1 kg/100 m^2 may be applied in two split doses at 15th and 30th day after planting. After 45 days, vines are ready for planting in the main field.

Selection of planting material

Vines of 20-30 cm length with at least 3-4 nodes are found to be ideal as planting material. The cuttings obtained from the apical and middle portion of vines are preferable to get higher sprouting percentages and better tuber yields.

Preparation of vines

The cuttings with intact leaves are stored under shade for two days before planting in the main field to promote better root initiation, early establishment and higher yield.

Time of planting

Sweet potato is grown as a rainfed crop during *Kharif* (June-August) and as an irrigated crop during *Rabi* (October-December). However, the major area under sweet potato is planted during rabi (autumn) season with warm sunny days, cool nights and moderate rainfall. In North Bihar and parts of Uttar Pradesh, it is also planted in February-March as spring crop with supplemental irrigation. In eastern part of U.P., sweet potato can be planted during September for maximum profit. In Chhattisgarh, Madhya Pradesh and Chotanagpur Plateau of Bihar, August-September is ideal where as in diara land of Bihar and Eastern Uttar Pradesh, planting may be taken in November after recession of flood. In Andhra Pradesh, planting can be taken up in August and November. In Tamil Nadu and in Assam, planting can be taken up in May-June as rainfed and September-October as irrigated crop.

Methods of planting

The different planting methods, which are practiced in sweet potato are mound, ridge and furrow, bed and flat method. Among the different methods, ridge-and-furrow method is mostly practiced which gives better yield and is helpful for convenient cultural operations, earthing up and economical water requirements. Land is ploughed to a depth of 15-20 cm to bring a fine tilth. The vines are planted at a spacing of 20 cm on ridges of 20 cm height made 60 cm apart. Only the middle of the vine with nodes is buried at 5 10 cm depth keeping both the ends exposed. Vertical planting on the ridge for sweet potato in sodic soil is recommended for UP. For Bihar, the flat method of planting and for Andhra Pradesh and North-Eastern India, the ridge method of planting is recommended.

Spacing and Seed rate

The recommended spacing for sweet potato is 60x20 cm, which will accommodate about 83000 cuttings per hectare. Spacings vary according to the planting methods and places. A spacing of 30x30 cm is recommended for the Indo Gangetic Plains, whereas 45x45 cm is adopted for diara land of Bihar and Dergaon areas in Brahmaputra rive near Assam. Closer spacings accommodate maximum number of plants and gives relatively higher yield, depending upon the soil and assured moisture.

Manures and Fertilizers/Biofertilizer

FYM or compost @ 10 ha^{-1} should be applied at the time of field preparation. In addition to FYM, fertilizers may be applied as recommended doses:

States	Nitrogen (kg ha^{-1})	Phosphorus (kg ha^{-1})	Potash (kg ha^{-1})
Bihar, West Bengal and Assam	40	40	60
Andhra Pradesh	60	60	60
Karnataka	60	90	60
Kerala	75	50	75

Full dose of P_2O_5 and K_2O and half dose of nitrogen should be given at the time of planting, whereas remaining half dose of nitrogen should be applied one month after planting along with first weeding and earthing up.

Application of recommended level of N along with Azospirillum vine dipping @2 kg ha^{-1} and soil application of 10 kg ha^{-1} was recommended for Tamil Nadu, Assam, Bihar, West Bengal and Kerala. For Ranchi, Andhra Pradesh, Jorhat and Assam, application of ?rd or ?rd recommended N (60 kg N ha^{-1}) + 2 kg Azospirillum ha^{-1} as vine dipping + 10 kg Azospirillum ha^{-1} as soil application has been recommended. Application of 45: 40: 60 kg NPK ha^{-1} with 15 kg N or 30:40:60 NPK with 30 kg N by vermicompost with as vermicompost was recommended for Faizabad areas of Uttar Pradesh.

In case of micronutrients deficiencies in sweet potato $ZnSO_4$ (1-2%), Amm. ferric sulphate (1-2%) and $MnSO_4$ (0.1%) spray are recommended.

Irrigation

A light irrigation after planting, to maintain sufficient moisture, gives proper establishment of the crop. Sweet potato is tolerant to drought but continuous long phase of drought reduces the tuber yield. If there is insufficient moisture, irrigation should be given at 10-15 days interval. The critical stage of moisture supply is 40 days after planting and water stress during this period reduces the marketable tuber yield.

Weeding and Earthing-up

The intercultural operation 30 days after planting along with weeding and earthing-up suppress weeds in the early stages of growth and improves the physical condition of the soil. The second earthing-up should be done 50 days after planting. Being a quick growing crop, it covers soil quickly and suppresses most of the weeds.

Training and Pruning

After proper establishment, sweet potato starts growing vigorously. Therefore, vines should be lifted at nodes 30 days after planting to prevent rooting and to facilitate better tuber development at the basal end.

Cropping systems

Sweet potato has been found to grow under different cropping sequences in eastern and southern region of India. The ideal cropping sequences in eastern India are mentioned below :

S.No.	States	Cropping sequences
1.	North Bihar	(i) Maize (Kharif)-Sweet potato (Autumn)-Wheat (Rabi)-Moong (summer)
		(ii) Maize (Kharif)-Sweet potato-Onion (Rainy
2.	West Bengal	(i) Moong (summer)-*Colocasia* (Pre-kharif)-Sweet Potato (Autumn)
		(ii) Moong-Elephant foot yam-Sweet potato
3.	Orissa	(i) Mid-season paddy (Rainy)-Sweet potato-Fallow
		(ii) Maize-Sweet potato-Fallow
4.	Kerala (low land)	(i) Paddy (1st crop)-Paddy (2nd crop)-Sweet potato
		(ii) Paddy-Sweet potato-Fallow
5.	Tamil Nadu	Sweet potato –Cereals
6.	Andhra Pradesh	Maize-Sweet potato-Vegetables
7.	Madhya Pradesh	Vegetable cowpea + Sweet potato
8.	Uttar Pradesh	Vegetable cowpea + Sweet potato
9.	Chhattisgarh	Vegetable cowpea + Sweet potato
10.	Maharashtra	Vegetable cowpea + Sweet potato

Intercropping

Planting sweet potato as intercrop in cashew orchard up to pre bearing period of 5 years with recommended dose of fertilizer (60:40:60 kg NPK ha^{-1}) was recommended for adoption in Konkan region of Maharashtra.

Harvesting and Post harvest Management

Harvesting

The crop matures in 100-135 days after planting depending on variety and environmental conditions. Harvesting sweet potato within 120 days of planting is normally recommended. Delay in harvesting increases sweet potato weevil infestation. Light irrigation 2-3 days before harvesting of tubers make easier digging. Early cultivars can be harvested 90-105 days after planting, whereas, medium and late maturing cultivar at 110 and 120 days after planting, respectively.

Yield

Yield of sweet potato tubers vary from 10-20 t ha^{-1} depending on varieties and cultural practices adopted. By adopting recommended varieties and improved cultural practices, an yield upto 30 t ha^{-1} may be obtained.

Marketing

Freshly harvested tubers of sweet potato are marketable. Sweet potato is highly prone to weevil damage hence delay in harvesting reduces the marketability.

Storage

If stored for longer period shrinkage results due to dehydration and several other forms of post-harvest losses. After harvest, tubers must be spread in partial shade for 5-6 days, for proper healing and curing and then can be stored in well-ventilated rooms.

2. Cassava

Soil and Climate

A well-drained loamy soil of pH 5.5-7.0 is best suited for the crop. It cannot tolerate water logging, high salt concentration and alkalinity but it can tolerate drought once it is established. It has been estimated that about 55% of cassava in Asia is grown on Ultisols, whereas in India it is widely grown on Ultisols (Kerala), Vertisols (Tamil Nadu), and Alfisols (Kerala).

Cassava is grown in areas ranging from humid (annual rainfall more than 2,000 mm) to semi arid (500-700 mm) conditions. It is grown up to an altitude of 2000 m near the equator, but performs better at lower altitudes. Ideal temperature for cassava cultivations is between 25°C to 32°C. It is a drought-tolerant crop. However, it cannot withstand frost, as its growth ceases at temperatures below 10°C.

Cassava is a perennial crop with indeterminate growth habit but is commonly cultivated as an annual crop between the latitudes 30° N and S, at altitudes from sea level up to about 2,000 m in equatorial regions, with annual precipitation ranging from 500 mm up to 6,000 to 8,000 mm. It tolerates hot climate, but at temperatures below 20°C, the plant does not grow normally and the storage root yield decreases drastically. Depending on the growing conditions and the cultivar, crop duration may vary between 7-12 months. In areas where a cool climate arrests plant growth and starch accumulation in the roots, cassava cultivars are being cultivated for a long duration, up to 2-3 years, although the eating quality declines with ageing due to increasing fiber content (Ravi and Mohankumar, 2004).

Propagation and Planting material

Cassava is vegetatively propagated through stem cuttings usually called 'setts'. True botanical seeds of cassava are generally used for breeding purpose. Freshly harvested cassava seeds germinate very rapidly at temperature range of 25°C - 35°C.

Stems collected from the previous crop are generally used to raise the new crop. Disease and pest free planting material of 8-12 months maturity having a thickness of 2-3 cm should be selected for planting. Stalks obtained from bottom and middle portion of the stem, after discarding one-third of the total length from the top, gives better establishment and root initiation. A smooth circular cut, rather than an irregular cut, gives uniform callusing and root initiation. A stalk length of 25-30 cm is economically beneficial.

Time of planting

Cassava can be planted through out the year under irrigated conditions. April-May with the onset of pre-monsoon showers or August-September, with the onset of north-east monsoon showers is the ideal time of planting as a rainfed crop.

Methods of planting

Different methods of planting stakes can be adopted such as vertical (90° to the ground), slanted (45° to the ground) and horizontal, but vertical planting results in more uniform callusing around the cut surface, which facilitates uniform distribution of tuber forming roots all round the base of plant. Shallow planting facilitates production of more number of roots, but a planting depth of 5 cm is optimum.

Spacing and Plant population

The spacing of cassava plantation depends upon the branching behaviour *i.e.,* semi-branching and non-branching types. Branching and semi-branching types require 90 cm x 90 cm while non-branched types require 75 cm x 75 cm for optimum yield. The optimum plant population on these spacings ranged from 12,000 to 17,000 plants per hectare. Normally one set is planted per hill, but planting two per hill could improve total tuber yield but reduce the size of tuber.

Manures and Fertilizer/Biofertilizer

FYM or compost @ 10 ha^{-1} is given at the time of land preparation and 1 kg manure per pit while filling up the pits is recommended. In addition to FYM, N:P:K: @100:50:100 kg ha^{-1} for long duration varieties and 75:50:75 kg ha^{-1} for short duration varieties, has been recommended. Green manuring *in situ* with fast growing vegetable cowpea varieties (B-61 or C-152) can substitute FYM.

Full dose of P_2O_5 and half dose of Nitrogen and K_2O should be given at the time of sowing, whereas remaining half dose of Nitrogen and K_2O is applied 45-60 days after planting along with first weeding and earthing up. In Gujarat, Maharashtra and Kerala, 100: 50: 100 kg NPK ha^{-1} fertilizer is recommended.

For acid laterite soils studies with biofertilizers has been very effective in increasing the yield as well as reducing the use of inorganic fertilizers. In Andhra Pradesh and Tamil Nadu application of half dose of P_2O_5 (30 kg ha-1) + full N and K_2O (60:60 kg ha^{-1}) + FYM (12.5 t ha^{-1}) + phosphorous solubilizing bacteria (PSB) or Arbuscular mycorrhizal (AM) fungi (10 kg ha-1) has been recommended for cassava.

In Tamil Nadu 75:25:75 NPK ha^{-1} is given under rainfed condition and 90:90:240 under irrigated conditions. In Andhra Pradesh 60:60:60 kg NPK ha^{-1} is recommended under rainfed and 100:60:150 NPK ha^{-1} under irrigated condition. For Chhattisgarh 75:50:75 kg NPK ha^{-1} is recommended.

In case of micronutrient deficiency of Mg, Zn and B application of $MgSO_4$ (20 kg ha^{-1}), $ZnSO_4$ (12.5 kg ha^{-1}) and Borax (10 kg ha^{-1}) is recommended. The lime induced chlorasis as seen in Tamil Nadu, can be cured by 1% $ZnSO_4$ and 1% $FeSO_4$ spray.

Irrigation

As a rainfed crop in Kerala irrigation is normally not undertaken whereas irrigation is common in Tamil Nadu and Andhra Pradesh conditions. Sufficient moisture should be ensured in the field at initial stage *i.e.,* the first 20 days after planting, for proper establishment of cassava stakes. Supplementary irrigation during drought period can result in higher yield. Irrigation at approximately 25% available moisture depletion level through out the growing season could double the tuber yield compared to control. It also improves the starch content and reduces the HCN content.

Weeding and Earthing-up

Weeding in the early stage of the crop is essential for proper control of weeds and for improving the soil condition. The first weeding and earthing up should be given 45-60 days after planting and the second a month later.

Gap filling

All the stakes planted may not establish due to poor quality planting material, adverse weather conditions and terminate infestation, hence, at the time of planting of stakes in the main field, about 5% of the stakes may be planted separately at a very close spacing of 4 cm x 4 cm in a nursery. At the age of 15-20 days the plants are uprooted from nursery and used for gap filling.

Thinning of shoots

The sprouts emerging from the buds are more vigorous than those emerging from the lower nodes of the stakes. Removal of excess sprouts by nipping at the initial stage of establishment (10-15 days after sprouting) helps to prevent mutual shading and competition between plants. Retention of 2 healthy shoots on opposite sides is better for the production of more number of tubers per plant.

Cropping system

Being a small (2-7 m high), perennial, woody shrub, cassava is suitable as an intercrop under different cropping system. Cassava can be grown well with coconut, *Eucalyptus* and other fruits and forest trees under agri-horti or agri-silviculture system.

In Kerala cassava is inter cropped in 20% of coconut stands and performs well in newly planted coconut gardens up to 5 years. Cassava varieties H-165, H-226, Sree Viskham and Sree Sahya can be grown satisfactorily under different cropping systems. Cassava variety KMC-1 has been reported as a suitable intercrop in coconut gardens of Kuttanad in Kerala.

Cassava is a long duration crop that takes 9-10 months for harvest. It is planted at a spacing of 90 x 90 cm and it takes about 3-4 months to develop enough canopy due to slow initial growth which provide a great opportunity to grow short-duration intercrops like, ground-nut, vegetable cowpea, frenchbean, black gram and other legumes. In Tamil Nadu black gram/ Bellary onion (*Alliun cepa)*, in Andhra Pradesh, green gram/black gram and in Assam soy bean and French bean are recommended as suitable intercrops in cassava.

Bunch type ground nut varieties (TMV-2 and TMV-7) and French bean (cv. "Contender") are promising and economical intercrops. Sree Prakash and Sree Jaya have been found to be suitable for low land cultivation in a paddy based cropping system *i.e.*, sequential cropping of cassava and paddy. The cultivation details of various intercrops, is given below.

Cultivation details of intercrops in Cassava (Varma *et al.*,1997)

Name of intercrop	Cultivar	Duration (days)	Spacing (cm)	No. of rows	Seed (kg ha^{-1})	N:P:K (kg ha^{-1})	Yield (kg ha^{-1})
Groundnut	TMV-2	100	30 x 20	2	40-45 (kernel)	10:20:20	1200 dry pod
Groundnut	TMV-7	100	30 x 20	2	40-45 (kernel)	10:20:20	1200 dry pod
French bean	Contender Top cross-4	70	30 x 20	2	40	20:30:40	2000
Cowpea (grain)	S-488	90	30 x 25	2	20	10:15:10	800
Cowpea (Vegetable)	B-61	65	40 x 20	1	8	10:15:10	3000

The gross net income was always higher in the following cassava intercropping systems with legumes/non-legumes than the pure stand of cassava. The gross income increased by 40-82% in a cassava + French bean system, 25% in a cassava + urd bean system, 29% in a cassava + soy bean or mung bean system, 13-20% in a cassava + green gram, or black gram system. Cassava can be successfully relay cropped following maize, peanut, sorghum, selected vegetables and banana (Ravi and Mohankumar, 2004).

Sequential cropping of cassava and vegetable cowpea has been recommended for Tamil Nadu with application of P @ 30 kg ha^{-1} along with recommended dose of N & K and for Chhattisgarh, with 5 t ha^{-1} FYM and 50 kg ha^{-1} P_2O_5.

Harvesting and Post harvest management

Harvesting

The time of harvesting in cassava depends on the cultivars and the use of harvest. Normally, short duration cultivars can be harvested at 6-7 months while long-duration ones can be harvested at 10 months after planting. Delay in harvesting may result in deterioration in cooking quality of the tubers.

Yield

Yield of cassava depends on the cultivars and stage of harvesting. High yielding varieties of long-duration cassava (10-11 months) yield 35-40 t ha^{-1}, whereas short duration ones (6-7 months) yields 25-30 t ha^{-1}.

3. Yams

Soil and Climate

Well-drained, loose friable sandy loam soil containing good amount of organic matter and pH 5-7 is ideal for growing yams. Yams are relatively tolerant to dry conditions, but gives better tuber yield with the supplement of water.

Warm climate, with a temperature 25°C - 30°C and adequate moisture are ideal for vegetative growth and tuber bulking. Yams cannot tolerate frost and growth is affected at a temperature below 20°C. Day length greater than 12 hours at the early stage promotes the vine growth, while short photoperiod favours satisfactory tuber production.

Propagation and Planting materials

Yams are propagated vegetatively through either small whole tubers or tuber pieces. The large whole yams are cut into setts (small pieces) consisting of tops, middles and bottoms while small tubers are planted as such. The small tuber (seed yams) and tops are preferred for their earliness and uniformity in sprouting. The average weight 200-300 g of tuber is optimum for the cultivation of *D. alata* and *D. rotundata* whereas 100-125 g is ideal for *D. esculenta.* In storage condition, yam tubers remain dormant for about 2-2½ months, hence quick dipping in 4-8% solution of ethylene chlorohydrin followed by dry storage is used to break the dormancy.

Yam can also be propagated through vine cuttings as it is a very useful technique for rapid multiplication of desirable cloned material, but tuber production by this method is slow.

Time of planting

Yams are planted mostly in the later part of the dry season from March-May and start sprouting with the onset of pre monsoon showers.

Methods of planting

Yams are planted by various methods in different places depending on soil types. In loose sandy type of soils, they can be planted in flat beds while raised bed or ridges or mounds over pits are practiced in the heavy soils. In mild sloppy land, the ridges are prepared to conserve soil.

Spacing and Plant population

The planting spacing of yams vary with the species, growth habit and purpose of planting. *D. alata* and *D. rotundata* require wider spacing due to their luxuriant growth and broad leaves. A spacing of 1 x 1 m with a plant population of 10,000 plants ha^{-1} or 90 x 90 cm with a plant population of 12,000 plants ha^{-1}, is recommended. For *D.esculenta,* a close spacing of 75 x 75 cm with a plant population of 18,000 plants ha^{-1} is optimum.

Manures and Fertilizers/Biofertilizer

FYM or compost @10 t ha^{-1} should be applied at the time of land preparation. In addition to organic manures, N: P: K @ 80:60:80 kg ha^{-1} for *D. esculenta,* whereas 100:60:80 kg ha^{-1} for *D. alata* and *D. rotundata* has been recommended.

Full dose of P_2O_5 and half dose of nitrogen and K_2O should be given at the time of sowing while remaining doses of nitrogen and K_2O should be applied again in two split doses along with first and second weeding and earthing up.

Irrigation

Yams are relatively tolerant to drought, but it gives better tuber yield with the supplement of water. Yams should be irrigated immediately after planting to ensure quick and uniform sprouting. In dry period, before the onset of the monsoon, it can be irrigated to maintain the soil moisture in the field. The crop is sensitive to water logging.

Weeding and Earthing-up

In initial stage, after a week of sprouting, weeding and earthing-up should be done sufficiently deep for the better growth. The second weeding and earthing up should be done one month after the first one. It takes up to 40 days for sprouting of yams. The critical stages for weed interference in yams coincide with growth phases when leaf development and tuber bulking were maximum.

Mulching

Mulching, after the planting of setts or tubers has its own importance. It provides protection from excessive temperature, conserves soil moisture, and ensures quick and uniform sprouting of the tubers and suppression of weeds, ultimately it results in the increase in tuber yield.

Staking

Staking or trailing of vines is an essential operation in yams for increase in tuber production. Staking exposes the leaves to sun light which encourages greater photosynthesis. The emerging shoots should be provided artificial supports to avoid any injuries for the tender shoots. *D. alata* and *D.rotundata* plants trail to a height of 3-4 meter, whereas *D. esculenta* requires comparatively lesser height of staking. In Andhra Pradesh, *D. alata* was found growing without staking as done in that of sweet potato. The farmers were happy with the yield of the crop.

Cropping system

Ideal crop sequences can generate more income. In Bihar, yam-wheat-green gram and yam-potato-green gram cropping sequence is followed. In mixed cropping system maize + yam is adopted in Bihar.

Intercropping

Lesser yam is grown as an intercrop in cashew plantation at Dapoli of Konkan region of Maharashtra. The variety Konkan Kanchan gave 20 t ha^{-1} tuber yield as an intercrop in cashew plantation in Konkan Region of Maharashtra. *D. alata* can be grown as an intercrop in banana and has been successful in Kerala.

Harvesting and Post harvest management

Harvesting

Yams are generally harvested at 7-9 months after planting, depending on the species and varieties. *D. esculenta* matures early as compared to other species. Yellowing of leaves and complete drying up of the vines indicates the maturity of crop. Two general practices are adopted for harvesting the yams *i.e.,* single harvesting and double harvesting. Double harvesting consists of removing the mother tubers after two months of growth and allowing for subsequent production of side tubers, but it is not economical as compared to single harvesting.

4. Colocasia

Soil and Climate requirement

As a tropical tuber crop, *Colocasia* require a continuous supply of water. In India, under lowland conditions, it is grown in alluvial and fertile soil while under upland conditions, it performs well on marginal land. Well-drained and fertile sandy loam soil with pH range 5.5-7.0 gives best result.

Colocasia prefers moist conditions for better growth and yield. It is best suited in swampy tropical climate with a temperature range of 25°C to 30°C and it is grown well in flooded conditions in Hawaii. In moist environment, it can be grown at a temperature of 20°C-21°C. A wide range of rainfall is found in taro growing areas but annual rainfall near 1000 mm is optimum for its growth and high yield.

Propagation and Planting materials

Eddoe type (Taro)

Colocasia is propagated vegetatively either by mother corms or cormels. It has been observed that the sucker production was significantly higher when mother corms were used for propagation where as the yield was found to be significantly higher when it was propagated by cormels. Cormels weighing 20-25 g are commonly used as planting material for eddoe in many taro growing areas under upland condition. Healthy, disease free planting materials having uniform size and no injury, should be selected and stored in a cool place at least for 3 months before planting. Cut piece of corm can also be used as planting material.

Dasheen type (Bunda)

Usually, the main corm of Dasheen is cut transversely into pieces of 4 to 5 cm thickness, containing one to two 'buds' and weight should not be less than 50 gms. The top portion of the main corm containing apical bud is the best for sprouting. Sometimes the daughter corms are also used as planting material. Healthy, disease free planting material having uniform size and no injury, should be selected and stored in a cool place at least for 3 months before planting.

Swamp taro

Long runners of roots are the edible starch part.

4.1 Eddoe type (Taro)

Time of planting

It is normally planted during the rainy season but it can be grown any time if irrigation facilities are available. In Kerala, April to June was the ideal time of planting under rainfed condition. However, February (irrigated) and June (rainfed) is recommended time for planting in North and Eastern part of India and April to June for Southern part.

Method of planting

Usually, eddoe is planted as flat bed method but ridge and furrow method can be adapted to conserve moisture. In flat bed method, the shallow trenches are made at the distance of 45 to 60, whereas, the cormels are placed at a spacing of 30 cm and then covered with soil. As plants increase in height, soil should be placed near the base during earthing up to make ridges.

Spacing and Seed rate

Spacing or planting largely depends on size of planting material, soil fertility and availability of water. A spacing of 60 x 45 cm was found to be optimum to get good economic return. Higher planting density at a distance of 45x30 cm is also adopted on marginal land.

Seed rate and plant population vary according to spacing. About 1 ton planting material is enough for planting one hectare land, at spacing of 60 x 45 cm while higher planting density (45 x 30 cm) requires about 1.5-2.0 t ha^{-1}. In the former spacing, the plant population will be 37,000 ha^{-1} whereas in the latter, plant population will be 74,000 ha^{-1}. The depth of planting varies between 5-10 cm depending on the size of planting material.

Manures and Fertilizers/Biofertilizer

Application of FYM or compost @ 10-15 t ha^{-1} at the time of land preparation makes soil friable, which results in better tuberization. In addition to FYM or compost, N.P.K. @ 80:80:120 kg ha^{-1} has been recommended for better tuber production. Full dose of phosphorus along with half dose of nitrogen and potash should be given in two split doses within first 7-10 days after sprouting and second dose 1 month later.

Available reports suggested combined application of the N, P and K as highly beneficial to the crop for its growth and yield @ 80:25:100 kg ha^{-1}.

Most of the investigators are of the opinion that Nitrogen and Potassium should be applied in split while phosphorous may be applied in full as basal. Ravi and Mohankumar (2004) reported significant higher yield for three split application of nitrogen and potassium at the rate of ?rd dose when fifty percent sprouting occurs, second ?rd one month later and the remaining ?rd one month after the previous application. The fertilizer requirement has been worked out for different states under AICRPTC and is presented here.

The fertilizer recommendation for different states

States/Region	Recommendation (kg NPK ha^{-1})
Bihar	80:60:80
Andhra Pradesh	120:60:120
Karnataka	60:60:90
Tamil Nadu	40:60:120
Uttar Pradesh	80:60:80
Assam	80:60:120

Biofertilizers

Studies on biofertilizers indicated that AM fungi help in the higher uptake of P by extending the area of absorption through fungal hyphae. In Andhra Pradesh, application of half the recommended dose of P_2O_5 (30 kg ha^{-1}) + full N and K_2O (120: 120 kg ha^{-1}) + FYM (15 t ha^{-1}) + AM fungi @10 kg ha^{-1} having a spore load of 400 spores $10ml^{-1}$ + phosphorous solubilizing bacteria @10 kg ha^{-1} has been recommended for taro.

Irrigation

Eddoe grows well under moist situations. Although soil moisture has least influence on sucker initiation, moisture stress adversely affects vegetative growth, which largely reduces the dry matter production and tuber yield. Usually 7-8 irrigations are required for best yield in summer season. Generally, eddoe type requires less amount of irrigation as compared to dasheen type, due to short life period. Eddoe type is relatively more tolerant to drought than dasheen type.

Weeding and Earthing-up

Clean and weed free field during initial periods of crop growth facilitate better crop growth. Generally, two manual weeding alongwith earthing up has been recommended for better growth and tuberization in eddoe. The first weeding along with earthing up is done after 7-8 days of sprouting and second a month later. Early chemical weeding by atrazine or simazine @ 1 kg ha^{-1} followed by one light earthing up during later stage of crop growth, gives higher yield.

Mulching

Mulching with dried leaves and other organic mulches, provides additional nutrients, conserves soil moisture, maintains soil temperature and reduces weed growth. In Meghalaya, black polythene mulching effectively controlled weed population giving an yield of 22.7 t ha^{-1} in *Colocasia* var. Kandha-5.

Desuckering

A large number of suckers per plant reduces the tuber size hence only 3 suckers/plant should be retained at second earthing up.

Cropping system

Colocasia is found to grow under different cropping systems. In Assam, green gram and okra have been suggested for adoption as viable intercrops in taro. In Uttar Pradesh, green chilly as intercrop in *Colocasia* and *Colocasia* + tomato crop sequence were recommended for the farmers. *Colocasia*-ridged gourd and *Colocasia*-cauliflower cropping sequence is followed in Bihar. *Colocasia* (Kadma local) + cowpea (Arka garima) combination is the most profitable intercropping system here. Intercropping of three rows of onion in between two rows of *Colocasia* has recorded on additional net return of Rs. 20,000/- under irrigated condition in Bihar. In Kerala, taro is recommended as intercrop in banana/coconut/aracenut.

4.2 Dasheen type (Bunda)

Time of planting

The time of planting greatly varies with region and availability of irrigation. It can be planted from February to June, but February is the best time of planting under irrigated conditions in North Bihar, parts of West Bengal and eastern U.P. As rainfed crop, it can be planted in June, where there is no frost during growing period.

Method of planting

Generally, planting is done by ridge and furrow method. Flat bed method of planting is also adopted under upland conditions with good drainage. Furrow method is better than others as it conserves moisture. In well prepared land, the trenches are usually made at a distance of 60 to 75 cm, whereas the plant-to-plant distance is kept at 30 to 45 cm.

Spacing and Seed rate

The dasheen type is usually taller, spreading and robust in structure than the plant of eddoe. Thus, the spacing for dasheen is more than the eddoe type. A spacing of 75 x 45 cm or 75 x 30 cm is optimum for dasheen to get good economic return. It can be planted at closer spacing *i.e.*, 60 x 45 cm or 60 x 30 cm when the soil is comparatively less fertile and heavy. The seed requirement varies with different spacing. About 1.5 tonnes planting material is enough for one hectare, at optimum spacing, whereas, at closer spacing, the seed rate increase up to 2.5 tonnes per hectare.

Fertilizer

There is no major difference in fertilizer requirement of eddoe and dasheen type. In bunda type of *Colocasia* (*Colocasia esculenta* var. esculenta) seed size of 40-50 g with application of nitrogen @100 kg ha^{-1} and 60 and 120 kg ha^{-1} of P_2O_5 and K_2O respectively was found to be optimum for getting higher tuber yield.

Irrigation

The life period of dasheen is comparatively longer than eddoe, therefore, it requires more amount of irrigation. Usually, 8-10 irrigations are required for best yield if planting is done in summer. Dasheen is comparatively more sensitive to drought than eddoe.

Weeding and Earthing-up

Initial weeding facilitates better crop growth, which results in higher tuber yield. The life period of dasheen is relatively longer than eddoe; therefore, it requires three weeding along with earthing up. As plants increase their height and spread, the soil on the ridges should be replaced in trenches.

In places where the weeds are serious, generally, the first weeding along with earthing up is done after 7-8 days of sprouting and other two at the interval of one month. Atrazine or simazine @ 1 kg ha^{-1} can be applied in early stage followed by two light earthing up when it is planted as flat bed method. Initial earthing up is a critical operation if dasheen is planted as ridge and furrow method.

Mulching

Use of organic mulches like dried-leaves, straw, grasses etc, is found to be beneficial for dasheen cultivation. Mulching provides additional nutrients, conserves soil moisture, ensures quick germination, maintains soil temperature, reduces weed growth and makes soil more bulky; which results in better tuberization and corm production.

Desuckering

In case of dasheen, there is no need of desuckering.

Harvesting and Post harvest management

Harvesting

The crop matures in about 120 to 150 days, but dry matter content of tuber increases up to 180 days after planting. *Colocasia* becomes ready for harvest within 5-7 months after planting. Generally eddoe type is harvested early while dasheen type requires larger duration. Drying up of leaves is the indicator of maturity.

Yield

Yields of *Colocasia* vary from 20 to 40 t ha^{-1} depending on variety, type, management and stage of harvesting.

Marketing

The mother corms and cormels should be separated before marketing. In dasheen type, mother corms are marketable where as in eddoe type cormels are marketable.

Storage

At the time of harvesting, damaged tubers must be removed to prevent rotting during storage. The tubers should be dried for a day under the shade by spreading on ground and the soil adhering to the tubers should be removed. It should be stored either in soil pits or in heap under well ventilated room.

4.3 Swamp taro

Grown in swampy conditions of East and North Eastern India. *Colocasia esculenta* var. stoloniferum is grown for use of its stolons in parts of Assam and West Bengal. Mother corms are used for planting during January. 300:100:200 kg ha^{-1} NPK is advocated by BCKV, Kalyani along with five irrigations. Periodical harvesting is done at fortnightly intervals from April onwards. Stolon length is about one metre. The total stolon yield ranges between 20-26 t ha^{-1} and fortnightly yield ranges from 1-3 t ha^{-1}. BCST-1, BCST-5, BCST-13 and BCST-14 are good selections for high yield. Application of Mahua cake (50kg N) + 225 kg ha^{-1} N produced significantly higher stolon production of 20.86 t ha^{-1} at Kalyani.

5. Tannia

Soil and Climate

Sandy-loam to clay and fertile well drained soil having pH range from 5.5-7.0 is ideal for the cultivation of *Xanthosoma*. Unlike *Colocasia, Xanthosoma* can adapt to varied soil conditions but cannot come up well in clay soils. It cannot withstand water logging and a long dry spell.

The hot and humid conditions with a temperature range of 25 to 30°C and a well-distributed rainfall of 1,500-2,000 mm during its growth period is ideally suited for its cultivation. It can be grown at 1,500 m above mean sea-level but it cannot withstand cool climate where frost is a problem.

Propagation and Planting material

Like *Colocasia*, both corm and cormels are usually used as planting material. Healthy cormels of bigger size (50-80 g), 20-25 cm long are generally recommended for planting. However, when corms are used as planting material, the setts from the top portion of the main corm with a thickness of 5-10 cm containing the apical bud is preferred. In general, the plants grown from the apical portion require shorter growing period than those plants propagated from the other portion of the main corm.

Time of planting

Xanthosoma is planted with the onset of monsoon, *ie.,*by April-May, under rainfed conditions but is can be grown throughout the year with irrigation.

Method of planting

Generally, *Xanthosoma* can be planted on ridges as well as in flat bed also, but ridge method of planting gives higher yield as compared to flat system.

Spacing and Seed rate

Planting is done at a spacing of 90 cm between plants on ridges. When it is grown for leaf purpose, it should be planted on both sides of raised bed of breadth 1 m and convenient length. The height of bed should be 30 cm. About 8 to 10 quintals of corm is required for planting one hectare field.

All the intercultural practices resemble that of *Colocasia.*

Fertilizer requirement

80:60:80 kg ha^{-1} NPK with 10-15 t ha^{-1} FYM is advocated.

Intercropping

Tannia is grown as intercrop in sapota orchards, for Patra in Maharashtra and is grown in banana and coconut gardens.

Harvesting and Post harvest management

Harvesting

The crops become ready for harvest within 9-10 months after planting, when older leaves start turning yellow. The corms can remain unharvested up to 3 months after maturity. If the crop is grown for its leaves, then leaves should be harvested immediately after they attain maturity. During its entire growth period about 40-50 leaves can be harvested from each clump.

In parts of Maharashtra, Goa and Gujarat where only leaves are harvested for Patra, the corms and cormels are used as planting material. The leaves are harvested daily after 45 days of sprouting.

Yield

It yields 20-30 t ha^{-1}, if the crop is grown only for corm production. The corms production is adversely affected by removal of leaves.

Storage and Marketing

Removal of damaged corms followed by curing in sunlight for 4-5 days enhance the marketability and storability of corms and cormels. The tubers can be spread on sand and stored on sand under well ventilated conditions having low relative humidity.

6. Giant taro

Soil and Climate requirement

Alocasia grows well in medium to heavy soil with adequate drainage. Naturally, this plant is grown along water bodies *viz.*, river banks and canals, but it cannot withstand waterlogging. It performs well in sandy loam soil with proper irrigation facility. But it cannot survive long periods of drought.

As it is a strictly tropical tuber crop, temperatures below 10°C check the vegetative growth and tuber development. The areas receiving high and well-distributed rainfall with a temperature range 25-30°C, is found to well suited for its cultivation.

Propagation and Planting material

Suckers are used to vegetatively propagate *Alocasia*, but shoot tips with a few inches of stem or sections of stem having two or three buds can also be used as planting material.

Time of planting

The planting is done before the onset of monsoon *ie.,* from March to May.

Methods of planting

Generally, planting is done in hole; cormels are planted 8-15 cm deep while suckers are planted at 15-20 cm depth.

Spacing of planting

Generally, it is grown at spacing of 0.6 - 1.5 x 0.6 - 1.5 m. As an intercrop with *Dioscorea*, it is planted at the spacing of 3.5 x 3.5 m. As a sole crop, all the intercultural operations should be adopted as it requires. When it is intercropped with *Dioscorea* then the cultural practices given to *Dioscorea* is enough for it. In general, all the intercultural operations resemble with *Colocasia.*

Harvesting

Alocasia takes about one to one and half years to reach maturity, and the corms can remain unharvested up to 3 months after maturity.

7. Elephant foot yam

Soil and Climate requirement

A well-drained, well-aerated, fertile, sandy loam to sandy clay loam soil with a pH of 6 to 7, rich in organic matter is ideal for the crop. The growth of corms in heavy soil reduces the yield greatly. Compost or FYM or organic matter should be used as amendment, which make the soil friable and light and increases the yield. It cannot survive water logging.

The warm humid climate with an average temperature of 30-35°C and well distributed rainfall of 1000-1500 mm promotes better shoot emergence and vegetative growth whereas dry condition with comparatively less temperature (20-25°C) is favourable for corm development.

Propagation and Planting materials

Amorphophallus is propagated vegetatively by using either whole tubers or cut tubers. Whole tubers should be preferred over cut tubers to minimize rotting. Whole or cut tubers weighing 500-1000 g are used for commercial cultivation, whereas for seed production 100-150 g tubers are used which give rise to seed corms weighing between 500 and 750 g. Some south Indian cultivars do not produce daughter corm for further multiplication, hence in such cases, the mother corms of small size or bigger corms cut into pieces are used as planting material. Generally, high yielding, non acrid varieties should be propagated through cut tubers or small tubers, whereas in medium yielding, acrid varieties, daughter corms (cormels) are used for planting. To enhance the sprouting, planting materials should be dipped in thick cow dung slurry and then dried in shaded place while to break the dormancy tubers should be treated with thiourea (0.1%) or smoking in confined room.

Agrotechniques

Time of planting

February-March is ideal time for planting the crop in south India whereas March-April is best time in eastern and other parts of India with adequate irrigation facilities. Under rainfed condition, April-June with onset of monsoon is better for planting the crops. For eastern Uttar Pradesh, March is found to be the best time for planting.

Method of planting

Amorphophallus is planted in pits then raised as mounds on germination to obtain higher yield. Planting is done vertically in the pits at the depth of 10-15 cm. Pit size may be 35 cm x 35 cm x 35 cm or 60 cm x 60 cm x 60 cm depending upon the size of planting material.

Spacing and Seed rate

Spacing and seed rate depends on the size of planting material. Corms weighing 500g, 600g, 750 g and 1 kg should be planted in pits of 60 cm x 60 cm x 60 cm size at the distance of 50 cm x 50 cm, 60 cm x 60 cm, 75 cm x 75 cm and 90 cm x 90 cm respectively. For a seed production, the planting material weighing 100-150 g should be planted in pits of 35 cm x 35 cm x 35 cm size at a distance of 30 cm x 30 cm. In eastern U.P., planting on 30th March with 75 x 50 cm spacing and 200:50:200 kg NPK ha^{-1} is recommended. Before planting the pit should be filled with 1-2 kg FYM or compost. Seed rate varies from 12 to 20 t ha^{-1} depending on spacing and size of planting material. In Gujarat, seed size of 1 kg at a spacing of 75 x 75 cm with a fertilizer schedule of 150:60:150 kg ha^{-1} in two equal split doses, one as basal and another at 60 days after planting is recommended.

Seed treatment

Cut surface leads to development of several saprophytic organisms, causing decay of seed pieces, leading to poor germination. Cut pieces should be cured to encourage rapid tuberization and wound periderm formation on the cut surface. The seed pieces should be treated with Agallol-6 or Aretan-6 or Dithane M-45 @2.5 gm per litre of water or cow dung slurry with *Trichoderma* fungus for 20-30 minutes before planting to protect them from infection by soil born diseases.

Fertilizer requirement

The FYM or compost dose 25-30 tha^{-1} for elephant foot yam has been found better when given at the time of land preparation. The NPK dose 80:80:120 kg ha^{-1} along with FYM compost has been recommended for better growth and tuber production. Full dose of phosphorus along with half dose of nitrogen and potash is applied at the time of sowing, whereas remaining doses of nitrogen and potash is given again in two split doses, first 7-10 days after sprouting and second 1 month later. Several reports have been given on the NPK doses for elephant foot yam in Table.

The fertilizer / biofertilizer recommendation for different states.

States/Region	Recommendation
Bihar	150:60:150 kg NPK ha^{-1}
Kerala (as sole crops)	100:50:150 kg NPK ha^{-1}
Kerala (as an intercrop in coconut plantation)	53:40:67 kg NPK ha^{-1}
Chhattisgarh	100:60:80 kg NPK ha^{-1}
Andhra Pradesh(Telangana)	150:60:150kg NPK ha^{-1}
Andhra Pradesh(Kovvur)	250:60:250 kg NPK ha^{-1}
Tamil Nadu	80:60:100 kg NPK ha^{-1}
Uttar Pradesh	125:60:100 kg NPK ha^{-1} +50 kg N by FYM or vermicompost
Uttar Pradesh	200:50:200 kg NPK ha^{-1}
Gujarat	150:60:150kg NPK ha^{-1}
Maharashtra	150:60:150kg NPK ha^{-1}
Jharkhand	120:60:120 kg NPK ha^{-1}+ cowpea green manuring
West Bengal	120:60:120 kg NPK ha^{-1}+ cowpea green manuring or 150:60:150 kg NPK ha^{-1}

Irrigation

A light irrigation immediately after planting should be done to get uniform sprouting; subsequent irrigation should be given on the requirement of plants. Water logging should be avoided.

Weeding and Earthing up

Generally, 2-3 weedings are essential during early stage of growth until the spread of the leaves. Manual weedings after 5 weeks and 10 weeks of planting followed by earthing up with the application of fertilizers results in better growth and tuberization.

In Kerala, West Bengal and Andhra Pradesh, application of straw mulch 2 kg sq m^{-1} at the time of planting of elephant foot yam has been recommended for management of weeds in the crop. In Bihar, Sesamum leaf mulch in elephant foot yam is recommended.

Mulching

Mulching the pits with organic sources, immediately after planting, results in better sprouting, moisture conservation and reduction of the temperature around the corms.

Cropping system

Elephant foot yam has been found to grow under different cropping sequences in Northern, Eastern and Southern states.

Ideal crop sequences identified for different States

States/Region	Crop sequences
Eastern parts of India NEH Region	Elephant foot yam -Wheat/ Bengal Gram/Pea or elephant foot yam -Okra/*Colocasia*
West Bengal	Moong - elephant foot yam - sweet potato

Intercropping

Vegetable cowpea as intercrop in elephant foot yam is profitable for Uttar Pradesh, Bihar, Assam, Tamil Nadu and West Bengal giving and additional income of about Rs.50,000/- ha^{-1}. In Bihar, bottle gourd or bhindi (lady's fingure) is also cultivated as profitable intercrops in elephant foot yam. Inter cropping elephant foot yam in banana was recommended for adoption in Coimbatore. Elephant foot yam is identified as an ideal intercrop in coconut gardens in Kerala, litchi orchards of Bihar and mango orchards of Chhattisgarh.

Weed Management

- In elephant foot yam cultivation straw mulching the crop @ 2 kg m^{-2} was found to be effective in the weed management with enhanced corm yield at Kalyani, Thiruvananthapuram and Rajendranagar.
- At Dholi sesame leaf mulching or black polythene mulching was effective.
- Cowpea mulching after harvest as green manuring also helped to reduce weeds and increase yield.

Harvesting and Post harvest management

Harvesting

The crop can be harvested from 6-9 months after planting, depending upon market demand. Yellowing and drooping of the leaves is considered as maturity indices of the crop. Early harvesting considerably reduces the yield of corms, which can be compensated by higher market price.

Yield

The corm yield is greatly influenced by size of planting material and stage of harvesting which ranges from 50-100 t ha^{-1}.

Storage and Marketing

Any type of injury during harvesting reduces the storability and marketability of corms. After harvesting, damaged corms should be removed and it should be spread in a shaded place for 3-5 days. Corms can be stored in well-ventilated rooms or under dry soil.

8. Coleus / Country potato

Soil and Climate

A fertile well-drained, sandy loam to alluvial soil with adequate organic matter is ideally suited for its cultivation. Water logging reduces tuber yield considerably. The best soil pH requirement ranges between 6.5-7.0.

Coleus grows well in hot-humid climate but the subtropical and hot temperature areas is also suitable for its cultivation. A comparatively lower temperature at night than day time favours better tuber development. Coleus requires an evenly distributed rainfall for its cultivation and cannot withstand drought conditions nor frost.

Propagation and Planting material

Coleus is vegetatively propagated through suckers raised from healthy tubers of previous harvest or stem cutting or tubers. About 0.5-1.2 tonnes of tubers are planted in 0.2 ha of land to produce enough planting materials (suckers) for one hectare crop.

The healthy tubers are planted at 5 x 15 cm spacing and 4 cm depth in well prepared nursery beds of 15 cm height and 1m widths. These tubers start sprouting within 15 days, which grow to a height of 15-20 cm in about three weeks and can be readily transplanted to the field. Stem cutting of 15-20 cm length from these sprouts are also used as planting material.

Time of planting

Coleus is mostly grown as a monsoon crop during August-October. It can also be planted in January and September as an irrigated crop. Under Hyderabad conditions the optimum time of planting was found to be between June and August but planting can be done from 15th June to 1st September. For UP best planting time is June to July end.

Methods of planting

There are three different methods, generally practiced for planting coleus;

(a) *Ridge method :* Stem cuttings of 15-20 cm length with 5 leaves are planted on ridges at 15-20 cm spacing.

(b) *Coiled planting method:* Stem cuttings of 22 cm length are used and about 12 cm of the mature portion is coiled and placed in holes of about 7 cm diameter and 5 cm depth.

(c) *Horizontal planting methods :* Two pieces of 30 cm long stem cuttings are placed side by side across the ridges in opposite directions. Planting pairs of cutting at 7-10 cm apart requires a lot of planting material, whereas dense planting produces smaller tubers but suppressed weeds. Among these three methods, coiled planting gives highest tuber yield.

Spacing and Plant population

The planting space varies according to the methods of planting as mentioned above. The optimum spacing for coleus cultivation is 60 x 30 cm, which accommodates about 55,500 plants/ha.

Irrigation

It is essential to give irrigation immediately after planting to ensure adequate soil moisture for proper establishment of stem cuttings. During the tuber development phase proper moisture must be maintained to encourage better growth and tuber development.

Weeding and Earthing up

A weeding along with earthing up and top dressing of fertilizers is done at about 3 weeks after planting to cover a portion of the stem with soil and to improve the physical conditions of soil, and promote better tuberization and tuber development. The second weeding and earthing up should be done at about 6 weeks after planting.

Manures and Fertilizer/Biofertilizer

About 10 tonnes FYM or compost must be added at the time of field preparation. In addition to FYM, 60 kg N, 60 kg P_2O_5 and 100 kg K_2O are applied for one hectare of land. Half dose of nitrogen and potassium along with full dose of phosphorus should be applied at the time of planting. Another half dose of nitrogen and potassium are to be applied along with second earthing up. In Tamil Nadu, 60:60:150 kg NPK ha^{-1} is recommended.

Harvesting and Post harvest management

Harvesting

Coleus is a short duration crop and can be harvested in about 5-6 months. With the maturity of tubers, leaves start falling from stems, which acts as an indicator of crop maturity. Harvesting should not be delayed as the mature tubers decay rapidly when left in the soil.

Yield

The tuber yield varies from 10-20 t ha^{-1} depending on the management practices.

Storage

The tubers can be stored in well-ventilated room after curing for 2-3 days under shade condition.

9. Yam bean

Soil and Climate requirement

Yam bean requires fertile well-drained, sandy loam soil of pH 6.0-7.0. It can do well in loamy and clay loam soil with good drainage and adequate humus content.

Warm (25°C-30°C) and humid climate is needed for initial vegetative growth, but cool night temperature of 18-20°C along with dry sunny days is required for tuber bulking and development. Cool climate during early growth period adversely effects the initiation of tuberization and also results in a prolonged vegetative phase. Yam bean requires long (14-15 hours) photoperiod for good vegetative growth, whereas shorter days are required for better tuberization.

Propagation and Planting materials

Yam beans are propagated usually through seed. The mature pods have about 8-10 brownish-yellow to red seeds, which contain a toxic substance 'rotenone' and are harmful for grazing cattles.

Time of planting

In North-Eastern India, June-July with the onset of rain is the ideal time of planting and for late sowing, August-September is ideal.

Methods of planting

Usually 2-3 seeds are placed on ridges made at the distance of 30 cm to a depth of 2 cm.

Spacing and Seed rate

Timely sown : 30 x 30 cm and Late sown : 30 x 15 cm or 15 x 15 cm.

Close spacing gives maximum yield but small tubers, which are free from cracks. Seed rate varies from 20-60 kg ha^{-1} depending upon the time of sowing, spacing and the purpose.

Irrigation

Rainfed crops, sown in June-July do not require irrigation but supplementary irrigation is essential for crops sown in August-September.

Weeding and Earthing-up

Weed infestation is maximum in June-July sown crops compared to September sown corp. One weeding is essential at the initial stage to check the weeds and promote the crop growth. First weeding at 40 days, along with nutrient application results in better growth and tuberization. Second may be done one month later if there is any need. Earthing up is not much effective in increasing tuber yield, but a light earthing-up around the stem gives better results.

De-flowering

Yam bean, normally, flowers at 75 days after sowing. Removal of flowers without allowing the plants to bear pods, results in better tuber yield and better quality. There is a significant negative correlation between tuber yield and pod formation. The flowers are manually removed or the flowering can be successfully prevented at bud initiation stage (60-70 days).

Manures and Fertilizers / Biofertilizer

FYM or compost @ 10-20 t ha^{-1} should be applied at the time of land preparation. In addition to this, the fertilizer dose of N:P:K@80:40:80 kg ha^{-1} is recommended for getting optimum yield. The full dose of P and K is applied as basal dose at the time of planting along with half dose of nitrogen. The rest half of N is applied as a topdressing 40-25 days after sowing along with weeding and earthing up.

Manure and fertilizer recommendations for Yam bean in different states of India

States	N (kg ha^{-1})	P (kg ha^{-1})	K (kg ha^{-1})	FYM (t ha^{-1})
Tamil Nadu	80	60	80	10
West Bengal	80	80	80	10
Bihar	80	40	80	10

Cropping sequence

In Bihar, maize-yam bean-onion and maize-yam bean-wheat-green gram cropping sequence is followed. Pigion pea or maize is grown as intercrop in yam bean in Bihar.

Harvesting and Post harvest management

Harvesting

Yam bean will be ready for harvest 150 days after sowing. The crops sown in June-July is usually being harvested in December-January, but can be harvested up to January-February. whereas those sown in August-September can also be harvested in December-January, which results in small-sized tubers free of cracking, and can easily fit in multiple cropping system.

Yield

The average yield of local cultivars varies from 18-20 t ha^{-1}, whereas Rajendra Mishrikand-1 yields 35-45 t ha^{-1}.

10. Winged Bean

Soil and Climate

Winged bean has no specific requirement of soil. It can withstand moisture to a reasonable extent and can be grown in heavy soil with poor drainage. For tuber production, well drained sandy loam or sandy soil is ideal. It is grown for various purposes in different types of soil like a green manure, cover crop, fodder crop and restorative fallow crop.

It thrives well in warm humid regions with partial shade. The plants require short day conditions for flowering and planting during short days would result in flowering within 8 weeks. The higher temperature (30°/22°C) promotes vegetative growth but inhibits tuberization, whereas the optimum temperature for tuberization is 24°/13°C.

Method of planting

The seeds are sown either in flat beds or ridges, the latter gives better tuberization. The seeds normally take about 10 days to germinate; the pre-soaking of seeds for 2 days before planting enhance germination per cent and induce early germination (within 5-6 days). Time of seed sowing is June-July as a rainfed crop.

Spacing and Seed rate

The spacing varies with respect to varieties purpose of growing and soil fertility. The optimum spacing recommended for better growth and yield vary from 60 x 60 cm to 100 x 60 cm. Generally 50-60 kg of seeds are sufficient for planting one hectare land.

Irrigation

The field must be irrigated. Lack of sufficient soil moisture may destroy the seedlings.

Weeding and Earthing-up

To keep field weed free and for proper tuber development, weeding and earthing up around the plant after 3 months of planting is essential.

Fertilizer requirement

N:P:K @40:100:40 gives better pod or tuber yield. Being a leguminous crop it does not require much amount of nitrogen, whereas K has only influence on initial stage of tuber development.

Harvesting and Yield

The green pods are ready for harvesting 10 weeks after sowing. Each vine yields about 25 pods every 5 or 6 days. The green pods yield varies from 1.7-3.9kg/1.5 sqm (10-25 t ha^{-1}), whereas the yield of tubers varies from 2.5-6.0 t ha^{-1}.

Selected References

Chadha, K. L. and Nair, G.G. 1994. *Advances in Horticulture Vol. 8. Tuber Crops* (Eds.). M.P.H., New Delhi. 714 pp.

Chadha, K.L. 2001. Hand Book of Horticulture. ICAR, New Delhi. 1031 pp.

Ravi, V and Mohankumar, C. R. 2004. Cassava based Multiple Cropping systems. *Horticultural reviews* (John Wiley & Sons, Inc) 30 : 355-500

Palaniswami, M.S. and Shirly Raichal Anil. 2006. Region specific technologies for tropical root ant tuber crops in India. All India Coordinated Research Project on Tuber Crops, CTCRI, Thiruvananthapuram. 67 pp.

Varma, S. P., Rajendran, P.G and Byju. G. 1995. Research highlights (1968-1995) AICRP Tuber Crops.. CTCRI, Thiruvananthapuram. pp. 27

Varma, S. P., Ravi,V and Suja, G.1996. Yam bean, Coleus, Arrowroot, *Colocasia* (Dasheen) and *Xanthosoma*. CTCRI, Thiruvananthapuram. 26 pp.

Varma, S.P., Suja, G., Rajendran, P.G. 1997. Production technology for cassava and sweet potato. CTCRI, Thiruvananthapuram. 24 pp.

❑❑❑

8

Pests and Diseases : Their Management and Plant Quarantine

Tropical tuber and root crops are subjected to the depredation of a number of insect and non insect pests. Comprehensive studies revealed that cassava is attacked by a wide range of arthropod and rodent pests. An exhaustive global account of cassava pests and their management was given by Anon (1978), Bellotti and Schoonhoven (1978) and Bellotti *et al.* (1999). Lal and Pillai (1982); Pillai and Palaniswami (1984), Pillai (1994) and Palaniswami (1999) reviewed the pests of cassava in India. Anon (1978), Sutherland (1986) and Jansson and Raman (1991) covered world sweet potato pests. Palaniswami (1999) detailed the pests attacking sweet potato in India. Pests of edible aroids, yams and chinese potato were comprehensively reported by Anon (1978); Mitchell and Maddison (1983) and Palaniswami (1994, 1999). In India about 34 species of insects and non-insect pests infest cassava, 80 species attack sweet potato, 46 species on edible aroids, 21 on yams and 10 on chinese potato. One of the main factors for the low productivity of these crops in India is attributed to the pest complex attacking them (Chadha, 2001).

Tropical tuber and root crops are also attacked by several diseases caused by fungi, bacteria, viruses and mycoplasma like organisms causing yield reduction in the crops. In India more than 80 diseases have been reported on these crops while the global figure is more than 200 (Anon, 1978; Thankappan, 1994; Thresh *et al.*, 1998; Kareyija *et al.*,1998). However, many of these diseases are of minor importance causing less damage and are limited in geographical distribution, while some are of considerable importance and cause economic damage.

Among all the diseases reported in India, only few diseases like cassava mosaic disease, tuber rot of cassava, leaf blight of *Colocasia* and collar rot of elephant foot yam are of major importance as they cause considerable reduction in the yield. Yield reduction of cassava mosaic disease (25-80%), leaf blight of *Colocasia* (25-50%) and collar rot of elephant foot yam (20-100%) varies depending upon the nature of cultivars and influence of various weather factors. The important viral diseases of tuber crops reported in India includes cassava mosaic (Indian cassava mosaic geminivirus (ICMV)), sweet potato feathery mottle (sweet potato feathery mottle potyvirus (SPFMV)), sweet potato mosaic (sweet potato mosaic potyvirus (SPMV)), dasheen mosaic on taro and tannia (dasheen mosaic potyvirus (DaMV)), *Amorphophallus* mosaic (*Amorphophallus* mosaic potyvirus (AmMV)) and yam mosaic potyvirus (YMV). Among these viruses Indian cassava mosaic virus is transmitted by whitefly, while others are by aphids. With the increase in the evolution of new hybrids and with the International exchange of genetic material, the number of diseases and pests are also on the increase in an alarming rate and in certain pathogens and pests new virulent strains are also developing. As these crops are vegetatively propagated and many of the diseases are spread through the planting materials / tubers, care should be taken during the import of seed materials from other countries so as to avoid introducing the non existing pathogen in to our country.

1. Sweet Potato

1.1 Pest of sweet potato

1.1.1 Sweet potato weevil

Among the pests that attack the crop, the most serious and ubiquitous one is the weevil *(Cylas formicarius* F.) which originated in India and was first recorded in 1792 from Tranquibar near Pondicherry. Even though many pests are recorded on this crop, none of them will come near to the weevil in inflicting direct damage, yield loss and at times total destruction of the crop. Both adults and grubs cause damage to the crop both in field and in store. Adult weevil feeds on all parts of the plant while1 the attack of the grubs is restricted to vines and tubers. Economic damage is caused by grubs. The feeding causes a characteristic terpenoid odour and a bitter taste on cooking, thus rendering the tubers unsuitable for human consumption. Other related species like *Cylas brunneus* F., *C. puncticollis* Boh. and the West Indian weevil *Eusceps postfasiatus* De Geer, which are considered to be important elsewhere, have so far not been recorded in India (Jannson and Raman, 1991). The weevil lays eggs on vines and maintain its population in the vines before tuber formation. The grubs bore and feed by making irregular tunnels. As a result of feeding by a number of grubs, the collar region usually bulges out due to hypertrophy and proliferation of tissues. At times of severe damage the region breaks off from the root zone. The emerging adult weevils get access into the tubers through the cracks and crevices formed in the soil during bulking. The pre oviposition, oviposition and post oviposition periods were 7.8, 81.3 and 4.0 days respectively. A single female on an average laid 180 eggs in its life span. The adult longevity is 95.8 days for female and 112.6 days for male. According to Jayaramaiah (1975) it was 90.5 and 91.1 days respectively. Subramaniam *et al.* (1977) observed 61 to 120 days adult longevity for males and 81 to 107 days for females. However the longevity was found higher in males than in females. Sex ratio found was 1.5 : 1 for male to female (Jayaramaiah, 1975). Seven to nine overlapping generations are completed in an year.

In India the yield loss assessment due to weevil damage ranged from 17 to 70% at Coimbatore (Subramaniam *et al.*, 1977), 16 to 40% in Bangalore (Jayaramaiah, 1975) and 27 to 82% in Trivandrum (Rajamma *et al.*, 1993). In farmer's field in Kerala the damage ranged from 0 to 22 in lowland and 4 to 50% in upland .The annual yield loss due to weevil damage in Kerala is estimated at Rs. 96 lakhs (Palaniswami *et al.*, 1991). The damage in Orissa, Bihar, U.P. and West Bengal varied from 25 to 45, 5 to 50, 7 and 5 to 11% respectively. In Orissa, the tribal people harvested tubers after five months and in such condition about 50% tuber damage was observed. In some villages of Bihar vine damage was more than 50% while the tuber damage was less than 5% indicating lack of relationship between tuber damage and vine damage. Yield loss due to weevil damage also varied from season to season, soil to soil, age of the crop, moisture and temperature of the soil. The population build up of weevil and tuber damage at different ages of the crop during the three seasons revealed that mean population increased from 2.5 at 30 days to 22.4 per plant at 120 days maturity in June to September season. The tuber damage at 105 days was 50% which increased to 72% at 120 days. In October-January season the weevil population increased from 2 per plant at 30 days to 27.3 at 120 days and the tuber damage 61% at 105 days and 82% at 120 days maturity (Rajamma *et al.*, 1993). Generally the weevil damage was higher in summer crop. There was a progressive increase in the weevil incidence and damage as the age of the crop increased. The crop grown in red loam and clay loam had higher incidence than that of sandy loam probably due to cracking nature of the soil (Palaniswami *et al.*,1991). In North India the crop raised in kharif and rabi seasons showed varying weevil damage. In rabi, the weevil damage was lower while it was higher in the kharif; this phenomenon can be mainly attributed to low soil temperature.

Management of sweet potato weevil

Before the advent of modern insecticides, clean cultivation like burning of plant residue after harvest, destruction of collateral hosts and cultural methods like deep ploughing of the field, earthing up of the crop etc., were adopted for the control of weevil. Even now these methods hold good to suppress weevil damage.

Alternate hosts

Quite a large number of Ipomoea species *viz., Ipomoea purpuria, I. prescapare, I. lateralis, I. trifida, I. trichocarpa, I. panduraria, I. biloba, I. graoilis, I. cepioria, I. learil, I. palmata, I. barlairiodes, I. acarica, Thumbergia* sp. and *Calonyction aculeate* were recorded as collateral hosts for sweet potato weevil in India. Even if sweet potato is not in the field, the weevils can survive on these collateral hosts. The destruction of collateral hosts may help to reduce population build up of weevil. Plant residues in the field often serve as the source of inoculum and the disposal of residues after harvest is therefore, essential to save the next crop from damage (Palaniswami *et al.,* 1991).

Planting material

Planting material is the primary source of infestation in many situations. The mature vines or at times over-mature vines used for planting are sure to contain different stages of weevil and the infestation thus starts from the very beginning leading to poor establishment of the crop and significant yield reduction. Planting materials must be selected from the nursery of 30-45 days old plants raised from pest free materials. Again only the tender portion of the vine is to be used for planting and mature basal portion is to be discarded.

Cultural methods

Adoption of different cultural methods like earthing up the crop, using deep rooting varieties, providing frequent irrigation, mixed cropping and crop rotation were found effective to reduce weevil damage on sweet potato. The adult weevils enter into the tubers through the basal portion of the plant and also through the cracks and crevices formed in the soil surface during tuber bulking especially in the case of shallow rooted varieties. Earthing up prevents the entry of weevils into the tubers through cracks and crevices and also makes the tubers deeper. Palaniswami and Mohandas (1991) and Palaniswami (1999) reported reridging or hilling up the mounds around 50 DAP to minimize the intensity of damage. When soil was sufficiently moist weevil damage was found decreasing. Trehan and Begal (1957) reported that irrigation at 10 days interval with heavy manuring resulted in highest yield and lowest infestation in Maharashtra.

While screening the varieties for weevil resistance, it was found that tuber damage was less in deep rooting varieties having the neck length of more than 10 cm. The varieties having short neck of less than 2.5 cm were found as highly susceptible as the weevil could easily enter into the tuber from the plant base. In short necked varieties the shape of tuber is more likely to be globular and as such more surface area is superficially exposed for the weevil to feed and breed. In deep rooting and long necked varieties the tubers are mostly elongate or spindle shaped and the weevil cannot that easily enter into the tuber from the plant base as long necked and deep rooting nature afford physical barrier. However, when there is great residual population of the pest in the soil (in plant debris) due to repeated monoculture, the long necked nature may not help much, as the weevil can directly reach the tubers from soil (Rajamma *et al.*, 1993).

Crop rotation

There was significant reduction in the population build up of weevil on sweet potato when crop rotation was adopted. The tuber damage on sweet potato under different rotations in three seasons (June-September, October-January and February-May) ranged from 8 to 19% while in monocropping (control) the damage was as high as 80% with very low mean tuber yield of less than 2 t ha^{-1}. The weevil population recorded in 5 kg of tubers taken from each treatment during harvest showed that the population ranged from 35 to 173 in different treatment against a mean of 382 in monocropping. The rotation like (i) Paddy-paddy-sweet potato (ii) Paddy-sweet potato-paddy (iii) Paddy-sweet potato-cowpea and (iv) *Colocasia*-sweet potato were highly superior to sweet potato-sweet potato-sweet potato (control) in reducing tuber damage and increasing marketable tuber yield (Rajamma *et al.*, 1993). In Bihar the sweet potato weevil damage was lower in intercropped sweet potato than in sole crop (Palaniswami, 1999).

Effect of organic materials against weevil infestation

Several organic substances reported to have repellent or insecticidal action were tried against sweet potato weevil in India. Soil application (before planting) of organic materials like *Calophyllum* cake and Mahwah cake (*Madhuka indica*) @2500 kg ha^{-1}, leaves of lemongrass (*Cymbopogan flexuosus*), *Clerodendron infortunatum* and *Chromolaena odorata* @ 5000 kg ha^{-1} were found effective to reduce weevil infestation (Johnson *et al.*, 1979, Rajamma, 1982; Rajamma *et al.*, 1993). Mulching with *C. odorata* @3000 kg ha^{-1} at 30 DAP at the base of plants or *Clerodendron* was found to reduce further weevil incidence by their repellent action and conservation of soil moisture (Palaniswami, 1999).

Chemical control

Insecticidal application was the most widely recommended control measure against sweet potato weevil. Spraying the crop at monthly intervals with fenitrothion 0.05% or methyldemeton 0.05% was highly effective to reduce weevil on sweet potato. There was 80-86% increased production of marketable tubers in fenitrothion treatment (Palaniswami,1999).

Subramaniam *et al.,* (1977) found that three application of fenthion 0.1% or carbaryl 0.1% at triweekly interval from the time of tuber formation reduced weevil infestation. Elaborate field experiments were done at CTCRI using eleven insecticides (E.C and Granules) and four soil amendments. Spraying the crop first after a month of planting and subsequently at an interval of three weeks with 0.05% conc. of fenthion or fenitrothion or monocrotophos (500 g a.i ha^{-1}) was found highly effective in reducing weevil infestation and to increase the marketable tuber yield. The residue analysis of the tubers after harvest showed that the residue was nil in monocrotophos, 0.09 ppm in fenitrothion and 0.07 ppm in fenthion which were far below the tolerance limit. Soil application of phorate of carbofuran (1.5 kg a.i ha^{-1}) at 45th day of planting also reduced weevil infestations. Soil drenching with any one of the chemicals *viz.,* fenthion, endosulfan, quinalphos, monocrotophos and fenitrothion 0.05% at 50 and 70 days after planting was found effective to reduce weevil infestation. The tuber damage was brought down from 44.4% in the control to 14.4 - 9.5% in the above treatments. Cost-benefit ratio was highest in fenthion (1:5.7) followed by endosulfan and fenitrothion. Cost-benefit ratio was lowest in monocrotophos due to its higher price (Rajamma, *et al.,* 1993). Residues of the above chemicals were below tolerance limit.

Different insecticides reported as effective for the control of weevil in various coordinated centers were selected for final evaluation and also to find out the cost-benefit ratio. Accordingly fenthion, endosulfan, fenitrothion and quinalphos each at 500 g a.i. ha^{-1} and carbayl @ 1 kg a.i. ha^{-1} (per spray) were tried against weevil for three seasons in 1983-84 adopting four spraying. The mean tuber yield in different chemical treatments ranged from 12 to 17.4 t ha^{-1} against 7.4 t ha^{-1} in the control. Though all the treatments were effective in reducing damage and producing more marketable tubers, carbaryl was less effective than other chemicals. Lowest damage was observed in fenthion treatment. The cost benefit ratio was highest in endosulfan (1:6.2) followed by fenthion, fenitrothion and quinalphos. Thus the efficacy of these chemicals reported by different centers for the control of weevil has been reconfirmed (Johnson *et al.,* 1979; Rajamma, *et al.,* 1993).

Resistant varieties

Isolation of clones that are resistant or less susceptible to weevil has been one of the important lines of work. Clones with long neck and deep rooting nature escape severe damage. Some of the clones with moderately long stalk and deep rooting nature *viz.*, S 3, S 13, S 234, S 238 and S 248 showed lower susceptibility to weevil. The resistance can be considered only in relative terms in sweet potato as purely resistant genotypes are not available in India or elsewhere. The cultivation of less susceptible or relatively resistant clones under good agronomic practices will help to reduce weevil damage leading to increased production (Palaniswami and Mohandas, 1992; Rajamma, *et al.*,1993).

Biological control of *Cylas formicarius*

Different parasites, predators and pathogens have been recorded on different stages of weevil in India. Two hymenopteran ecoparasitoids (*Rhaconotus* sp and *Bracon* sp) and a dipteran predator (*Drapetis* sp) were identified on grubs. Forty percent of the grubs collected from aerial vines were found parasitized by *Rhaconotus* sp and *Bracon* sp. *Metarhizium anisopliae*, a commonly prevalent fungus in the field was found to infect the adults. Weevil could be infected by *Bacillus thuringiensis* (60-64%). Among the biotic agents attacking the weevil, larval mortality factors were maximum and the parasitoids itself contributed upto 45 per cent, while others contributed only 5 per cent. *Rhaconotus menippus* and *Metarhizium anisopliae* were identified as potential ones in the biological control of *Cylas formicarius*. An entomopathogenic nematode was found frequently occurring on the weevil in the field. It was identified as *Heterorhabditis indicus* (Palaniswami, 1994, 1999, 2000).

Parasitoid

Rhaconotus menippus Nixon is an endemic solitary ectoparasitoid of *C. formicarius*. The field parasitism is restricted to grubs found in the vines and exposed roots. Adult parasitoids are honey brown in colour. Females can be identified by its long ovipositor. The egg to adult developmental period ranged from 13 to 23 days. The incubation period was about two days. Eggs are laid singly on the posterior side of grubs. Newly hatched larva is white and apodous and it turns light pink at maturity. Larval period ranged from 5 to 11 days. When full grown, the larva stops feeding and entered prepupal phase (one day) after which it finally moults and becomes pupa. By this time, the host larva becomes completely shrunken and black due to sucking and desapping by the parasitic grub. The pupation is in a dirty white papery cocoon, which is usually found attached in the larval galleries. Adult emerged in 6 to 9 days. The male-female sex ratio was 1:2. Adult longevity varied depending upon food availability. Adults without food lived for two days but when food was available, males lived for 7 to 18 and females for 11 to 30 days. Pre-oviposition period of two days was noted. Mated females oviposited after paralyzing the host larva. Oviposition period lasted for 27 days. Females preferred 10 to 15 days old host grubs and did not oviposit on early instars and pupae. In the laboratory, adult females occasionally found to feed on the body fluid oozed out due to ovipositor stingings, but not such feeding was observed by males. A mated female could parasitize up to 36 host grubs during its life time. At times, a female could oviposit 2-5 host grubs in a day and at times none. Usually 15-25 days after oviposition by *C. formicarius,* i.e., 10-15 days old grubs, are used for maximum parasitization.

The field parasitism ranged from 2 to 45 per cent. Higher larval parasitism was observed during July-October 20-36 per cent) with a peak during August. The population assessment of *R. menippus* in field revealed the prevalence of the parasitoid in the range of 4 to 30 numbers per ten plants. Pruning of sweet potato crop affected the population of the parasitoid significantly. The population was also found affected by insecticidal spraying. The fields without any treatment had a higher population (24-38 numbers per ten plants), while it was nil in the case of pruned and insecticidal sprayed fields and 4-8 in the granular insecticides applied fields.

Cylas formicarius infested vines and tubers can be used for mass rearing of parasitoid. More or less same old age host grubs reared in the vines or tubers are made available for parasitization. Usually 15-25 days after oviposition by *C. formicarius,* i.e., 10-15 days old grubs, are used for maximum parasitization (Palaniswami, 1994, 1999).

Palaniswami *et al.* (1991) observed the parasitoid in all the upland and low land cultivations of Kerala during the survey and reported that soil drenching with fenthion or endosulfan (0.05 per cent a.i) did not significantly influence parasitoid population in the field.

Entomopathogenic nematode *(Heterorhabditis indicus)*

H. indicus infected grubs, pupae and adults of *C. formicarius* found in sweet potato crop. The infection varied from 20-90 per cent depending upon season and soil condition (Ekanayakae, 2000; Palaniswami, 2004).

Entomopathogenic fungus (*Metarhizium anisopliae*)

Metschnikoff described this fungus from the beetle *Anisopliae austriaca* Herbst. *Metarhizium anisopliae* belongs to 'Fungi Imperfecti'. The disease caused by the fungus is commonly known as 'green muscardine'. The fungus has been reported to infect more than 200 insect hosts. Mode of infection is through outer integument, digestive tract, respiratory openings or through wounds. Infection unit is conidium, which on germination produced an infection peg called 'appressorium' and enters host by mechanical pressure or by enzymatic action or by both. It is facultative internal parasite. Infected adults showed loss of appetite, sluggishness, gradual loss in function resulting in general or partial paralysis and dissolution of various systems in the body. Fungal growth is observed at jointed parts and is dull white in colour. When adults are treated with spores of fungus in the laboratory, the mortality was observed from fourth day onwards. Maximum mortality of adults was noted on eighth day (63 per cent). Soil drenching with fungal spores (3×10^9 spores per ml) was found to be encouraging. The number of adult emergence (30 days after harvest), in the treated field was almost one-third that of the control. The recovery of fungus could be made both from the dead weevils as well as from soils. The micro-climate of sweet potato crop is highly favourable for the growth of the fungus which requires high humidity and low temperature. Release of the parasitoid and soil drenching of the fungus at 50 DAP and a reridging at 60 DAP were quite effective in containing the SPW (Palaniswami, 1994).

Sex pheromone

The synthetic sex pheromone for the control of SPW was tried and Z(3)-dodecen 1-ol-E(2) butenoate pheromone attracting males was standardised. One mg was useful for mass trapping of male adult weevils. 10 μg was enough for monitoring and the results were spectacular. Field experiments were conducted for two crop seasons (June-September and November-February) and the pheromone traps were installed @ 1 trap /100 m^{-2} area from first day of planting till harvest and there after for 1 month to collect the residual population. The male weevils were collected and killed in alternate days from the traps. During June-September season nearly 12,000 male weevils were collected from the experimental field of 400 m^2 (within 15 weeks) and 8300 weevils in November-February season. The residual populations in June-September and November-February were 6300 and 1500 respectively. In the experimental fields, the tuber damage was less than 10% in both the seasons while in the control the damage ranged from 33 and 39% respectively. The population build up of the weevil was drastically reduced in the experimental field by continuous elimination of male weevil. The mean weevil population/

stem and tuber were 1.3 in the treatment and 9.5 in the control. Thus synthetic sex pheromone has been found as an efficient tool to contain the population build up of weevil and consequent damage of the crop (Rajamma *et al.*,1993; Palaniswami and Chattopadhyay, 2006).

Kairomone

Tubers are attractant to adult weevils. Boehmeryl acetate has been identified as the attractant. More females were found to be attracted by the kairomone (Palaniswami *et al.*, 2000). Use of kairomone and sex pheromone in the IPM has been reported as useful against the weevil (Palaniswami and Chattopadhyay, 2006).

Integrated Pest management for sweet potato weevil

Sweet potato weevil incidence is a complex phenomenon influenced by several factors such as seasons, type of soil, soil moisture, age of the crop, varietal susceptibility and cultural practices including crop sanitation, crop rotation etc. Therefore, a proper technology of integrated pest management utilizing mechanical, cultural, chemical and biological methods in a systematic fashion would be an ideal one for the pest management. Thus we can integrate the effective mechanical and cultural practices, use of resistant/tolerant variety and or chemical control with due consideration to the parasitoids, predators and pathogens of the weevil. Integrated approach for the control of weevil using a new sex pheromone developed at BARC was found to be cost effective and sustainable.

- Collection and destruction of crop residues.
- Removal of *Ipomoea* species weeds which will otherwise serve as alternate hosts.
- Selection of pest free planting materials from nursery.
- Proper earthing up whenever there is soil crack especially after tuber bulking.
- Basal application of mahwa cake or calophyllum cake @2500 kg ha^{-1} or leaves of *Clerodendron* or *Chromolaena* @5000 kg ha^{-1}.
- Adoption of crop rotation utilizing only one sweet potato crop in the sequence.
- Retention of soil moisture in the field.
- Use of resistant/tolerant varieties.
- Encouraging the naturally occurring biotic agents.

- Installation of pheromone traps in the field from first DAP to trap and kill the male weevil at a distance of 10 x 10 m using BARC sex pheromone. Continue trapping till a fortnight after harvest.
- Harvesting at 110 days
- Destruction of weevil infested crop residues.

This IPM programme resulted in the increased production of marketable tubers (over 53%) (Palaniswami, 2000; Palaniswami and Chattopadhyay, 2006). Onfarm trials of IPM using the suggested components indicated drastic reduction of sweet potato weevil damage in tubers and enhanced marketable tuber yield. AICRPTC trials at Hyderabad, Coimbatore, Bhubaneswar, Kalyani, Dholi, Dapoli, Ranchi, Jorhat and Faizabad clearly gave enhanced marketable tuber yield with very low weevil incidence (Palaniswami and Shirly, 2006).

Yam bean has been identified as barrier crop to the weevil. Yam bean as border crop in sweet potato or as an inter crop in between two rows of sweet potato has helped in the reduction of weevil damage in the crop.

1.1.2 Vine borer (Omphisa anastomosalis)

Generally wide spread and is of sporadic nature. White stout caterpillars bore into the vines causing terminal drying and wilting. Larvae occasionally bores into tubers. Larval period is about 28-50 days. Egg and pupal periods are about one and two weeks respectively. The adult moth has straw coloured wings with wavy markings. Crop hygiene and crop rotation reduces the incidence. Spraying 0.05% fenthion, fenitrothion or endosulfan in the case of severe infestation is advocated. Another borer *(Metasia coniotalis)* is also found in India (Pillai and Palaniswami, 1984).

1.1.3 Defoliators

Leaf folder (Brachmia convolvuli Wals.)

Cream and black larvae spin webs close to the main veins of the leaves, on the underside, while older larvae spin webs across the upper surface of the leaf and draw the two sides together. The leaves eventually dry up, die and drop off. The adult is a small slender moth greyish brown in colour. It lays eggs in small groups at the base of the radiating veins on the underside of the leaves. They hatch in about three days. The larvae feed on the leaves and become fully grown in about two weeks undergoing five instars. It pupates within the leaf fold for about a week. Sprays of fenitrothion control the pest infestation. *Brachmia engrapta* Meys., *B. arotraea* Meys and *B. macroscopa* are also reported to cause similar damage in India (Palaniswami, 1999).

Horn worms (Agrius convolvuli)

Horn worms are voracious eaters of foliage. Eggs are spherical and laid singly on the upper surface of the leaf. The horned caterpillar is dark brown or green in colour with oblique red yellow strips laterally. There are five larval instars and larvae mature in 3-4 weeks and grow up to a length of about 100 mm. Pupation is in the soil and may take 3 weeks to several month depending upon the climate. The moth has pale-grey wings and has violet bands on the abdomen. Ploughing the land between crops to expose the pupae thus reducing infestation. Hand picking of the larvae is also found effective in small areas. It is only a sporadic pest. *Theretra oldenlandiae* and taro horn worm *Hippotion celeria* also feed on sweet potato (Subramaniam *et al.*,1977).

Bihar hairy caterpillar (Diacrisia obliqua)

It is a sporadic pest. Numerous larvae covered in yellow hairs skeletonise the leaves, entire vines may become defoliated. Eggs are laid on the leaves in rows or clumps. About 600 eggs are laid by a single female. Egg period is 4-9 days, larval period is 28-34 days and pupal period is 16-22 days. Moths are pinkish buff coloured with black dots on wings. The pest hibernates in pupal stage in India. The larva is parasitized by *Apantiles obliquae* and *Sarcophago unisera*. Endosulfan or fenitrothion spray 0.05% is suggested (Palaniswami, 1999).

Syntomid caterpillar (Euchromia polymena)

Hairy caterpillars feed on the leaves. The adult is a colourful moth measuring 4.2 cm in wing span and 1.6 cm in body length. Wings are black with golden yellow patches and the body has black-blue and crimson stripes. Eggs are laid in masses of 80-100 on the underside of leaves. They hatch in 8-10 days. The hairy caterpillars grow feeding on foliage and become full grown in 3-4 weeks undergoing six instars. The fully grown caterpillar is 35 mm long. It pupates in a cocoon of silk and felted hairs attached to the stem. Moth has a longevity of 15-17 days (Subramaniam, *et al.*, 1977).

Tortoise beetles

Several species are reported as foliage feeders in India. *Aspidomorpha miliaris, A. indica, A. furcata, A. dissentanea, Chirida sexnotata, Metriona circumdata, M. varians, Cassida indicola, Glyphocassis trilineata, Chiridopsis ornate* and *Conchyloctaemia punctata*. Adults and larvae feed on the leaves making large round holes and eventually skelentonising the leaves. Eggs are laid on the under side of the leaves in batches. The larvae are flattened, spiny and usually carry excreta and the skins of earlier instars on anal projections. Adult tortoise beetles are broadly oval in shape. They are bright sometimes patterned or metallic in colour. Convolvulaceae weed plants should be controlled as tortoise beetles can breed on them. They are minor pests (Subramaniam *et al.,* 1977; Pillai and Palaniswami, 1984).

Spiny beetle (Oncocephala tuberculata)

Both grubs and adults mine the leaves and cause them to wither. It is a sporadic pest. Adult is a small beetle with blunt projections all over the body. It lives up to 50 days. Eggs are laid inside the leaf tissues. When full grown it constructs a short tunnel in the leaf and pupates inside. Egg, larval and pupal stages last for 7-10, 17-23 and 8-15 days respectively (Pillai and Palaniswami, 1984).

1.1.4 Sap feeders

Halticus minutus is a minute sized mirid bug, sucking sap from shoot in large numbers. It is a minor one. Mealy bugs *(Pseudococcus* sp. and Planococcus sp) and thrips *(Thrips apicatus, Dendrothrips ipomeae)* also feed on the crop (Subramaniam *et al.,* 1977, Palaniswami *et al.,* 1999).

1.1.5 Flower and Fruit feeders

Many insects prevent seed setting by feeding on floral parts. Important among them are *Mylabris thunbergi, Oxycetonia versicolor* and *Macroura orientalis.* The weevil *Protocylas coimbatorensis* feeds on fruits and seeds (Subramaniam *et al.,* 1977).

1.1.6 Soil insects

The ground beetles *Gonocephalum oivicum* and *G. depressum* feed on the tubers by making irregular shallow galleries. They are found in the soil in large numbers when the crop matures. Damage to the extent of 10-15% was recorded (Pillai and Palaniswami, 1984).

1.1.7 Non insect pests

A few non insect pests including mites, nematodes and rodents also attack and cause damage to sweet potato.

Spider mites

Tetranychus cinnabarinus, Eutetranychus orientalls colonise on the leaf surface and the leaf curl and dry up. Eriophyd mites *Oxypleurites convolvuli* Channa. causes severe rusting of the leaves which turn pale brown. These are sporadic pests.

Nematode

Nematodes - *Meloidogyne incognita* and *Rotylenchus reniformis* were found attacking roots and tuber. Root knot nematode *(Meloidogyne icognita)* cause swelling or knotting of roots and heavy infection can inhibit apical growth. The most obvious symptoms of damage on tubers are longitudinal cracking and general rough appearance of the skin. Studies conducted at CTCRI indicated availability of very high degree of resistance to the nematode in all CTCRI released high yielding varieties Sree Nandini, Sree Vardhini and H 268, (Mohandas and Palaniswami, 1996). *Rotylenchulus reniformis* a semi endoparasitic nematode is found in high numbers (4000 to 9000 per 100 cc soil) in sweet potato fields. The nematode penetrate roots and tubers causing injury. Normally this nematode does not produce galls but on sweet potato the nematode was surrounded by galls formed from the host. The infection causes chlorosis in the foliage and reduction in the yield and size of tubers.

Many varieties of sweet potato are resistant to root knot nematode and resistance has been used either alone or in combination with other measures for the biological suppression of the nematode. The identified resistant varieties can be included for cultivation in the root knot problem areas in order to reduce the population and there by increasing the yield (Mohandas and Palaniswami, 1996).

Rats

In India rats damaging sweet potato in the field are mainly field rat *(Bandicota bengalensis)* and field mouse *(Mus booduga)*. Under storage, house rat *Rattus rattus* and house mouse *Mus musculus* cause damage. Yield loss was reported to be 1-10% in India under field conditions and 10-20% under storage (Rajamma *et al.*,1989). An integrated approach utilizing the following methods are suggested.

Field Sanitation : Removal of bushes and weeds in and around the crop field is important.

Trapping : Different types of traps are available and suitable one can be used for each species.

Catching and killing : Live burrow can be located by plugging the burrows with moist soil and observing for opening the next day. Traditional rat killers can be utilized for this purpose.

Smoking : Dry coconut leaves made into small bundle about 50 cm long can be thrust into the live burrow, fired and fanned in such a way that smoke is pressured down into the burrow. Rats die inside the burrow.

Poison baiting : Zinc phosphide is the effective poison. It is mixed with some attractive food material like cereal flour, dry fish or groundnut to get 2-5% concentration. When cereal flour is used some sugar and oil can be added. Prebaiting using food material alone should be done for 2 to 3 days before poison baiting as the rats tend to develop shyness.

Biological method : Predators like rat snake, owl, etc. need to be encouraged with least ecological disturbance. An owl can kill a rat every five minutes during nesting period, while a cat devours only 30 per year.

1.1.8 Storage pests

Keeping sweet potato tubers in sand helped to prevent weevil damage in storage. Storing tubers as heaps on floor and covering with red earth or wood ash prevented the entry of the weevils which is an important storage pest and thus protected the tubers for two months. Studies at CTCRI showed that different methods like keeping uninfested tubers in earthern pots or in sand or keeping in heaps covered with straw and plastering with mud paste could protect the tubers from weevil damage up to 45 days. The tubers in such storage retained their freshness without any deterioration. Recent studies showed that keeping the tubers in wooden box in layers of sand was effective for the control of storage pests and post-harvest deterioration up to 2½ months and beyond which the tubers got spoiled by rotting either due to microbial or physiological degradation (Rajamma, *et al.,* 1993).

1.2 Diseases of Sweet Potato

Several viral, fungal, bacterial pathogens and other micro organisms attack sweet potato and reduce the crop yield and quality. Some diseases occur in the field; others occur during storage, handling, and marketing. Among these viral diseases are the most important and about 20 virus diseases have been reported to infect sweet potato. Sweet Potato Feathery Mottle Virus (SPFMV) is most important and most commonly occurring virus. SPFMV alone cannot damage the crop much but mixed infection with other minor viruses inflict sweet potato virus disease (SPVD) syndrome which causes severe damage to the crop. Sweet potato chlorotic stunt virus (SPCSV) (family Closteroviridae) and sweet potato feathery mottle virus (SPFMV) (family Potyviridae) are whitefly- and aphid-transmitted, respectively, which in double infection cause sweet potato virus disease (SPVD) in Africa, a serious sweet potato disease. Sweet potato plants showing symptoms similar to SPVD have been observed in most areas of India. Nevertheless, not much information is available about the identity of the viruses infecting this crop in India. NCM-ELISA, DAS-ELISA and PCR detection of viruses revealed predominance of SPFMV in the states, Kerala and West Bengal in India.

Variable symptoms in sweet potato plants with foliar mosaic, stunting, leaf malformation, chlorosis, and ring spot symptoms were observed in several farms in West Bengal and NCM-ELISA gave many positive results of sweet potato chlorotic stunt virus (SPCSV), C-6 (unknown virus) and seldom occurrence of sweet potato cauliflower like virus (SPCaLV). Plants obtained from vine cuttings and tuber lets collected from SPFMV, leaf curl, severe mosaic symptomatic plants of different varieties/cultivars in West Bengal gave identical symptoms. The important virus disease of sweet potato are discussed.

1.2.1 Chlorotic leaf distortion of sweet potato

This disease is caused by *Fusarium lateritium* Nees Fr. The prominent symptoms are bright chlorosis on the young leaves, twisting or deformation of leaves in advanced stage, white powdery growth of fungal structures on leaf surface and margin or leaf blade. It affect the leaf opening and reduces the chlorophyll content. Spraying of Dipotassium hydrogen phosphate or Disodium hydrogen phosphate @0.001% for 3-4 times contains the disease (Thankappan, 1994).

Stem Rot or Fusarium Wilt (caused by fungus: *Fusarium oxysporum f. batatas*)

This disease is more prevalent in the nursery beds during vine multiplication. The infected plants become very weak, thin and leaves become yellow. Initially wilting appears in a patch and infected plants gradually wilt and dry. Stems are blackened and the roots are dried. Cultivation of sweet potato in the same field for several years increases the disease incidence. The infected plants fail to produce the storage root and dry at premature stage. If the disease appears at later stage of the crop, the storage roots become pencil shaped, shriveled and black.

1.2.2 Virus diseases

Sweet potato feathery mottle disease is caused by sweet potato feathery mottle potyvirus. The important symptoms are ringspot, pink veinbanding, mosaic and puckering. Spots enlarge and intermix with pink veinbanding and entire leaf becomes pinkish. The chlorophyll content is reduced which affect photosynthetic process resulting in yield loss. Chlorotic patterns developed on the leaves, sometimes with pigmented borders. Some infected cultivars express internal (internal cork) or external (russet crack) necrosis on the roots. The disease is non-persistently transmitted by a large number of aphid species and also by vegetative propagation. The virus is a flexuous rod with helical morphology, about 13 x 850 nm (potyvirus group). This disease can be contained using virus free planting material and rouging out of infected material from field. Diagnosis can be confirmed by serological tests and graft transmission on *I. setosa.* Meristem tip culture is the method for obtaining disease free material .

Sweet potato chlorotic leaf spot virus, sweet potato chlorotic stunt virus, sweet potato leaf curl virus, C-6 virus, sweet potato caulimo virus, sweet potato mild mottle virus, Sweet potato sunken vein virus, sweet potato vein clearing virus, sweet potato vein mosaic virus, sweet potato yellow dwarf virus, sweet potato virus A are other viruses reported on sweet potato. Sweet potato mild mottle virus and sweet potato leaf curl virus are found in Kerala, Bihar, Orissa and West Bengal (Jayanta, 2006; Palaniswami and Shirly, 2006).

Sweet potato latent virus (SPLV) belongs to potyvirus group. It causes latent or mild chlorosis in sweet potato. No known vector. The virus is a flexuous rod with helical morphology, about 13 x 750 nm. Graft transmission to *I. setosa.* Diagnosis can be confirmed by serological tests of *I. setosa.*

Sweet potato mild mottle virus (SPMMV) infected sweet potatoes exhibiting stunting and leaf mottling. The disease is efficiently transmitted by the whitefly *Bemisia tabaci* and by vegetative propagation material. Virus are flexuous rod shaped particles, 800-900 nm in length (potyvirus group). Graft transmission to *I. setosa.* Diagnosis can be confirmed by serological tests of *I. setosa.*

Russet crack (caused by SPVD)

The sweet potato plants infected with virus diseases (SPVD) showed mixed symptoms on the foliage. The severely infected plants produce deformed storage root, splitting and internal cork tissues. The flesh of the roots produces dark, hard, corky patch with dark brown to black in color. The central portion of corky tissues produce hollow zone with shrinkage of tissues. Sometimes the roots do not show external symptoms and occasionally produce tiny corky growth on the skin (Jayanta, 2006).

1.2.3 Bacterial Wilt and Root Rot (caused by bacterium: Erwinia chrysanthemi)

This disease is also referred bacterial soft rot. The disease manifests with yellow leaves and water-soaked lesions on stems and petioles. The infections occur in more localized area of the plant and in most cases only one or two branches collapse and die. In advanced condition the leaves and petiole turn black and whole plant dies. Roots decompose completely. The disease is more prevalent in rainy season mainly during vine multiplication. The heavy rainfall with poor drainage and water logged condition predispose this disease.

1.2.4 Leaf spots

Two species of *Cercospora* have been reported to cause leaf spot disease on sweet potato. *Cercospora batatae* Zimm., a synonym of *C. timorensis* Cooke, reported by Sydow and Mc. Rae in 1929 (from Tamil Nadu and Bihar), Salam and Rao in 1957 (from Hyderabad) and Chiddarwar in 1959 (from Maharashtra). Later in 1982, *C. ipomoeae* Wint. was also reported to cause leaf spot disease on sweet potato in Bihar (Sinha and Mehta, 1982).

Other leaf spots are *Cercospora batatae* Zimm., *Cercospora timorensis* Cooke, *Cercospora Ipomoee* Wint. *Alternaria* sp. *Alternaria capsici*- annui Saval and Sandhu, Colletotrichum capsici (Syd.) Butler and Bisby, *Helminthosporium euphorbiae* Hans.

Use of field resistant varieties alone can give a satisfactory control for this disease. Fifteen sweet potato lines out of ninety five tested have been identified as resistant at Rajendra Agricultural University, Bihar.

1.2.5 Vine fasciation

Striking diagnostic symptom in the appearance of thick flattened transformation of vines which remain green and grow till harvest. This new disease is of unknown etiology occurring sporadically (Jayanta, 2006).

1.2.6 Diseases in storage

Black Rot (caused by fungus: *Ceratocystis fimbriata*) can cause severe loss in storage, in the plant bed, and in the field. The infected storage root of sweet potato shows black, sunken circular spots that often cover the surface. The infected tissues may spoil and make it non-marketable.

Rhizopus Soft Rot (caused by fungus: *Rhizopus stolonifer*) is found in the storage and is most destructive disease. Once the disease appears it spread very quickly in the storage and may cause heavy toll to the harvested tubers. The infected sweet potato becomes soft, watery with offensive odour and producing fluffy white fugal growth. The disease is more prevalent in high humidity with poor ventilation in the storage.

Scurf (caused by fungus: *Monilochaetes infuscans*) turns the skin of the sweet potato blackish patch, sunken and shrivelled later. The disease is usually worse during wet weather and late harvesting during summer scurf occurs in the storage and whole marketable roots become hard and shrivelled. The disease spreads when healthy sweet potatoes come in contact with diseased roots (Jayanta, 2006).

1.2.7 Management of the diseases

As sweet potato is infected by large number of diseases both in the field and storage, it is very difficult to eliminate the diseases. Most of the diseases caused by viruses initially spread through vine cuttings and even through the tuber slips. Clean cultivation and use of disease free planting materials and careful post harvest management of the harvested storage roots may minimize the disease.

Field Diseases

- Disease free healthy quality planting materials to be used. Vine cuttings should be collected from the healthy and disease free plants from the field free from any infection.
- Crop rotation at least for 2-3 years in the SPVD endemic areas
- Pre-planting soil drenching with suitable fungicides and dipping of vine cuttings/tuber slips in fungicides reduce the rot diseases.
- Use resistant varieties to virus diseases if available.
- Control insect vectors that may spread the virus diseases in the field.

Storage Diseases

- Store only clean and fresh sweet potato tubers.
- Sweet potato tubers are to be kept in fresh, clean container or in well ventilated storehouse.
- Careful handling and movement of seed stock is necessary to avoid injuries that may cause decay (Jayanta, 2006).

2. Cassava

2.1 Pests of Cassava

Among the pests that attack cassava, spidermites, white grubs, termites, thrips and rats, are important and whitefly as a vector of cassava mosaic is significant. The pests of cassava in India can be broadly grouped into (i) pests attacking stems (ii) pests damaging plant setts and young seedlings (iii) pests infesting foliage (iv) pests attacking tubers and (v) storage pests.

2.1.1 Pests attacking stems

Scale insects

Scale insects of different species attack the stem in the field and during post harvest storage of planting material. The most important scale insect is white mussle shaped soft scale (*Aonidomytellus albus* CkII.). The insect sucks and desaps the stem. It multiplies rapidly covering the stem in large numbers and it also attacks the side shoots and at times the petioles. In severe infestation the stems become weak and dry due to the prolonged infestation, resulting in the loss of viability of planting material and poor establishment of the crop. The weakening and breaking of the main stem causes profuse branching which gives bushy appearance. The severity of attack increases during prolonged dry seasons. The problem is perpetuated through planting materials which serves as a source of multiplication and spread. Eggs are laid beneath the scales. The young nymphs which come out in 3-4 days are active and spread over and settle. Once they start feeding they lose their legs and become sedentary. It develops white waxy secretion over which it develops into a scale. In 20-25 days the nymphs become full grown. The male is winged and female is sedentary and wingless.

For the effective control, scale free stems should be collected for storing and planting. Stacking the stems in horizontal position encourage multiplication of scales due to higher temperature and humidity. Therefore healthy stems are to be kept in vertical position under shade to facilitate easy aeration and diffuse day light. As a prophylactic measure the stems may be sprayed with dimethioate 0.05% at the time of storing. If further infestation is observed one more spraying may be done. The infested stems are to be rejected and burned at the time of planting. In case of acute shortage of planting material and the scale attack is mild, them the setts can be given a dip in the above insecticidal solution for 10-15 minutes before planting. Other scale insects observed are *Saissetia nigra* N. and *Lepidosaphes* sp.. Larvae and adults of coccenellids - *Chilocorus nigritus* F. *Syemnus quimeti* Muls., *Chilomenus sexmaculatus* F. and *Stethorus pariepunctatus* Kapur prey on the scales (Pillai, 1994, Palaniswami, 1999).

2.1.2 Foliage feeding pests

Spidermites

Among the foliage feeding pests the most serious ones are spidermites. There are two distinct groups of spidermites. One group feeds on the lower surface of the leaves (*Tetranychus cinnabarinus* (Boisd.) and *T. neocaledonicus* (Andre) causing elongated streaks, chlorosis, withering and drying of leaves. The other group which prefer to infest on the upper surface (*Eutetranychus orientalis* (Klien) and *Oligonychus biharensis* (Hirst)) cause depletion of chlorophyll, resulting in the characteristic bronzing and typical rusted bathery appearance; such leaves begin to curl from margins and later fold into one twisted roll. The teranychid mites usually infest lower surface but during epidemics may cover both sides of the leaf. The predominant species are *T. cinnabarinus* and *E. oreintalis* and are found in cassava growing areas of Kerala and Tamil Nadu. They appear together and multiply rapidly during dry periods. Both groups infest from bottom leaf upwards and in periods of heavy infestation the population density ranged from 200-1000 / leaf in almost all leaves. The lower surface feeding mites causes more severe leaf fall than the other group. The biology and binomics of *T. cinnabarinus* and *E. orinetalis* reveal that the total life cycle from egg to adult ranges from 11-20 and 10-14 days respectively (Pillai and Palaniswamai, 1985). The spidermites incidence increases year after year depending upon the climate. The data during 1977 to 1982 revealed that rainfall was the most limiting factor. Heavy rainfall accompanied by wind was deleterious, while a fall in RH and an increase in temperature were highly conducive for rapid multiplication of mites. An out break can be expected during rainless summer where temperature is 32°C and above and RH below 75%. The yield loss due to spidermite complex alone ranged from 17.8 to 33.1 per cent in different varieties under natural infestation in Kerala (Pillai and Palaniswami, 1981). Economic threshold levels could be fixed for spidermites; for Sree Vishakham 25% leaf infestation and for Sree Sahya and M4 10%. Spraying of dimethoate or methyldemeton 0.05% during severe infestation (January to April) at monthly interval is highly effective against the pest or spraying water at run off level is also effective in controlling mite damage. Spraying of water can be resorted to where labour is cheap. This has an added advantage of preserving the biotic agents of mites. They include *Coccenellids- Pharoscymnus hornii* Weise, *Stethorus gilvifrons* Muls., thrips - *Scolothrips indicus* P. and phytoseiid mites - *Amblyseius longispinosus* Evans (Pillai and Palaniswami, 1985). Thrips is an effective predator but the thrips population in the field is very low. However even though there are a number of natural enemies their population in general is very low under field conditions. If these predators are multiplied and released they will be effective biological agents for mites. Cassava accessions at CTCRI *viz.,* CE-4, CE-14, CE-38 and CE-139 are found highly resistant to spidermites. The host plant resistance is the most satisfactory and long range solution to control mites on cassava. Foliar application of urea followed by spraying of water at the beginning of mite infestation or spraying of dimethoate @0.05% in severe situation suggested to contain the pest outbreak (Pillai and Palaniswami, 1985).

Thrips

Thrips *(Retithrips syriacus* (Mayet) another foliage pest is serious during dry periods when spidermites also occur and accentuate the problem. Infested leaves lose chlorophyll and turn pale with deposits of dark brownish excreta.

Infestation starts on the older leaves and spread to the younger ones. The control measures adopted for spidermites will also take care of this pest (Pillai, 1994).

White fly

Bemisia tabaci (Genn.) is the common white fly on cassava. It is more important as a vector of CMD than as a pest. It is abundant throughout the year Palaniswami *et al.* (2001) Lisha *et al.* (2003) identified two biotypes of *B. tabaci* and reported that only cassava biotype acted as a vector of cassava mosaic virus and not the other biotype *viz.*, sweet potato biotype. There were differences in the gel-banding pattern between the two biotypes. The ability of cassava biotype to transmit ICMV from cassava to cassava was confirmed through nested PCR primers and electron microscopy. *B. tabaci* feeding on cassava leaves induced pathogenesis related proteins (Binu and Palaniswami, 2006a).

Aphelimid parasitoids found often regulating the field population with 6.2 to 12.1% parasitoids (Palaniswami *et al.*, 2001). Parasitoids recorded on *B. tabaci* include *Encarsia transvena* (Timberlake), *E.bimacualata* (Heraty and Polazek), *E. nigriphala* (Dozier), *E. strenua* (Silvestris), *E. haitiensis* (meritoria) (Dozier), *E. Guadeloupe* (Viggiani), *E. azimi* (Hayat), *E. pergandiella* (Howarl), *Eretmocerus munduls* (Mercet) and *Eretmocerus* sp. (Antony *et al.*, 2006b).

Spiral whitefly

Spiral whitefly *Aleurodicus dispersus* Russel was found to seriously infest cassava since 1993. It is highly polyphagous in nature. It causes yellowish specks of leaves, resulting in crinkling and curling and sooty mould development. Adults are much larger than *B. tabaci* (2 mm in size). Severity of damage increases during summer. The rainfall had deleterious effect on the pest (Palaniswami, 1999).

2.1.3 Pests attacking setts and seedlings

Termites

Termites *(Odontotermes obesus)* attack the planted cuttings and young seedlings resulting in the poor establishment. The pest problem becomes serious in areas with prolonged dry spell. In certain periods the seedlings mortality ranged from 20 to 40%. It is a sporadic and localized pest. In the case of severe incidence, soil application of insecticides may be resorted to (Palaniswami, 1999).

White grubs

White grubs *(Leucopholis coneophora* Burm.) are polyphagous and are important in certain localities. The attacked plants become pale, wither and dry up. The grubs feed on the developing roots and on the bank of the planted stems. Eggs are laid in the soil and first instar grubs which emerge out first feed on the organic matter and then they migrate to the cassava roots and cause damage. The pest can be effectively controlled through collection of adult beetles from harbouring plants and trees, deep ploughing to expose the grubs and soil application of insecticides (Palaniswami, 1999).

2.1.4 Pest attacking tuber

Rats

Rats are serious and wide spread pest on tubers. The enormity of the damage to tubers is not realized until harvest, an attack is often not recognized until lodging occurs because most of the activity is underground. Field rat *(Bandicota bengalensis),* Indian gerbil *(Tatera indica)* and field mouse *(Mus booduga)* are important species noted in the fields, while *Rattus rattus* (house rat) and *Mus musculus* (house mouse) are problems in the houses and godowns.

Bandicota bengalensis also known as Indian mole rat or lesser bandicoot makes extensive and deep burrows which extent up to 30 m with a depth of 1 m. Each burrow had 1-3 food cum bed chambers with 6-10 outlets. Male and female live separately. Usually only one adult per burrow is observed. They plug the burrow openings with heaps of mud. On an average there are 13 live burrows per ha. The Indian gerbil is nocturnal and live in groups. Burrows are simple but deep (1-2 m) with 3 or more openings and these burrows are usually found in or nearby bunds covered with weeds and bushes. The field mouse lives in shallow burrows with only one or two openings.

The extent and intensity of damage to cassava was assessed and the actual tuber loss was found in the range of 1 to 45% with a mean of 9.4% in upland conditions. The percent of damage done to the plants was found to be in the range of 2 to 90% with a mean of 18% (Rajamma *et al.,* 1989).

2.1.5 Pests of stored chips

Among a dozen insects that infest cassava chips in store, the most serious one is the beetle *Araecerus fasciculatus* Da Geer. Others like *Denoderus bifoveolatus, Tribolium castaneum, Lasioderma serricorne* also damage chips. Storing chips in parboiled form is an effective method to protect from these pests (Pillai, 1994).

2.2 Diseases of Cassava

2.2.1 Cassava Mosaic Disease

Cassava mosaic disease is caused by Indian cassava mosaic geminivirus. The symptoms are chlorotic areas intermixing with normal green tissue which gives mosaic pattern. In severe cases leaves are reduced in size, twisted and distorted. It affects the growth of the plant resulting in significant reduction in plant height, stem girth, petiole length and leaf size. It causes 25-80% yield reduction depending up on the varieties and grade of ICMV infection (Thresh *et al.,* 1998).

Primary spread of the disease was effected through the indiscriminate use of infected planting material and the secondary spread is through the vector whitefly *Bemisia tabaci* G.-cassava biotype (Lisha *et al.,* 2003). Studies on true seed transmission indicated that the disease was not transmitted through true seeds. Sap transmission of the disease from cassava to cassava could not be achieved. However, the disease could be transmitted through sap from cassava to *Nicotiana benthamiana* and *N. glutinosa.* CMD could be easily transmitted through grafting also.

The disease causing agent (virus) was purified and found to belong to the group Gemini viruses. The particles measured 18-24 nm in diameter. The antisera with adequate titre strength, which can detect successfully even latent infection of the diseases, have also been developed (Thresh *et al.,* 1998, Antony *et al* 2006b). Nested primer and few sequencing primers were designed from AUI coat protein (CP) gene of ICMV. This nested primer when used in the diagnosis of ICMV infection in different varieties of cassava was found to be precise and useful (Antony, *et al.,* 2006b).

The symptoms of the disease were masked during dry and hot months and make it impossible to identify the diseased plants. Therefore, the disease free planting materials should be selected before the beginning of the dry/hot season.

Resistance to CMD has been studied since 1968 in the country. No cassava variety is immune to CMD. Even the related species *viz., M. glaziovi* is susceptible to artificial inoculation. However, a great variation in the degree of susceptibility was observed. Few field resistant/tolerant lines such as M4, H 97, H 165, H 2304, H 1687, S 1315, S 1310, S 2381, S 2380 MNga-1 were identified for cultivation.

The integrated management of CMD including selection of disease free planting material, use of field tolerant varieties like H-97, H-165 and Sree Visakham, selection of disease free meristem derived planting material followed by clonal multiplication with periodical screening and roguing of freshly infected plants, will be useful to raise a good disease free crop. The disease free planting materials can be multiplied in large scale at higher altitude where cassava whitefly population is low. Raising the plants in the nursery at closer spacing prior to transplantation into the main field is an useful step to prevent the primary spread of the disease in the main field. Adherence to strict sanitary practices such as timely harvest and prompt disposal of crop residues, and eradication of self sown plants and weeds which may harbour both the disease and vectors (Thresh *et al.,* 1998).

Development of transgenic plants against CMV through ICMV replicase mediated resistance is in progress. Use of barrier crops around cassava crop and spray of antiviral plant extracts of *Boerhavia diffusa* and *Mirabibis jalapa* were found to be useful in containing the ICMV (Palaniswami *et al.,* 2004).

2.2.2 Cassava Tuber Rot

Cassava tuber rot is caused by *Phytophthora palmivora.* There are no external symptoms seen. Infected tubers show brown discoloration of internal tissues, rotten and exhibit foul smell. Tubers become unfit for consumption or marketing. Causes heavy yield loss when it occurs. Integrated Management for tuber rot recommended include the following. Provide proper drainage in the field. Remove infected tubers from field and burn. *Trichoderma viridae* is a good biological control agent and can be incorporated in the problem areas (Thankappan,1994).

In case of severe incidence, deep ploughing, planting cassava on ridges and incorporation of neem cake (@500 kg ha^{-1}) and *Trichoderma viridae* (1 kg culture in 100 kg FYM) with well drained conditions could reduce the tuber rot and increase the yield.

2.2.3 Leaf spot

Disease	Causative Organism
Brown leaf spot	*Cercospora henningsii* Allescher
White leaf spot	*Cercospora caribaea* Chupp and Ciferri
Snow white leaf spot	*Cercospora manihotae* Viegas
Diffuse leaf spot	*Cercospora vicosae* Muller and Chupp
Phyllosticta leaf spot	*Phyllosticta manihotae* Viegas
Leaf spot	*Periconia shyamala* Roy
Leaf spot	*Epicoccum nigrum* Link

Severe spotting causes premature defoliation resulting in significant loss of leaf area available for photosynthesis. Some of the simple inexpensive cultural practices, if followed at the right time can reduce the severity of the disease and the loss caused by it.

Wider spacing: Cultural practices such as wider spacing would reduce excess humidity in the crop stand, and thereby the multiplication and spread of the disease. Removal of severely infected leaves and burning them with a view to reduce inoculum source has also been recommended. H.-97, Sree Visakam and Sree Sahya are moderately resistant, and Sree Prakash is highly resistant to this disease. Spraying of Bordeaux mixture (1%) has been recommended. However, spraying of fungicides should be recommended only when it is absolutely essential and with caution.

Phomopsis leaf spot disease on cassava has been reported from Kanpur (U.P), Bangalore (Karnataka) and Kerala. Two species of *Phomopsis viz., P. manihotis* Swarup, Chauhan and Tripathi, and *P. manihot* (Speg.) Chavaugeon have been recognised as the causal organisms of the disease.

Phomopsis blight widespread throughout the cooler cassava growing areas at high altitudes, or in the lowlands during the rainy season. Large, watersoaked spots on young leaves, often at the tips and margins. They are pale green at first, rapidly enlarging and turning brown. Spore- containing pycnidia are formed in concentric rings on the upper leaf surface or on stems. Severe attack leads to defoliation, stem girdling and dieback. Caused by *Phomopsis manihotis*. Meristem-tip culture eliminates the disease.

2.2.4 Disease affecting stem and tuber (Jayanta, 2006)

Sett rot (Diplodia natalensis P.)

Sett rot is an important and wide spread disease of cassava found in the stored stems as well as in the new plantings in drier areas where planting is done after irrigation. Rotting of stems (setts) results in failure of germination and gap formation in the field. Sett rot is caused by the pathogen *Diplodia natalensis* Pole-Evans. (=*Botryodiplodia theobromae* Fab.)

Use of healthy planting material is the best means of control. Fungicides, such as benlate (0.1%), bavistin (0.1%), vitavax (0.1%) and thiram (0.2%) have been found to be very effective in checking the disease in field. The cuttings are to be dipped in one of the fungicidal solutions for 15 minutes, dried in shades and then to be planted.

Dry rot (*Fusarium solanii* (Mart).)

A dry rot of cassava tuber has been found to be prevalent in Tamil Nadu. The infected tubers are usually hard and the colour of the pulp instead of being white varies from brown to reddish brown and dark brown streaks are seen inside the tubers; one or more tubers are affected in each plant and the diseased tubers are unfit for marketing and consumption.

3. Yam

3.1 Pests of Yams

3.1.1 Leaf feeders

Spotted Beetle

Galerucida bicolor Hope attacks and defoliates the leaves of yam. It is also observed on elephant foot yam. *Crioceris impressa* Fb. is a pest in Kashmir. Grubs feed on the tender leaves and succulent tissues of shoot tips. Adult feeds on mature leaves by perforating leaf blade without affecting leaf veins. Adult is stout 7-9 mm, fulvous beetle. Female lays 250-300 eggs in groups of 3 to 5 on leaves or on stem. The grubs emerge out in 3 to 4 days and pass through three instars in 5 to 8 days. It pupates inside the soil and adults come out after 10-13 days. This species was reported damaging *D. bulbifera* in Bihar during September-November season.

Lema lacordairei Baly, attacks *D. alata* in Kerala. It feeds on leaves and tender vines. Adult is a small (8 mm long) beetle with blue elytra and yellow body. Eggs are laid in groups of 10-12 on leaves or vines. On an average, the incubation, larval and pupal periods last for 2.5, 6.5 and 8 days respectively. The grub is yellowish with thick fleshy abdomen and it carried its excrement on its back. Pupation takes place in a white frothy cocoon. Longevity of adult is 90-95 days. It appears in May and is found up to November. The damage caused by defoliation is seen often serious. It is often regulated by a tachinid parasitoid *Pseudalsomiya* sp (Palaniswami, 1999).

Caterpillars

The common cut worm *Spodoptera litura* is a leaf feeder causing severe defoliation. The damage and biology are similar to that in aroids. *Pericallia ricini* Fb. is a leaf eating hairy caterpillar causing severe damage when the crop is left untrailed. Hornworms-*Hippotion celerio* Linn., *H. oldenlandiae* Fb. and *Theretra gnoma* Fb. are leaf eating ones. These caterpillars are sporadic on yams. Pillai and Palaniswami (1984) observed hairy caterpillars and cut worms in the yam growing areas of Kerala and Andhra Pradesh. They are of minor importance.

Sawfly

In South India larvae of sawfly *Senocolidia dioscorea* R. was reported as a leaf feeder. A number of larvae defoliate by feeding together often in rows. It is sporadic in nature. The pest was later reported as *S. purpurata* (F. Sm.)

Mealy Bugs

Palaniswami (1999) reported the incidence of *Ferrisia virgata* (Ckll.) on *D. dumetorum*. The mealy bugs cause damage by desapping the leaf. During heavy incidence, the leaves dry up and wither and the damage caused the drying of entire population of *D. dumetorum* in Trivandrum.

Spidermites

Dioscorea was noted as a new host for *Tetranychus telarius* Linn. which was later on identified as *T. cinnabarinus*. All the four species found on aroids infest yams too as pests during dry hot periods (Palaniswami, 1999).

3.1.2 Pests on Root and Tuber

Scale Insects

The white scale insect *(Aspidiella hartii* Ckll.*)* is an important pest of yams in India. Among the different species of *Dioscorea, D. alata, D. esculenta* and *D. rotundata* were found attacked by the pest, while *D. hispida* and *D. bulbifera* were relatively free from attack. This insect also infests ginger, turmeric, taro and elephant foot yam. The field infestation ranged from 1 to 100 per cent among different varieties and species of *Dioscorea.* In severe cases of infestation, attacked plants show drying due to continuous sucking and desapping.

The males are winged and free living. The females are wingless, covered by a dirty white scale and parasitic. It lays eggs inside the scale itself. The emerging out nymphs are known as crawlers. They move about in search of a suitable feeding site. Once a crawler starts feeding, it settles down and looses its legs and becomes sedentary. Soon a white secretion covers the body and becomes hard to form a scaly cover. Male has one instar less than the female. The life cycle is completed in 11-20 days. The scale insects are small (1-2 mm) in size. Nymphs and females infest the basal vines and tubers. During January to April, they multiply very fast in large numbers covering the entire tubers.

The infested tubers start shrivelling affecting the quality, viability and marketability. The pest problem is perpetuated through seed tubers. The nymphs are parasitized by an aphelinid *Physcus comperi* H. and an encyrtid *Adelencyrtus moderatus* H. (Palaniswami, 1994, 1999).

White Grubs

The white grub, *Leucopholis coneophora* Burm., attacks root and tuber. White grubs attack on yams grown in coconut gardens. The extent of infestation was such that the cultivation of yams has to be abandoned for years. Incidence as serious pest in certain parts of Kerala, especially in Nileshwar and Wynad was noted. During severe incidence of white grubs, the intensity of damage was assessed as 10 to 35 pits per tuber and each gnawed pit measured 15-20 mm deep. The tuber yield loss was in the range of 4 to 9 per cent in greater and lesser yams. The damaged tubers are adversely affected and loose the quality and culinary taste. Field infestation in the range of 4 to 46 per cent was reported (Palaniswami, 1999).

Termites

The species identified attacking yams are *Odontotermes obesus* (Rambur) and *O. escherichi*. These occur during prolonged rainless dry hot periods. It attacks basal vine and tubers and is an occasional pest of local importance. It is noted in Orissa, Bihar and Andhra Pradesh. During 1980-81, the termite incidence was severe in Trivandrum with 20-100 per cent attack on yams.

Nematodes

The association of yam nematode *Scutellonema bradys* Andrassy with dry rot of *Dioscorea alata* was reported in India as early as 1967 by Nadakal and Thomas. This spiral nematode is a migratory parasite invading root and tuber through root tips and cracks of skin. Eggs are mainly laid in the plant tissues where they hatch and develop into adults and all stages of the life cycle are infective. It causes a disease of tuber commonly known as 'dry rot'. It feeds intracellularly causing cell destruction and necrosis. The nematode infection is restricted to 1-2 cm of tuber in the peridermal and sub-peridermal layers. In the early stage of infection, small yellow lesions can be observed just below the tuber skin and these become brown resulting into a continuous dark dry rot layer just beneath the surface in heavily infected tubers. The infected tuber exhibit cracks in the skin and at times shows skin flakes off exposing dark brown dry rot. Due to nematode infection, the quality, consumption and market values are lost. *Meloidogyne* spp. causes malformation and reduction in size of tubers. It is found infecting the root and tuber in large numbers.

Rat

The tubers of yams are reported to be attacked by field rat *Bandicota bengalensis* and it is only of minor importance.

3.1.3 Storage pests

Scale Insect

When *Aspidiella hartii* infested tubers are stored along with uninfested tubers, it multiplies and spreads rapidly. If the infested tubers are used as planting material, it results in poor germination and growth and establishment. The pest multiplies in the field perpetuating the problem. Two parasitoids *Adelencyrtus moderatus* and *Physeus comperei* often play significant role in regulating the scale insect population on yam tubers (Palaniswami, 1994).

Arecanut Beetle

Araecerus fasciculatus De Geer is often found damaging the tubers as well as dried yam chips. It was observed that the damage due to this pest render the tubers unfit for consumption as well as for seed purpose (Palaniswami, 1994).

Nematodes

The yam nematode *Scutellonema bradys* causes storage loss. Whenever the infected tubers are stored, the nematode continue feeding, causing severe dry rot and wet rot. Moisture loss is significantly greater in infected tubers during storage than the non-infected ones.

3.1.4 Management of Pests /Resistant/Tolerant Varieties

Among the different species of *Dioscorea, D. bulbifera, D. hispida* and, to certain extent, *D. rotundata* are relatively tolerant to scale insects.

About 190 accessions of *D. alata* available in the germplasm collection were evaluated for the scale insect resistance/susceptibility during five cropping seasons (1978 to 1983). Of which, 97 were found infested with scale insect. Among these, 38 accessions had recurring infestation for more than three seasons. The remaining 93 accessions were identified as resistant to tolerant to scale insect. Depending upon the grade of scale infestation on tubers, the infested 97 accessions were grouped into mild (70), moderate (9), severe (9) and very severe (9) insusceptibility to the scale insect.

The white grub damage was free in as many as 174 accessions of *D. alata,* while 16 accessions were identified as susceptible with maximum damage (46 per cent) in Da 172. During 1980-81, the termite attack was very serious in the farm of CTCRI, Trivandrum. About 156 accessions did not show any infestation, 34 accession showed termite attack with cent per cent infestation on DA 15, Da 35, Da 58, Da 102, Da 135, Da 176, Da 177 and Da 187 (Palaniswami, 1994,1999).

Chemical control

In case of severe incidence of leaf feeding beetles, caterpillars, mealy bugs and/or saw fly can be controlled with malathion or carbaryl (0.05 per cent) spray. Application of sevidol at the rate of 20-25 kg ha^{-1} was recommended against white grubs and termites.

The scale insect can be controlled in the field using pest-free seed tubers and dipping seed tubers before planting in 0.1 per cent monocrotophos if scale infestation was noted during storage.

Use of nematode-free tubers for propagation effectively reduces field infestation. Fallowing of land meant for yam cultivation also reduces yam nematode infection.

3.2 Diseases of Yams

3.2.1 Anthracnose (dieback)

Widespread throughout the tropical countries. The disease is characterized by brown spots on young leaves, which enlarge and sometimes coalesce, as leaves approach maturity. Spots may have pale yellow margins. Epidemics occur during prolonged rains; young growth is infected and destroyed by rapidly expanding blacken in response to sunlight and the presence of numerous *Colletotrichum* spores which germinate but rarely penetrate the leaf surface. Stems also blacken. Repeated re-growth of vines between epidemics leads to multi-stemmed plants and production of several small tubers. *C. gloeosporioides* attacks many crops, and spores from these sources may infect yams. Fungicidal spray is suggested in case of epidemics (Jayanta, 2006).

3.2.2 Mosaic

Yam mosaic disease is common in all species of *Dioscorea.* In case of severe infection mosaic symptoms appear with leaf distortion. Potyvirus is the causal agent. It is non persistently transmitted by aphids (Jayanta, 2006).

4. Aroids

4.1 Pests attacking Aroids

4.1.1 Sucking pests

Aphids

One of the important sucking pests of taro is *Aphis gossypii* Glov. It sucks and desaps the chlorophyll content of leaves causing yellowish, speckling and curling. Aphid attacks both sides of leaves and petioles. Aphids are often attended by ants for its sugary secretion which manifests into sooty mould. Palaniswami (1994) observed the presence of aphids on taro throughout the year with higher levels of population during August to October and in March (around 830 to 1050 aphids per leaf). They also noted the aphid population was more during early stage of the crop and the infestation spread from young to older leaves. The activity of aphids is maximum during cloudy days with high humidity and temperature. It reproduces parthenogenetically and viviparously. Aphids are yellowish green to black in colour and females are alate. An adult may give rise to 12 or more nymphs per day which becomes adult in 7 to 9 days through four instars. An aphid may give rise to 8 to 22 nymphs per day. The multiplication rate is very high. It is widely prevalent in Andhra Pradesh, Kerala, Maharashtra, Orissa, Tamil Nadu and Bihar. The aphid is reported to be a potent vector of 'dasheen mosaic disease' of aroids.

Banana aphid, *Pentalonia nigronervosa* Coq. is reported to infest aroids. It is usually seen whenever the crops are grown in banana fields as intercrops or near banana fields. The nymphs become adult in 8 to 9 days through four instars. An adult produces 32 to 50 nymphs at the rate of 1-3 per day.

Other species of aphids reported in India are *Pentalonia galadii* D., *Hydronaphid colocasiae* Ray, *Aphis durranti* Das. and *A. punicae* Passerini. Among all these aphids, *A. gossypii* and *P. nigronervosa* also infest tannia and elephant foot yam.

Aphelinus mali H., *Aphelinus* sp., *Coccophagus* sp. nr. *cowperi* Gir. and *Aphidencyrtus aphidivorus* May. are reported as parasitoids. Aphidophagous predators reported are *Menochilus sexmaculatus* F., *Verania discolor* F., *Pseudospidimerus circumflexus* Mots., *V. inops* Muls., *Ischiodon scutellaris* F. and spiders (Palaniswami, 1994, 1999).

Spidermites

Tetranychus cinnabarinus Boisd., *T. neocalydonicus* Andre., *Oligonychus biharensis* Hirst. and *Eutetranychus orientalis* Klein. attack taro, tannia and elephant foot yam. The former two species are important ones. Spidermites are polyphagous. Both nymphs and adults cause damage by sucking and desapping the chlorophyll content of the leaves. The damage is chracterised by the appearance of yellowish white specks and chlorotic symptoms. Usually they feed from the lower surface of leaves but during severe infestation, both sides are attacked. The mite population is in serious proportion on dry land rainfed crops. A rainless, hot dry spell during March-April and November-December is found conducive for an outbreak. On the other hand, heavy rainfall accompanied by windy weather suppresses mite population. The life cycle is completed in a fortnight. They are distributed in Andhra Pradesh, Kerala, Maharashtra, Tamil Nadu and in Orissa. Palaniswami (1994) noted that the attack starts from older leaves to younger ones and the population build up is more during later stage of the crop.

Scolothrips indicus F., *Stethorus gilvifrons* Muls., *Oligotta* sp., *Amblyseius longispinosus* E. and *Trypoxylon errans* S. are observed to prey upon spidermites (Palaniswami, 1994, 1999).

Thrips

The creamish white *Caliothrips indicus* Bagnall. is seen infesting on both sides of taro leaves. Both nymphs and adults lacerate and sucks the sap, causing characteristic silverfish white specks with black dots. It is often found in peak (17.2 number per leaf) during February-May. It is common in Kerala, Tamil Nadu and Andhra Pradesh. It is prolific in its rate of multiplication and passes through several generations in a year. It takes 13-33 days to complete its life cycle depending upon weather conditions. It is common pest on groundnut.

Other thrips attacking taro and elephant foot yam are *Helinothrips kadaliphilus, Heliothrips haemorrhoidalis* Bouche and *Sciothrips cardamomi* (Ram.). *H. kadaliphilus* is the banana thrips and the leaves turn yellowish brown due to infestation. *Astrothrips asiaticus* (Bhatti) and A. *globiceps* (Karny) also feed on taro in India.

Tingids

Stephanitis typicus D. is a pest of taro. The nymphs and adults congregate on the lower surface of leaves and desap the leaf content causing yellowish white specks and chlorotic symptoms. It is a small bug with transparent lace like shiny reticulate wings. Occasionally, it assumes serious proportions especially when taro is grown in banana/coconut gardens.

Jassids

Green jassids (*Empoasca* sp.) is found breeding on taro in Kerala, Tamil Nadu and Andhra Pradesh as a minor pests. *Hecalus* sp. is found to feed on taro.

Fulgorids

Nisia atrovenosa Leth. and *Anigrus* sp. were seen feeding on taro.

White fly

Bemisia tabaci Genn. is found to breed on taro. Adults, nymphs and pupae are seen on the lower surface of leaves. It feeds on the sap and is a minor pest. It is often parasitized by an aphelinid parasitoid *Encarsia transvena.*

4.1.2 Leaf Eating Pests

Tobacco caterpillar or common cut worm

The tobacco caterpillar or common cut worm, *Spodoptera litura* Fb. is an important pest of taro and elephant foot yam. The adult moth has greenish brown forewings with white markings and hyaline hind wings with pale brown margins. Eggs are laid in batches covered with plumes/hairs on the lower side of leaves. A female lays up to 200 eggs. On hatching, the caterpillars feed the leaf by scraping gregariously for few days and later on disperse and move to other leaves to feed voraciously causing significant leaf damage. Caterpillars are stout, cylindrical (30-40 mm) and pale greenish brown with dark markings when full grown. They hide during the day time in crevices or among plant residues and become active during dusk to dawn. Pupation is in soil in an earthern cell and it is reddish brown in colour. Egg, larval and pupal periods last for 3-4, 13-19 and 6-10 days respectively. It is a polyphagous pest and is common in taro growing areas of Kerala, Tamil Nadu, Andhra Pradesh, Maharashtra and West Bengal. This is a pest of economic importance in Ganjam and Bolangir districts of Orissa. They reported the extent of damage as 16 to 80 per cent and the population in some cases was as high as 63 caterpillars per leaf.

Hornworms

Hornworms, *Hippotion oldenlandiae* Fb., *H. celerio* L., *Rhyncholaba acteus* C. and *Agrius convolvuli* (L.) Hb. are leaf feeders. The eggs are laid singly. The emerging hornworms feed on taro and elephant foot yam. Hornworms are stout (60-90 mm) and greenish to brown in colour with stripes. Pupa is reddish brown and pupation is in the soil. Adult is a hawk moth. The egg, larval and pupal periods vary from 6-11, 20-28 and 17-26 days respectively depending upon species and weather. These are minor pests of sporadic occurrence. Hornworms are parasitized by a parasitoid *Charops hersei* G. and M.

Hairy caterpillars

Hairy caterpillars *(Pericallia ricini* Fb.) are seen infesting taro and is sporadic in occurrence.

Grasshoppers

The common grasshopper that attacks taro in *Gesonila punctifrons* Stal. It burrows into petiole and lays eggs within tissues. Up to 13 eggs were recorded in a single petiole. Egg period is one week. The emerging out black nymphs feed on the leaf gregariously. They scrape the greenish part of the leaf leaving only a parchment membrane. The black nymphs turn greenish with black stripes from third to fourth instars. The total life cycle is about five nymphal instars. The mean duration of different instars are 4.5, 6.5, 5.5, 5.0 and 7.5 days respectively for first to fifth instar. Mean adult longevity for female and male are 14 and 11 days respectively. It is prevalent in Kerala, Tamil Nadu, Andhra Pradesh, Orissa, Bihar, Assam and West Bengal. The intensity of damage in Cuttack district of Orissa was as high as 50 per cent. In severely infested fields, most of the petioles were found bored by adults causing early lodging of leaves. This species is also a beneficial biotic agent in the biological control of water hyacinth *(Eicchornia crassipes),* which is a menace in ponds and tanks. The eggs and nymphal periods on water hyacinth ranged from 21-30 and 28-38 days respectively with five nymphal instars (Palaniswami, 1994).

Spotted beetles

The white spotted beetle, *Monolepta signata* Ol. is an important pest which feeds on leaves of taro. It also attacks a number of other crops including cabbage, cauliflower, chillies, beet root, radish, mango and grape vine. Adult is reddish brown with two big white spots on each elytra. This polyphagous pest is more active in south India than in the north. Whenever it occurs, it becomes a serious pest.

Oides affinis Jacoby is a pest on elephant foot yam and is seen in Kerala, Tamil Nadu and Andhra Pradesh. The grubs and adults feed on the leaves and pseudostem and severely infested plants wither away. The grubs bore into the pseudostem and cause stem splitting and decay. The pest was also found to breed on a weed *Bauhinia* sp. Infestation ranged from 15 to 25 per cent during July-August. Besides the direct feeding damage, it also predisposes to sclerotium rot. The grubs are often preyed upon by *Eocanthecona (Cantheconidia) furcellata* in field.

Galerucida bicolor Hope attack on the leaves of elephant foot yam in south India. The larvae are gregarious, skeletonizing the leaves. It gnaws into the leaf stalk and main stem. Eggs are laid in clusters in soil near plants. The incidence was noted during May-June. The grubs after a month pupate in the soil which extends for a fortnight. The adults are red and blue black beetles (Palaniswami, 1999).

4.1.3 Pests Attacking Root and Tuber

Scale Insect

White scale insect *Aspidiella hartii* (Ckll.) attacks tubers of taro and elephant foot yam. It is common in rainfed upland crop. Palaniswami (1994) reported shrinking and shrivelling of cormels of taro due to scale insect attack. The problem is perpetuated through seed tubers. The bio-ecology of the pest is explained under yams.

Mealy Bugs

Palaniswami (1999) recorded two species of mealy bugs, *viz., Pseudococcus citriculus* Green and *Rhizoecus* sp. Both are found together, and the nymphs and adults suck and desap the cell contents of tuber. The field infestation ranged from 6 to 45 per cent. The mealy bugs multiplied faster during high temperature and humidity. They cover the tuber surface along with dirty white powdery mealy substances. Severely infested tuber shrivelled, adversely affecting the quality and marketability.

White Grubs

Incidence of a white grub *Leucopholis coneophora* Burm. was noted on young seedling of taro in certain localities where it is grown as an intercrop in coconut plantations (Palaniswami, 1999).

Nematodes

As early as 1959, the infection of root knot nematodes was noted on taro in India. *Meloidogyne icognita* Chitwood and *M. javanica* Chitwood cause severe losses to taro and elephant foot yam. Foliage of seriously infected plants at first become yellow and then turn to brown and ultimately withers. The corms and cormels are deformed with galls of 2-15 mm size. Rotting of corms and cormels occurs in severe cases of attack. *Pratylenchus mutablis*, two new species of *Aphelenchoides* and two species of *Longidorus* are reported to attack taro (Jayanta, 2006).

4.1.4 Other Pests

Field rat *Bandicota bengalensis* feed on tubers of taro occasionally. A small snail *Kaliella barrackporensis* Pr. is noted to attack elephant foot yam. Snails are becoming pest on elephant foot yam and *Colocasia* in Maharashtra. Spraying of 5% soap nut and yam bean seed extract is effective against snails. Yam bean as border crop also contain snail attack on tuber crop. Waste leaves of cauliflower acted as a trap by luring them (Palaniswami and Shirly, 2006).

4.1.5 Storage Pests

Palaniswami (1994) reported two mealy bugs *(Pseudococcus citriculus* and *Rhizoecus* sp.) on elephant foot yam under storage. Corms and cormels of taro and tannia are also attacked by mealy bugs. Mealy bugs infested tubers showed poor germination and establishment in the field.

Scale insect *(A. hartii)* attack the tubers of taro, tannia and elephant foot yam. Damage is similar to that in yams.

Root knot nematodes cause rotting of corms and cormels under storage when infected ones are stored.

The arecanut beetle *Araecerus fasciculatus* De Geer was found damaging the corms of taro, tannia and elephant foot yam. *A. fasciculatus* infested tubers are unfit for consumption as well as for seed purpose. The beetle lays eggs on tubers and grubs bore inside forming zig-zag galleries and feed the contents. Pupation is inside the tuber and adult emerges making emergence holes on the tuber surface. The information on the storage losses due to this pest is scanty. Occasionally, field rat *Bandicota bengalensis* and house rat *Rattus rattus* are reported to feed on tubers (Palaniswami, 1994).

4.1.6 Management of pests

The pest infestation on taro mainly affected the size and weight of cormels but not the number of cormels. The pest control and the cormel yield showed a positive significant correlation. The loss was in range of 13-29% with a mean of 21.5 per cent depending upon the intensity of pest population (Palaniswami, 1994).

The availability of resistance/tolerance to pests in these crops is very important as it will be one of the effective and economic tools against the pests. Among the accessions/varieties of taro, Thamarakkannan (C9), Kovvur, Panchamuki were observed to be susceptible to major pests (leaf eating caterpillars, aphids, mites and thrips), while the accessions C82, C135 and C266 were fairly resistant to these pests. Mohandas and Palaniswami (1996) identified Thamarakkannan (C9) and Sree Rashmi as resistant to root knot nematode attack and suggested the use of such field resistant varieties in the problem areas.

Among the germplasm collections of elephant foot yam in CTCRI, 14 were grouped as susceptible to mealy bugs with 12-45 per cent field infestation and about 17 accessions were found to be field resistant (Palaniswami, 1994).

Endosulfan or carbaryl (0.05 per cent) spray was suggested against taro pests. The major pests of taro and elephant foot yam can be controlled using dimethoate, quinalphos or fenthion at the rate of 0.05 per cent spray (at an interval of 30 and 60 days after planting). The pest control measures had significantly increased the cormel yield.

The mealy bugs and scale insects of elephant foot yam can be controlled by dipping the seed tubers in 0.05 per cent monocrotophos before planting. The storage pests can be controlled if the tubers are stored in sand medium.

The major pests of taro and elephant foot yam can be controlled using need based application of dimethoate, quinalphos or fenthion 0.05 per cent spray at 30 and 60 days after planting. There are a number of parasitoids, predators and pathogens on the pests of taro. The role of natural enemies in the suppression of pests are to be given importance in the pest management programme (Palaniswami, 1994).

4.2 Diseases of aroids

The important aroids are *Colocasia* (taro/eddoe, swamp taro and bunda types) tannia, *Alocasia, Xanthosoma* and elephant foot yam (EFY) and giant taro. Among these, taro and elephant foot yam are cultivated in many tropical countries. Upland taro and EFY are commercially cultivated in many states of India. Taro leaf blight caused by *Phytophthora colocasiae* is the most devastating disease and wrecks heavy damage of the crop. EFY is also attacked by several fungal and viral diseases which seriously interfere its cultivation and commercial seed corm production. Some important disease of taro and elephant foot yam are described below (Jayanta, 2006).

4.2.1 Taro leaf blight

The *Phytophthora* leaf blight caused by *P. colocaciae* Racib. is the serious menace in taro cultivation. The disease is widespread in tropical and sub-tropical region. The blight may appear at any stage of the crop but in India the disease appears at the onset of monsoon when the crop is at an active growth stage or at maturity time. The disease first appears on the leaves with typical round shaped lesion and rapidly spreads over whole leaf lamina. Many lesions may coalesce to form patches and ultimately the whole leaf become blighted. The symptoms varied greatly with the cultivars/varieties of taro and weather conditions. The conspicuous symptoms includes small dark round spots on upper surface which enlarge, coalesce together and forms characteristic rings of yellow and brown zones. Petiole infection leads to collapse of entire plant. It causes 25-50% yield reduction.

The blight pathogen, *Phytophthora colocasiae*, has a complex life cycle and can attack taro and other aroids in several ways. The assault usually begins when airborne spores land on the foliage and either germinate directly or release swimming zoospores. In either case, the pathogen penetrates the leaf or stem and grows into a dark lesion that releases thousands of new spores. Carried by wind or rain-splash, the spores can infect the same or other plants, sometimes miles away. Zoospores can be washed from the foliage to the taro corms in the soil, causing them to rot immediately or in storage. Infected seed tubers can start a new cycle of infection in a following season.

Durable genetic resistance provides the best option for effective management of this disease. Taro varieties show diverse response to *Phytophthora* leaf blight. The resistance might be occurred intra-species or varietals level i.e. cultivar specific or race specific resistance and Inter-species or genus level resistance i.e. non-host resistance. Factors related to resistance reactions in taro are associated with quantitative phenotype, hypersensitive response subjected to cytological and biochemical changes which is triggered up by signal transduction system in host-pathogen interaction (Mandi *et al.*, 2006, Jayanta, 2006). The pattern of disease development in resistant cultivars highly depends on the leaf characters, angle of divergent with petiole. The leaves with narrow angle of divergent can not hold free water on the leaf surface which is needed for germination of sporangium. The spread of leaf blight of taro is more in the broad ovate leaves with wide angle of divergent. The resistant varieties of taro show Hypersensitive Reaction (HR) due to pathogen invasion and induce extremely fast and plant cells rapidly sacrifice to check fungal growth inside tissue, especially in highly resistant taro against *Phytophthora*, which causes small necrotic spots on the leaves. The cytological evidence in infected resistant cultivars of taro shows containment of hyphae in HR lesion causing trailing necrosis, hyphal coagulation, cytoplasmic granulation with high intensity of stain of Trypan blue due to irreversible membrane damage. Several mesophyll tissue of partial resistant plant is involved in HR. Sometimes, the *Phytophthora* invaded tissues of resistant cultivars induce series of cell death, condensed protoplast which give barrier to the pathogen spread. Penetrated cell of susceptible taro cultivars do not show visible plant response and pathogen hyphae expand through the plant tissue leads to the external symptoms manifestation with massive gum deposition.

Several biochemical changes occur due to pathogen invasion and plant peroxidases have many putative functions including lignification, and likely to play indispensable roles in plant defense system against leaf blight pathogen attack in taro. Moreover, Super Oxide dismutase, Esterase, Glucose -6 phosphate dehydrogenase exudates may work together as a multicomponent defense system to protect plant tissue during pathogenic assault in taro. Isozyme steadily increased in activity during pathogenesis. The diversity of peroxidase, EST in individual cultivar expressed in leaf tissue of taro which is quick migrating isozyme, was expressed at a slightly higher level in infected plant tissue than in healthy tissues. Nitrate reductase is also an important signalling and defense molecule, plays a key role in activating disease resistance in plants. Hypersensitive is always associated with the resistance response of taro to *Phytophthora colocasiae*; the timing, severity and extent of the HR varied considerably, depending upon the examined cultivars. As observed in taro leaves *P. colocasiae* interactions, various level of pathogen ingress correlated with the different necrotic responses (Jayanta, 2006).

In resistant varieties the blight appears along the mid-vein of the leaf and progress on towards the leaf tip. In many varieties small blight lesion are produced and cover the whole leaf lamina. In the advance stage of the disease the blight also appeared on the petiole and severe attack of the pathogen cause dropping of the leaf at the junction of the petiole. The blighted portion of the leaves shows heavy oozing out of phenolic gummy substances which dry up later in the hot climate. Sometimes white fungal growth appeared around the lesion during high humidity. In severe infection the plants could stand for long and ultimately whole plant become blighted and very few small cormels are formed. In the later stage, the corms are also infected and the corms start to rot before harvest. The pathogen also attacks the corms in storage. In the storage condition, the inner tissue of the corms become pale brown colour; in high humidity the corms are rotten but in case of late infection - the corms are firmed and dry with distinct margin.

The *Phytophthora* blight is predisposed with continuous high humidity (90-95%) and temperature (30-35°C) during July to October and heavy rainfall with high temperature spreads the disease very quickly and blights the whole plant in no time. The secondary shoots are also infected at late stage. The blight infection continues up to October - November and a short phase development of disease occurs in the old plants which are favoured by the heavy dew deposition and low night temperature (Mishra, 1996, Jayantha, 2006).

Growing of resistant / tolerant varieties like Muktakeshi, Jhankri, Topi-1, Nadia local etc. (Mandi *et al.*, 2006), early planting (Feb-March) to escape from disease, two to three sprays of Ridomil @0.05% or Mancozeb @ 0.25% and strict field sanitation (Misra, 1996) are found to contain the disease. In the place of synthetic fungicides, application of biocontrol agents (*Trichoderma viridae)* and spraying of neem leaf extract (1 kg in 10 litres water) proved to be helpful in suppressing the disease (Palaniswami and Shirly, 2006).

4.2.2 Collar rot of elephant foot yam

Collar rot of elephant foot yam is caused by *Sclerotium rolfsii* Sacc. The most important symptoms are water soaked lesions on the stem, yellowing at the tip of leaves which moves downward and collapse of stem due to rotting at collar region. It causes 20-100% yield loss. It can be controlled following crop rotation for 2-3 years, field sanitation and if necessary drenching the soil around the plant with 0.2% captan or 0.5% benzimidazole (Thankappan, 1994). Seed tuber treatment with cowdung slurry incorporated with *Trichoderma viridae* (100 g in 20 kg FYM) is highly effective (Palaniswami and Shirly, 2006).

4.2.3 Dasheen mosaic virus (DMV)

There is evidence for the occurrence of a severe strain of the virus in French Polynesia. DMV is not lethal; its chief effect is to retard plant growth and reduce corm yield. Foliar mosaic, mottle and 'feathering' symptoms are apparent. Foliar symptoms are intermittently expressed. Transmission by aphids *Pentalonia negronervosa*, non-persistently, under natural conditions is the vector. Flexuous rods, about 11 x 750 nm. Induces cylindrical inclusions, as do other potyviruses (potyvirus group). Plants free of this virus can easily be obtained through meristem-tip culture when small (0.5 mm or less) meristem-tips are used (Jayanta, 2006).

4.2.4 Storage diseases

The other fungal pathogens like *Pythium* spp., *Fusarium* spp., *Sclerotium* sp., *Botryodiplodia* sp., *Rhizopus* sp. etc are the important pathogens which attack the plant in the field but mainly in the storage.

Pythium spp : Infected tissues of corm become pale and soft. The lesion is bounded by irregular but sharp margin. In advance stage decay is accompanied by unpleasant odour.

Fusarium spp : Decay usually begin at the base or side of the Spongy whitish grey with distinct brown margin. In case of infection with *F. solani,* the dry corky or powdery rot with white mould develop on the surface.

Sclerotium rolfsii : Caused tuber rotting and the tissue becomes pale brown to pinkish, soft but not watery. In humid condition there is development of white mould which given rise to small spherical sclerotia.

Botryodiplodia theobromae : Showed no external symptoms even when decay is advance. Internal tissue is initially greyish or pinkish and subsequently turn black distinct margin. Flesh growth may turn spongy.

Rhizopus stolonifer : The damage starts at one end of the corm. In dry condition firm flesh growth appear on sunken lesion. In humid condition affected tissue is soft and watery with a pronounce odour. The periderm develops a white mycelia.

Aspergillus niger : The tuber tissue is affected with typically fawn or brown colouring.

Pythium aphanidermatum, P. splendens, P. myriotylum, Fusarium solani, F. moniliforme, Black rot (*Ceratocystis fimbriata),* Bacterial soft rot (*Erwinia carotovora),* Blue mold rot (*Penicillium crustosum),* Cylindrocarpon rot (*Cylindrocarpon lichenicola),* Pink mould rot (*Trichothecium roseum)* and Rhizoctonia rot (*Rhizoctonia solani)* are other storage diseases on corms of aroids (Jayantha, 2006).

5. Chinese Potato

5.1 Pests Attacking Leaf and Shoot

5.1.1 Leaf Feeders

The caterpillars of *Hymenia recurvalis* (Fab.) damage leaves. They form a silken thread and web and leaves either folding the same leaf or leaves together and feed on the leaf within. It feeds on the green matter leaving a thin parchment tissue. Only one caterpillar is noted in a fold. The damage results in reduced leaf area by feeding and due to folding of leaves, which in turn, affects photosynthesis of plants. The eggs are laid singly or in small groups and they hatch in 2-4 days. Larval period lasts for 7-10 days through five instars. Pupation is in a silken cocoon inside the folds or in the dried leaves. Moths emerge out in 7-10 days and has blackish brown wings with white wavy markings. The pest is very common in Tamil Nadu and west coast area. The larvae are parasitized in nature by braconid, *Apanteles delhiensis* and *Cardiochiles* sp.

As early as 1938, a caterpillar of economic importance was noted in south India, *viz.,* leaf folder *Pycnarmon cribrata* Fb. Moths are white in colour. Eggs are laid on young leaves singly or in groups. Eggs, larval and pupal period is 2-5, 7-10 and 6-7 days respectively. The caterpillars of this species also fold the leaves and feed within, similar to *H.recurvalis*. In the field, it is often regulated by a braconid parasite *Microgaster psarae* W.

Top Shoot Folders : Phostria piasusalis Wlk. is a pale brown small moth. The larva folds top shoots and leaves with silken webs and feeds within on the green matter of the leaves. Damage is similar to leaf folders both by means of leaf feeding and reduction in photosynthetic efficiency of the plant. Egg, larval and pupal periods lasts for 2-4, 9-12 and 6-7 days respectively. The activity of the pest is more during September-December. It is often regulated by a parasitoid *Leptobatopsis indica* Cameron (Palaniswami, 1994, 1999).

Bihar hairy caterpillar, *Diacrisia obliqua* Wlk. is observed as a pest on coleus. Young larvae feed on the young leaves gregariously. They scrape the chlorophyll content and skeletonize the leaves. Full grown larvae feed on the leaf and, to some extent, vine also. The field infestation reported was 14.5 per cent. The caterpillars possessed long thick hairs and fully grown ones measure 35-40 mm in length. The larvae pupate in cocoons of silk and hairs in dried leaves and branches. It is found as a sporadic pest during September to March. Adult moths are yellowish brown and lay eggs in groups. The total life cycle is completed in about 58 days. Egg, larval and pupal periods are about 7, 34 and 9 days respectively. Adults live for a week.

Tussock caterpillar *Dasychira mendosa* Hubn. lay eggs in masses on young leaves. The caterpillars emerge in nine days and they scrape and skeletonize the leaves. The larvae are greyish brown, 30-35 mm in length with tufts of hairs on prothoracic and pre-anal ends. Larval period lasts for a month. Pupation occurs in cocoons of yellowish brown hairs in dried leaves. Pupal period lasts for a week. Moths are yellowish brown with white patches. The total life cycle completes in 55 to 65 days. When compared to Bihar hairy caterpillar, this pest is less serious (Palaniswami, 1994, 1999).

Black Tingids : Cochlochila bullita S. is a pest in Tamil Nadu and Kerala causing substantial damage to the leaves. Adult lace wing bugs are black in colour with hyaline wings and hollow globular out growth on the dorsal side of the thorax. Adults are active, short distance fliers and live for 14 to 17 days. Pre-oviposition, oviposition, egg, nymphal and total life cycle periods are in the range of 2 to 3, 11 to 15, 6-8, 12-14 and 18 to 22 days respectively. It has five nymphal instars. The eggs are laid either on the edges of leaf lamina or on leaf veins. Due to desapping, the leaf margins dry up which leads to curling, cupping, withering and drying. During heavy infestation, the infested plants show a burnt up appearance. Population was maximum during August to December. Higher minimum temperature and lower maximum temperature favoured the population build up. It also attacks *Ocimum sanctum* (Tulsi) and *Mentha spicata* (Mint) (Palaniswami, 1994).

5.1.2 Pest Attacking Stem/Vine

Palaniswami (1994) reported a cerambycid stem borer (vine borer), *Nupserha vexator* P. It was observed as pest in severe form during October-February. The grubs feed in the stem by boring and tunnelling which result in the arrest of nutrient transmission and destruction of tissues. Infested plants showed yellowing of leaves and terminal drying. At times, the grub also attack the stem and tubers found underneath soil. As many as 6-9 grubs per plant are noticed. Nearly half the plant population in the field was affected by the borer. The total life cycle was completed in 49-60 days and adults live for 7-12 days. All stages of pest could be recorded in the same plant indicating the presence of overlapping generations.

5.1.3 Pests Attacking Root and Tuber

Root Knot Nematodes : Meliodogyne incognita (Kofoid and White) Chitwood is a pest of economic importance. The infection by the nematode causes suppressed root growth and heavy gall formation resulting in the stunted growth and wilting. The infected tubers are malformed and hypertrophied in size due to heavy galling rendering them unsuitable for consumption and marketing. In case of heavy infection, the tubers rot even before harvest. Cracking on the surface of infected tubers and blackening have been observed. The starch content of the tubers were considerably reduced. This nematode has a wide range of host and adaptability (Palaniswami, 1999).

Mealy Bugs : Geococcus coffeae Green, at times, infest roots and tubers in field. This is minor pest.

White Grubs

Leucopholis coneophora Burm. is also reported as a pest of local importance. They feed on the tubers and roots in field. They are sporadic in nature.

5.1.4. Storage Pests

Root knot nematode infected tubers start shrivelling and drying in storage. The quality and marketability of tubers are affected.

5.1.5 Pest Management

Insect pests can be controlled using a spray of fenthion or malathion (0.05 per cent). For effective nematode management, use of nematode-free seed tubers for raising nursery, keeping the field free from weeds and rotating the crop with non-host or resistant varieties sweet potato variety Sree Bhadra (Mohandas and Palaniswami, 1991) are recommended. The leaf folders and shoot folders are often regulated by parasitoids in the field.

The practice of chemical control on tuber crops has certain limitations and hence emphasis should be given to maximum utilization or preservation of available natural enemies, use of field tolerant varieties and cultural methods with a need based application of bio-pesticide.

5.2 Diseases

Leaf spot disease caused by *Cercospora* and a mosaic disease are occasionally seen.

6. Yam Bean

Spotted pod borer *Maruca vitrata* is a problem in seed setting in yam bean. Dimethoate or endosulfan 0.04% spray was useful in checking the infestation of pod borer. Aphids and mealy bugs area occasional pests. Yam bean leaves had been noted to be infected with mosaic (Palaniswami and Shirly, 2006).

7. Plant Quarantine

Many of the economically important pests and diseases found in different parts of the world have not so far gained entry into India. Therefore we need to guard our tuber crops against the entry of such devastating exotic pests and diseases. This is very important especially in the context of liberalization of export/import policies. The movement of vegetative/ reproductive pests of different tuber crops has to be done scientifically with caution. Prevention is better than the cure. Hence international movement of planting materials, tubers, etc. has to be under strict plant quarantine.

Fortunately some of the pests of international importance which do great havoc to the crop *viz.*, green mite *(Mononychellus* spp.), thrips *(Frankliniella williamsi)*,mealy bug *(Phenacoccus mainihoti* Matile-Ferrero) and hornworm *(Erynnis ello)* have not so far gained entry into our country. Strict plant quarantine measures are to be adopted to prevent their entry.

The economically important exotic pests which are not prevalent in India are given below.

7 a. Exotic diseases of economic importance to be taken care during exchange of germplasm of tuber crops and while importing planting materials from other countries into India (Anon 1978, Thakappan 1994, Palaniswami *et al.,* 2004, Jayanta, 2006).

Disease / Geographic distribution	Symptoms/transmission/particle morphology
1. Cassava	
1.1 African Cassava Mosaic disease (African Cassava Mosaic Virus)	
Tropical Africa (Uganda), Cape Verde, Sao Tome and Principe, Malagasy and Seychelles	Cassava plants show mild to severe mosaic, yellowing distorted leaves, and stunted growth. ACMV is transmitted by vegetative propagation and by the whitefly *Bemisia tabaci* in a persistent manner. ACMV has not been shown to be transmitted by seed. The particles typically are bipartite, approximately 30 x 20 nm. Thermotherapy followed by meristem-tip culture. The use of serological techniques (preferably ELISA, using monoclonal antibodies) is recommended. As there are serological differences between isolates of ACMV from East and West Africa and between ACMV and ICMV, it is important to use monoclonal antibodies that detect the virus isolates occurring in the country of origin.
1.2 Cassava Brown Streak Virus disease (CBSV)	
Cassava growing regions in East Africa; Kenya, Malawi, Tanzania and Uganda.	Yellow blotches and mottling develop on mature leaves, but not on young leaves. Small thin lesions on young stems which coalesce to form blotchy patches and extend into the cortex. Unripe fruit may have black necrotic spots which extend to the fleshy tissues beneath. Roots show longitudinal fissures surrounded by discoloured tissues and internal lesions also develop. In young plants, stem symptoms are often observed before leaf symptoms. CBSV is transmitted by vegetative propagation.Possibly whitefly transmitted (*Bemisia* spp.). Thermotherapy followed by meristem-tip culture. The virus can only be detected in plants showing symptoms. Mechanical inoculation with sap from older, lower leaves will produce

Contd...

Table 7.a. Contd...

	local lesions in *Chenopodium quinoa* and vein yellowing, mottling and distortion in *Nicotiana benthamiana*. The virus can be detected in symptom-bearing cassava leaves by ELISA . using polyclonal antiserum. Symptom development takes several months in cassava.
1.3 Cassava Common Mosaic Virus disease (CCMV)	
CCMV occurs throughout tropical America from Mexico to Paraguay.	Mosaic symptoms and chlorotic areas that are often limited by veins. No known vector. Semiflexuous rod, approximately 495 x 15 nm (Protexvirus group). Thermotherapy followed by meristem-tip culture. The virus is readily detected by serological methods including ELISA and ISEM.
1.4. Cassava Green Mottle Virus disease (CGMV)	
CGMV disease has been reported only from Choiseul, Solomon Islands.	Faint or distinct mottling and mosaic, often on puckered leaves with distorted margins. Plants often recover to give slightly stunted but otherwise symptomless plants. Occasionally, plants are severely stunted. CGMV is transmitted by vegetative propagation. In cassava symptoms develop in 2-5 months. The virus has been shown to be transmitted through seeds of *N.clevelandii*, but seed transmission in cassava has not been tested. Affinities to the nepovirus group imply possible transmission by nematodes. Isometric, approximately 26 nm in diameter. Indexing by elisa using a polyclonal antiserum. In mechanically inoculated plants, the virus can be detected only after symptoms develop, but can still be detected after the symptoms disappear.
1.5 Cassava Vein Mosaic Virus disease (CVMV)	
The virus has been reported from many states in Brazil and is prevalent in the northeastern states.	Symptoms include a chlorosis of the veins which can either appear as a chevron pattern or coalesce to form a ringspot pattern. Some leaves show a mosaic pattern over the entire

Contd...

Table 7.a. Contd...

	leaf. There is often leaf distortion and sometimes the young leaves show epinasty. The virus is transmitted by vegetative propagation. There is no conclusive evidence about the vectors of the virus. Isometric, approximately 50 nm in diameter (Caulimovirus group). Thermotherapy followed by meristem-tip culture. The virus can be detected by standard serological methods.
1.6 Cassava American Latent Virus (CALV)	
The virus was isolated from cassava collected in Brazil and Guyana.	The virus can be mechanically transmitted to cassava. Isometric, approximately 28 nm in diameter. The virus can be detected by serological methods including ELISA.
1.7 Cassava Colombian Symptomless Virus disease (CCSV)	
The virus was isolated from cassava collected in Colombia.	CCSV is transmitted by vegetative propagation. No vectors of CCSV are reported. Semiflexuous rod, approximately 495 x15 nm. The virus is readily detected using ELISA or ISEM.
1.8 Cassava Ivorian Bacilliform Virus disease (CIBV)	
The virus was isolated from cassava collected in cote d'Ivoire.	Vector unknown. Transmitted mechanically. Bacilloform particles of three predominant lengths: 42, 48 and 75 nm (Tentative member of the alfalfa mosaic virus group). The virus can be detected in test plants using ELISA with polyclonal antiserum.
1.9 Cassava X Virus (CsXV)	
CsXV is found in the north coast of Colombia.	CsXV is transmitted through vegetative propagation. Vector unknown. Not transmitted through seed. Semiflexous rod, approximately 15 x 495 nm (Tentative member of the potexvirus group). Thermotherapy followed by meristem-tip culture. The virus is readily detected by ELISA or ISEM.
1.10 Frogskin disease (FSD)	
FSD is found in the Andean regions of central and southern	No symptoms on the stems or leaves, but the roots are stunted and do not fill with starch.

Contd...

Table 7.a. Contd...

Colombia. It is also endemic in the Amazon regions of Colombia and Caribbean.	The lower stem may become enlarged. The root periderm and corky layer enlarge and form raised lip-shaped fissures. When the root symptoms are mild, the fissures often form a constricted ring around the root. A few cassava clones develop mosaic symptoms. The whitefly *Bemisia tuberculata* is suspected to be the vector of both FSD and CMD. Thermotherapy followed by meristem-tip culture. Plants can be indexed for FSD and CMD by grafting cuttings of an indicator clone of cassava (Secundina) onto them. The buds should be removed from the plant to be tested and the grafted plants should be maintained below 30°C. Plants which index positively show a severe mosaic on the new leaves, approximately a month after grafting.
1.11 Cassava antholysis	
Brazil (especially in the Northeast region), the Caribbean, Central America, Colombia and Venezuela.	Symptoms are first observed as virescence of the normally pinkish-cream tepals, followed by the hypertrophy and phyllody to produce a syndrome known as antholysis. Elongation of the floral axis is common in these flowers. Deformed flowers are sterile and abort prematurely. Caused by MLOs.
1.12 Cassava Witches' Broom	
Important in Brazil (States of Ceara, Pernambuco and Sao Paula) and southern Mexico.	Three different types of symptoms have been reported : a) stunting and excessive proliferation of branches: shoots have small leaves and shortened internodes, without distortion or chlorosis; b) proliferation of shoots from the cutting with generally weak growth; c) a few weak, stunted shoots germinate from the cutting which never reach normal size. Transmitted by vegetative propagation. Thermotherapy followed by meristem-tip culture. Caused by MLOs.
1.13 Cassava bacterial blight (CBB) Caused by *Xanthomonas campestris* pv.	
CBB has been reported in almost all countries of Africa, Asia and	*Mahihotis* .Initially, angular, water-soaked leaf spots, more clearly seen on the undersurface,

Contd...

Table 7.a. Contd...

Latin America. It is present in the Federated States of Micronesia, Guam, Indonesia and the Philippines.	sometimes with yellow halos above, rapidly expanding, and turning brown. Leaves wilt, desiccate, roll and fall off. Petioles are also attacked, leading to vascular infection on young shoots, stem rot and dieback. Yellowish exudates often collects in droplets on leaf spots or is exuded from cracks that develop on young infected stems and petioles. Roots rots around dead vascular tissues. To prevent movement of CBB, cuttings or seeds should only be taken from plantations that are free of the disease.
1.14 Cassava anthracnose Caused by *Colletotrichum* spp..	
Anthracnose occurs in most cassava growing areas where annual rainfall is higher than 900 mm.	Initial infection is characterized by leaf spots and distortion of young leaves. Later the pathogen invades petioles and green stems, inducing partial or total necrosis. On the stems, it produces cankers and dieback. Meristem-tip culture eliminates the disease.
1.15 Dry root stem rot Caused by *Diplodia manihotis.*	
Cassava growing regions in Asia and tropical America.	The root system becomes blackish and deteriorates. Stem and root rot lead to partial or total wilting and dieback. Meristem-tip culture eliminates the disease.
1.16. Superelongation disease Caused by *Sphaceloma manihoticola.*	
Brazil, the Caribbean, Central America, Colombia and Venezuela. *Sphaceloma manihoticola* has also been recorded from Cook Islands associated with scab-like lesions and leaf distortion, but without stem elongation.	White, irregular spots on the leaf blades, but more commonly along midribs and veins, and on petioles, often joining together to form cankers several centimeters long. Yields of heavily infected plants are severely reduced.
1.17. East African cassava mosaic virus	
Africa	Geminiviridae :begomo virus
1.18. South African cassava mosaic virus	
Africa	Geminiviridae :begomo virus

Contd...

Table 7.a. Contd...

1.19 Cassava kumi virus	
Africa	Unassigned
1.20 Cassava 'Q' virus	
Africa	Unassigned
2. Sweet Potato	
2.1 Cucumber Mosaic Virus (CMV) disease	
CMV occurs worldwide, but is only reported in sweet potato from Israel and West Africa.	Stunting, chlorosis and yellowing of plants. CMV apparently only infects sweet potato plants previously infected with SPFMV. The virus is transmitted through vegetative propagation and by a number of aphid species. Transmission requires the presence of SPFMV in the acceptor plant. Isometric, about 30 nm in diameter (cucumo virus group).
2.2 Yellow dwarf	
Taiwan	Mottling,chlorosis and dwarfing. Sweet potato yellow dwarf ipomovirus.
2.3 Ipomoea leaf crinkle	
Israel	Leaf crinkling and puckering. Ipomoea leaf crinkle virus.
2.4 Sweet potato little leaf, or witches' broom disease	
Japan (Ryukyu Islands), Korea, New Caledonia, Niue, Palau, Papua New Guinea, Solomon Islands, Taiwan, Tonga and Vanuatu.	Vein-clearing of otherwise healthy leaves is the first sign of infection, followed by the development of progressively smaller, chlorotic leaves sometimes rolled upwards at the margins, puckered and in some cultivars more rounded than normal. Diseased stems are short, erect, and multibranched due to growth of normally dormant axillary buds, giving a bushy appearance, further increased by gradual shortening of internodes to 1 cm or less. Storage roots are either absent or pencil-thin. *Orosius lotophagorum ryukyuensis* (in Pacific countries)

Contd...

Table 7.a. Contd...

	and *Nesophrosyne (Orosius) ryukyuensis* (in Japan) are vectors. The MLO is not seedborne. Caused by MLOs. Only meristem-tip culture is efficient. Tetracycline will give remission of symptoms but will not eliminate the MLO from the plants.
2.5 Sweet potato scab	
Brazil, Brunei, China, Indonesia, Japan, Malaysia, Philippines, Taiwan and the following Pacific island countries: Cook Islands, Federated States of Micronesia (excluding Kosrae), Fiji, French, Polynesia, Guam, New Caledonia, Niue, Palau, Papua New Guinea, Solomon Islands, Tonga and Vanuatu.	Small, brown, scabby areas mostly among the leaf midrib and veins and on the petioles, 1-5 mm at first, later joining together for several centimeters. Numerous pinpoint size spots occur in patches between the veins. Severe distortion of leaves occurs as early infection of veins prevents normal leaf expansion. Leaves are small, curled, with deeply torn edges and petioles are short and twisted. Caused by *Elsimoe batatus*. The movement of vines and tubers between countries should be avoided. Preference should be given to importing seeds or sterile, pathogen-tested, tissue cultures growing in a tissue-culture medium.
3. Aroids	
3.1 Alomae	
Papua New Guinea and Solomon Islands.	Leaves are short and the tissues are thickened and severely distorted but remain greenThe bobone virus is transmitted by the taro plant hopper *(Tarophagus proserpina)* whereas the smaller bacilliform virus is transmitted by the mealybug, *Planococcus* sp. Virions of bobone are bacilliform in outline (50 x 330 nm) as are those of the smaller virus (30 x 125 nm). Plants free of alomae can readily be obtained by meristem-tip culture when small (0.5 mm or less) meristem tips are used. Negatively stained leaf extracts examined by electron microscopy will reveal bacilliform virus of two size ranges. A post-entry quarantine period for one crop cycle is essential to assure that the 'male' taro plants are free of the disease.

Contd...

Table 7.a. Contd...

3.2 Bobone virus	
Papua New Guinea, Solomon Islands. There is evidence for a mild strain of bobone in several Pacific island countries, causing vein clearing without severe leaf distortion.	Severe stunting of recently planted stock, leaves are puckered, distorted, brittle and thickened. Plant eventually produces leaves with few or no symptoms Original leaves with symptoms remain green. Transmission is by the delphacid planthopper *Tarophagus proserpina*, in a persistent manner. Bacilliform, 50 X 330 nm (rhabdovirus group). Caused by mixed infection of rhabdovirus and unnamed bacilliform virus. Plants free of this virus can easily be obtained by meristem-tip culture when small (0.5 mm or less) meristem tips are taken. Detection of diagnostic symptoms of actively growing, recently planted material is the only other reliable indexing method for this virus.
3.3 Small bacilliform virus disease	
Cook Islands, Papua New Guinea, Samoa and Solomon Islands.	Vein clearing, especially along leaf margins. Leaf blade curved downward along margins, not thickened or distorted as in bobone. Transmitted by the mealybug, *Planococcus* sp. Bacilloform, 30 x 125 nm. Plants free of this virus can easily be obtained through meristem-tip culture.
4. Yams	
4.1 Chinese yam necrotic mosaic vi ıs disease (CYNMV)	
Japan	Conspicuous chlorotic and/or necrotic spots and severe interveinal chlorosis and necrosis. The virus is transmitted efficiently in the non-persistent manner by aphids (especially *Aphis gossypii* and *Myzus persicae*. Filamentous particles of about 12-13 x 660 nm. The virus can be rapidly detected and identified by electro-blot immonassay.
4.2 Cucumber mosaic virus (CMV)	
Worldwide in a wide range of plant species, but reported in	Severe leaf chlorisis. Efficiently transmitted in the non-persistent manner by a wide range of

Contd...

Table 7.a. Contd...

D. alata only in West Africa.	aphid species. Isometric, about 30 nm in diameter (cucumovirus group). It is readily identified by standard serological procedures (ELISA and ISEM).
4.3 *Dioscorea bacilliform* virus	
Widespread in *D. bulbifera* in the Caribbean area, and occurs in *D. alata* cv. White Lisbon in Barbados.	Severe interveinal leaf chlorosis in *D. bulbifera* and associated with internal brown spotting of tubers in *D. alata*. Vector unknown. Bacilliform, about 28 x 30 nm. The virus is sap transmissible to seedlings of *D. bulbifera*. It is readily identified by standard serological procedures (ELISA and ISEM).
4.4 *Dioscorea* latent virus	
Common in *D. floribunda* and *D. composita* in Puerto Rico.	Potexvirus group. No known vector; through planting material, probably by mechanical means. Slightly flexuous filaments, about 11 x 485 nm. The virus is readily detected and identified by standard serological procedures (ELISA and ISEM).
4.5 Yam mosaic virus (YMV)	
In *D. rotundata-cayenensis* and *D. esculenta,* the virus is widespread in West Africa and detected occasionally in *D. alata* from the South Pacific.	Severe leaf chlorosis and leaf distortion. Non-persistently transmitted by aphids (especially *Aphis gossypii*). Flexuous filaments, about 12 x 750 nm (potyvirus group). The virus is mechanically transmissible to indicator plant species (eg. *Nicotiana benthamiana)* and is readily identified by standard serological assays (ELISA and immunosorbent electron microscopy).

7 b Exotic pests of economic importance to be taken care during exchange of germplasm of tuber crops and while importing planting materials from other countries into India (Palaniswami, 1994, 1999).

1. Cassava	
1.1 Coreid bugs (*Amblypelta* spp.) Northern Australia, Indonesia	Initially a wilt of the terminal shoot which, if it survives, develops cankerous swelling and

Contd...

Table 7.b. Contd...

(Java and Timor), New Caledonia, Papua New Guinea, Solomon Islands and Vanuatu. Coreidae:Homoptera	cracks. Alternatively, shoots die and laterals develop. Continued attack of regrowth results in stunted, flat-topped plants. Cracks also occur in petioles with characteristic drooping of the leaf. Levels of damage are high relative to Amblypelta numbers, hence the assumption that toxins are injected during feeding. There are 18 species and subspecies. For *A. coccophaga,* eggs are laid singly on leaves or stems, they hatch after 8 days and there are five nymphal instars. Development from egg to adult takes about 35 days. Adults are 20-30 mm long, pale reddish brown with yellowish legs. Both sexes have an unpleasant smell when handled. *Amplypelta* spp. are polyphagus and has been recorded from species of many families especially *Euphorbiaceae.*
1.2 Mealy bugs	
Phenacoccus manihoti Mat. Ferr. found throughout most of the cassava growing regions of Africa, while in the Americas it is confined to Bolivia, southwestern Brazil and Paraguay. *P. herreni* is reported from northeastern Brazil, Colombia, the Guyanas and Venezuela. Neither species is present in Asia. Pseudococcidae : Homoptera	Damage symptoms due to *P. manihoti* and *P. herreni* are very similar and are characterized by leaf yellowing, curling and deformation resulting in a tight cabbage-like appearance of the apical buds. Large numbers of mealybugs cause extensive leaf necrosis, defoliation, stem distortion and dieback. Eggs, placed in ovisacs where several hundreds may be present, hatch in 5 to 8 days at 25-30°C. First instar nymphs are highly mobile and disperse throughout the plant. *P. manihoti* is parthenogetic and no males have been found; *P. herreni* has both males and females. Mealybugs can be disseminated by wind, or carried by animals or on vegetative planting material. Tissue culture plantlets eliminates the mealy bugs.
1.3 Green Mites	
Mononychellus mites are found throughout Africa, and South Americas. They have not been reported in Asian countries.	The *Mononychellus tanajoa* (Bondas) mite is usually found on growing points of the plant, on buds, young leaves and stems. Newly emerged leavers are marked with yellow spots

Contd...

Table 7.b. Contd...

Tetranychidae : Acarina	or appear mottled, bronzed and deformed. Not known for *Mononychelles*, but over 100 plant species are host for Tetranychus.
1.4 Scale insects	
Scales have been reported attacking cassava stems in many cassava -growing regions of Africa, the Americas and Asia. *Pseudaulacaspis pentagona*, probably native to China and Japan, has been reported on various fruits trees and ornamental plants in Africa, the Americas, Asia, Australia, the Caribbean, Mediterranean Europe and the Pacific. Diaspiditae : Homoptrea	High scale populations may cover the stems and lateral buds. Leaves yellow prematurely, wilt and fall. Severe infestations retard plant growth. Cuttings from infested stems do not sprout well, initial growth is slow and shoots may desiccate and die. Adults of *A. albus* are grey, elongated and mussel-shaped (2-3 mm long). *P. pentagona* has a white, roughly circular scale (2-2.5 mm in diameter) with a reddish-yellow nipple towards one side. *A. albus* females produce an average of 47 eggs deposited beneath the scale. Eggs hatch in 4 days; the first instars (crawlers) are mobile and can disperse to other leaves and stems of the same or adjacent plants, or to distant plantings by wind. Once the crawlers have settled, their stylets are inserted into the plant, where they remain throughout the feeding period of the immature stages. The total life cycle is 22-25 days. *P. pentagona* females moult twice and produce an average of 140 eggs during their life span of 27-23 days. The total developmental period ranges from 24-27 days. Male scales have eyes, long antennae, legs and a pair of forewings. Important pests of a wide range of other economic plants and weeds.
1.5 Stem borers	
Coelosternus species are confined to the Americas, except for *C. manihot* which is reported from West Africa. *Logochirus* species are primarily reported from the Caribbean, Central America and northern South America. *C. clarkei* has been reported	*Coelosternus* spp. and *Lagochirus* spp. are coleoptera, whereas *C. clarkei* is a lepidopteran. Larvae of all species damage cassava by penetrating the stem and tunneling to the pith; this weakens stems and branches which may eventually wither and break. Coelosternus oviposits on various parts of the plant but mostly in succulent tissues. Eggs of C.

Contd...

Table 7.b. Contd...

from Colombia, Paraguay and Venezuela. Cerambycidae : Coleoptera	granicollis are white and larvae are curved, with a yellowish white to pale brown body. The larval period ranges from 30 to 60 days. Adults are light to dark brown. Adults of Lagochirus spp. oviposit 2.5 mm below the bark and eggs hatch in 5-6 days. The larvae which take about 2 months to develop, measure up to 29 mm and feed at the base of the plant. The pupal period, which lasts about 1 month, takes place in the larva chamber in the stem. Adults are brown, about 17 m long and feed on leaves and bark. Chilomima clarkei lays eggs on the stems around the lateral buds of leaves that have already fallen. The larva, during its first four instars, is found feeding under a fine web around the buds. After the fifth instar it penetrates the stem and completes its life cycle in galleries. The duration of the life cycle depends on the variety and varies from 62 to 68 days.
1.6 Thrips *(Frankliniella williamsi* and other species)	
F. williamsi is reported on cassava only from the Americas and Hawaii (Bellotti and van Schoonhoven, 1978). *Euthrips manihoti* is reported only from Brazil (Belloti and van Schoonhoven, 1978). Thripidae : Thysanoplera	Damages the terminal bud of the plant. Leaflets are deformed and show irregular chlorotic spots. Brown wound tissue appears on stems and petioles, and internodes are shortened. Growing points may die, causing growth of lateral buds which may also be attacked, giving rise to a witches' broom appearance. Grasses, especially Maize, are alternative hosts.
1.7 Flea Beetles	
Pale-striped fleabeetle, *Systena frontalis*, feed on sweet potatoes (both foliage and roots).	Foliage damage caused by the beetles may vary from holes eaten in leaves, to narrow channels, or to grooves eaten into the upper leaf surface. Early harvest of the sweet potato is recommended to avoid storage root infestation
1.8 Weevils and wire worms	
Cylas puncticollis (Boheman). Africa	Attack tubers/vines Serious Pest

Contd...

Table 7.b. Contd...

C. brunneus (F). Africa	Attack tubers/vines Serious Pest
Eusceps postfasciatus (Fairmaire) West Indian weevil	South Pacific, Caribbean, Central and South America. Attack tubers/vines Serious Pest
Conoderus spp. Elateriidae : Coleoptera Wire worms	Americas. Attack tubers/vines Serious Pest
2. Aroids	
2.1 Taro beetles, *Papuana armicollis* (Fairmare), *P. huebneri* (Fairmaire), *P. inermis* Prell, *P. uninodis* Prell, *P. woodlarkiana* (Mont)	
There are many species occur in Fiji, Kiribati, Indonesia (Irian Jaya and the Moluccas Island), Papua New Guinea, Solomon Islands, Vanuatu. Australia and Indonesia Dynastidae : Coleoptera	Adult beetles bone and tunnel into the corms to feed. Infestation can totally destroy the corm and even low infestations reduce marketability. The unrestricted movement of taro between countries should not be permitted. Imports should preferably be in the form of sterile, pathogen-tested plantlets growing in tissue culture medium.
2.2 Taro planthopper, *Tarophagus proserpina* (Kirkaldy)	
Australia, Hawaii, Indonesia, Japan, Fiji, Malaysia, Philippines Samoa, New Guinea, Soloman Islands Delphacidae : Homoptera	Feeds on the sap of plants and heavy infestations can cause wilt. *Tarophagus* is the vector of the taro rhabdovirus (bobone virus).
2.3 Taro Thrips, *Organothrips bianchi* Hood. Thripidae : Thysanoptera	
Hawaii, Polon Is., Samoa	Usually found in stems from the water line.
2.4 Weevil, *Elytroteinus subtrundatus* (Fairmaire). Curculionda : Coleoptera	
Hawaii, Fiji, Pacific Islands	Tunnel in corms
2.5 Taro beetle, *Ligyrus ebenus* (De Geer) *L. cuniculus* (F). Dynastidae : Coleoptera	
North South America, West Indies	Adults and larvae bore and feed the corms serious pest.
2.6 Taro nematode, *Hirschmanniella miticausa*	
Papua New Guinea and Solomon Islands	Internally, a dry brown rot extends in narrow irregular bands upwards from the base of the

Contd...

Table 7.b. Contd...

	corm, Often the basal parts of the corms are completely decayed by secondary soft rot
3. Yams	
3.1 Greater yam beetle, *Heteroligus meles* (Bill)	
Tropical Africa. Dynastidae:Coleoptera	Tubers are attacked; Further attack occurs just before harvest when the beetles again feed voraciously and then migrate to the breeding sites. The unrestricted movement of tubers between countries should be avoided.
3.2 Lesser yam beetle, *Heteroligus appius*	
Occurs in southern Nigeria	The beetle is smaller than H. meles, but the damage is similar.
3.3 Yam weevil, *Paleopus costicollis* Marshal.	
Occurs in Tropical Africa	Both weevil and grubs tunnel the tubers.

References

Anonymous, 1978. Pest control in tropical root crops. *Pans pest manual* No. 4, Centre for Overseas pest research, ODA, London. 235 pp.

Antony, B., Lisha, V.S., Palaniswami, M.S., Sugunan, V.S., Makesh Kumar, T. and Henneberry, T.J., 2006a. *Bemisia Tabaci* and Indian Cassava mosaic Virus transmission. *Int. J. Tropical Insect Science*, 26(3): 176-182.

Antony, B., Palaniswami, M.S., Kirk, A.A and Henneberry, T.J. 2006b. Aphelinid parasitoids of *Bemisia tabaci* G. in India. *Entomon* 31: (3):1-8.

Belloti, A.C. and Schoonhoven, A.V. 1978. Mites and insect pests of cassava. *Ann. Rev. Entomol.* 23:39-67.

Bellotti, A.C., Smith, L. and Lapointe, S.L. 1999. Recent Advances in cassava pest management. *Ann. Rev. Entomol.* 44:343-370.

Binu Antony and Palaniswami, M.S 2006. *Bemisia tabaci* feeding induces pathogenesis – related proteins in cassava (*Manihot esculenta* Crantz) *I. J. Biochemistry and Biophysics* 43: 182-185.

Chadha, K.L. 2001. Hand Book of Horticulture. ICAR, New Delhi. pp. 499-522 and 806-812.

Ekanayake, H. 2000. Potential of entomopathogeic nematodes as biocontrol agents of sweet potato weevil *Cylas formicarius*. *International News letter. CIP SWA* New Delhi, 3(2):9.

Jansson, R.K and Raman, K.V. 1991. *Sweet potato pest management – a global perspective.* West view Press. Oxford. 458 pp.

Jayanta, T. 2006. Virus diseases of sweet potato and *Phytophthora* leaf blight. Report submitted to AICRP Tuber Crops, Trivandrum. 41 pp.

Jayaramaiah, M. 1975. Bionomics of sweet potato weevil, *Cylas formicarius* Fab. Oliver (Coleoptera:Curculionidae). *Mysore J. Agricultural Science,* 9, 99-109.

Johnson, J., Visalakshi, A., George Koshy and Nair, M.R.G. 1979. Studies on the use of some insecticides and organic materials for the control of sweet potato weevil, *Cylas formicarius* Fab. *J. Root Crops,*5 (1&2):132-135.

Kareyija, R.F., Gibson, R. W. and Valkonen, J.P.T. 1998. The significance of sweet potato feathery mottle virus in subsistence sweet potato production in Africa. *Plant Disease* 82(1): 4-15.

Lal, S.S and Pillai, K.S. 1982. Pests attacking cassava *Manihot esculenta* Crantz. in South India. *Tropical Pest Management* 27:480-491.

Lisha Vijayan, A., Antony, B., Palaniswami, M.S. and Henneberry, T. J. 2003. *Bemisia tabaci* (Gennadius) Biotypes (Homoptera: Aleyrodidae) in India. *J. Econ. Entom* 96 (2):322-327.

Mandi, S., Sinha, B. and Tarafdar, J. 2006. Diverse response of promising Taro (*Colocasia esculenta* var antiquorum) Cultivars to *Phytophthora* blight. *Indian J. Plant protection* 34(2): 236-238.

Misra, R.S. 1996. Prevalence and assessment of yield losses caused by *Phytophthora* leaf blight in *Colocasia* in Northern and Eastern parts of India. In Tropical tuber crops : Problems, Prospects and Future Strategies. Ed. G.T. Kurup, M.S. Palaniswami *et al.,* Oxford and IBH Publishers. pp. 380-387.

Mitchel, W.C and Maddison, P.A. 1983. Pests of taro. In: Taro – a review of Colocasia esculenta and its potentials. (Ed. Wang, J.), University of Hawaii Press, Honolulu. pp. 180-225.

Mohandas, C. and Palaniswami, M.S. 1996. Resistance in Tuber Crops to *Meloidogyne incognita (In:)* Tropical Tuber Crops. Oxford IBH, New Delhi pp.360-369.

Palaniswami, M.S. 1994. Pests of edible aroids, yams and Chinese potato and Biological control of pests of tropical tuber crops. *(In:)* Advances in Horticulture volume 8. Tuber Crops (Eds. Chadha, K.L. and Nayar, G.G) Malhotra Publishing House, New Delhi. pp. 485-522.

Palaniswami, M.S. 1999. Major pests of tropical tuber crops and their management. *(In:)* IPM system in Agriculture. Vol. VI (Eds. R.R. Udpadhyay, K.G. Mukerjee and O.P. Dubey), Aditya Books Ltd., New Delhi.

Palaniswami, M.S. 2000. Biological and parabiological approaches for sweet potato weevil management in India. *International potato centre New letter* 3(2):6-8

Palaniswami, M.S. 2004. Final Report on Utilization of Kairomone for sweet potato weevil management and its potential as a biocontrol Components of IPM, ICAR-NATP-CGP Report. 81 pp.

Palaniswami, M.S., Antony, B., Lisha, V.S. and Henneberry, T.J. 2001. Sweet potato whitefly *Bemisia tabaci* : Ecobiology, host interaction and natural enemies. *Entomon* 26: 256-262.

Palaniswami, M.S. and Chattopadhyay, S. 2006. Ecology based integrated management of the sweet potato weevil in India. *Acta Hort.* 703: 183-187.

Palaniswami, M.S., Makeshkumar, T and Hegde, V. 2004. Management of vector borne disease in Horticultural crops. *(In:)* Crop Improvement and Production Technology of Horticultural crops. Horticultural Society of India, New Delhi, vol. I. pp. 698-726.

Palaniswami, M.S. and Mohandas, N. 1991. Incidence of *Cylas formicarius* F. on sweet potato at different stages of crop growth. *J. Root Crops.* 17:60-66.

Palaniswami, M.S. and Mohandas, N. 1992. Search for resistance to *Cylas formicarius* F. in sweet potato. *J. Root Crops* 18:41-52.

Palaniswami, M.S., Mohandas, N. and Saraswathy, P. 1991. On farm survey on the incidence, extent and intensity of *Cylas formicarius* F. damage on sweet potato in Kerala. *J. Root Crops,* 17(Special), 180-196.

Palaniswami, M.S and Shirly Raichal, A. 2006. Region specific technologies for tropical root and tuber crops in India. Tech. Bulletin 47, AICRPTC, CTCRI, Trivandrum. 67 pp.

Palaniswami. M.S, Sreeja, V.G. and Roshan, C.J. 2000. A potent kairomone for management of sweet potato weevil *Cylas formicarius. Entomon,* 25: 269-280.

Pillai, K.S. 1994. Pests of Cassava, *(In:)* Advances in Horticulture Vol. 8. Tuber Crops (Eds.) K.L. Chadha and G.G. Nair. M.P.H., New Delhi pp. 429-456.

Pillai, K.S. and Palaniswami, M.S. 1984. Pests of tuber Crops. *Indian Farming* 33(12): 58-66.

Pillai, K.S. and Palaniswami, M.S. 1985. Spidermites of cassava. Tech. Bull. No.1, CTCRI, Trivandrum. pp17.

Rajamma, P., Palaniswami, M.S and Mohandas, C. 1989. Rodent pests of tuber crops. Technical Bulletin 25, CTCRI, Trivandrum. 28 pp.

Rajamma, P., Pillai, K.S and Palaniswami, M.S. 1993. Sweet potato weevil Research in CTCRI. Technical Bulletin 16, CTCRI, Trivandrum, 43 pp.

Sinha, R.K.P. and Mehta, P.P. 1982. A new host, *Ipomoea batatas,* of *Cercospora ipomoeae* from India. *Indian Phytopath.,* 35: (4): 679-80.

Subramaniam, T.R., David, B.V., Thangavel, P. and Abraham, E.V. 1977. Insect pest problems of tuber crops in Tamil Nadu. *J. Root Crops,* 3(1), 43-50.

Sutherland, J.A. 1986. A review of the biology and control of the SPW *Cylas formicarius* (F.). *Trop. Pest Management* 32: 304-315.

Thankappan, M. 1994. Fungal diseases of aroids and yams. *(In:)* Advances in Horticulture vol.8, Tuber Crops Eds. K.L. Chadha and G.G. Nayar, Malhotra Publishing House, New Delhi. pp. 397-417.

Thresh, J.M, Otim-Nape, G.W., Thankappan, M and Muniyappa, V. 1998. The mosaic disease of cassava in Africa and India caused by whitefly borne gemini viruses. *Rev. Pl. Path.* 77(9):937-945.

Trehan, J.N and Begal, S.R. 1957. Life history, bionomics and control of sweet potato weevil (*Cylas formicarius*) with short notes on some other pests of sweet potato in Bombay State. *Indian J. Entomology* 19:245-252.

❐❐❐

9

Tools and Equipments for Tuber Crops

The major tuber crops cultivated in tropical countries are: cassava (*Manihot esculenta* Crantz), sweet potato (*Ipomoea batatas* L.), yams (*Dioscorea sp.*), aroids (*Amorphophallus, Alocasia, Xanthosoma, Colocasia*), *Canna edulis*, arrowroot, Chinese potato, winged bean and yam bean. In addition to this, there are a number of under exploited and under utilized tuber crops grown as wild cultivars having medicinal and pharmaceutical applications. Processing and utilization of tuber crops as food are traditional with both primitive as well as modern equipments being in use (Nanda and Kurup, 1994). One of the oldest traditional processing tools is a stone pounder for making poi from taro in Hawaii.

The importance of tuber crops is realized as a high energy food, animal feed mix and as an industrial raw material. Cassava and sweet potato are the most important among tuber crops and they find extensive use in food, feed and industrial sector. Other tubers are grown as vegetable crops in homestead or in semi commercial scale. Development of cost and energy effective production and processing equipment is an essential requisite for tuber crops which are invariably associated with labour intensive farming and low cost produce. Homestead cultivation, longer duration to mature, inconsistent supply of marketed surplus, irregular shape and bulky size of tuber crops are factors which render mechanization of their production and processing operations rather difficult (Nanda *et al.,* 2002).

Harvesting Tools

Tuber crops are traditionally harvested by hand after digging the soil. In cassava, the soil around its stem is dug up by a spade and the plant is then uprooted. Harvesting by such method is quite strenuous, slow and leads to damage in the form of cuts, bruises or complete breakage of roots causing infection by fungus and bacteria and affecting its post harvest life adversely. Harvesting requires maximum labour in tuber crops cultivation. It represents, for example, about 30% of cassava production costs.

Two types of cassava uprooting tools were developed at Central Tuber Crops Research Institute (CTCRI), Trivandrum, India, which function on the lever principles. These harvesting tools reduce the effort of lifting to about one-fourth and one-third, respectively. The first order lever type harvester (Nanda *et al.*, 1996) has a mechanical advantage of four, a total weight of 14 kg and the production cost is Rs 1000. The lever is supported on a fulcrum at the top of a stand. The shorter arm has a stem holding mechanism consisting of a fixed jaw, a pivoted jaw and a spring to keep the two jaws in tension. During operation, the jaws open to get around the cassava stem. As the longer arm is lowered, the jaws grip around the stem tightly. Further lowering of the longer arm uproots and raises the cassava plant out of the soil.

The second order lever type harvesting device (Nanda *et al.*, 1996) from CTCRI has a mechanical advantage of 3.4, total weight of 8 kg and the production cost is Rs 700. The height of the fulcrum at the far end of lever can be adjusted which facilitates uprooting of cassava plants raised on flat bed as well as on mounds or ridges. A self-tightening mechanism is used to grip the cassava stem between the two jaws. Initially the jaws are opened by means of a metallic wire rope to get a hold around the stem. After gripping the stem, the plant is uprooted by raising up the effort end of the lever.

To harvest one hectare of cassava by the harvesting tools require 14 to 15 man days whereas 30 to 34 man days are necessary for traditional manual operation. There can be a saving of about Rs 200 ha^{-1} by using the harvesting tools in lieu of the manual harvesting. However, the primary objective of these mechanical devices is to reduce drudgery involved with traditional manual harvesting. Their low initial cost and economic viability make them an efficient practical solution to cassava harvesting.

To decrease the amount of labour involved in the manual uprooting, a universal harvester is being manufactured and marketed by Kobashi Kogyo Co. Ltd., Japan propelled by a water-cooled diesel 46 PS engine. Its field capacity is 7, 10, 5 and 6 acres per hour for harvesting potato, sweet potato, taro and carrot respectively. A tractor-mounted cassava harvester was developed at CIAT, Colombia (Diaz-Duran, 1979) for use with a 3-point hitch from a 75 hp tractor. The implement consists of a horizontal blade to harvest individual rows or two rows at a time depending on length of blade. This implement can be used for other root and tuber crops as well since the desired depth of penetration of blade into the soil can be controlled by the hydraulic system of the tractor.

Size Reduction

The simplest and the most common mode of processing cassava is conversion of tubers into chips. The hydrogen cyanide is also reduced from root tissues during slicing and drying operations. Cassava chips are used for edible purposes as such and for preparation of flour. Dried chips/flour are used in cattle feed formulation. In Industry, it serves as raw material for manufacturing of starch, dextrin, glucose and ethyl alcohol.

Chips and crisps are conventionally produced from cassava, *Dioscorea, Colocasia, Amorphophallus* and *Xanthosoma* tubers (Nanda, 1984). Under the conventional practice, tubers are sliced with help of hand-knives, usually after peeling the outer skin/ rind and dried in the sun for 3 to 5 days depending upon weather conditions. The method is tedious, time-consuming and also leads to uneven and delayed drying. The output by manual chipping of cassava varied from 11 to 37 kg hour^{-1} when the chip thickness varied from 2.7 to 12.5 mm. Several manually operated and also motorised chippers were developed in Asian, African and Latin American countries to increase operational convenience and output (Nanda *et al.,* 2000, 2002).

Manually operated chipping machines

To remove the tedium of operation and produce chips of uniform shape and thickness, CTCRI, Trivandrum developed first a hand-operated type chipping machine for cassava tubers (Nanda, 1983). The machine consists of two concentric mild steel drums, the annular space between which is divided into compartments for feeding tubers, supported on four m.s. legs. A rotating disc at bottom of drum carries the knives assembly. A pair of H.S.S. bevel gears is provided to operate the machine manually with a crank arm. Tubers are fed into compartments from the top and the chips are collected at the bottom. The average outturn of the hand-operated cassava chipping machine is up to 120 kg hour^{-1} for 6.9 1mm thick chips, which is 3 to 5 times more than traditional method. Mechanical chipping results in 2-5% breakage of chips. The capital cost of the machine is approximately Rs 6000 (2005). Machine chipping is highly economical where labour charges are higher or when family labour cannot be employed (Nanda, 1987).

Subsequently developed pedal-operated chipping machine from CTCRI is a modified version of the earlier prototype with additional provision of a pivoted pedal for transmitting power to the cutting disc through suitable belt and pulley drive mechanism (Kurup *et al.*, 1995; 1996). A trimming knife is also provided on the frame to remove the woody neck portion of tubers before feeding into compartments. Four castor wheels are fixed to legs of machine to make it portable. Two persons are required for the most efficient operation of machine, one to trim and feed the tubers and another to rock the pedal. Height of the operator's seat can be altered according to convenience of operator. Thickness of chips can be adjusted from 0.9 mm to 10 mm. Blades can be easily removed for sharpening or replacement.

The capacity of the machine increases from 83 to 768 kg h^{-1} with increase in chip thickness from 0.9 mm up to 6.9 mm. For further increase in chip thickness up to 10.0 mm, average output gradually decreases to 529 kg h^{-1}. In case of chipping *Dioscorea rotundata* (white yam) tubers, which resemble cassava tubers in shape and size, the average outturn of the machine 471kg h^{-1} for chips of 4.73 mm average thickness,. The cost of the machine is Rs 9000 (1997). The economic analysis of the chipping machine showned a profit of Rs 5376/- per year (Sheriff and Nanda, 2000).

Motorised cassava chipper

The motorised chipper developed is run with a 0.5 hp single phase motor through suitable belt drive. The feed hopper consists of two concentric rows of 25 cm high ms cylinders. The outer row of cylinders are of 10 cm diameter while the inner row of cylinders meant for thinner tubers are of 7 cm diameter. A ms circular disc of 87 cm dia and 10 mm thickness carries two pairs of stainless steel blades. A brick masonry foundation with a sloping chute serving as chips outlet is constructed with the motor and the chipper installed over it. The rotational speed of cutting disc is optimised between 80 - 100 rpm to overcome the jolting of tubers within feed cylinders while producing chips thicker than 5 mm at the reduced angular velocity (Nanda *et al.*, 2005).

The output of the machine was 286, 655 and 1091 kg $hour^{-1}$ for chip thicknesses of 2.5, 5.3 and 9.9 mm respectively. The cost of the machine is Rs 25,000/- (1997) inclusive of the prime mover.

Malaysian Chipping machine

The cassava chipping machine used in Malaysia (Manurung, 1974) consists of a rotating circular steel plate of about 12 mm thickness and 100 cm diameter, to which 4 or 6 blades are attached. The blade consists of 1 to 1.5 mm steel plate that is corrugated at the cutting edge. The machine produces strips about 6 mm wide and 3.6 mm thick, with a small percentage of fine particles. Length of the strips depends on position of roots at the time of cutting. The rotary disk is vertically mounted on driving shaft and driven by a 3 bhp gasoline engine/motor with a belt-drive mechanism at a required disk speed of 500 rpm. This chipper has a turnout of 1 tonne $hour^{-1}$ when operated by two men and fed manually. Since the chips produced from the Malaysian chipping machine are finer in size, the drying process becomes faster and the pounding and pelleting operations are easier.

Thai Chipping machine

The cassava chipper used in Thailand (Manurung, 1974) consists of a thin circular instead of a thick heavy plate as used in Malaysia. The plate, which is normally made out of the cover of an oil drum, is cut and formed to produce cutting elements throughout its surface. This rotating notched disc is mounted on a wooden frame equipped with a hopper. The chips or shreds produced are irregular in shape and take usually 2 to 3 days to reach a moisture content of 13 to 15% when dried on a concrete surface with periodic turning of the chips (Thanh, 1974).

Cassava Chipping machine in Nigeria

The machine consists of an assembly of three knives held in slots of a metallic cylinder in a fashion similar to wood planer (Odigboh, 1979). Roots cut 15 to 20 cm length are fed into a hopper placed above horizontally mounted cylinder. The knives assembly rotated clock-wise at 375 rpm, belt-driven by a 1 hp electric motor, while another cylinder of smaller diameter placed parallel at a gap of 2.5 mm from it rotated counter-clockwise at 95 rpm. The machine is powered by a 1 hp electric motor. Chips of 30 to 75 mm length are produced by this machine, with finer particles ranging up to 8%. Average turnout of this machine is 225 kg h^{-1}.

Besides the above, some pedal-operated bicycle and treadle typed as well hand-operated chipper-cum-grater units developed in the Phillippines (Orias and Cruz, 1981) can be suitable for the small- scale cassava processors and farmers.

Cassava Pelleting

Thailand and Indonesia are the world's largest exporters of dried cassava, largely in the form of pellets as a source of energy in compound animal feed. Cassava pellets were introduced to make handling easier, to maintain uniformity in size and shape and to help lower the shipping costs due to reduced volume-to-weight ratio. Pelleting is a process of particle size upgradation, utilizing a principle of compaction and extrusion. Two types pf processes are in operation in the pelleting plants used in Thailand and Indonesia (Thanh *et al.,* 1978). One press uses a horizontal circular platedie with a large number of holds driller to form a honeycomb pattern. A set of six rolls is resting on the die which rotates by friction when the die is rotated around a vertical axis. The other type of press uses a vertical ring die with a set of two rolls inside the ring. Holes are drilled radially throughout the face of the die. When the ring is rotated, the rolls also rotate with it by friction.

Chips are first screened to remove sand and soil, and to separate oversized ones, which are passed through a hammer mill. Cassava particles are then conditioned to correct moisture content with the use of a water spray and fed in front of the press rolls, which crush and force the material through the holes. As the compressed material emerges from other side of die, a knife shears the pellets to adjusted length. Cassava pellets are usually less than 1 cm in diameter and 2 cm in length. The pellets from the presses are quite warm and soft. These are sent to coolers where the temperature and moisture content are reduced, after which pellets become reasonably hard. The air-cooled pellets are sent to a screen from which oversized are sent to a bagging machine or storage.

Cassava strips having a bulk density of 0.288 g/cc, pressing after addition of steam produced pellets with a bulk density 0.808 g/cc (Thanh *et al.,* 1978). The savings in space, transportation cost, storage, and packaging cost from reduction in volume are obvious from these values.

Drying

Conventionally, cassava chips are dried in open air under sunlight by spreading in a single layer on cemented flour, bamboo mat, rock surface or sometimes even on bare earth. Depending upon weather conditions, it takes 2 to 5 days to dry the chips to 13-15% moisture content. Unreliable climatic conditions, however, make continuous sun drying difficult, particularly in Kerala where the monsoon is of long duration and coincides with time of harvesting. Artificial drying, therefore, becomes essential besides offering saving of time and floor space requirements (Kurup and Nanda, 1996). Keeping this in view, a thorough-circulation batch type dryer with a holding capacity of one m^3 of fresh chips was developed.

Natural drying

At the Asian Institute of Technology in Bangkok, a three-tier shelf-tray dryer was developed and studied as an alternative to concrete floor drying

(Thanh and Lohani, 1978). Superiority of shelf-dryer was attributed to better circulation of the ambient air through the chips. Experiments conducted on the influence of shape and size on the rate of drying showed that strips and slices required less drying time than cubes and conventional chips, but the difference was not substantial except drying on a plain cement floor.

Experiments at CIAT, Colombia (Best, 1978) showed improved circulation of air and faster natural drying when chips were spread on wood-framed trays with 1 in. chicken wire supported fine plastic mosquito netting base. The trays are placed at an angle of 25 to 30^0, which just prevents chips from sliding down, being supported on bamboo posts and rails facing direction of prevailing wind. Before a rain or at night, the trays can be stacked horizontally, with a canvas or a corrugated iron sheet covering top tray. Average drying time was observed 6 hr and 11 hr at loading rates of 5 and 10 kg m^{-2} of fresh chips. Inclined tray drying was faster when compared with methods of drying on plain concrete, black concrete, and horizontal trays placed 30 cm above the ground.

Best (1978) observed that if the moisture content is reduced to 50% on the first day, whether dried on concrete floor or on an inclined mesh tray, the cassava would not deteriorate for 3 days after chipping. This may be of considerable importance when a large amount of chips have to be processed during peak seasons of harvest.

To take the maximum advantage of drying power of the wind, vertical trays could be used. Chips are sandwiched in between two parallel wire panels and positioned against the most predominant direction of wind. When covered by a roof, drying may continue overnight or during rains.

Electrical dryer

The system consists of a motorised centrifugal blower with ducts, an insulated heater box and the drying chamber. A thermostat was provided to regulate temperature of drying air within 65-70 °C range to avoid browning and case-hardening of chips. The drying chamber has a perforated GI sheet floor of one sqm area and four sliding walls of AC sheet to facilitate easy loading and unloading of cassava chips. Maximum holding capacity of this dryer is 500 kg cassava chips on fresh weight basis. The overall cost (including blower) is about Rs 30,000 (1997). The dryer can be used to dry batches of 100 kg cassava chips in 8-10 hours to 500 kg in 20-24 hours from an initial moisture content of 65-73% to a final moisture content of 12-14% (Nanda,1991).

Taking the practical feasibility into account, a 300 kg batch size will be optimum for a 'one batch per day' operation with this dryer so that the said batch of cassava tubers can be harvested, peeled (optional), chipped and also dried - all on the same day. The dryer is used for drying of chips produced from other tuber crops too. It has the following avantages: permits harvesting, chipping and drying of cassava on the same day, shorter drying time, very little floor space requirement compared to sun drying, can be used on rainy days too, drying can be continued at night time during peak periods of harvest, discolouration and excessive drying are avoided due to relatively low temperature of the drying air, mechanically dried chips store better and remain free from insect infestation longer than sun dried chips (Nanda,1996).

Renewable energy dryer for cassava chips

Mechanical drying system may produce a higher quality product but are more likely to be associated with high costs. Hence a renewable energy oriented dryer for cassava chips was developed which makes use of solar energy and/or combustion of agricultural waste for heating the air. The dryer is of batch type and has a holding capacity of two cubic meters or one tonne of freshly chipped cassava tubers (Nanda and Kurup, 2000). The system consists of an agricultural waste fuel burner (internally insulated with ceramic fibre wool and white clay), a centrifugal ash separator, a motorised centrifugal blower coupled to a three-way distributor duct (connecting the blower with the solar collector and the waste burner) and a double-cylinder drying bin housed a shed with corrugated GI sheet roofing painted black to serve as a solar collector. Glass panel covers, though a common feature in solar collectors for attaining higher temperature, were adopted due to economic considerations.

Test runs were conducted with 500 to 1000 kg cassava chips of about 5mm chip thickness with the drying air temperature regulated around 85 ^{0}C before entry into the drying bin plenum. Fuel (partially dried coconut husks, cassava stems, etc.) feed rate in the waste burner was adjusted to 5-6 kg h^{-1} to achieve desired temperature limits. A drying time of about 20 hours was required to bring down moisture from 67-69% to 12-13%.

Production of Starch and Sago

Tuber crops are the richest sources of starch. Although sweet potato is commercially utilized for starch extraction in Japan, USA, China and Vietnam, such practice is not common in India. Cassava starch production, however, is carried out on a rudimentary cottage scale, semi-mechanised small scale enterprises and modern large scale factories. It is a thriving agro-industry in Tamil Nadu and about 1100 factories are operated to produce starch and sago (Pal *et al.,* 1987; Sheriff *et al.,* 2005).

The process consists of peeling, rasping, screening, settling and drying. Peeled roots of cassava are disintegrated into pulp by a rasper which releases the starch granules from fibrous matrix. Resulting slurry is pumped onto a series of vibratory screens (80, 150, 260 and 300 mesh). While the screens retain the fibrous waste (thippi), the starch milk is channeled into sedimentation tanks. After at least 8 hours, the supernatant liquor is run off and the starch cake settled at the bottom is scooped up for sun drying on a cement floor. Sago (Saboo dana) is manufactured from partially dehydrated (35-40% m.c) starch cake. The lumps are broken in a spike mill and then globulated on a gyratory shaker. The globules are graded according to size and then partially gelatinized by roasting on shallow metal pans. Finally the sago pearls are dried in the sun on cement flour for 6-8 hours. The agglomerates are separated by means of a spike beater and polished before bagging. Large space requirement, high water consumption, long retention time, absolutely weather-dependent drying and process equipment of obsolete design are major bottlenecks of the cassava starch industry in India (Trim *et al.,* 1996).

Cassava peeling knife

Peeling is the most labour intensive unit operation in the process of starch extraction. A cassava peeling knife of novel design was developed and tested at CTCRI, Trivandrum and subsequently evaluated on-site at a Starch and Sago Factory in Salem by the professional peeling workers. Results show that the average output of peeling knife is 113.56 kg h^{-1}, comparable to that of the traditional knife used by professional workers. Flesh loss with improved knife is only 1.38% compared to 5.70% flesh loss by traditional knife (Nanda, *et al.,* 2005). The traditional knife costs Rs 5/- each, and two to three knives are disposed by a labourer each week, with the minimum cost of operation being Rs 10/- . Cost of the improved knife is Rs 15/- and its blade can be replaced and/or sharpened at same interval with a cost of operation of Rs.1.50 per week.

Machineries for small scale processing of tubers

Commercially available machines for starch extraction are of high capacity and the cost is prohibitive for small farmers. Extraction of starch from tubers at production site itself will be a boon to farmers as it avoids middlemen to exploit the peasant. The high return also ensures promotion of cottage and small-scale industries. In order to facilitate *in situ* starch extraction, especially in the remote villages, two starch extraction machineries were developed at CTCRI.

Cassava rasper

A power operated rasper for small scale processing of cassava roots into starch was developed at CTCRI. The main components of the machine are hopper, top and bottom covers, rotating drum, chute and pulp collecting box , fabricated in mild steel. The overall dimensions of the machine are 760 X 395 X 1092 mm. The capacity of the rasper for two industrially popular varieties of cassava viz. H-165 and H-226 is 360.17 and 384.94 kg h^{-1} respectively with corresponding rasping effects of 75.99 and 78.81% for two industrially popular varieties of cassava viz. H-165 and H-226. For the tubers of curcuma, arrowroot and sweet potato, the capacity of this rasper is 323.56, 218.74 and 356.33 kg h^{-1} respectively with corresponding rasping effects of 76.66, 79.37 and 72.74 % (Nanda *et al.,* 2004).

Hydrocyclone

Cassava starch extraction by wet process requires large quantity of water (4.5-7.3 m^3 of water /tone of roots) and separation of starch from water used is a great problem. One of the earliest, but still widely practiced methods to separate starch from its suspension in water is the gravity settling in tanks or tables. Settling tanks held large volume and the process took several hours (10-12 hr). The longer detention time of starch with fresh water causes fermentation, producing alcohols and organic acids which give foul smell, polluting the entire atmosphere around the plant, besides resulting undesirable changes in the starch qualities. Hydrocyclone is a solid -liquid separator working on the basis of centrifugal separation whereas accelerated form of sedimentation takes place by centrifugal force, which is several hundred times greater than the gravitational force.

Hydrocyclones are used in large scale starch manufacturing since long to separate starch from nitrogenous impurities and to concentrate starch milk obtained from different crops. Though these are widely used, the low throughput capacity of single hydrocyclone makes them unpopular. Works has been initiated at CTCRI to optimize the operating and machine parameters of hydrocyclone to get maximum efficiency of starch separation.

Multipurpose mobile starch extraction plant

The major components of the machine are hopper to feed the tubers, crushing disc or cylinder with nail punched protrusions rotating inside crushing chamber to crush the tubers, sieving tray to remove the fibrous and other cellulosic materials, stainless steel or plastic tanks to collect the sieved starch suspension, tuber storage chamber, handle and wheels for easy transportation from place to place and a frame to support these components. Addition of water during processing can be controlled through a water pipe with holes fixed inside the hopper along its length and during sieving by a shower attachment connected to the water line. An electric motor (¾ hp) or a generator (kerosene–petrol) attached to the frame can be used as the energy source to operate the machine.

Out of the 24.1 % starch in cassava, about 84.2% could be extracted by using this machine while for sweet potato having 21.0% starch, the recovery is 75.1%. *Amorphophallus* has a starch content of 13.4%, however, the recovery is very low (40%) due to presence of mucilaginous content. Purity of starch extracted by the machine is also fairly good (80-85%) in all the samples. Capacity of the machine is 200 kg h^{-1} for cassava, 135 kg h^{-1} for sweet potato and 120 kg h^{-1} for *Amorphophallus*.

Extraction of starch from dried chips

Rapid deterioration of fresh tubers hindering year-round supply has induced attempts for preparation of starch from dried root. Extraction of starch from cassava tubers is a simple process. Rapid microbial decay of tubers hinders their storage beyond a few days and any delay in processing adversely affects recovery of starch both quantitatively and qualitatively. Transport of cassava tubers is also not economical due to their bulk and high moisture content. A process for extraction of starch from the dried chips of cassava can overcome these difficulties besides ensuring year round supply and surmounting glut. Possibilities of starch production from dehydrated cassava chips through wet milling and dry milling process were investigated (Nanda, 1994). Dried cassava chips are crushed to a long-fibered mass by rollers or hammer mill. The mass is then soaked in water and the starch is extracted by washing and sieving. The qualitative characteristics such as moisture, dry matter, starch, sugar, fibre, viscosity and colour were analysed and compared with those of the starch extracted from fresh tubers serving as control. Brabender viscosity characteristics of the starch extracted from cassava chips by wet milling process were comparable to that of starch extracted from cassava tubers. However, the viscosities of starch factions extracted by coarse as well as fine dry milling of cassava chips were significantly lower than viscosity values for starch extracted from fresh tuber. Pasting temperature of starch extracted from chips, whether dry milled or wet milled, was always higher. The starch faction from dry milled chips exhibited better stability after cooking the paste at 95°C. Chromameter whiteness of the starch extracted from cassava chips by wet milling was quite close to that of starch from the tuber, whiteness of the wet milled starch from unpeeled chips was 96-97% of the whiteness of wet milled starch from unpeeled tubers, and 98-99.5% in case of peeled. The dry milled starch factions from unpeeled chips were more reddish while the dry milled starch factions from peeled chips were more yellowish in comparison to the starch extracted from tubers. A settling time 4 to 20 hours was necessitated for starch extracted from unpeeled and peeled tubers. Uptake of moisture slowed down at around 24hr of soaking in the tests to determine optimum soaking time of dehydrated cassava chips. Results indicate that it would be possible to obtain starch from dry chips with quality similar from fresh tubers. However, improper drying and delayed processing would easily have adverse effect on whiteness and the viscosity characteristics of the starch so prepared. It is also obvious that the preliminary operations of dehydration and disintegration make the process highly energy consuming and expensive (Nanda, 1994; Nanda *et al.,* 2005).

Granulator for cattle / poultry feed

The growing demand for nutritionally and economically viable food and feed preparations necessitated increased production of tuber crops based formulations. Tubers, being rich source of carbohydrate, alone or incorporated with protein rich ingredients stand atop in this regard. The low palatability of cassava diet due to powdery nature of flour gave way for particle size up gradation by the process of pelleting, granulation or globulation. Owing to the high energy consumption of roller or screw type pelletising machines, small scale processing becomes difficult due to low investment capacity of majority of farmers.

A drum type centrifugal granulator operated both manually and by an ¾ hp electric motor was designed and fabricated. Capacity of machine is about 20kg/hr. Cost of operation to produce one kilogram of feed material was worked out to be Rs.1.6/- for manual operation and Rs. 0.92/- for mechanical operation (Sajeev *et al.,* 2004).

Research Priorities

The industrial potential of tuber crops has not been tapped except for cassava to some extent. Development of cost and energy effective pre- and post-harvest equipment is an essential requisite considering the low investment capacity of farmers raising tuber crops and low cost of produce. Improved technology needs to be developed for small and medium scale extraction of starch from cassava, sweet potato, taro and yams. Growing demand and trend for new types of snack foods by urban and affluent section of society should be recognised and enchased by processing tubers into various convenience and fast food products. Extrusion processing is one such technology to get top priority in value addition/ product diversification front.

Interventions that aim for an increase in productivity and production areas to stimulate commercialization should be made with prior analysis of market demand development.

References

Best, R. 1978. Cassava Processing for animal feed.(*In:*) Proceedings of Workshop on Cassava Harvesting and Storage. Weber,E. S., Cock, J. H., and Chouinard, A.(Eds.) .International Dev. Res. Centre, Ottowa, Canada. pp. 12.

Diaz-Duran, A. 1979. A cassava harvesting Aid. Series 05 Een-3. Centro Internacional de Agricultura Tropical (CIAT), Cali, Colombia. 14pp.

Kurup, G.T., Sheriff J.T. and Nanda S.K. 1995. Development and testing of a pedal-operated cassava chipping machine. *J. Root Crops* 21 (1) : 7-11.

Kurup, G.T. , Sheriff J.T . and Nanda S. K.1996. Pedal-operated cassava chipping machine. Central Tuber Crops Research Institute, Trivandrum. 6pp.

Kurup, G.T. and Nanda S. K. , 1996. Drying and storage system for tuber crops.(*In:*) Issues in Horticulture and Plantation Crops, Vol. 5 : Post Harvest Management, K.L. Chadha, (Ed.). Malhotra Publishing House, New Delhi. pp.71-489.

Manurung, F.1974. Technology of cassava chips and pellets in Indonesia, Malaysia, and Thailand. (*In*:) Proc. Interdisciplinary Workshop on Cassava Processing and Storage. Araullo, E. V., Nestel, B., and Campbell, M.(Eds.).Pattaya, Thailand. pp. 89-112.

Nanda, S.K. 1983. Hand-operated cassava chipping machine. Central Tuber Crops Research Institute, Trivandrum. 6pp.

Nanda, S.K.1984. Post-harvest practices of tuber crops. *Indian Farming*. 33(12):64-65.

Nanda, S.K. 1987. Field evaluation of hand-operated cassava chipping machine. Tropical Tuber Crops : Production and Utilization, ISRC, Trivandrum. pp. 225-228.

Nanda, S.K. 1991. Development and evaluation of a through circulation batch dryer for cassava chips. *J. Root Crops* 17: 274-279.

Nanda, S.K. 1994. Extraction of cassava starch from dehydrated chips by dry and wet milling. (*In:*)Abstracts of International Meeting on Cassava Flour and Starch, 11-15 Jan. 1994, Cali, Colombia. pp 134.

Nanda, S.K. 1996. Electrical dryer for cassava chips. Central Tuber Crops Research Institute, Trivandrum. 6 pp.

Nanda, S.K., Balagopalan, C., Padmaja, G., Moorthy, S.N.and Sajeev, M.S. 2002. Post Harvest Management of Cassava for Industrial Utilization in India. VII Asian Cassava Workshop held during 28th October - 1st November 2002 at Bangkok, Thailand.

Nanda, S.K., Hemasankari, P. and Sheriff, J.T. 2004. Development of a primary rasper for starch extraction form tuber crops. National Seminar on Root and Tuber Crops - NSRTC I, held at Reg. Centre of CTCRI, Bhubaneswar during 29-31 October 2004.

Nanda, S.K., Kurup, G.T., and. Sheriff J.T.1996. Harvesting tools for cassava. Central Tuber Crops Research Institute, Trivandrum. 6 pp.

Nanda, S.K. and Kurup, G.T. 2000. Non-conventional energy aided drying of cassava chips : an experimental study. Paper presented in International Symposium on Tropical Root and Tuber Crops, ISRC, 19-22 Jan. 2000, Trivandrum, India.

Nanda, S.K. and Kurup, G.T. 1994. Processing and process equipments for tropical tuber crops.(*In:*)Advances in Horticulture, Vol. 8 : Tuber Crops. Malhotra Publishing House, New Delhi. pp. 703-714

Nanda, S.K., Sajeev, M.S., Sheriff, J.T. and Hemasankari, P. 2000. Primary Processing Equipment for Cassava. Central Tuber Crops Research Institute, Trivandrum. 18p.

Nanda, S.K., Sheriff, J.T. and Sajeev, M.S. 2005. Starch Extraction Machinery for Tuber Crops. Central Tuber Crops Research Institute, Trivandrum. 14p.

Odigboh, E. U. 1979. Cassava chips processing and drying: a cassava chipping machine (*In:*) Proc. Workshop on Small-Scale Processing and Storage of Tropical Root Crops. Plucknett, D. K.(Ed.). Westview Press, Colo., pp 243.

Orias, R. R. and Cruz, R. 0. 1981. Root crop processing: assessment for village level adaptability, (*In:*) The Radix, Root Crop Research and Training Centre, Leyte, Philippines. 2:3.

Pal, T.K., Nanda, S.K., Anantharaman,M., and Gadewar, A.U.1987. Techno-economic survey of small scale cassava based starch units in Kerala. *J. Root Crops* 13 (2) : 67-74.

Sajeev, M.S., Padmaja, G. and Nanda, S.K. 2004. Development of a centrifugal granulator for cassava based feed. National Seminar on Root and Tuber Crops - NSRTC I, held at Reg. Centre at CTCRI, Bhubaneswar during 29-31 October 2004.

Sheriff, J.T. and Nanda, S.K.2000. Field evaluation of pedal-operated cassava chipping machine. Paper presented in International. Symp. on Trop. Root & Tuber Crops, 19-22 Jan. 2000, Trivandrum, India.

Sheriff, J.T. , Nanda, S.K. and Sajeev, M.S. 2005. Current Status of Cassava Processing Industries in South India. Central Tuber Crops Research Institute, Trivandrum, 78pp.

Thanh, N. C. 1974. Technology of cassava chips and pellets processing in Thailand, (*In:*) Proceedings of Interdisciplinary Workshop on Cassava Processing and Storage, Araullo, E.,V., Nestel, B., and Campbell, M., Eds., Pattaya, Thailand. pp 113-122.

Thanh, N. C. and Lohani, B. N.1978.Cassava chipping and drying in Thailand (*In:*) Proceedings of Workshop on Cassava Harvesting and Storage. Weber, E. S., Cock, J. H., and Chouinard, A.(Eds.). International Dev. Res. Centre, Ottawa, Canada, 21.

Thanh, N. C., Muttamara, S., and Lohani, B. N. 1978. Technological improvement of Thai tapioca pellets,*Thai J. Agric. Sci.*, 2 :75.

Trim, D.S., Nanda, S.K., Curran, A., Anantharaman, M and Nair, J.S. 1996. Investigation of starch and sago production technology in India (*In:*) Tropical Tuber Crops Problems, Prospects and Future Strategies. G.T. Kurup *et al* (Eds.). Oxford and IBH Publishing Co. Pvt. Ltd. , New Delhi, India, pp 397 - 415.

❑❑❑

10

Root and Tuber Crops for Food and Feed in North Eastern States of India

The North Eastern region stretches from 21°25' to 29° 28' N and 80° 4' to 97°25' E and covers a total geographical area of 2.55 lakh square kilometer sharing about 8 percent of the country's geographical area. The NE states, otherwise known as eight jewels, which are Arunachal Pradesh, Assam, Manipur, Meghalaya, Mizoram, Nagaland, Tripura and Sikkim. This region is characterized by inaccessibility, marginality, fragility, cultural heterogeneity, ethnicity, rich bio-diversity, low livestock and crop productivity. These states are the reservoirs for genetic diversity of tuber crops like *Colocasia*, *Xanthosoma*, *Alocasia*, elephant foot yam, *Dioscorea* and other minor tuber crops. The annual precipitation in the region is about 2450mm and maximum rainfall is received from South West monsoon. The annual average rainfall in the region was recorded at a minimum of 1637mm and a maximum of 6317mm. This region shows great variation in temperature, ranging from 15 to 20°C in the temperate zone and 25 to 32°C in plains while during winter it ranges from 6 to 18°C and 15 to 26°C in above two zones, respectively. The prevailing agro climatic conditions (six agro climatic zones), surplus rainfall, good soil depth and mild acidic soil favour cultivation of cassava, sweet potato, elephant foot yam, greater yam, lesser yam, white yam, taro, swamp taro, giant taro, giant swamp taro and tannia. These crops are food crops for tribal and poor masses in the region. The per capita availability of food grain in this region is low compared to national average. The region has a population of 39.04 million (3.8% of India's population). Assam, the most populous North Eastern state accounts for 70% of this population. There are about 400 ethnic groups which constitute the major population in the region, the tuber crops are very much popular among the people as food and medicine. Apart from human food, these crops have potential for the agro based industries for value addition and to produce concentrate feed for livestock, poultry and fishery. Tuber crops (cassava and sweet potato) also play a vital role to substitute as fodder for livestock, duckery (prefers tapioca and elephant foot yam leaves) and fishery in the region.

Genetic Resources of Tuber crops in North Eastern States of India

Table 1. Percentage distribution of area at different elevation in North Eastern India

State	< 150 M	150-300M	300-600M	600-1200M	> 1200M
Arunachal Pradesh	1.67	7.20	10.70	16.43	64.00
Assam	78.73	11.18	5.90	0.86	0.33
Manipur	5.29	1.38	12.49	53.47	27.37
Meghalaya	16.53	16.11	19.05	30.96	17.34
Mizoram	17.78	24.89	24.18	32.55	-
Nagaland	4.15	11.82	17.68	30.58	35.77
Tripura	83.67	15.55	0.78	-	-
Average	38.01	11.91	11.99	8.76	29.33

Das and Dutta, 1986

Tuber crops have wide range of adaptability starting from tropical wet land to temperate dry regions. The genetic diversity and adaptability of tuber crops in the region are classified based on the altitudinal differences, which are as follows:

(i) *Altitude < 150m:* The crops like *Colocasia* (dasheen type and eddoe type), *Alocasia* (giant taro and swamp giant taro) *Xanthosoma* and elephant foot yam.

(ii) *Altitude 150-300m: Colocasia* (All types), *Alocasia, Xanthosoma,* elephant foot yam and sweet potato.

(iii) *Altitude 300-600m:* Cassava, Sweet potato, *Dioscorea* spp. and elephant foot yam

(iv) *Altitude 600-1200m:* Cassava, Sweet potato, *Dioscorea* spp. and elephant foot yam

(v) *Altitude >1200m :* Cassava, Sweet potato, *Dioscorea* spp. and elephant foot yam

Rich genetic diversity of edible aroids (*Colocasia, Alocasia, Xanthosoma* and elephant foot yam) are found in low lands of Tripura, Assam, Mizoram and Meghalaya (83.67, 78.73, 17.78 and 16.53%, respectively) (Table1.).

Table 2. Geographical area, recorded forest and actual forest cover of NE Region (km^2)

State	Geographical area (km^2)	Recorded forest		Forest cover	
		Area (km^2)	Percent (km^2)	Area (km^2)	(%)
Arunachal Pradesh	83,743	51540	61.54	68847	82.21
Assam	78,438	30708	39.15	23688	30.20
Manipur	22,327	15154	67.87	17384	77.87
Meghalaya	22,429	9496	42.34	15633	69.70
Mizoram	21,081	15,935	75.59	18338	86.99
Nagaland	16,579	8629	52.04	14164	85.43
Tripura	10,486	6296	60.01	5745	54.79
Sikkim	7,096	NA	NA	NA	NA
All India	3,287,263	765,253	23.28	637293	19.39

Source: Anonymous, 1999

The Eastern Himalaya is one of the biodiversity hot spots recognised by the world (Table 2). There is huge genetic diversity of edible aroids, Dioscorea sp and lesser known tuber crops like winged bean, yam bean, arrow root and typhonium. In Arunachal Pradesh, Meghalaya, Mizoram and Nagaland the tribal population ranging from 80- 90%. However, in remaining states (Assam, Manipur, Sikkim and Tripura) the tribal population is 20-30%. The tribal and rural masses used to collect petiole (*Colocasia, Alocasia, Xanthosoma* and elephant foot yam), runners (swamp taro), corms (*Colocasia, Alocasia, Xanthosoma* and elephant foot yam) and tubers of various *Dioscorea* sp. for food (*D.alata, D.esculenta, D.rotundata, D.bulbifera, D.glabra, D.pentaphylla, D.hamiltonii*) and for medicinal purposes *(D.floribunda and D.opposita)* from forest for the food and medicinal purposes.

Out of a total geographical of 2, 62,040 km^2, 88.50% of area is under land utilization. About 63.30% area is under forest, 5.38% area is in non agricultural usc, 9.06% is barren and uncultivable land, 6.64% is uncultivated land excluding fallow and 15.31% area is net sown area (Table.3).This offers scope to introduce the low input and high volume crops like cassava, sweet potato, greater yam and lesser yam in cultivable waste lands. The tuber crops (cassava, elephant foot yam greater yam and lesser yam) are ideal for intercropping in coconut, arecanut and rubber gardens (*Alocasia* and *Xanthosoma* are shade loving plants)

Table 3. Area under different land use ('000ha)

State	Net Sown area	Area under current fallow	Net Cultivated area	Other uncultivated Area (excluding fallow)	Fallow land other than current waste land	Cultivable waste land	Total uncultivated waste land	Available area for cultivation
Arunachal Pradesh	166	28	194	44	36	33	113	43
Assam	2800	37	2837	108	34	43	185	182
Manipur	160	3	164	3	2	2	7	4
Meghalaya	298	2	300	1	Neg	1	2	Neg
Mizoram	82	2	84	-	-	-	-	-
Nagaland	218	100	318	197	248	148	593	57
Tripura	244	5	249	5	17	8	31	28
All India	140413	12246	152659	2828	3798	3290	9916	2332

Source: Anonymous, 2002

Status of tuber crop production in North Eastern States

Among tuber crops, cassava, sweet potato and elephant foot yam are cultivated commercially in considerable area. Area and production of edible aroids, yams and other minor tuber crops are not well documented.

Cassava

Cassava is one the major root crops being grown in North Eastern states. These states contribute approximately 3% of total area and 1% of total cassava production in India. In 1984-85, area and production were about 44,400 t and 10000 ha, respectively. Area under cassava cultivation went down to 7000 ha with a production of 55,600 t during 1996-97. There has been instability in area and production of cassava in the region (Table.4). The average global productivity of cassava is about 10.0 t ha^{-1}, India stands in first position with respect to productivity of cassava (24.60 t ha^{-1}). North Eastern states, productivity is always low and it ranges from 2.00 to 17.50 t ha^{-1}. The lowest productivity was from Arunachal Pradesh during 1988-89 (1.98 t ha^{-1}), where as higher productivity (17.50 t ha^{-1}) was recorded in Mizoram in 1993-94.

Sweet potato

Sweet potato is the most important tuber crop next to cassava in India. Sweet potato plays vital role as animal feed and fodder in the region as compared to cassava. The national average productivity of sweet potato is about 8.30 t ha^{-1}, where as in North Eastern states, the productivity is far away from national average in all states except Tripura and Nagaland (12.0 t ha^{-1}). During 1997-98, area and production of sweet potato was 1700 ha and 70000 t, respectively (Table 5). The North Eastern states contribute 1.41% of area of total cultivation and 0.70% of total production in the country. Instability in area and production of sweet potato are mainly due to non availability of high yielding varieties and non tuberisation.

Table 4. Area and Production of cassava in North Eastern states

Year	North Eastern States		All India	
	Area ('000ha)	Production ('000t)	Area ('000ha)	Production ('000t)
1984-85	10.60	44.40	305.30	5662.10
1985-86	11.00	44.80	269.90	4474.40
1986-87	9.90	44.10	265.30	4814.40
1987-88	10.40	47.00	269.70	5212.30
1988-89	7.60	36.50	241.60	4962.00
1989-90	7.30	37.20	243.20	5111.20
1990-91	7.00	36.40	250.90	5832.50
1991-92	7.00	40.90	226.90	5021.90
1992-93	6.90	47.30	234.90	5412.80
1993-94	6.80	38.20	245.80	6028.90
1994-95	6.80	38.10	231.70	5856.80
1995-96	7.70	41.60	228.20	5443.20
1996-97	7.50	53.40	256.10	5662.80
1997-98	7.80	55.60	244.30	6681.90

Source: Anonymous, 2002

Table 5. Area and production of sweet potato in North Eastern states

Year	North Eastern States		All India	
	Area ('000ha)	Production ('000t)	Area ('000ha)	Production ('000t)
1984-85	16.60	65.30	192.20	1627.70
1985-86	14.30	57.40	183.40	1508.20

Contd...

Table 5. Contd...

1986-87	15.00	58.70	178.30	1490.60
1987-88	14.70	58.40	164.20	1346.70
1988-89	14.80	60.00	157.10	1251.50
1989-90	15.70	62.60	151.40	1202.30
1990-91	16.70	63.60	136.50	1130.70
1991-92	16.50	65.90	139.70	1185.20
1992-93	16.50	65.80	141.40	1215.50
1993-94	16.50	63.70	143.90	1220.60
1994-95	16.40	62.90	144.60	1127.80
1995-96	16.30	64.20	140.70	1138.10
1996-97	16.20	65.70	129.90	1061.20
1997-98	17.00	70.70	120.60	1048.10

Source: Anonymous, 2002

Yams

The genus *Dioscorea* has 600 tropical and subtropical species of which 11 are grown for tuber production and 50 species are distributed throughout India except in dry areas of North Western India. In north eastern states, there are about 28 species and 25 varieties are found in wild and cultivated form, respectively (Sharma and Hore, 1995) (Table.6). *Dioscorea* grows in all states, the important species are *D. alata, D. esculenta, D. rotundata, D. bulbifera, D. puber, D. glabra, D. hamiltonii, D. pentaphylla, D. oppositifolia, D. spicata, D. wightii, D. wallachi, D. floribunda* and *D. deltoidea* found in North Eastern states. Among *Dioscorea* species, *D. floribunda* is cultivated commercially in Tripura, which is promoted by the TFPDC (Tripura Forest and Plantation Development Corporation) under Forest Department, Govt.of Tripura. The *Dioscorea* species like *D. floribunda* and *D. deltoidea* are grown for extraction of "diosgenin" alkaloid, used for production of oral contraceptives. In *D. alata,* about 12 accessions were collected and characterised at Jorhat. Out of 12 accessions, accessions DA-3, DA-10 and DA-11 had good cooking quality (Anonymous, 2005).

Edible Aroids of NE states

(i) Taro

- *Colocasia esculenta* (L.) Schott var. *esculenta-* Dasheen type – found in low land of all states but the maximum diversity can be seen in Assam, Meghalaya, Tripura and Manipur.

Table 6. Edible yams of North Eastern Region

Botanical Name	Vernacular Name	General description	Use
Dioscorea glabra Roxb.	Pisala Alu (T)	The plants are annual (or) biennial, climbing to the right, glabrous and prickly at base. Leaves are deeply cordate or sagitate, opposite or lower alternate glaucous or nearly white beneath; petiole long, sometimes, as long as leaf blade. Tubers descending deep into soil on a long stalk gradually increasing in size; cylindrical and flesh is white	The tubers are eaten as vegetables along with dry fish.
D.pentaphylla L.	Kanta Alu (T)	Vines climbing to the left; stem prickly at the base. Leaves trifoliate or pentafoliate. Male and female flowers appear on separate racemes.	Flowers and leaves are used as vegetable Tubers possesses properties like antihelmintic and useful against leprosy, piles, gonorrhoea and swelling.
D.pubera Blume.	-	Vines climbing to right; densely pubescent; tubers are long and yellow flesh. Leaves cordate, ovate or orbicular and pubescent .Male flower axes 1-2 on leafless branches, where as female flowers 1-3 in leaf axis.	Tubers are used as vegetable
D.bulbifera L. var.*bulbifera*	-	Vines twining to left. Bulbils appear in leaf axils. Tubers and bulbils are small and warted	Tubers are used as vegetable Tubers are useful for piles and dysentery.
D.bulbifera L. var.*sativa*	-	The bulbils are large and smooth. The tubers are almost absent	Used as vegetable
D.hamiltonii		Vines twining to the right. Leaves ovate or lanceolate, acuminate, more or less cordate. Tubers are long and white flesh.	

- *C. esculenta* (L.) Schott.var. *antiquorum*- Eddoe type – found in upland of all states but maximum diversity can be seen in Assam, Meghalaya, Tripura and Manipur.

(ii) Giant taro

Alocasia indica- found in low land of Assam, Meghalaya and Tripura

(iii) Swamp giant taro

A. macrorrhizha- found in low land of Assam, Meghalaya and Tripura

(iv) Tannia

Xanthosoma sagittifolium- found in upland of all states but the maximum diversity can be seen in Assam, Meghalaya, Tripura and Manipur

(v) Elephant foot yam

Amorphophalus companulatus- local germplasms are available in Meghala and Tripura.

Minor tuber crops

Tannia *(Xanthosoma sp.)*

A herbaceous perennial, several large leaves sprout from main stem, which are sagittate and erect with long, ribbed petioles; inflorescence sprouts between leaves in a spadix, with a white 12 to 15 cm spathe which closes at its base in the form of a spherical chamber and opens at the top into a concave lamina; the spadix is cylindrical, slightly longer than the spathe, with female flowers on the lower portion, male flowers on upper portion and sterile flowers in middle portion. Spadices are rarely fertile and produce a few viable seeds. It has a corm or main underground stem in the form of a rhizome from which swollen secondary shoots, or cormels, sprout. Duration is 9 to 11 months, during the first six months, the corms and leaves develop and in the last four months, the foliage remains stable and, when it begins to dry, the plants are ready for harvest. The tuber and petiole are edible. The approximate nutrient composition of 100 g tuber is as follows: water (65%) protein (2.0g) Fat (0.3g) Carbohydrate (31g), Fibre (1.0g), Calcium (20mg), Phosphorous (47mg) and Ascorbic acid (10mg).

Winged bean (*Psophocarpus tetragonalobus* (L.) D.C.)

Family: Fabaceae, **Common name**: Khamranga Bean

The winged bean plant grows as a vine with climbing stems and leaves, 3-4 m in high. It is a herbaceous perennial; but can be grown generally as annual. Inflorescence raceme, flowers lilac colour; peduncle upto 15 cm long. Calyx glabrous, teeth shorter than the tube, upper two connate. Corolla: petals are equal in length, keel petal incurved. Stamens: monodelphus, anthers uniform, ovary subsessile. The pods are 4-sided with fringed wings and can be 6-30 cm in length and consisting of 5-20 seeds/pod. The plants take 50 days to flower, young pods are ready to harvest about 20 to 30 days after anthesis. After this, winged bean can be used as grain legume. The tubers are harvested 5 to 12 months after sowing when they reach 2 to 4 cm in diameter. The seeds are used for industrial purposes for commercial exploitation, they are also useful as food when steamed, boiled, fried, roasted, fermented, made into milk and other methods. The plant also produces underground tubers of varying sizes, rich in carbohydrates (25-30%) and proteins (10-15%). Almost all parts of winged bean are used as vegetable. It produces edible immature pods, mature seeds, leaves, flowers, tuberous roots, succulent shoots and edible seed oil. Seeds are rich in crude proteins (33.70%), fat (17.74%) and the minerals like Fe and Ca content 13.78 and 240.11 mg/100g.

Vigna vexillata (L.) A.Rich (Family: Leguminaceae)

It is a climbing herb found in tropical and subtropical regions.In India , about 15 species have economic importance in terms of protein supplementation and a few are useful as green manure or cover crop(CSIR, 1976). The plant produces fusiform tubers with or with out branches.Both seeds and tubers can be used as vegetable (Vimala and Balanambisan, 2005). It is found grown wild in Assam, Meghalaya and Tripura.

Costus specious (Koenig) Sm.

It is a common plant in Assam, Meghalaya and Tripura. The plant attains 1.5-2.0 m height and plant bears tuberous rhizome. The root has purgative and tonic properties (CSIR, 1959, 1976).

Yam bean (*Pachyrrhizus erosus*) (Family: Leguminaceae)

This crop is grown in Assam and Meghalaya. The tubers are tap root modification and are turnip shaped (or) conical. The underground portion of the plant is perennial, where as aerial part withers and dies after flower and seed set. It can be propagated through seeds and tubers.The tubers can be used as raw (or) salad and cooked.

Tuber Crops as Animal Feed and Fodder in NE India

Animal Husbandry is an inseparable part of economy, as this region is dominated by mono cropping and crop failure very often pushes the farmer's economy at the verge of collapse. The region is inhabited by considerable tribal population (>35%) who are mostly non-vegetarian and hence, the demand for animal protein is much higher than other parts of the country. There is an acute shortage of dry matter and green roughage in NE regions. Grasses, tree leaves, herbaceous weeds, shrubs, herbs straw and stubbles are the main indigenous feed resources available in the region. The feed and fodders have great impact on productivity of livestock, poultry and fishery.

North Eastern Region of India has a repository of livestock and poultry (Table 7). Among meat-producing animals, pig occupies a unique position in the region, as pig rearing is socially intermingled with livelihood of people of the region. This region accounts for about 28% of the total country's pig population and about 25 % of the total pork produced in the country is being consumed in this region. In spite of having a rich animal biodiversity, region has not yet been able to make a desired headway in its production status. The per capita availability of milk is around 77 g against the recommended level of 280 g. Similarly, annual availability of meat and egg was only 5.6 kg and 33 numbers against the recommended levels of 12 kg and 180 eggs. The region heavily depends on outside supply of meat and egg. The deficiency in meat production accounts to about 49%.

Table 7. Livestock and poultry population in NE India *vis-à-vis* India

State	Cattle ('000)	Buffalo ('000)	Sheep ('000)	Goat ('000)	Pig ('000)	Poultry ('000)
Arunachal Pradesh	453	11	28	183	275	1267
Assam	8000	728	84	2677	1082	18210
Manipur	508	95	8	33	388	3055
Meghalaya	738	18	17	280	351	2152
Mizoram	33	5	1	16	163	1307
Nagaland	383	36	2	161	571	2444
Sikkim	295	4	11	96	30	252
Tripura	1228	18	6	639	237	3595
Total North East	11538	915	157	4085	3097	32282
All India	197708	88832	56766	120774	1358	346693
% in NE	5.84	1.03	0.28	3.38	22.80	9.31

Source: Anonymous, 2002

At present, number of pigs/100 persons in the region comes to 18, while it is 20, 14.2, 12.3 and 3.7, respectively in Eastern, Southern, Northern and Western regions of the country. It is estimated that pork contributes 72 % of total meat production in Mizoram. Although every household rears at least one or two pig(s) in backyard, a wide gap exists between need/ demand and availability of pork in the region. Total pork production in the region in 2003 was around 48, 000 tonnes against approximate requirement of 58, 000 tonnesleaving a production short fall of 10, 000 tonnes (Table 9). Major reason for this insufficiency is that a major chunk of livestock and poultry population in this region is of local/indigenous type, whose productive performance is poor besides the traditional unscientific management systems followed. The tribal farmers have evolved their own livestock and poultry production system depending upon infrastructure available, which is almost a zero grain animal production system. Pigs reared in the region mostly depend on local vegetations, agro industrial by products, household and kitchen waste. The farmers themselves evolved, based on traditional knowledge, feeding system for their animals with locally available materials like plants suitable for feeding, left out rice and kitchen wastes. The average CP content of the cooked pig feed offered by farmers was only 6.66±2.14%, lesser than the recommended level (12 - 18 %). Similarly, quantity of dry matter offered to pig/day is much lower than the recommended level. In this feeding system, it has been claimed that local pigs proved more prolific than exotic breeds. On contrary, growth of exotic pigs were comparatively higher than nondescript pigs under tribal pig production system suggesting that exotic pigs are also capable of attaining moderate growth rate with a relatively low nutrient density feeding system. Rational exploitation of local feeds supports much more sustainable production systems in the medium and long term. Owing to lack of quality feed resources, small-scale farmers in these areas cannot afford enough feed to maintain their livestock in good condition over dry season. Throughout the region, cost and availability of feed are probably the most important constraint to increase livestock production since feed costs are major part of production cost that accounts 60-80%.

The gap between demand and availability of feed resources is wide and in alarming situation (Table 8). There is around 75% deficiency in availability of concentrate feeds in this region (Table 9).

Table 8. Demands and availability of livestock feed and fodder (In million tonnes)

Types of feed	NEH		All India	
	Demand	Availability	Demand	Availability
Dry roughages	12.00	12.00	799	551
Green fodder	74.60	34.40	558	430
Concentrate	5.13	0.90	117	62

Table 9. Production, requirement and deficit of food of animal origin (1999-2000)

State	Production			Requirement			Deficit		
	Milk ('000t)	Egg (Millio n nos)	Meat ('000t)	Milk ('000t)	Egg (Millio n nos)	Meat ('000t)	Milk ('000t)	Egg (Millio n nos)	Meat ('000t)
Arunachal Pradesh	45	35	19	87	199	24	42	164	5
Assam	733	488	16	2139	4862	583	1406	4374	567
Manipur	67	71	18	192	436	52	125	365	34
Meghalaya	62	85	29	185	420	50	123	335	21
Mizoram	18	43	12	71	162	19	53	119	41
Nagaland	49	51	20	160	363	44	111	312	24
Tripura	77	106	7	256	582	70	179	477	63
Total	1051	879	121	3090	7024	842	2039	6146	755

Source: Anonymous, 2002

Cassava

Cassava is a carbohydrate source with considerable potential as feed in subtropical humid areas. It is also known as yucca, manioc, cassava or manioca. The world average yield of cassava tubers is 10 t ha^{-1}. It is possible to obtain 68 t ha^{-1} by use of improved varieties, fertile soil and good management. Cassava meal is obtained from tuber and is high in starch and 1ow in protein and fibre. The proximate composition of cassava meal (7 varieties evaluated at ICAR Research Complex NEH are presented in Table 10). The average DM content was 37.00 ± 1.00 (32.48-39.82%). The values for DM, CP, EE, CF, NFE and total ash were 97.64 ±0.19, 3.33 ± 0.19, 0.55 ± 0.04, 2.28 ± .016, 91.47 ± 0.36 and 2.36 ± 0.19% respectively. There is not much variation among different varieties. Similar values are reported by Ravindran, *et al.* (1982).

Table10. Biochemical composition of different varieties of cassava (Estimated from the peeled tubers)

Varieties	Dry matter (%)	Dry Matter basis (%)					
		Organic matter	Crude protein	Ether extract	Crude fibre	Nitrogen free extraction	Total ash
Megh-61	39.82	98.05	2.96	0.63	1.92	92.54	1.95
Sree Vishakam	36.99	97.44	3.49	0.61	2.29	91.05	2.56
H-226	38.53	97.78	3.53	0.58	1.85	91.85	2.22
H-165	36.24	96.94	3.92	0.38	2.04	90.60	3.06
Sree Prakash	32.48	97.13	3.89	0.52	2.44	90.28	2.87
Sree Vijaya	35.23	97.72	2.85	0.60	3.06	91.21	2.28
Sree Jaya	39.72	98.41	2.68	0.59	2.36	92.78	1.59
TRC-T-1	38.95	98.12	3.26	0.34	2.98	92.36	2.16
Mean ±SE	37.55 ±1.00	97.94 ±97.64	3.37 ± 0.19	0.55 ±0.04	2.28 ±0.16	91.34 ±0.36	2.30 ±0.19

Sankaran *et al.*, 2006a

Raw chopped cassava tubers supply major source of energy for growing finishing pigs if properly supplemented with protein, minerals and vitamins. Pigs given chopped raw cassava tubers with a protein supplement gained less weight but FCR was similar when compared to those fed on maize-SBM ration. Under humid conditions, it is difficult to dry cassava tubers. Root silage can be prepared. This silage feeding had similar performance in growing finishing pigs as those given fresh cassava tubers (Buitrago *et al.*, 1978).

One of the major factors for restricted use of cassava tuber in swine diets is presence of HCN. Various processing methods like boiling, chopping/ crushing followed by sun drying remove HCN (Maner, 1973; Tewe, 1992). The low HCN varieties could be used in feed preparations.

Cassava meal is prepared by chopping whole tuber and drying it. Cassava meal can substitute maize in 16% CP diets. A slight decrease in ADG was noticed as level of cassava meal increased. Intake was similar indicating palatability is not a problem in cassava meal based diets. A level of 680-770 g ADG could be reduced using cassava meal based ration. Piglets from weaning to 20-25 kg BW, given diets containing 20-25% cassava meal showed similar performance as those given maize-SBM ration.

Results show that sows given cassava /SBM produced smaller litters compared to those given maize/SBM diets mainly due to deficiency of Methionine (Maner, 1973,). Supplementation with 0.3% Methionine in the diet improved reproducing performance. cassava meal possessed DE and ME values 3.54 and 3.15 MCal/kg DM, respectively for swine and ME of 3.52 MCal/kg for poultry (Ranjan, 1980). A value of 3.3 MCal/kg DM was reported for poultry (Reddy and Bhosale, 2001). Cassava waste, a byproduct of starch extraction from cassava tubers, can be successfully used in growing and lactating cattle which possesses 2.0% DCP and 64% TDN (Ranjan, 1980).

Leaves

Yield of cassava leaves ranging 2-8 tonnes of DM/ha/yr. could be obtained depending on variety and soil fertility (Oke, 1978). The leaves are high in CF and CP compared to tubers. Cassava leaf meal is deficient in methionine but rich in lysine. Inclusion of cassava leaf meal at 20-40% level in diet resulted in lower body weight gains and feed efficiency (Alhassan and Odoi, 1982).

Sweet Potatoes in Animal Feeding

About 43% of the world's annual sweet potato production is used as animal feed. Feeding sweet potato roots to pigs and other livestock is a common practice in China, North and Central Vietnam, Eastern islands of Indonesia, Philippines, Papua New Guinea, Cuba and Uganda. Sweet potato (*Ipomoea botatus*) has higher biological efficiency as food and shows the highest rate of production (35-45 t ha^{-1}) per unit area. In India, it is grown over an area of 0.14 million ha producing 1.17 million tonnes. Under hill agro-eco-system, North Eastern region has tremendous potential for cultivation and this crop occupies 17,000 ha with a production of 70,700 tonnes.

Sweet potatoes store starch and sugar in the edible underground tuber. The parts above ground – the vine and the leaves – are also edible, but they contain relatively a little starch. Starch is a carbohydrate that when eaten becomes an energy source. The dry matter content of tubers ranges 20 to 40 % with protein content of <10%. Skin and leaf tips contain comparatively higher protein, 50- 90% and 18-21%, respectively. Tubers also contain essential amino acids, with exception of sulfur-containing amino acids, especially cystine/ cysteine. From 60 to 85% of nitrogen in sweet potato is proteinaceous, with a protein efficiency ratio (PER) equal to casein (milk protein). Sweet potatoes contain compounds known as trypsin inhibitors that inhibit action of protein-digesting enzymes in the gut. Heat treatment/silage making can inactivate these compounds. Fermented silage, containing a mixture of different locally available food sources is the ideal.

Sweet potato vines, commonly left unused, can also be used as a protein feed for animals. They do not have as much protein as cassava leaves, but they do have more protein than several grasses. One hundred kg of dried sweet potato vines supply 10-12 kg of protein. Grasses which commonly grow in tropical climates supply much less protein than this. Sweet potato vines have no substances, bad for animals and hence they can be safely fed to animals.

Table 11. Proximate composition of sweet Potato varieties (Whole tuber)

Variety	DM (%)	% on DM basis					
		Organic matter	Crude protein	Ether Extract	Crude fibre	Nitrogen free Extract	Total Ash
Gouri	24.66	96.49	4.99	2.12	2.38	87.00	3.51
Sree Gouri	22.92	97.72	4.97	1.78	2.58	88.39	2.28
440038	21.70	96.54	6.41	1.95	2.30	85.88	3.46
4400127	30.32	97.71	4.70	2.03	2.86	88.12	2.29
Sankar	29.50	97.79	4.13	1.88	1.92	89.86	2.21
442074	31.28	97.83	4.04	1.65	2.20	89.94	2.17
187071	32.61	97.69	3.90	1.82	2.11	89.96	2.31
420027	34.78	97.11	5.30	1.96	1.83	87.95	2.89
Mean ± SE	28.47 ± 1.69	97.36± 0.20	4.81 ± 0.29	1.90 ± 0.05	2.27 ± 0.12	88.38 ± 0.52	2.64 ± 0.20

Source: Sankaran *et al.*, 2006b

The dry matter content of sweet potato varieties from 21.70 to 34.78%. The dry matter content was high in 442074(34.78%) and the low (21.70%) was in 440127. The dry matter content is directly related to starch content and biochemical composition varies with genotypes (Ramanujam, 1985). The nutritional parameters like organic matter, crude protein, ether extract, crude fibre, nitrogen free extract and total ash of different varieties ranged from 96.54 to 97.83%, 3.90 to 6.41%, 1.65 to 2.12%, 1.83 to 2.86%, 85.88 to 89.99% and 2.17 to 3.51% respectively. The values obtained are similar to those observed by other workers (Cullison, 1978, Morrison, 1984). Values of 0.70 and 74.00% for digestible crude protein and total digestible nutrients respectively was observed in cattle (Morrison, 1984).

Pig ration

Pork is the preferred meat of this region and pigs enjoy a protected status in tribal society because they are important in economic terms. Pigs have played an important role in NE India, as valuable commodities and an important source of animal protein. Importance of pigs is increasing because they have the potential to be a source of income and meeting more immediate food needs. The only non utilized parts of the pig are gall bladder, bristles, and bones too large to be chewed. There are several problems while trying to expand pig population, of which inadequate availability of feed occupies an important position.

In China, farmers knew the value of sweet potatoes for a long time. They fatten their pigs mainly on sweet potatoes. Farmers also make dried chips from sweet potatoes, store and use them throughout the year. Some make silage with sweet potato tubers to feed their pigs. Dry matter intake for older pigs of mean live weight of 90 kg is around 900 g/day, while the younger pigs (<30 kg) can consume around 450 g/day. It is claimed that feeding sweet potato reduces parasitic load, thus having an additional advantage in body weight gain of pigs.

Fresh tubers can be chopped and fed to pigs. Pigs readily eat them. Fresh sweet potato tubers are better feeds than fresh cassava roots because they have more protein and have no cyanide. Tubers from some cultivars have high levels of trypsin inhibitors and hence they can be cooked to increase digestibility and availability of protein. Sweet potato meal can also be prepared and fed to the pigs. Sweet potato meal is a good energy feed. It can be used in compounded rations instead of cereal grains. It must be mixed with protein feeds, mineral supplements and vitamin supplements. The energy that pigs can get from sweet potato meal is similar to that they get from cassava meal and maize. Sweet potato meal can completely replace maize in pig rations. Trypsin inhibitor may be a problem in the meal and hence, it is safe to replace only 50% of the maize in pig ration with sweet potato meal. Sweet potato tubers can also be preserved as silage. Pigs readily eat silage.

Sweet potato vines can be fed to pigs either in fresh form or after drying. Pigs readily eat the vines. Vines can be dried and ground into a meal. By drying 100kg of fresh vines, about 30kg of dried vines can easily be obtained. Sweet potato vine meal can be used in compounded pig rations, but only at low levels. It should not be used more than 5% level in pig rations.

Poultry ration

Level of inclusion of sweet potato meal into a ration depends on the age of the birds. In rations of young birds that are less than eight weeks old, it should not be used beyond 20%. As the birds grow bigger, it can be increased gradually and in rations for laying chickens it can be used up to 30%. Fresh green leaves of sweet potato can be chopped and given to birds in addition to mash. It supplies the pigments, minerals and vitamins to poultry. Giving green leaves to chickens is a good practice. Amount of mash given to chickens can be reduced when additional greens are given. Fresh sweet potato vines can also be fed to chicken but to a level of less than 3%. Tubers are good sources of energy (ME = 3500 K.Cal/kg) for poultry (Reddy and Bhosale, 2001).

Ruminant ration

Sweet potato tubers can be given to all ruminants. They can be fed as fresh, chopped tubers, dried chips and silage. Sweet potatoes can be fed to ruminants as energy supplements along with locally available grasses during dry season for both fattening and milking animals. Fresh sweet potato vines make good feed for all ruminants. They can be given without any restriction. When sweet potato vines are fed to milking animals or fattening animals, there is no need to give any other protein supplement as vines can alone supply all the protein needed by these animals. Fresh vines can be preserved as sweet potato vine silage and fed during lean season when fodder availability is inadequate. Method of making silage from vines is similar to that used for making silage from any other forage. Good sweet potato vine silage will be brownish green. It has a pleasant fruity smell and can be fed free choice to all ruminants.

Yams

The greater yam *(Dioscorea alata)* tubers are also rich in carbohydrates which are mainly in form of amyl pectin. Tubers contained 34.2% DM. The values for CP, CF, Ash, EE and NFE are 8.1, 2.6, 5.2 0.8 and 83.3% respectively (FAO, 1992). It contains high level of trypsine but deficient in lysine and methionine.

The leaves are good urea of protein (10.82%) but the level was lower than in tapioca and sweet potato leaves. The proximate composition of leaves is similar to that reported by FAO (1992). The CWC content was similar to those of tapioca and sweet potato leaves. IVDMD and IVOMD values were 55.28 and 56.82%, respectively. The leaves had ME value of 7.66 MJ kg^{-1} DM.

Tuber crops in relation to Fisheries

Table 12. Fish Production, requirement and deficiency in NE States

NE States	Production ('000 t)	Requirement ('000 t)	Deficit ('000 t)
Arunachal Pradesh	2.65	12.08	9.43
Assam	180.95	293.21	112.26
Manipur	17.60	26.21	8.67
Meghalaya	5.15	25.51	20.36
Mizoram	3.68	9.71	6.09
Nagaland	5.56	21.89	16.33
Sikkim	0.14	5.95	5.81
Tripura	17.98	35.19	17.21
Total	233.71	429.87	196.16

Source: Annymous, 2002

The data pertaining to the fish production, requirement and deficit are given in Table 12. The percapita fish availability is about 5.98 kg/annum, but, the actual demand is 11kg/capita/ annum (5.02 kg deficit/annum). This is mainly due to low productivity of ponds and tanks, non availability of feed and high feed cost. The fish production in North Eastern states is mainly from the culture fisheries (tanks and ponds). In the integrated farming systems, the composite fish culture is popular in all North Eastern states. In composite fish culture, the phytophagous fishes like grass carp, silver carp, common carp and catla relish leaves of tapioca, sweet potato and elephant foot yam. The cultivation of tuber crops would reduce feed cost by substituting the concentrate feed level in fish feed.

Fish feed is an important constraint in the aquaculture of this state. Availability of fish seed and water resources and natural productivity are very good in the state. But fish production is still lesser than national average. Almost 100% of people are fish eaters. Demand of fish is very high and the region is struggling to meet it. One of the important ingredients in fish food is starch. Starch acts as an energy source and an excellent binder for the fish feed. Tuber crops are the cheapest source of starch. Cassava is being utilized as fish feed ingredient in several parts of the world.

Fish feed generally consists of a starch source, an oil cake, an animal protein source, a fiber source and vitamin mineral mixture. The ingredients can be different but prepared in a way that overall 15-50% of protein should be there depending on species of fish, life stage and purpose of fish culture. Brooder fishes and carnivorous fishes require diet with high protein content.

Composite fish culture with higher stocking density and intensive feeding of grass carp with green leaves can effectively reduce cost of supplementary feeding (Grygerik, 1973; Manissery and Verghese, 1988). Identification and cultivation of different fodder plants and incorporation of grass carp at a higher ratio help the resource poor farmers of North East India to have better yield with lesser inputs.

Tuber crops as fish feed

Though all tuber crops can be utilized for fish feed preparation, tapioca got maximum popularity compared to almost all other tubers. Tapioca is easy to cultivate, easy to harvest and easy to process. Due to its low cost of production, tapioca became a popular ingredient in fish feed throughout the tropical region. In fact, all parts of tapioca are useful in fish culture. Tapioca leaf is an excellent fodder and can be utilized for feeding fishes like grass carp and *Barbus gonionotus*. Tapioca stem is an excellent material for periphyton development and fishes like tilapia, rohu and mrigal browse the partially decomposed skin with periphyton. Dried and powdered tubers are utilized as a starch source and a binder to have good quality pelletted feed.

Falayi *et al.,*(2001) evaluated the economic losses due to leaching of feed nutrients in fish farming at Nigeria and reported that cassava tuber was the most efficient ingredient tested in terms of nutrient retention, higher growth, feed utilization and economic evaluation indices compared to all other binders tested. Adebayo, *et al.*, (2003) reported that cassava meal can be used at 2% level to reduce nutritional loss in fish feed. Boscolo, *et al.*, (2002) reported that cassava meal can be incorporated up to 24% in the feed of Nile Tilapia for reducing feed cost without any adverse effect on its growth and performance. El-Baki *et al.*, (1999) reported that the cassava root meal can be incorporated upto 50% replacement of corn meal without affecting performance of Nile Tilapia in Egypt.

The 30% of the starch ingredient can be replaced by cassava peel with an addition of 0.4% methionine in pelletted feeds to maintain the serum and tissue thiocyanate concentration for Tilapia (Oresegun and Alegbeley 2002). Akegbejo-Samsons, (1999) reported that cassava flour can replace the yellow maize totally in the fish feed for African catfish. Fontana *et al.*, (2001) developed the technology for converting cassava starch to high-value single-cell protein (pigmented yeast biomass of *Xanthophyllomyces dendrorhous* (*Phaffia rhodozyma*) which can increase protein supplement and also enhance colour of fishes. Tapioca flour can be used for replacement of starch source without any noticeable change in fish production in common pond fish culture in Tripura (Santhosh *et al.,* 2006).

Dried and powdered sweet potato can be utilized for fish feed along with other ingredients. Sweet potato leaves also can be used as a fish feed for grass carp. Mokolensang, (2003) reported that sweet potato distillery byproduct increased growth of common carp more than conventional feed in experimental condition.

Cassava is the second most commonly used fodder for integrated fish culture in Malaysia (Ahmad, 2003). The palatability tests conducted using grass carp in Tripura also showed that tapioca leaves are preferred by grass carps than other plant materials tested. Laboratory level experiments with feeding of green tapioca leaves to grass carp showed better growth rate of grass carp fingerlings than five commonly used grasses (Santhosh, 2006).

Cassava stem with its soft coating easily disintegrates in pond water and forms a thick mass of periphyton that fishes prefer to browse upon. It is popularly advised to small farmers to go for cassava cultivation around the ponds and feed the fishes with the green fodder produced out of it. Even in lean season, cassava produces good vegetation and can be used for feeding fishes. The varieties which can grow luxuriant foliage are ideal for this purpose.

Nandeesha *et al.,* (2002) reported that *Colocasia esculenta* can be effectively utilized to replace protein source in the diet of common carp (*Cyprinus carpio*) without much negative impact in their performance.

Tuber crops as a cover crop

Tubercrops can be used to strengthen bunds and can be used as a cover crop to prevent soil erosion. These crops can check runoff of soil and silt in drainages. A few sweet potato varieties are identified which can produce maximum vegetation within a short span. These can cover entire area for years together and reduce soil loss and check siltation in earthen ponds. Some of the varieties are identified as suitable for this purpose are 187071 and 420027, which are producing very small quantity of tubers and producing foliage upto80-100 t ha^{-1} in 120 days. Sweet potato leaves can also be fed for grass carp. One of the major problems in the region is the heavy siltation of ponds and subsequent shallowing of the fish culture ponds. A few sweet potato varieties are fast growing and ideal to protect bunds. If it is planted in slops of bunds even near the water region for newly excavated ponds or renovated ponds within three months, the total bunds will be covered by vegetation. It is a duel purpose crop which gives fodder and tuber for fish feed and also helps to strengthen bunds. Unlike *Ipomea aquatica* and other marshy grasses which absorb nutrients from water column and reduce nutrient availability in water and thus fish production, above recommended varieties are ideal for cultivation in ponds bunds and slops. Harvesting of tubers in these cover crops should be restricted because it may loosen soil. Some non-edible tapioca varieties also can be used to strengthen bunds.

Direct Integration approach

Low land *Colocasia* and *Alocasia macrorhiza* can be directly integrated with aquaculture practice. Hardy fishes like tilapias and air breathing fishes like catfishes, murrels and climbing perches can be cultured in this system. Northeast India with vast area of swamps can have potential for integrated aquaculture of low land colocasia with air breathing fishes. Cruz *et al.,* (2003) described the farmer's own innovative integrated practice of growing low land *Colocasia* and Nile tilapia in Triala Village, Guimba, Philippines.

Conclusion

The agro climatic conditions of NEH region are suited to root and tuber crops cultivation. The huge genetic diversity of root and tuber crops be used for creation of gene banks especially for edible aroids. As the NE Region of India is very much deficient in food grain green fodder and feed, the sweet potato and tapioca could be a good feed and fodder resource for animals as they can be produced in better way in hill agro-eco-system. Enormous potentials are available in these eco-system to produce quality feed and fodder, which can be utilized for feeding the livestock to decrease dependence on external inputs. Intensification of root and tuber crop production and development of storage systems to ensure continuity of supply may result in improvements in animal feed availability. Evolving cassava and sweet potato cultivars suitable for hill agro ecosystem and their use in animal feed. These crops are suited to grow as intercrops in coconut, arecanut, fruit orchards and integrated farming system. More over, there is scope for exporting products of root and tubers to neighbouring countries like Nephal, Bhutan, Bangladesh, China and Burma.

References

Anonymous, 1999. State of forest report. FSI, GOI, Dehradun.

Anonymous, 2002. Basic statistics of North Eastern Region. NEC, GOI, Shillong, Meghalaya.

Anonymous, 2005. Annual report .AICRP on Tuber crops. CTCRI, Trivandrum.

Adebayo, O. T., Falayi, B. A. and Balogun, A. M. 2003.Comparative evaluation of some binding agents for water stability and nutrient retention in aquaculture diets. *Tropical Agriculture* 80, 128-131.

Ahmad, R., 2003. Fodder-fish integration practice in Malaysia. *FAO Document technique sur les peches [FAO Doc. Tech. Peches]*, no. 407, 33-37.

Akegbejo-Samsons, Y. 1999. The use of cassava flour as a substitute for yellow maize in diets for *Clarias gariepinus* fingerlings. *J. Aquaculture in the Tropics*, 14; 247-253.

Alhassan, W. S. and Odoi, F. 1982. Use of cassava leaf meal in diets for pigs in the humid tropics. Tropical Animal Health and Production 14 : 2160218.

Boscolo, W. R., Hayashi, C. and Meurer, F. 2002. Cassava by-product meal (Manihot esculenta) on feeding of Nile tilapia (*Oreochromis niloticus* L.) fingerlings. *Revista Brasileira de Zootecnia*. 31; 546-551.

Buitrago, J. A., Gomez, G., Portela, R., Santos, J. and Trujillo, C. 1978. Yuca ensilada para alimentation de credos. CIAT Cail, Colombia, Mimeo. 36 pp.

Cruz, C. D., Sevilleja, R.C. and Torres, J. 2003. Rice-fish system in Guimba, Nueva. Ecija, Philippines. FAO Document technique sur les peches [FAO Doc. Tech. Peches], no. 407, pp. 91-95.

CSIR, 1959. The Wealth of India. Raw Materials. CSIR, New Delhi. 5:25-27.

CSIR, 1976. The Wealth of India. Raw Materials. CSIR, New Delhi. 9:257-258.

Cullison, A. E. 1978. Feeds and Feeding: Animal Nutrition Prentice Hall of India Pvt. Ltd. New Delhi.

Das, M.M. and Dutta, L .1986. Regional variation in land use and agriculture in NE India. Vol: 18(1&2).

El-Baki, S. M. A., Ghoneim, S. I., El-Husseiny, H. M., El-Gendy, K. M. and Marghany, M. 1999. Cassava as a new animal feed in Egypt. 9. Cassava root meal in Nile tilapia (*Oreochromis niloticus*) diets. *Egyptian J. Nutrition and Feeds.* 2: Special issue; 753-763.

FAO, 1992. Roots, Tubers, plantains and bananas in animal feeding (D. Machin and S. Nyvold, editors). Animal Production and Health paper No. 95. Rome. pp 289.

Falayi, B. A., Balogun, A. M., Adebayo, O. T., Madu, C. T. and Eyo, A. A. 2001. Leaching of feed nutrients, economic losses to fish farming. *J. Aquatic Sciences (Nigerian Association for Aquatic Sciences)* 18; 119-123.

Fontana, J. D., Passos, M., Baron, M., Mendes, S. V. and Ramos, L.P. 2001. Cassava starch maltodextrinization/monomerization through thermopressurized aqueous phosphoric acid hydrolysis. *Applied Biochemistry and Biotechnology* 91: 469-478.

Grygierek,K. 1973. The influence of phytophagous fish on pond zooplankton. *Aquaculture,* 2: 197-208.

Manissery, J.K. and Varghese, T. J. 1988. Role of grass carp (*Ctenopharyngodon idella* Val.) fecal matter in composite fish culture. (*In:*) *Proceedings of the First Indian Fisheries Forum,* 31-37. M. Mohan Joseph (Ed.), Asian Fisheries Society, Mangalore.

Maner, J. H. 1973. Effect of processing method on the nutritional value of certain feeds for wine in Colombia and Ecuador. (*In:*) Effect of processing on the Nutritional Value of Feeds. *Nat. Acad. Sci,.* Washington D. C. pp 428-442.

Mokolensang, J. F., Yamasaki, S and Onoue, Y., 2003. Utilization of sweet potato distillery byproducts as a feedstuff for red carp *Cyprinus carpio. J. World Aquaculture Society* 34: 512-517

Morrison, F. B. 1984. Feeds and feeding. 22nd edition. CBS Publishers and Distributors. New Delhi, India.

Nandeesha, M. C., Gangadhara, B. and Manissery, J. K., 2002. Further studies on the use of mixed feeding schedules with plant- and animal-based diets for common carp *Cyprinus carpio* (Linnaeus). *Aquaculture Research.* 33; 1157-1162,

Oke, O.L. 1978. Problems in the use of cassava as animal feed. *Anim. Feed. Sci.Technol.* 3 : 345-380.

Oresegun, A. and Alegbeleye, W. O. 2002. Serum and tissue thiocyanate concentration in Tilapia (*Oreochromis niloticus*), fed with cassava peelbased diets supplemented with DL-methionine. *J. Aquaculture in the Tropics.* 17: 93-100.

Ramanujam, T. 1985. Extractable starch in six released varieties of Cassava and its relationship with tuber dry matter. *Natl. Symp. on Tropical tuber crops.* pp.125-127

Ranjan, S. K. 1980. Animal Nutrition in Tripura. Vikas publishing House Pvt. Ltd. New Delhi.

Ravindran, V. 1990. Feeding value and digestibility of cassava leaf meal for growing pigs. Proc. Fifth Anstralasian Animal Production Congress. Vol. 3. Taipei, Taiwan. 20 pp.

Ravindran, V., Kornegay, E.T., Wabb Jr. K.E. and Rajaguru, A.S.B. 1982. Nutrient Characterization of some feedstuffs of Sri Lanka. *J. Nut. Agric. Soc. Ceylon.* 19 : 19-32.

Reddy, V. R. and Bhosale, D.T. 2001. Handbook of Poultry Nutrition. International Book Distributing Co. Lucknow, India. Published in association with American Soy bean Association and United Soybean Board.

Sankaran, M., Singh, N. P., Chander Datt, Santhosh, B and Jai Prakash. 2006a. Studies on the performance of high yielding tapioca varieties and their nutritional parameters. Proceedings of the National symposium on improving input use efficiency in Horticultural crops. Organised by IIHR, 9-11th August, Bangalore.

Sankaran, M., Singh, N. P., Chander Datt, Santhosh, B and Jai Prakash. 2006b. Studies on the performance of high yielding sweet potato varieties and their nutritional parameters in uplands of Tripura. Proceedings of the 14th Triennial International symposium on Tropical tuber crops. Organized by CTCRI and ISRTC, 21-27 November, Thiruvananthapuram.

Santhosh, B., 2006. Utilization of Tuber crops as fish feed in Tripura (*In:*) Tuber crop production techniques and their utility in Tripura. Technical bulletin, 1; ICAR Research Complex for NEH Region, Tripura Centre, Lembucherra, West Tripura. (In press).

Santhosh, B., Singh, N. P., Datta, M., Dhiman, K. R. 2006. Production technology for composite fish culture in Tripura, Publication No. 14.,ICAR Research Complex for NEH Region, Tripura Centre, Lembucherra, West Tripura. 16 pp.

Sharma, B. D. and Hore, D. K. 1995. Genetic Resources of Yams in NE India with special reference to Garo Hills (Meghalaya). *Indian J. Hill.Farming* 8 (2): 145-151.

Tewe, O. O. 1992. Detoxification of cassava products and effects of residual toxins on consuming animals. (*In:*) Roots, tubers, plantains and bananas in animal feeding. FAO. Animal Production and Health paper No. 95. pp 81-98.

Vimala, B. and Nambisan, B. 2005. Tropical minor tuber crop. Technical Bulletin Series: 44, CTCRI, Trivandrum.

11

Tuber Crops Starches

1. Tuber crops Starches

Tropical root and tuber crops are rich sources of starch (Ghosh *et al.*, 1988; Chadha and Nair, 1994; Moorthy, 1994) besides many vitamins, minerals etc (Table 1). In addition to food applications, they have important industrial uses; especially, cassava in view of the considerable starch content. Cereals form the major source of commercial starches in most developed countries, but cassava (*Manihot esculenta* Crantz) and to a lesser extent, sweet potato (*Ipomoea batatas* Lam.) tubers are used for starch extraction in India, Brazil, Thailand, Indonesia, Philippines and China. Physical and chemical modifications of cereal starches are often carried out to overcome certain undesirable characteristics and make them suitable for specific end uses. Chemical modification is not desirable especially in food applications. Detailed studies carried out on different tuber starches revealed wide diversity in starch characteristics of tuber crops and the possibility of using these native starches instead of chemically modified starches (Moorthy, 2001). The isolation and characterisation of tuber crop starches *viz.*, cassava (*Manihot esculenta* Crantz), sweet potato (*Ipomoea batatas* Lam.), aroids (taro, *Colocasia esculenta*; tannia, *Xanthosoma sagittifolium* and elephant foot yam, *Amorphophallus paeoniifolius*); yams (*Dioscorea* spp.) and minor tuber crops like arrowroot (*Maranta arundinacea*), yam bean (*Pachyrrhizus erosus*), canna (*Canna edulis*), coleus or Chinese potato (*Solenostemon rotundifolius*), *Curcuma* sp. etc. have been elaborated, highlighting their advantages and potential uses.

2. Extraction techniques

Extraction of starch from cassava is simple and the resultant starch is pure white in colour and relatively free from other chemical impurities like fats, proteins, fibre etc. Moreover, the extraction of starch from other tuber crops is not so simple. The settling of starch granules is often slowed down by presence of various components like mucilage and latex leading not only to lower yield of starch, but also affecting quality of extracted starch. The long residence time can also bring about microbial growth bringing about breakdown of starch. Hence maximum recovery of starch with good physicochemical and functional qualities coupled with economical extraction of starches from tuber crops other than cassava is thus relevant. Use of ammonia (0.03M), during starch extraction, improved the yield and functional characteristics like colour, viscosity and swelling (Moorthy, 1991). Ammonia complexes with the mucilagenous material, releases starch granules and enables faster settling of starch. Lactic acid and citric acid are used to improve yield and colour of starch from sweet potato tubers (Balagopalan *et al.*, 1996). An enzymatic method for enhancing recovery (26% increase) of starch from cassava tubers using pectinase and cellulase enzymes was also reported. These enzymes break the pectin-cellulosic matrix of cell membranes and thereby facilitate release of starch granules. Using same technique, starch recovery from sweet potatoes could be enhanced by 20% without affecting starch properties (Kallabinski and Balagopalan, 1991). Mathew *et al.*, (1991) used a mixed culture inoculum to enhance yield of starch from cassava tubers. The extracted starchy flour possessed better attributes like puffing ability which makes it suitable for specific food applications like confectionery and bakery products (Moorthy *et al.*, 1993a; Mathew *et al.*, 1995).

3. Other components in starch

Starch extracted is invariably contaminated by minor quantities of fibre, lipids, proteins and minerals depending on method of extraction, age of crop, environmental conditions etc (Table 2). Some of these impart desirable qualities to starch, while others have detrimental effect on quality.

3.1 Moisture content

The moisture content of starch varies from 6-16% depending on the process used for drying starch. Higher levels of moisture result in microbial damage and also deterioration in quality. The moisture content suggested for safe storage of starch is 13% (ISI, 1970; Radley, 1976 b), but among tuber starches, large variation was found (Kay, 1987; Takeda *et al.*, 1986; Melo *et al.*, 1994; Soni *et al.*, 1990).

3.2 Fibre content

The fibre content in starch varies to a great extent depending on sieve used for removal of fibrous material and also on varietal variation and age of crop. Wide variation in fibre and ash contents in different tuber crops is reported (Asoaka *et al.*, 1991; Woolfe, 1992; Gebre-Mariam and Schmidt, 1998). Cassava flour (containing 2-3% fibre) had different properties compared to isolated starch (having 0.1-0.15% fibre) showing influence of fibre on starch properties (Moorthy *et al.*, 1996). The total dietary fibre in cassava flour is from 4.7 to 5.5 %.

3.3 Lipid content

Lipids in various forms have strong effect on starch properties (Eliasson and Gudmundson 1996; Hoover, 1998). Formation of starch-lipid or starch-surfactant complexes can affect textural properties of various foods (Krog, 1973; Evans, 1986; Gudmundson and Eliasson, 1992). The starch-lipid interaction is particularly important in native cereal starches, which contain lipids to noticeable extent. The tuber starches contain much lower quantities of lipids in them and hence effect is not so pronounced. In fact, the tuber starch is superior due to their bland taste. The lipid content in five cultivars of cassava varied from 0.11 to 0.22% in starch and 0.27-0.45% in flour. Widely varying lipid contents are reported for different tuber starches. Viscosity stability of cassava starch could be enhanced by treatment with surfactants (Moorthy, 1985a). Since the root starches contain much smaller quantities of native lipids in them, addition of lipids or surfactants would be desirable to obtain quality improvement and it was found that there is no hindrance for tuber starches to form complex with surfactants or lipids added externally. The specific complexing ability of amylose with surfactants was utilized for determination of amylose content in various starches (Kugimiya and Donovan, 1981; Sievert and Holm, 1993; Moorthy *et al.*, 2006).

3.4 Phosphorus content

Phosphorous is another important component invariably present in starch, since it is associated in synthesis of starch in chloroplasts. Wide variation occurs in the phosphorus contents of different starches. The phosphorus content in sweet potato starch and cassava starch (Tian *et al.*, 1991; Woolfe 1992) is much lower than that in potato starch. The P content in starch of different *Colocasia* cultivars varied from 0.006 to 0.013% (Moorthy *et al.*, 1993) while P content in six accessions of *D. rotundata* showed only negligible variability (0.011-0.015%) (Moorthy and Nair, 1989). For *D. dumetorum* starch, Nkala *et al.*, (1994) reported that almost all of phosphorus exists as bonded to starch. Starch from three cultivars of *Canna edulis* contained P in the range of 0.05 - 0.08%, even higher than that found in potato starch (Moorthy *et al.*, 2002). Curcuma starch also contained high P content (0.045%) (Jyothi *et al.*,

2003). The high phosphorous content leads to high viscosity and improved gel strength. Hence high P starches have use in food applications requiring high gel strength like in jellies. Cookies made using *Canna* starch are very popular in Latin American countries.

4. Colour and appearance

Colour is an important criterion for starch application in sago and textile industries. In food applications, the starch paste should be clear and free from any off-colour for better acceptability (Radley, 1976a). Starch from cassava tubers has white colour, if the skin and rind are removed prior to crushing or (Balagopalan *et al.,* 1988) addition of chemicals like inorganic and organic acids, sulphur dioxide, bleaching powder and ammonia. Mechanisation of starch extraction process, replacing conventional method of sun drying with centrifugal separation instead of settling tanks are other techniques to improve colour and appearance of starch.

5. Granule shape and size

Starch granules exhibit wide variability in size and shape (determined using light microscopy and SEM) depending on starch source (Table 3, Plates 1 and 2).

5.1 Cassava starch

Cassava starch granules have mostly round shape (Rickard *et al.,* 1991; Swinkels, 1985) with a flat surface on one side containing a conical pit, which extends to a well-defined eccentric hilum. Under polarised light, a well defined cross can be noticed. Starch granule size increased up to 6th month with age of crop and then remained steady (Moorthy and Ramanujam, 1986). Defloor *et al.,* (1998) found that the granule size was lower during dry season and attributed it to higher percentage of small granules during this period.

5.2 Sweet Potato starch

Sweet potato starch is polygonal or nearly round in shape (Tian *et al.,* 1991; Woolfe, 1992; Shin and Ahn, 1983; Bouwkamp, 1985) and has a centric distinct hilum. Polarisation crosses are comparatively less distinct. Granule size of sweet potato starch is almost similar to that of cassava. The granule size affects functional properties like swelling, solubility and digestibility and Bouwkamp (1985) reported negative correlation between particle size and susceptibility to amylase and acid degradation for sweet potato cultivars.

5.3 Other starches

Yam starches are characterised by a large variability in shape *viz.*, round, triangular, oval and elliptical. Other tuber starches also exhibit different granular shapes (Table 3). *Colocasia* granules are among the smallest of starches in plant kingdom making them useful in various applications eg. as a filler in biodegradable plastics, toilet formulations and aerosols (Griffin, 1979; Wang, 1983). Strauss and Griffin (1994) examined a large number of taro cultivars and found maximum granules having size of 5.10 mm and the least in the size 1.79 mm (mean value of 3.34 mm). *Colocasia* starch also exhibited varietal difference and 10 taro varieties revealed a significant difference in average granule size (Fig. 1, Table 4) (Moorthy *et al.*, 1993). The average granule size and distribution of the granule sizes showed only minor difference between corms and cormels of four cultivars of *Colocasia*. Varietal changes were not observed in *Dioscorea alata, D. rotundata* and *D. esculenta* starches. These starches exhibit an increase in granule size upto 5 months and thereafter remain steady. *D. esculenta* also has small granules (2-15 mm), useful in applications similar to *Colocasia* and cereal starches. Starch granules of *D. alata* and *D.rotundata* are much bigger. Farhat *et al.*, (1999) examined the distribution in size of different yam species. *D. rotundata* and *D. cayensis* starches showed similar patterns with a single symmetrical distribution centred at 32 and 35 mm respectively. *D. alata* starch however exhibited a non symmetrical wider particle distribution around 31mm and *D. dumetorum* starch had uneven distribution. Variation in granule size was minor among the three yam species *viz D. rotundata, D. alata* and *D. esculenta*. There was only small variation among 10 accessions of *elephant foot yam* (Table 5) (Moorthy *et al.*, 1994). The largest granule size was observed in *Canna edulis* starch (>35 mm) (Moorthy *et al.*, 2002). The granule size was in the range of 16-58 mm for *Curcuma zedoaria* and 14-46 mm for *C. malabarica* starches (Jyothi *et al.*, 2003).

6. X-ray diffraction (XRD) pattern

The crystallinity of starch is assigned to the well-ordered structure of the amylopectin molecules inside the granules. Different starches possess either 'A', 'B' or 'C' pattern (Table 6).

Cassava starch possesses 'A', 'C' or a mixed pattern with three major peaks at 2Ø=15.3, 17.1 and 23.5° (Rosenthal *et al.*, 1974; Gallant *et al.*, 1982; Zoebel 1988; Rickard *et al.*, 1991; Defloor *et al.*, 1998;). The absolute crystallinity of starch from four varieties of cassava lay between 8-14% (Moorthy *et al.*, 1996). Asaoka *et al.*, (1991) determined the diffraction pattern and absolute crystallinity of Colombian cassava varieties and reported that all possessed a

Table 1. Biochemical content of different Tuber crops

Tuber crop	Moisture (g %)	Protein (g %)	Fat (g %)	Crude fibre (g %)	Ash (g %)	Nitrogen Free extract (g %)	Vitamin C $(mg\ 100g)^{-1}$	Starch (%)
Cassava	59.4	0.7	0.2	0.6	1.0	38.1	25.2	20-25
Sweet potato	68.5	1.2	0.3	0.8	1.0	28.2	24.0	10-29
Yams	70.0	1.8	0.1	0.1	0.8	27.0	6.0	18-30
Taro	73.1	3.0	0.2	1.0	1.7	25.0	8.0	13-30
Elephant foot yam	78.7	3.0	0.4	0.8	0.8	18.4	–	16-18

Table 2. Biochemical content of different tuber starches

Starch	Moisture	Fibre/Ash	Lipid	Phosphorus
Cassava	10-14.8	0.01-0.8	0.07-1.54	0.007-0.012
Sweet potato	9.8-17	0.05-1.3	0.006-0.26	0.009-0.022
Taro	16.6-17.4	0.81-0.92	-	0.006-0.01]
Tannia	15.5-16.5	0.19-0.22	0.39	
Yam bean	10.9	0.06	0.33	
Arrowroot	10.10			
Elephant foot yam	10.55	0.58	0.088-0.1	0.045
Canna	*11.0*	*0.061*	*0.30*	0.01-0.08
Greater yam	13.6-18.2	0.22-0.26		
Lesser yam	16.8	0.46 [21]		
White yam	8-18.6	0.19-0.46		0.011-0.015
D. dumetorum	12-16.5	0.16-0.30	3.73-3.96	0.003
D. ballophylla	-			0.005
D.abyssinica		0.1	1.0	
Coleus	12.2-15.1	0.183- 0.4	-	
Curcuma sp.	-	-	-	0.045

Table 3. Granular properties of different root starches

Starch	Granule shape	Granule Size (μm)
Cassava	Round, Truncated, Cylindrical , Oval, Spherical , Compound	3-43
Sweet potato	Round, Polygonal, Oval, Bell, round, polygonal	2-72
Taro	Round	0.3-10
Tannia	Round	2.8-50
Yam bean	Round, Cupoliform or convex biconcavepolyhedral	6-35
Arrowroot	Round, polygonal	5-50
Elephant foot yam	Round, Polygonal,	3-30
Canna	Oval, polyhedral	13-145
Greater yam	Oval, Shell shaped, ellipsoid	5-140
Lesser yam	Round, Oval, Polyhedral, compound	1-15
White yam	Oval, Polyhedral Oval,round, triangular	10-70
D. dumetorum	Round, polygonal	1-16
D.cayenensis	Polyhedral,ellipsoid	10-70
Coleus	Round, oval	2.5-20

Table 4. Starch granule size, total and soluble amylose contents in starch from different varieties of *Colocasia* -Taro

Variety	Granule size [μm]	Total amylose content [%]	Soluble amylose content [%]	Viscosity(BU) (6%)	Pasting temperature (°C)
C-9	5.19	19.4	8.8	600	81-83
C-62	2.96	14.5	6.6	380	81-83
C-46	4.27	16.1	9.3	230	83-84
C-149	3.06	15.1	8.8	220	83-84
C-189	3.30	14.0	4.4	440	81-83
C-216	3.51	14.5	6.5	460	82-83
C-218	3.39	15.7	8.3	320	82-83
C-220	3.55	17.6	10.7	240	84-86
C- 266	3.16	15.7	9.8	320	81-83
C-304	3.20	16.1	7.5	300	83-84

Table 5. Yield of starch, granule size and amylose content of starch from different varieties of Elephant foot yam

Variety	Yield [%]	Average granule size [μm]	Total amylose [%]	Viscosity (BU) (6%)	Pasting temperature (°C)
Am 2	7.0	13.03	23.2	450	82-85
Am 5	10.0	12.49	23.5	480	82-85
Am 14	8.1	10.32	22.9	480	82-85
Am 15	12.3	11.12	23.3	620	82-85
Am 27	11.1	9.62	23.3	575	81-84
Am 32	9.9	10.62	23.2	570	82-85
Am 34	12.2	11.69	22.9	580	82-84
Am 36	10.5	10.35	23.9	600	82-85
Am 43	10.5	10.19	21.9	560	81-84
Am 51	14.3	9.85	23.2	500	81-84

Table 6. XRD patterns of different tuber starches

Starch source	XRD pattern
Cassava	A
Taro	A
Elephant foot yam	A
Yam bean	A
Tannia	A
Arrowroot	A
Greater yam	B
Lesser yam	B
White yam	B
Canna	B

'C_A' pattern. Sriburi and Hill (2000) found that extrusion of starch led to destruction of crystallinity.

Sweet potato starch also has 'A' (Szylit *et al.*, 1978; Gallant *et al.*, 1982; Takeda *et al.*, 1986) 'C' (Shin and Ahn, 1983; Zoebel, 1988; Chiang and Chen, 1988) or intermediate pattern (Tian *et al.*, 1991). Takeda *et al.*, (1986) observed 'A' pattern for two varieties while it was 'C_A' for another variety with absolute crystallinity of 38%.

6.1 Aroids, yams and other minor starches

Taro, tannia, yam bean, arrowroot, elephant foot yam and *D. dumetorum* starches possess 'A' pattern (Gallant *et al.,* 1982; Moorthy, 2001) and the starches of greater yam, lesser yam, white yam, *D. abyssinica* and *D. cayensis* possesss 'B' patterns. The XRD pattern of extracted starch is the same throughout the growth period of *D. rotundata* (Fig. 5). *Canna edulis* and *Curcuma* sp. starches exhibit 'B' XRD pattern. The absolute crystallinity of *Canna* starch was 26% (Zoebel, 1988). Heat-moisture treatment of potato starch brings about change in the XRD pattern of potato starch from 'B' to 'A '(Sair, 1967) but such a change was not noticed with *D. rotundata* (Moorthy, 1999). Lorenz and Kulp (1982) also observed change in XRD patterns of cassava and arrowroot starches by heat-moisture treatments. The treatments compress starch granules leading to change in properties.

7. Molecular weight

7.1 Cassava starch

The molecular weight of amylose from cassava starch was 232,000 (Ciacco and D'Appolonia, 1977), 431,000 (Takeda *et al.,* 1984), 522,000 and 1,250,000 (Susuki *et al.,* 1985) and for amylopectin, the molecular weight was 450 x 10^6 (Banks *et al.,* 1972). The average chain length was reported as 21(Susuki *et al.,* 1985), and a weight average of 26 (Hizukuri, 1969). Takeda *et al.,* (1987) showed that cassava amylose was composed of nearly equal numbers of branched and unbranched molecules. Susuki *et al.,* (1985) reported a value of 7710 for weight average degree of polymersiation and 3220 for number average degree of polymerisation. The molecular parameters of cassava amylose and amylopectin were as follows, mean chain length - amylose 100 to 10,000, amylopectin 20 to 30; DP (Glucose residues) amylose –100 to 10,000 and amylopectin- 10, 000 to 100,000.

7.2 Sweet Potato starch

Takeda *et al.,* (1986) reported trimodal pattern for sweet potato amylopectin and Hizukuri (1969) a bimodal distribution. They concluded that sweet potato has a higher proportion of 'A' chains and short 'B' chains compared to potato. Studies on sweet potato starch revealed amylopectin to have peaks at DP=12 and DP=8. The concentrations of the peaks at DP=6 and DP=7 were 7.1-7.5% and 6.7-7.0% respectively. Chain length varied in some varieties based on low viscosity and high reducing values (Woolfe, 1992).

7.3 Aroid, yam and other starches

The starches from different varieties of taro and elephant foot yam had similar molecular weights as exhibited by their reducing values. For *Coleus* starch, the reducing value was 1.71, while it was between 1.7 to 2.1 for *Curcuma* starch (Jyothi *et al.*, 2003) indicating the same range as for other tuber starches (Moorthy, 1986).

8. Amylose content

The linear component of starch, *viz.*, amylose plays an important role in determining characteristics of starch. Amylose content varies considerably among different starches. However genetic engineering has been used to obtain starch of amylose contents varying from 0 - >75% amylose. Part of amylose can exist as more labile soluble amylose in the amorphous regions of starch granules.

8.1 Total amylose

8.1.1 Cassava starch

The total amylose content in cassava ranged from 8-27% (Rickard *et al.*, 1991). Only non-significant differences in the amylose content were noticed during the growth period of six cassava varieties (Moorthy and Ramanujam, 1986). The gel permeation chromatographic analysis of five varieties of cassava carried out after debranching with iso-amylose also did not show any minor difference among the varieties (Moorthy *et al.*, 1996). In a comparative study of five varieties from Nigeria, Olorunda *et al.*, (1981) reported that tuber mealiness could be correlated to higher amylose. Defloor *et al.*, (1998) reported that the amylose content of five cassava varieties harvested at different seasons could not be related to the genotype. Asaoka *et al.*,(1991) reported the effect of season on amylose content. Added surfactants affected the amylose content in cassava starch. Maximum reduction in blue values was observed with cetyl trimethyl ammonium bromide while the lowest reduction for potassium stearate. Sodium lauryl sulphate and cetyl trimethyl ammonium bromide, having bulky hydrophilic groups can block the entry of iodide into the amylose helix causing reduction in blue values.

8.1.2. Sweet Potato starch

For sweet potato starch also, considerable variation in amylose content is found (Table 7). Garcia and Walter (1998) obtained values from 20-25% (by potentiometric titration) for a few Peruvian cultivars and location did not show any effect. Effect of fertilisation on amylose content was minimum. Similarly, there was no change in the amylose content during growth period of two sweet potato varieties (Noda *et al.*, 1992). Starches having a fewer amylose molecules and amylopectin molecules with higher content of short chains (DP 10) retrograded slower compared to others. Curing had only minor effect on amylose content (Bertoniere *et al.*, 1966) or a slight increase (Hammet and Barrentin, 1961).

Table 7. Physicochemical properties of cassava and sweet potato starches

	Cassava	**Sweet potato**
Amylose content [%]	8-27	8.5-38
Pasting temp. [°C]	55-90	65-90
Swelling Vol. [ml g^{-1}]	42-71	32-95
Solubility [%]	25-48	11.4-79
T_{onset} [°C]	50.7-69	58-75
T_{max} [°C]	54.7-78	63-79
T_{end} [°C]	60.2-100	78.83-88.5
ΔH [Jg^{-1}]	4.8-22	13.6-18.6

8.1.3 Other starches

Taro starch showed considerable variability in the amylose content and relationship between amylose content and granule size was observed. Studies on a number of cultivars of taro by Strauss and Griffin (1994) found maximum value for amylose content was 43% and minimum 3% with a mean value of 24.04%, but could not observe any good correlation between amylose content and granule size. Amylose content in yam starches also varied considerably according to various reports. The amylose content in starches of different *Dioscorea* species were greater yam – 25%, white yam and *D. cayenensis* – 23.8%, *D. dumetorum* -12.6%, *D. abyssinica* -29.7% (Gebre-Mariam and Schmidt, 1998), while *Soni et al.*, (1990) obtained 24.1% for *D. ballophylla.* Only very little variation in the amylose content was observed with age of crop for lesser yam, greater yam and white yam starches. Tannia starch also had a similar amylose content *viz.*, 15-25% and very little varietal variation was observed. The amylose content of ten cultivars of elephant foot yam showed very little variation (Moorthy, 2001). Starch from three accessions of *Canna edulis* had an amylose content of 24-30%, the highest being observed for purple accession (Moorthy *et al.*, 2002). Soni *et al.*, (1990) reported a still higher value of 38%. Coleus starch had an amylose content of 33-34 % (Moorthy, 1986). Based on all reports, it is inferred that among various tuber crops, *Canna edulis* and *Coleus* starches have the highest amylose contents.

8.2 Soluble amylose

Soluble amylose which exists in the amorphous region plays a significant role in deciding textural properties, since it is easily leached out and hence considered responsible for cohesiveness in cooked tubers (Hoover and Hadziyev, 1981). The soluble amylose contents in tuber crop starches normally ranged from 10-40% of total amylose. Soluble amylose content in different varieties of cassava did not vary during the growth period. Leelavathi *et al.*, (1987) obtained soluble amylose of 4.8% out of total amylose content of 17.9% in cassava starch. Similar trend was observed for *Colocasia, D. alata* and *Xanthosoma* starches. The soluble amylose content in starches of different accessions of elephant foot yam ranged between 9-11% forming nearly 45% of total amylose content (Moorthy *et al.*, 1994). The soluble amylose content in *Coleus and Canna* starch was 12.8% (Moorthy, 1986) and 10-12% respectively (Moorthy *et al.*, 2002).

8.3 Effect of surfactants on soluble amylose

Various surfactants suppressed the soluble amylose content to different levels in different starches and surfactants. The anionic surfactants, potassium palmitate and potassium stearate and the neutral surfactant glyceryl monosterate had little suppressive action, whereas the cationic surfactants cetyl trimethyl ammonium bromide and sodium lauryl sulphate had significant suppressive effect on cassava starch. Increase in concentration of these surfactants led to reduced blue values for soluble amylose (Moorthy, 1985a).

The cationic surfactant cetyltrimethyl ammonium bromide reduced the blue value for soluble amylose of all the starches (Table 8), more prominent in taro and lesser yam starches indicating that the soluble amylose of these two starches may be more anionic in nature or that the amylose helix of these two starches is able to sterically favour the complex formation.

9. Thermal characteristics

Differential Scanning Calorimetry (DSC) is an important tool to investigate starch gelatinisation (Biliaderis, 1983, 1990; Eliasson, 1994; John and Shastri, 1998). Most of these DSC studies were carried out on cereal starches and to some extent on potato and cassava starches (Stevens and Elton, 1971; Wootton and Bamunuarachchi, 1979; Asaoka *et al.*, 1992; Moorthy *et al.*,1996; Defloor *et al.*,1998; Farhat *et al.*,1999) whereas information on the DSC of other root starches is comparatively limited. Table 7 gives the DSC characteristics of tuber starches. A typical DSC curve of cassava starch is shown in Fig. 2.

9.1 DSC Gelatinisation temperatures

9.1.1 Cassava starch

Starch extracted from five varieties of cassava possessing different organoleptic quality (Moorthy *et al.*, 1996) had affect on DSC patterns which showed that varietal differences manifest themselves in the DSC patterns (Table 9). The characteristic peak shape could be traced to structural differences among the varieties (Fig. 3). Onset of gelatinisation as indicated by T_{onset} was the earliest for H-165 starch (65.35°C) and the latest for H-97 starch (69.35°C). While T_{end} was the highest for M4 showing the largest range of gelatinisation for M4 starch. Examination of gelatinisation characteristics of four cultivars harvested at different seasons indicated that both genetic constitution and environmental conditions affected the DSC parameters. Starch of one particular cultivar displayed the highest gelatinisation temperature through all the harvesting seasons. They found that although within each planting season for each genotype and for each harvest time, significant differences in gelatinisation temperatures were noticed, no systematic changes as a function of genotype or harvest time was noticed. Starch samples from tubers harvested at six months stage in dry season had higher onset, peak and temperatures compared to that from rainy season, but the opposite trend was true for the 12, 15 and 18 months harvested tubers. The temperature of drying also enhanced the DSC gelatinisation temperatures. Heat-moisture treatment enhanced the gelatinisation values of cassava starch considerably (Lorenz and Kulp, 1982).

9.1.2 Sweet Potato starch

Collado *et al.*, (1999) obtained considerable variation in all the DSC parameters of 44 sweet potato varieties. The mean T_{onset} was 64.6°C and ranged 61.3-70° C, mean T_{peak} 73.9°C (range 70.2-77°C) and mean T_{end} 84.6°C, range being 80.7-88.5°C and the mean gelatinisation was 20.1°C with a range of 16.1 to 23°C. Garcia and Walter (1998) found the range to be between 58-64°C for T_{onset}, 63-74°C for T_{peak} and 78-83°C for T_{end} for the two varieties cultivated at different locations. While selection index did not affect the values, location influenced the parameters (Tian *et al.*, 1991). Noda *et al.*, (1996) did not find effect of fertilisation on DSC characteristics of two sweet potato varieties. During growth period, the T_{onset} was the lowest at the latest stage of development.

9.1.3 Other starches

The DSC data on other tuber starches showed considerable variability in values. Compared to cassava and sweet potato starches, these starches generally had higher T_{onset} and T_{end} values. The highest values were noticed for *taro* starch and the other starches had values in between. For taro starch, T_{onset} values were 83.2°C and 79.9°C. The corresponding T_{end} values were 90.0 and 85.0°C respectively, being the highest among the starches. For elephant foot yam starch, the values were 77.8°C for T_{onset} and 83.5°C for T_{end}.

DSC data of starches from the two *Curcuma* varieties indicted that the gelatinisation peak of *C. malabarica* starch was a doublet (Jyothi *et al.,* 2003). The onset of gelatinization was lower for *C. malabarica* starch whereas T_{end} was nearly similar. The higher T_{onset} for *C. zedoaria* starch also suggests curcumin forming complex with starch molecules. The range of gelatinization was higher (17-22°C) for *Curcuma* starch resembling yam and potato starches which can be attributed to presence of phosphate linkages in these starches. T_{end} values and the ranges of gelatinization were higher for *Curcuma* starch.

The difference in gelatinisation temperatures among various tuber starches can be traced to variation in intermolecular bonds. High temperature of gelatinisation can be attributed to the higher stability of starch crystallites in starch molecules, which implies that more heating is required to swell the granules. A number of other factors like varietal differences, environmental conditions and the experimental protocols like level of moisture, sample preparation, rate of heating and instrument used contribute to the differences in values.

9.2. Gelatinisation range

Range of gelatinisation is also quite different among different starches. The highest range was obtained for cassava starch (12.90°C) and the lowest for *tannia* starch (4.7°). Higher range indicates to higher level of crystallinity, which imparts higher structural stability so that the water molecules need longer time to penetrate the crystalline areas. Biliaderis *et al.,* (1990), Tester and Morrison (1990) and Leszkowiat *et al.,* (1990) suggested that higher transition temperatures indicate more stable amorphous regions and lower degree of chain branching. Granule size and gelatinisation range also do not show any relationship.

9.3 Gelatinisation enthalpy

Gelatinisation enthalpy depends on crystallinity and intermolecular bonding. For cassava starch, values range from 4.8 to 16 J g^{-1}. Gelatinisation enthalpy was also found to depend on genetic and environmental factors as illustrated by Asaoka *et al.,* (1992) and Defloor *et al.,* (1998) from studies on different varieties, time of harvest and seasonal variations. For sweet potato starch, the value for gelatinisation enthalpy lies between 10-18.6 J g^{-1} (Tian *et al.,* 1991; Garcia and Walter, 1998; Collado *et al.,* 1999). Effect of variety and environmental conditions was also evident (Garcia and Walter, 1998; Noda *et al.,* 1996). During growth period, the ΔH was the lowest at the earliest stage of development in two sweet potato cultivars and the enthalpy ranged between 11.8-13.4 J g^{-1} (Noda *et al.,* 1992).

The gelatinisation enthalpy for other starches also illustrates considerable variability among the starches. There does not appear to be any definite relationship between enthalpy and factors like amylose content, granule size and XRD patterns. Since the amylopectin has a more crystalline nature, higher amylopectin content contributing to higher enthalpy of gelatinisation but such an effect is not evident. The differences do not reflect the structural differences among the tuber starches. The results also show that unlike the gelatinisation temperatures, which show wide variation, the enthalpy of gelatinisation is within in a small range for tuber starches.

10. Gelatinisation and Pasting temperatures

Gelatinisation temperature is indicative of temperature at which the starch granules start gelatinising. Here also considerable variability exists among different tuber starches.

10.1 Cassava starch

Cassava starch has the lowest gelatinisation temperature among the different tuber starches. Gelatinization temperatures of starch of seven cassava varieties examined microscopically showed that starch of H 165 and H 1687 gelatinised earlier and the gelatinization range was relatively higher, *viz.,* over 12°C. No relationship between granule size and gelatinization temperature was observed (Defloor *et al.,* 1998). The gelatinisation temperatures of cassava starch determined microscopically by various workers ranged from 49-64°C to 62-73°C (Rickard *et al.,* 1991). When cassava starch was subjected to steam pressure treatment at different pressures, increase in pasting temperature by 2 to 9°C occurred depending on time of treatment and pressure used. The pasting temperature rise was higher at longer time of treatment and higher steam pressures. Surfactants led to increase in pasting temperatures (Moorthy, 1985a), and the most pronounced effect was noticed on treatment with potassium palmitate and potassium stearate. On esterification to acetate, the pasting temperatures were lowered (Moorthy, 1985a). Gelatinisation temperatures were enhanced tremendously in glycerol and ethanol, while in DMSO and formalin, only slight increase was noticed and results can be traced to steric factors (Moorthy, 1982b).

10.2. Sweet potato starch

The pasting temperature of sweet potato starch obtained using Brabender Viscograph varied between 66.0 and 86.3°C while microscopic determination gave values between 57-70 to 70-90°C. Noda *et al.*, (1996) observed the pasting temperatures of starch from two sweet potato cultivars grown at different fertiliser levels to be 70.8 –73.9 °C.

10.3 Aroids, yam and other starches

Strauss and Griffin (1994) found the gelatinisation temperature of different varieties of taro starch to be between 69 to 74°C. Pasting temperatures of different cultivars of taro and tannia were distinctly higher than those of cassava and sweet potato starches. Perez *et al.*, (1998) obtained the gelatinisation temperatures (obtained by Rapid Visco Analyser) and DSC of Arracaccia, tannia, canna, arrowroot and cassava starch as 68-95 °C, and 56-73°C, 85-95°C and 74-87°C, 73-95°C and 61-75°C, 79-92°C and 68-85°C, 73-90°C and 62-84°C respectively. Soni *et al.*, (1990) found gelatinisation temperature of *Canna* starch as 65-70°C. The gelatinisation temperature of starch of three accessions of *Canna* was 74-85 to 80-95°C by the Brabender Amylograph and 74-75°C by the RVA (Moorthy *et al.*, 2002). Pasting temperatures of starch of elephant foot yam extracted from 10 accessions were nearly equal and in the same range as other aroid starches (Moorthy *et al.*, 2006). The range was lower compared to cassava or yam starches. For yam bean starch, values of 74-79°C were obtained and only minor variation existed among varieties (Moorthy *et al.*, 2006). Values for *Curcuma* starch were 81°C by RVA (Jyothi *et al.*, 2003) and 65-85°C for *Coleus* starch determined microscopically (Moorthy, 1986). The gelatinisation temperature of *Coleus* starch was 75 - 81°C depending on concentration used.

Yam starches gelatinised over a temperature of 20°C and gelatinisation continued even after 95°C showing strong intermolecular linkages. The RVA results on pasting of different yam species indicate the values to range between 75 and 83°C, the highest for *D. dumetorum* and lowest for *D. cayenensis* starch (Farhat *et al.*, 1999). *D. abyssinica* starch had a gelatinisation temperature of 73°C (Gebre-Mariam and Schmidt, 1998) . Nkala *et al.*, (1994) found the values to be similar, ie. 83°C for starch of *D. dumetorum* from dry and wet seasons. The large range for yam starches may be attributed to presence of phosphate linkages in these starches.

11. Viscosity

Use of starch in textile, paper, adhesive and food industries depends on viscosity of starch paste. Different tuber starches exhibit considerable variation in their viscosity characteristics.

11.1 Cassava starch

Peak patterns obtained using Brabender Viscoamylograph are different for different varieties of cassava. Wide variation among viscosity data for cassava starch from various sources is obtained (Srivatsava *et al.*, 1970; Rosenthal *et al.*, 1974) (Table 9). Slight reduction in peak viscosity and breakdown are observed in cassava starch extracted by an enzymatic method. The viscosity of cassava starch extracted from inoculum provided fermentation was lowered due to presence of fibrous matter (Moorthy *et al.*, 1993). The breakdown was also correspondingly reduced due to cementing of granules by fibrous materials and its rheology was similar to cassava flour.

Cassava starch, when subjected to steam pressure treatment showed reduction in viscosity which fell steadily with increase in pressure and time of treatment. It was also observed that reduction in viscosity was linearly related to the severity of treatment. Lorenz and Kulp (1982) observed changes in viscosity by heat-moisture treatment at different moisture levels. Different salts affected viscosity characteristics of cassava starch. It was observed that the influence of ionic solutes of chloride and sulphate was nearly same, but the effect of sulphate was more evident. Disodium orthophosphate produced different effect compared to sulphate or chloride. Non ionic solutes like glucose and sucrose brought about only slight effect (Raja and Ramakrishna, 1988). Blending of cassava starch with maize and potato starches modified viscosity properties with the resulting blend having properties in between those of starches. The Peak Viscosity fell from 475 to 415 BU while setback was lowered from 283 to 231 BU. The breakdown was not affected to any major extent. Extrusion cooking led to reduction of apparent viscosity of cassava starch.

Surfactants affect viscosity of cassava starch considerably (Moorthy, 1985a). Surfactants like sodium lauryl sulphate increased the peak viscosity, especially at higher concentrations, while potassium stearate and potassium palmitate had only minimum effect. Glyceryl monostearate lowered the peak viscosity at higher concentrations. Non-aqueous solvents like ethanediol reduced the Redwood viscosity from 55 seconds to 17.5 seconds, while glycerol enhanced it to 175 seconds. In addition to peak viscosity, the breakdown in viscosity is another important criterion that decides the suitability of starch in food and industry. Cassava starch is considered inferior to maize starch, because its viscosity is considerably lowered on heating under shear leading to a 'long' and cohesive texture for its paste, which is not desirable in many food and textile applications. However, surfactants have a more desirable effect and are easy to handle as well. On incorporation of potassium stearate or potassium palmitate, even at 0.02% mole concentration 100 g^{-1} starch, the viscosity was stabilised (Moorthy, 1985a).

In view of the presence of a large number of hydroxyl groups in starch, cross-linking using di- and tri functional chemicals has been used to stabilise starch viscosity. The most common reagents for crosslinking are epichlorhydrin, phosphorus oxychloride and sodium metaphosphate. These chemicals, by binding starch molecules in granules, increase the associative forces, rendering starch granules stronger and preventing breakdown during heating and stirring (Srivatsava *et al.,* 1973).

11.2 Sweet Potato starch

Sweet potato starch behaves almost similarly to cassava starch in all its viscosity characters. In a study of 44 different sweet potato genotypes using Rapid Visco Analyser, Collado *et al.,* (1999) have worked out the correlations among RVA parameters. They observed wide variation not only in the PV but broadness of peak. A negative correlation between PV and amylose content was noticed. The rheological properties of sweet potato starch extracted using an enzymatic process did not vary among the different concentrations of enzyme (Moorthy and Balagopalan, 1999). The rheological properties of sweet potato starch have been examined in detail. During heating, the G′ and G″ increased while phase angle decreased indicating change from sol to gel. The initial increase in G′ and G″ was attributed to progressive swelling of starch granules leading to close packing. When the starch granules became very soft, deformable and compressible, decrease has been observed. The rheological properties of various tuber starches were compared using the Bohlin rheometer and wide variability in the values of G′ and G″ was observed. However, all the different tuber starches exhibited uniformity in their elastic behaviour predominating over viscous nature.

11.3 Aroid, yam, and other starches

The viscosity of taro starch is less than that of cassava starch and close to cereal starches. Considerable difference in peak viscosity values among accessions was observed. There was only nominal breakdown in viscosity even at the highest concentration pointing to the possibility of using this starch in various applications, which require paste stability. The viscosity values were nearly the same for the corm and cormel starches.

Viscosity properties of elephant foot yam starch extracted from ten accessions were quite similar (Moorthy *et al.,* 1994). The viscosity breakdown was very low, similar to other aroid and cereal starches. There was only a slight difference in viscosity values of tannia starch among the varieties. Breakdown varied from 0 to 30 BU at 7% concentration, showing stability of starch paste.

The yam starches are usually characteristised by a slow rise in viscosity even after 95°C. This indicated that all the granules do not gelatinise at 95°C thus resembling potato starch. Starch of four varieties of lesser yam extracted using water and ammonia solution showed only minor difference in viscosity values between the varieties (800-950 BU), but the viscosities of starch extracted using ammonia solution had higher values than that obtained by water extraction. As observed for lesser yam starch, there was no noticeable breakdown in viscosity during holding. The viscosity breakdown in yam starches was quite low in spite of high viscosity levels. Yam starches usually have three to four times more phosphorus than cassava and aroid starches. It is reported that the phosphate linkages in potato starch are responsible for its high viscosity and similar effects are observed in yam starches. The intrinsic viscosity of yam bean starch was 2.23ml g^{-1}, similar to cassava (2.45 ml g^{-1}) and corn (1.81 ml $g^{-1)}$. For coleus starches, reported value obtained is 40 sec for this starch in a Redwood Viscometer while the Brabender values were: Peak Viscosity (6%)-890 BU, Peak Viscosity (10%)-2670 BU; breakdown being 260BU and 430 BU respectively.

Farhat *et al.,* (1999) found peak viscosities ranged from 2028 mPas for *D. dumetorum* to 3893 mPas for *D. cayenensis* compared to 8900 mPas for potato and 3134 mPas for cassava starches. The breakdown is lowered compared to cassava and potato starches. Retrogradation tendency was also clear. *D. dumetorum* had all the viscometric parameters lower than the other three (Gebre-Mariam and Schmidt, 1998). Yam bean starch had similar viscosity to arrowroot starch with only little difference among 10 varieties. Among cassava, arrowroot and Canna starches, the highest peak viscosity was for canna starch followed by arrowroot and cassava. Breakdown was the highest for cassava starch among the three and consistency, the highest for Canna starch. Similarly Arracaccia starch had higher peak viscosity compared to tannia, but the latter had higher consistency. The peak viscosity values for starches from three accessions of Canna ranged between 3887 to 4187 cPs. The *Curcuma malabarica* starch had a higher peak viscosity of 3387 cP compared to *C. zedoaria* starch. The breakdown in viscosity was quite low for curcuma starches more specifically for *C. zedoaria* starch (66 cP), showing that the granules are quite strong and resist breakdown under shear and heat, thus resembling yam starches rather than cassava. A reasonable amount of reassociation of granules during cooling was also observed. The high setback viscosity and low breakdown viscosity of *Curcuma* starch can be attributed to the high phosphorus content resembling potato and *Canna edulis* starches (Jyothi *et al.,* 2003).

12. Swelling power

Swelling power and solubility indicate the strength of non-covalent bonding among starch molecules which depends on factors like amylose-amylopectin ratio, chain length and molecular weight distribution, degree / length of branching and conformation (Rickard *et al.,* 1991).

12.1 Cassava starch

Cassava starch has swelling power, medium between potato and cereal starches and values vary from 42-71 ml g^{-1}. During growth period, starch of two varieties H-2304 and M4 maintained their swelling volumes within narrow ranges, while that for some of the varieties showed wide variations which indicate that these varieties are very much susceptible to environmental influences (Moorthy and Ramanujam, 1986). Asoaka *et al.,* (1992) found that swelling power was higher in samples harvested in November compared to those harvested in August. Swelling volumes are affected by presence of various chemicals and treatments carried out on starch. High amylose content and presence of stronger or more number of intermolecular bonds lead to lower swelling. Formation of lipid-starch complex can also affect swelling volumes (Swinkels, 1985) as also presence of naturally occurring carbohydrates and non-carbohydrates along with starch (Swinkels, 1985; Eliasson and Gudmundson, 1996). The fibre and lipids act as barrier to free swelling of starch (Moorthy *et al.,* 1996). The starchy flour extracted from fermented tubers also exhibited lower swelling volume (Moorthy *et al.,* 1993). Sodium sulphite and thiosulphate suppressed swelling volume to very low values at different concentrations (Fig. 4 and Fig. 5). The effect is attributed to oxidative-reductive depolymerisation brought about by the sulphite ions. Another salt that brought about a similar effect was sodium thiosulphate at the same concentrations. Surfactants also affect swelling volume of starches. Swelling volume was reduced by potassium palmitate and potassium stearate even at the lowest concentrations, while sodium lauryl sulphate and cetyl trimethyl ammonium bromide enhanced swelling volume and glyceryl monostearate affected only to a small extent. Steam pressure treatment and heat-moisture treatments also lowered swelling volume by compressing starch molecules and thus restricting free swelling of starch (Moorthy, 1982a).

13.2. Sweet potato starch

Swelling power of sweet potato starch varies considerably not only among varieties, but also at different temperature (Table 7). Delpeuch and Favier (1980) reported a two stage swelling but Madamba (1975) found a single stage swelling for same starch. The lower swelling volume of sweet potato starch is attributed to a higher degree of intermolecular association compared to cassava or potato starch. There was no significant correlation between amylose content and swelling volumes.

12.3 Other starches

Considerable variation in swelling volume of different varieties of *Colocasia* was also reported (26.5 to 60 ml g^{-1}). Swelling volume of starch of six clonal selections of white yam showed only slight differences among selections. Starch from different varieties of lesser yam, greater yam, tannia and elephant foot yam had much lower ranges. *D. ballophylla* and elephant foot yam starches had lower swelling volumes compared to cassava, but among these, the elephant foot yam starch had higher value. The relatively lower swelling of yam starch compared to potato starch was attributed to higher lipid content in the starch and also higher inter-associative forces compared to potato starch. For *Coleus* starch, the value is around 25 ml g^{-1} (Moorthy, 1986). For Curcuma starch, values obtained are 9 ml g^{-1} and 30 ml g^{-1} for *C. zedoaria* and *C. malabaricum* starches respectively.

13. Solubility

Solubility of starch depends on source, inter-associative forces, swelling power, presence of other components etc.

13.1 Cassava starch

Higher solubility of cassava starch compared to other tuber crop starches can be attributed partly to the high swelling it undergoes. The solubility data indicate that starch is more soluble in polar solvents or solvents with affinity towards water (Moorthy, 1982b).

13.2 Sweet Potato starch

Solubility of sweet potato starch is similar to cassava starch. The high swelling volume of starch is reflected in its solubility. It is presumed that the bonding forces might be tenuous but comparatively extensive, immobilising the starch within the granules even at high levels of swelling.

13.3 Other starches

Solubility of other yam and aroid starches varied from 10-30% and in general, the aroid starches had usually lower solubilities.

14. Clarity and Sol stability

The high clarity of starch has much relevance in food and textile applications. Clarity characterizes the associative bonds between starch molecules in granules.

14.1 Cassava starch

The weaker associative forces in cassava starch compared to cereal starches impart better clarity to starch paste. Acetylation and propylation derivatives lead to higher clarity. Esterification weakens available hydroxyl groups (Moorthy, 1985b). In contrast, heat moisture treatment reduced clarity by strengthening associative forces.

14.2 Yam, aroid and other starches

Yam starches have almost equal clarity as cassava starch, pointing that their associative forces are similar. Variation was minimum among different varieties of greater yam, lesser yam and white yam starches. Steam pressure treatment decreased the clarity of yam. The clarity of aroid starches is poor and similar to that of cereal starches. The clarity of three accessions of Canna was much higher than that of the aroid starches. Generally the 'B' starches appear to have higher clarity compared to 'A' starches.

15. Sol stability

Sol stability or paste stability reflects retrogradation tendency of starch pastes. Cassava and sweet potato starches have low retrogradation tendency and therefore exhibit high paste stability. It is suggested that the lower retrogradation tendency of cassava may be due to the higher weight-average molecular weight of amylose fraction in cassava.

Sol stability was also affected by added surfactants, by derivatisation etc. Yam starches have good stability, while the aroid starches have poor stability. The paste stability of starch of different varieties of greater yam, white yam, lesser yam, tannia, and elephant foot yam were nearly the same. Heat-moisture treatment (steam pressure treatment) of yam starches lowered paste stability.

16. Digestibility

Starch digestibility by enzymes is of importance for evaluating nutritive value and in industrial applications. When compared to other tuber starches cassava starch is one of the least resistant root starches (Madamba *et al.*, 1975; Tian *et al.*, 1991; Rickard *et al.*, 1991). Sweet potato starch was highly susceptible to degradation by α-amylase and glycoamylase (Cerning-Beroard and Le Dividich, 1976). Digestibility of raw starch of 8 sweet potato varieties by glycoamylases was compared by Noda *et al.,* (1992). The digestibility was >80% after 24 hours. Gallant *et al.,* (1982) found that 'A' type starches showed high susceptibility to α-amylase. Gallant *et al.,* (1982) found that pelletisation increased the raw starch digestibility with bacterial α-amylase from 17% to 45%. SEM studies indicated that enzymatic corrosion occurred mainly at surface of the granules. Deep radial corrosion channels were not observed on starch granules (Franco *et al.,* 1988).

For raw starches, digestibility of cassava, sweet potato, taro, tannia and elephant foot yam starches was quite high (65-75%), compared to corn starch (76%) but that of the yam starches was low (15-25%) as in potato (10%). Cooking increased the digestibility substantially and is attributed to the change in starch structure on gelatinisation (Moorthy and Padmaja,1991). The digestibility could not be related either to the amylose content or the soluble amylose contents in these starches. Popping of sweet potato improved starch availability and nitrogen digestibility. Digestion of elephant foot yam starch by α-amylase carried out by Wankhede and Sajjan (1981) obtained 98% digestibility for gelatinised starch compared to 21% for raw starch.

17. Applications of tuber starches

Tuber crops contribute significantly to nutrition and economy of developed and developing nations. Cassava tubers are used primarily for food purpose. Root and tuber crops are also considered ideal raw materials for animal feed as they are concentrated sources of energy. The tuber crops are rich sources of starch and hence are used widely for the extraction of starch. Among tuber crops, cassava and sweet potato are the most important from industrial point of view as they are rich sources for a number of value added products. Starch has tremendous application in various fields like food, feed and industry.

In addition being a major component in human diet, starch also forms an excellent raw material for modifying texture and consistency of foods, owing to its ability to form viscoelastic gel when heated in water. Besides their food uses, starch derived products have considerable applications in pharmaceuticals, textiles, alcohol based fuels and adhesives (Kempf, 1985). Emerging uses of starch include as low calorific fat substitutes, biodegradable packing materials, thin films and thermoplastic materials with improved thermal and mechanical properties (Griffin, 1994).

Tuber crop starches widely differ in their properties. Depending on type of application, different tuber crops can be selected. In addition there is scope of modification by physical and chemical means so that the products can be further improved for specific applications. A large amount of work was done for production of value added products from tuber crops. The most important products are cassava macaroni, cassava rava and biscuits. The pastiness of cassava flour was reduced by various treatments including cross linking (Emilia Abraham *et al.*, 1979), phosphoric acid treatment (Moorthy, 1983) and fractional sieving (Raja *et al.*, 1978).

17.1 Food industry

Native and modified starches are used in food industry as thickener, filler, binder and stabilizer. For example, in instant pudding, soups, gravies, custards and jellies starch are used as thickener. The bland taste, higher paste clarity and less tendency to retrograde compared to cereal starches make it the most preferred starch for food purpose. The gel stability of cassava starch is much higher compared to cereal starches and hence preferred for many food products especially when the food has to be stored for a long time. In addition to native starch, modified ones are used in food products. Starch modification is therefore a common practice for development of processed foods, having both textural and nutritional advantages.

17.2 Modified starches

Cassava starch is linked to improve its viscosity stability to suit specific applications especially in textile industries. The chemicals used commonly are phosphorus oxychloride, epichlorohydrin metaphosphates, acrolein, cyanuric chloride etc (Srivastava and Patel, 1973). Cross linking reduces stringiness without interfering with desirable properties. In addition, the cross linked starches are resistant to acids. Formaldehyde is a simple reagent used in starch modification and by manipulating the amount of formaldehyde used, the properties can be modified according to the requirement (Fig. 6). Carboxymethyl starch is an important ester derivative whose properties depend on the degree of substitution (Fig. 7). The other methods include direct reaction with acids, reaction with acid anhydride in presence of alkali, reaction with acids in presence of catalyst and reaction with acylating agents. Of all the chemical modifications, acid modification has the single advantage that it alters the functional properties of starch without affecting its granular nature. Acid modified starches have wide applications in various starch based industries (Rohwer and Klem, 1984). For esterification of cassava starch various catalysts are tried of which perchloric acid catalysis gave good results

(Moorthy, 1985b). Oxidized starch was also prepared by using permanganate and chromic acid oxidation. All these reactions were done to improve the already existing properties of tuber starches. Dialdehyde starch was prepared by oxidation of starch using potassium meta periodate. Dialdehyde starches are also used to produce water resistance adhesives

17.3 Sweeteners

Maltodextrins, glucose syrup, crystalline dextrose and fructose syrup are used widely in sweetener industry. All these products are widely accepted as sweetener in confectionaries, colas and the demand for them is steadily increasing. High fructose syrup has almost displaced glucose syrup in soft drink industry in many developed nations. Attempts were made to prepare high fructose syrup from cassava and sweet potato starch. When alkali was used as isomerising agent it gave only low levels of fructose in the syrup (Moorthy, 1982b,c). However when glucose isomerase was used, conversion was facile and the product was free from colour and had the desired sweetness. It is used in production of carbonated and non carbonated beverages.

17.4 Alcohol production

The production of alcohol from cassava flour was also an important out come in starch research. In addition, a number of researchers carried out extensive work on improving yield of glucose from starch and flour (Srikanta *et al.,* 1987, Emilia Abraham *et al.,* 1987, 1988, Vijayagopal and Balagopal, 1989) and from cassava factory waste (Kunhi *et al.,* 1981, Ahmed *et al.,* 1983, Sreekanta *et al.,* 1987, Jaleel *et al.,* 1988).

17.5 Animal feed

Cassava flour was also tried in animal feeds and many workers analyzed the suitability of starch and flour in various formulations indicating possibility of large scale utilization of cassava starch and flour in animal feed industry.

17.6 Biodegradable plastics

Production of biodegradable plastics using starch assumes a lot of importance. Patel *et al.,* (1986a,b) grafted cassava and sweet potato starch to polyacrylonitrile and studied their characteristics. There exists vast potential for exploitation of tuber starches in food and other industries.

17.7 Textile industry

Starch is used in textile industry mainly for textile sizing, finishing and printing operations. The major advantages of cassava starch compared to maize starch are mainly lower price, higher strength, higher viscosity and transparency of its film. Textile industry used starch in finishing giving brightness and stiffness to textiles and cassava starch is superior for this purpose. However, it has also disadvantages. It possesses unstable viscosity, tendency to stick to machinery since the swollen granules are not very strong and disintegrate under shear. Lots of modifications were done to improve all its properties. The availability and properties of various starches decide their use in different areas.

17.8 Adhesives

Among the modified starches, starch acetate is used in gummed tape formulation because of its flexibility high gloss and rewetting ability. Hydroxy ethylated starches are used in adhesives such as bag pastes, case sealing and label and envelop adhesives.

17.9 Dextrins

Dextrins and British gums are used as thickeners or textile printing pastes. Dextrins with cold water solubility are used as carriers for active ingredients like food flavours. Pharmaceutical companies use dextrins in the fermentation process.

17.10 Pre-gelatinized starches

The textural qualities of food are derived from starch as a result of changes during and after cooking. Gelatinisation of starches, swelling, viscosity, paste clarity and retrogradation, all affect textures of food. Pre-gelatinized starches are receiving much attention because of their solubility, digestibility and ease of preparation. The most popular food based on pre-gelatinized starch is instant pudding.

In addition to major industries which used starch, there are a number of other applications of starch, encountered in many fields. They are soap and detergent industry, laundry starches, cosmetic uses, pharmaceutical uses, plastic industry, optical whiteners and in production of poly hydroxy compounds which are industrially important. Thus we find a variety of uses for starch in different fields ranging from food to medicine, effluent treatment to soil conditioning, biodegradable plastics to explosives and a very large spectrum of other uses. In addition, the higher paste clarity and bland favour are desirable properties of cassava starches. Cassava starch is the only starch used for sago manufacture.

Table 8. Effect of surfactants on the Blue Values of different tuber starches

Starch	Total amylose [Blue Values]	Soluble amylose [Blue Values]
Cassava	0.32	0.18
Cassava + CTAB	0.27	0.13
Taro	0.28	0.18
Taro+CTAB	0.20	0.07
Lesser yam	0.29	0.14
Lesser yam +CTAB	0.22	0.04
Greater yam	0.43	0.18
Greater yam +CTAB	0.38	0.11
White yam	0.38	0.18
White yam +CTAB	0.35	0,12
Sweet potato	0.38	0.13
Sweet potato+CTAB	0.34	0.09
Tannia	0.38	0.21
Tannia+CTAB	0.33	0.15

Table 9. Effect of Varietal differences on the physicochemical properties of cassava starch

Varieties	Pasting temp. [°C]	Visc. 2% paste	Sw. Vol. [ml/g]	Solubility [%]	DSC Data		
					T_{onset} [°C]	T_{end} [°C]	ΔH $J g^{-1}$
M-4	60.70	58.0	30.5	22.8	68.20	75.5	11.8
Kalikalan	63.70	58.0	38.8	24.8	-	-	-
H 1687 (Sree Visakham)	55.68	58.0	25.5	23.6	67.2	73.1	12.6
H-2304 (Sree Sahya)	52.68	55.0	30.5	24.8	-	-	-
H-226	55.66	56.0	33.8	27.6	-	-	-
H-97	58.70	55.0	30.5	17.2	69.4	77.1	13.7
H-165	52.65	54.0	37.8	27.2	63.4	74.7	13.1
S- 856 (Sree Prakash)	-	-	-	-	65.6	74.9	10.6

Conclusions

Detailed studies on different tuber crops reveal the vast variability available among them, which is not generally observed in of cereal starches. The high viscosity of cassava and canna starches makes these starches very useful in many food and industrial applications, especially where high thickening power is desired. The low viscosity of aroid starches can be exploited in paper industries which require lower viscosity and good film forming capacity. The small granular size of Taro and lesser yam and *D. dumetorum* starches make these starches ideal as filler in biodegradable plastics, in aerosols and talcum powders. The clarity and gel strength of cassava, canna and yam starches can be useful in many food applications. The easy gelatinisation of cassava and sweet potato starches makes them suitable in manufacture of starch derived hydrolytic products. The range of characters observed makes tuber starches suitable for different applications in place of chemically modified starches. An awareness of their potential uses can help in large-scale cultivation of these crops and extraction of starch from them. It is also possible to modify starch properties by simple physical methods like hydrothermal or steam pressure treatments. Latest developments in biotechnology can also be used to modify starches. These include fermentation of starch by use of selective organisms or enzymatic modifications, which can bring about specific substitutions. A lot of work has been done on fermentation of cassava and its effect on starch quality whereas use of enzymes in starch derivatization has not been exploited and it offers very good scope for value addition.

References

Ahmed, S.V., Ghildyal, N.P., Kunhi, A.A.N. and Lonsane, B.K. 1983. Confectioner's syrup from tapioca processing waste. *Starch/Stärke,* 35 : 430-432

Asaoka, M., Blanshard, J.M.V.and Rickard, J.E. 1992. Effect of cultivar and growth season on the gelatinisation properties of cassava (*Manihot esculenta*) starch. *J. Sci. Food Agric.* 59 : 53-58

Asaoka, M., Blanshard, J.M.V. and Rickard J.E. 1991. Seasonal effects on the physicochemical properties of starches from four cultivars of cassava. *Starch/ Stärke* 43 : 455-459.

Balagopalan, C., Padmaja, G.,.Nanda, S.K. and Moorthy, S.N. 1988. Cassava in Food, Feed and Industry, CRC Press Inc. Boca Raton, Florida. pp 212.

Balagopalan, C., Ray, R.C. and Sheriff, J. T. 1996. Enzymatic separation of food grade industrial starch from sweet potato (*Ipomoea batatus* (L.) Lam) *(In:)* Tropical Tuber Crops, Eds G.T.Kurup, M.S.Palaniswami, V.P.Potty, G.Padmaja, S.Kabeerathumma and S.V. Pillai, Oxford-IBH, New Delhi. pp 471-474.

Banks, W., Geddes, R., Greenwood, C.T. and Jones, G. 1972. Physicochemical studies on starches. *Die Stärke,* 24 : 245-251.

Bertoniere, N.R., McLemore, T.A., Hasling, V.C., Catalano, E.A. and Deobald, H.J. 1966. The effect of environment variables on the processing of sweet potatoes into flakes and on some properties of their isolated starches. *J. Food Sci.,* 31 : 574-582.

Biliaderis, C.G. 1983. DSC in food research- a review. *Food Chem.,* 10 : 239-265.

Biliaderis, C.G. 1990. Thermal analysis of food carbohydrates. *(In:)* Thermal analysis of foods, Eds V.R.Harwalker and C.Y. Ma, Elsevier Science Publishers Ltd, Essex. 168-220.

Bouwkamp, J.C. 1985. Sweet potato products: A natural Resource for the tropics, CRC Press, Florida.

Cerning-Beroard and Dividich, J. Le. 1976. Feeding value of some starchy tropical products- *in vitro* and *in vivo* study of sweet potao, yam, malanga, breadfruit and banana, *Ann. Zootech,* 25 : 155-168.

Chadha, K.L. and Nair, G.G. 1994. Advances in Horticulture, Vol.8, Malhotra Publications, New Delhi.

Chiang, W.C. and Chen, K.L. 1988. Comparison of physicochemical properties of starch and amylolytic enzyme activity of various sweet potato varieties. *Shih Pin Ko Hsueh* (Tapei), 15 : 1-11.

Ciacco, C.F., and D'Appolonia, B.L . 1977. Characterisation of starches from various tubers and their uses in breadmaking. *Cereal Chem.,* 54 : 1096-1107.

Collado, L.S., Mabesa, R.C., and Corke, H. 1999. Genetic variation in the physical properties of sweet potato starch. *J. Agric. Food Chem.,* 47 : 4195-4201.

Defloor, I., Dehing, I. and Delcour, J.A. 1998. Physicochemical properties of cassava starch. *Starch/Stärke,* 50 : 58-64.

Delpeuch , F., and Favier, J.C. 1980. Characteristics of starches from tropical food plants; α-amylase hydrolysis, swelling and solubility patterns. *Ann. Technol. Agric.,* 29 : 53-67.

Eliasson, A.C . 1994. Interactions between starch and lipids studied by DSC. *Thermochimica Acta,* 246 : 343-356.

Eliasson, A.C. and Gudmundsson, M. 1996. Starch: Physico-chemical and functional aspects (In:) Carbohydrates in Food. (Ed.) A.C. Eliasson. Marcel Decker *Inc.* New York. pp 431-503.

Emilia Abraham, T., Krishnaswamy, C. and Ramakrishna, S.V. 1988. Hydrolytic depolymerisation of starch raw materials. *Starch/Stärke,* 40 : 387-392.

Emilia Abraham, T., Krishnaswamy, C. and Ramakrishna, S.V. 1987. Effect of hydrolytic conditions of cassava on the oligosaccharide profile and alcohol fermentation. *Starch/Stärke,* 39 237-240.

Emilia Abraham, T., Raja, K.C.M., Sreemulanathan, H., and Mathew, A.G. 1979. Improvement of texture of cassava flour by chemical treatment. *J.Root Crop.* 5 : 11-18.

Evans, I.D. 1986. An investigation of starch/surfactant interactions using viscosimetry and differential scanning calorimetry. *Starch/Stärke,* 38 : 227-235.

Farhat, I.A., Oguntona, T. and Neale, R.J. 1999. Characterization of starches from West African Yams. *J. Sci. Food Agric.,* 79 : 2105-2112.

Franco, C.M.L.,.Ciacco, C.F., and Tavares, D.Q. 1988. Studies on the susceptibility of granular cassava and corn starches to enzymatic attack, Part 2. Studies on the granular structure of starch. *Starch / Stärke,* 40 : 29-32.

Gallant, D.J., Bewa, H., Buy, Q.H.., Bouchet,. B., Szylit, O. and Sealy, L. 1982. On ultra structural and nutritional aspects of some tropical tuber crops. *Starch/Stärke,* 34 : 255-262.

Garcia, A.M. and Walter, W.M. 1998. Physico-chemical characterization of starch from Peruvian sweet potato selections. *Starch/Stärke,* 50 : 331-337.

Gebre-Mariam, T. and Schmidt, P.C.1998. Some physicochemical properties of *Dioscorea* starch from Ethiopia. *Starch/Stärke,* 50 : 241-246.

Ghosh, S.P., Jos, J.S., Ramanujam, T., Moorthy, S.N., and Nair, R.G. 1988. Root Crops, Oxford-IBH, New Delhi. 602 pp.

Griffin, G.T.L. 1979. Non-food applications of starch, especially potential uses of taro, *(In:)* Small scale processing and utilisation of tropical root crops, Westview Press, Colorado.

Griffin, G.T.L. 1994. In chemistry and Technology of biodegradable polymers, Blackie Academic and Professional, London. pp. 134-149.

Gudmundsson, M. and Eliasson, A.C. 1992. Comparison of thermal and viscoelastic properties of four waxy starches and the effect of added surfactants. *Starch/Stärke,* 44 : 379-385.

Hammet, H.L. and Barrentin, B. 1961. Some effects of variety, curing and baking upon the carbohydrate content of sweet potatoes. *Proc. Amer. Hort. Sci.,* 78 : 421-426.

Hizukuri, S. 1969. Effect of environment temperauture of plants on the physicochemical properties of their starches. *J. Jap. Soc. Starch Sci.,* 17 : 73-88.

Hoover, R. 1998. Starch-lipid Interactions (In:) Polysaccharide association in foods, (Ed) R,H Walter,. Marcel and Decker Inc. New York. pp 227-256.

Hoover, R. and Hadziyev. 1981. The effect of monoglycerides on amylose complexing during a potato granule process. *Starch/Stärke,* 33: 346-355.

ISI 1970. Specifications for tapioca starch for use in cotton textile industries. IS 1605-1960, Indian Standards Institution, New Delhi.

Jaleel, S.A., Sreekanta, S., Ghildyal, N.P. and Lonsane, B.K. 1988. Simultaneous solid state fermentation and saccharification of cassava fibrous residue or production of ethanol. *Starch/Stärke,* 40 : 55-58.

John, A. and Shastri, P.N. 1998. Studies on macromolecules by differential scanning calorimetry, A critical appraisal. *J. Food Sci. Technol.,* 35 : 1-14.

Jyothi, A.N., Moorthy, S.N. and Vimala, B. 2003. Physicochemical and functional properties of starch extracted from two species of *Curcuma. Int. J. Food Properties 6, 135-145.*

Kallabinski, J. and Balagopalan, C. 1991. Enzymatic starch extraction from tropical root and tuber crops, *(In:) Proc. Ninth Symp. Internat. Soc. Trop. Root Crops,* (Ed. *F. Ofori* and *S.K.Hahn).* pp 83-88.

Kay, D.E. 1987. TDRI Crop and Product Digest No.2, TDRI, London. pp 166-173.

Kempf. W. 1985. New possibleoutets for starch and starch products in chemical and technical industries. *Food Technology in Australia,* 37 : 241-245

Krog, N. 1973. Influence of food emulsifiers on the pasting temperature and viscosity of various starches. *Starch/Stärke,* 25 :22-27.

Kugimiya, M. and Donovan, J. W. 1981. Calorimetric determination of the amylose content of starches based on formation and melting of the amylose-lysolecithin complex. *J. Food Sci.* , 46 : 765-770,777.

Kunhi, A.A.M., Ghildyal, N.P., Lonsane, B.K.,m Ahmed, S.Y. and Natarajan, C.P. 1981. Studies on the production of alcohol from saccharified waste residue rom cassava processing industries. *Starch/Stärke,* 33 : 275-279

Leelavathi, K., Indrani, D. and Sidhu, J.S. 1987. Amylograph pasting behaviour of cereal and tuber starches. *Starch/Stärke,* 39 : 378-381.

Leszkowiat, M. J., Yada, R,Y., Coffin, R.H. and Stanley , D.W. 1990. Starch gelatinisation in cold temperature sweetening resistant potatoes. *J. Food Sci.,* 55 :1338-1340, 1420.

Lorenz, K. and Kulp, K. 1982. Cereal and Root starch modification by heat-moisture treatment. *Starch/Stärke,* 34 : 50-54.

Madamba, L.S.P., Bustrillos, A.R. and San Pedro, E.L. 1975. Sweet potato starch; Physicochemical properties of whole starch. *Philipp. Agric.,* 58 : 338-350.

Mathew George, Moorthy, S.N. and Padmaja, G. 1995. Biochemical changes in cassava tuber during fermentation and its effect on extracted starch and residue. *J. Sci. Food Agric.,* 69 : 367-371.

Mathew George. Padmaja, G. and Moorthy, S.N. 1991. Enhancement in starch extractability from cassava tubers through fermentation with a mixed culture inoculum. *J. Root Crops,* 17 : 1-9.

Melo, E.de A., Krieger, N. and Stamford, T.L.M. 1994. Physicochemical properties of Jacatupe (Pacchyrrhizus erosus L. Urban) starch. *Starch/Stärke,* 46 : 245-247.

Moorthy, S.N. 1982a. Effect of steam pressure treatment on nature and properties of cassava. AFST (I), Trivandrum, pp. 68-71.

Moorthy, S.N. 1982b. Behaviour of cassava starch in various solvents. *Starch/ Stärke,* 37 : 372-374.

Moorthy, S.N. 1982 c. Production of high fructose syrup from cassava starch. *(In:)* Proc. Internat.symp. on food AHARA-82.

Moorthy, S.N. 1983. Effect of some physical and chemical treatments on cassava flour quality. *J. Food Sci. Tech.,* 20 : 302-305.

Moorthy, S.N. 1985a. Effect of different types of surfactants on cassava starch properties. *J. Agric. Food Chem.,* 35 : 1227-1232.

Moorthy, S.N. 1985 b. Acetylation of cassava starch using perchloric acid catalysis. *Starch/ Stärke,* 37 : 307-308.

Moorthy, S.N.1986. Properties of Coleus starch. *Starch/Stärke,* 38 : 77-78.

Moorthy, S.N.1991. Extraction of starches from tuber crops using ammonia. *Carbohydrate Polymers,* 16 : 391-398.

Moorthy, S.N. 1994. Tuber Crop Starches, Tech. Bull. No 18, CTCRI, Trivandrum. 32 pp.

Moorthy, S.N.1999. Effect of steam pressure treatment on the physicochemical properties of Dioscorea starches. *J. Agric. Food Chem.,* 47 : 1695-1699.

Moorthy, S.N. 2001. Tuber crop starches. Tech. Bulletin, CTCRI, Trivandrum

Moorthy, S.N. and Ramanujam, T. 1986. Variation in properties of starch in cassava varieties in relation to age of the crop. *Starch / Stärke,* 38 : 58-61.

Moorthy, S.N. and Nair, S.G. 1989. Studies on Dioscorea rotundata starch properties. *Starch / Stärke,* 41 : 81-83.

Moorthy, S.N. and Padmaja, G. 1991. Comparative study on digestibility of raw and cooked starch of different tuber crops. *J. Root Crops,* 17 (Sp) : 255-258.

Moorthy, S.N., Mathew George and Padmaja, G. 1993a. Functional properties of the starchy flour extracted from cassava on fermentation with a mixed culture inoculum. *J. Sci. Food Agric.* 61 : 443-447.

Moorthy, S.N., Thankamma Pillai, P.K and Unnikrishnan, M. 1993. Variability in starch extracted from taro. *Carbohyd. Polymers,* 20 :169-173.

Moorthy, S.N., Unnikrishnan, M. and Lakshmi, K.R. 1994. Physico-chemical properties of some accessions of Amorphophallus paeoniifolius. *Trop. Sci.,* 34 : 371-376

Moorthy, S.N., Wenham, J.E. and Blanshard, J.M.V.1996. Effect of solvent extraction on the gelatinization properties of starch and flour of five cassava varieties. *J. Sci. Food Agric.,* 72 : 329-336.

Moorthy, S.N. and Balagopalan, C. 1999. Physicochemical properties of enzymatically separated starch from sweet potato.*Trop. Sci.*, 39 : 23-27

Moorthy, S.N., Vimala, B. and Archana Mukherjee. 2002. Physicochemical and functional properties of Canna edulis starch. *Trop . Sci.*, 42 : 75-77.

Moorthy, S.N., Andersson, L.,Eliasson, A.C., Santacruz, S. and Ruales, J. 2006. Determination of amylose content in different starches using modulated differential scanning calorimery. *Starch/Stärke*, 58 : 209-214.

Nkala, D., Sibanda, S., Tomasik, P. and Palasinski, M. 1994. Isolation and properties of starch from wild yam from Zimbabwe. *Starch/Stärke,* 46 : 85-88.

Noda, T., Takahata, Y., and Nagata, T. 1992. Developmental changes in properties of sweet potato starch. *Starch / Stärke*, 44 : 405-409.

Noda, T., Takahata, Y., Sato, T., Ikoma, H. and Mochida, H. 1996. Physicochemical properties of starch from purple and orange fleshed sweet potato roots at two levels of fertiliser. *Starch / Stärke*, 48 :395-399.

Olorunda, A.O., Aworh, O.C. and Numfor, F.A.1981. Technical note: predicting quality of cassava products with the aid of chemical and rheological properties. *J. Food Technol.*, 16 : 447-450.

Patel, A.R., Patel, N.R., Suthar, J.N., Patel, K.G. and Patel, R.D. 1986 a. Morphology of some legume and tuber starches grafted with polyacrylonitrile. *Starch/Stärke,* 38 :160-167

Patel, A.R., Patel, N.R., Patel, M.R., Patel, K.G. and Patel, R.D. 1986 b. Graft copolymers of starch and polyacrylonitrile. *Starch/Stärke,* 38 : 235-237.

Perez, E.E., Breene, W.M. and Bahnassey, Y.A. 1998. Variation in the gelatinization profiles of cassava, sagu and arrowroot native starches as measured with different thermal and mechanical methods. *Starch/Stärke*, 50 : 70-72.

Radley, J.A. 1976 a. Examination and analysis of starch and its derivatives. Applied Science Publishers Ltd, London.

Radley, J.A. 1976 b. Starch production technology. Applied Science Publishers Ltd, London pp.189-229.

Raja, K.C.M. and Ramakrishna, S.V. 1988. Influence of salts on the pasting characteristics of cassava. *Food Chem.*, 28 : 311-318.

Raja, K.C.M., Sreedharan, V.P., Emilia Abraham, T., Sreemulanathan, H. and Mathew, A.G. 1978. Effect of fractioal sieving on some quality aspects of cassava flour. *J.Root crops,* 4 :17-21.

Rickard, J.E., Asaoka, M. and Blanshard, J.M.V. 1991. The physicochemical properties of cassava starch. *Trop. Sci.,* 31 :189-207.

Rohwer.R.G and Klem.R.E. 1984. Acid modified starch, production and uses, (*In:*) Starch Chemistry and Technology, Eds, Whistler, R.L., Bemiller, J.N.and Paschell, E.F. IInd edition, Academic Press. INC, NewYork. pp 537-539.

Rosenthal, F.R.T., Nakamura, T., Espindola, A.M.C. and Jochimek, M.R. 1974. Structure of starch granules. *Starch/Stärke,* 26 : 50-55.

Sair, L. 1967. Heat-moisture treatment of starch. *Cereal Chem.,* 44 : 8-26.

Shin, M. and Ahn, S.Y. 1983. Physicochemical properties of several sweet potato starches. *J. Korean Agric. Chem. Soc.,* 26 : 137-142.

Sievert, D. and Holm, J. 1993. Determination of amylose by Differential Scanning Calorimetry. *Starch/Stärke,* 45 : 333-338.

Soni, P.L., Sharma, H., Srivatsava, H.C. and Gharia, M.M. 1990. Physicochemical properties of canna edulis starch- comparison with maize starch. *Starch/Stärke,* 42 : 460-464.

Sriburi, P., and Hill, S.E. 2000. Extrusion of cassava with either variation in ascorbic acid concentration or pH. *Int. J.Food Sci. Technol.,* 35 :141-154.

Srikanta, S., Jaleel, S.A., Ghildyal, N.P., Lonsane, B.K. and Karanth, N.G. 1987. Novel technique for the saccharification of cassava fibrous waste for alcohol production. *Starch/Stärke,* 39 : 234-237.

Srivatsava, H.C., Harshe, S.N., Gharia, M.M. and Mudia, G.P. 1970. Physicochemical studies on some starches. *Starch/Stärke,* 22 : 162-166.

Srivatsava, H.C. and Patel, M. M. 1973. Viscosity stabilisation of tapioca starch. *Starch/ Stärke,* 25 :17-20.

Stevens, D.J. and Elton, G.A.H. 1971. Thermal properties of Starch/Water systems. *Starch/* ***Starch/Stärke,*** 23 : 8-11.

Strauss, M. S. and Griffin, G.J.L. 1994. Variability in taro, (*Colocasia esculenta*) starches-size, gelatinisation and amylose content *(In:)* Proc. Symp. International Soc. Trop. Root Crops, 165-170.

Susuki, A., Takeda, Y. and Hizukuri, S. 1985. Relationship between the molecular structures and retrogradation of tapioca, potato and kuzu starches. *J. Jap. Soc. Starch Sci.,* 32 : 205-212.

Swinkels, J.J.M. 1985. Composition and properties of commercial native starches. *Starch/ Stärke,* 37 :1-5.

Szylit, O., Durand, M., Borgida, L.P., Atinkpahoun, H., Prieto, F.and Devort-Lavel, J. 1978. Raw and steam-pelleted cassava, sweet potato and yam *cayensis* as starch sources for ruminant and chicken diets. *Anim. Feed Sci. Technol.,* 3 : 73-87.

Takeda, Y., Hizukuri, S., Takeda, C. and Suzuki, A. 1987. Structure of branched molecules of amylose of various origins and molar fractions of branched and unbranched molecules, *Carbohyd. Res.,* 165 :139-145.

Takeda, Y., Shirasaka, K. and Hizukuri, S. 1984. Examination of the purity and structure of amylose by gel permeation chromatography. *Carbohyd. Res.,* 132 : 83-92.

Takeda, Y., Tokunaga, N., Takeda, C. and Hizukuri, S. 1986. Physicochemical properties of sweet potato starches. *Starch/Stärke,* 38 : 345-350.

Tester, R.F. and Morrison, W.R. 1990. Swelling and gelatinisation of cereal starches, II Waxy rice starches. *Cereal Chem.*, 67 : 558-563.

Tian, S.J., Rickard, J.E. and Blanshard, J.M.V. 1991. Physicochemical properties of sweet potato starch. *J. Sci. Food Agric.*, 57 : 459-491.

Vijayagopal, K. and Balagopal, C. 1989. Fermentation of cassava starch hydrolysate with immobilised cells of Saccharomyces cerevisae. *Starch/Stärke,* 41 : 271-275.

Wang, J.K. 1983. Taro. University of Hawai Press, Honolulu.

Wankhede, D.B. and Sajjan, S.U. 1981. Isolation and physicochemical properties of starch extracted from yam, elephant *(Amorphophallus compalunatus). Starch/ Stärke,* 33 :153-157.

Woolfe, J.A. 1992. Sweet Potato: an untapped food resource, Cambridge University Press, Cambridge. 643 pp.

Wootton, M., and Bamunuarachchi, A. 1979. Application of differential scanning calorimetry to starch gelatinisation I. Commercial native and modified starches. *Starch/Stärke,* 31 : 201-204.

Zoebel, H. F. 1988. Molecules to granules - A comprehensive starch review. *Starch/Stärke,* 40 : 44-50.

Figures Contd....

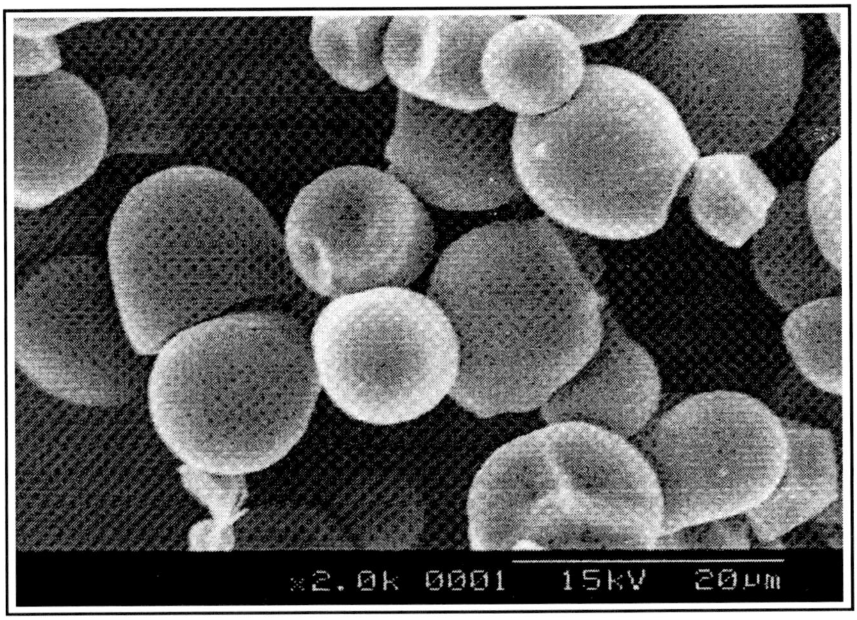

Plate 1. SEM photograph of native cassava starch

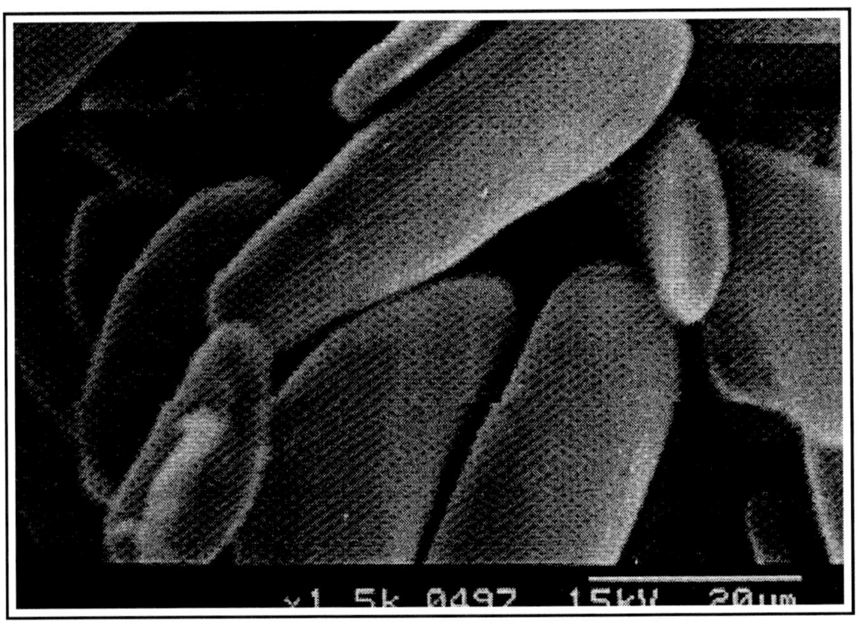

Plate 2. SEM photograph of *Curcuma* starch

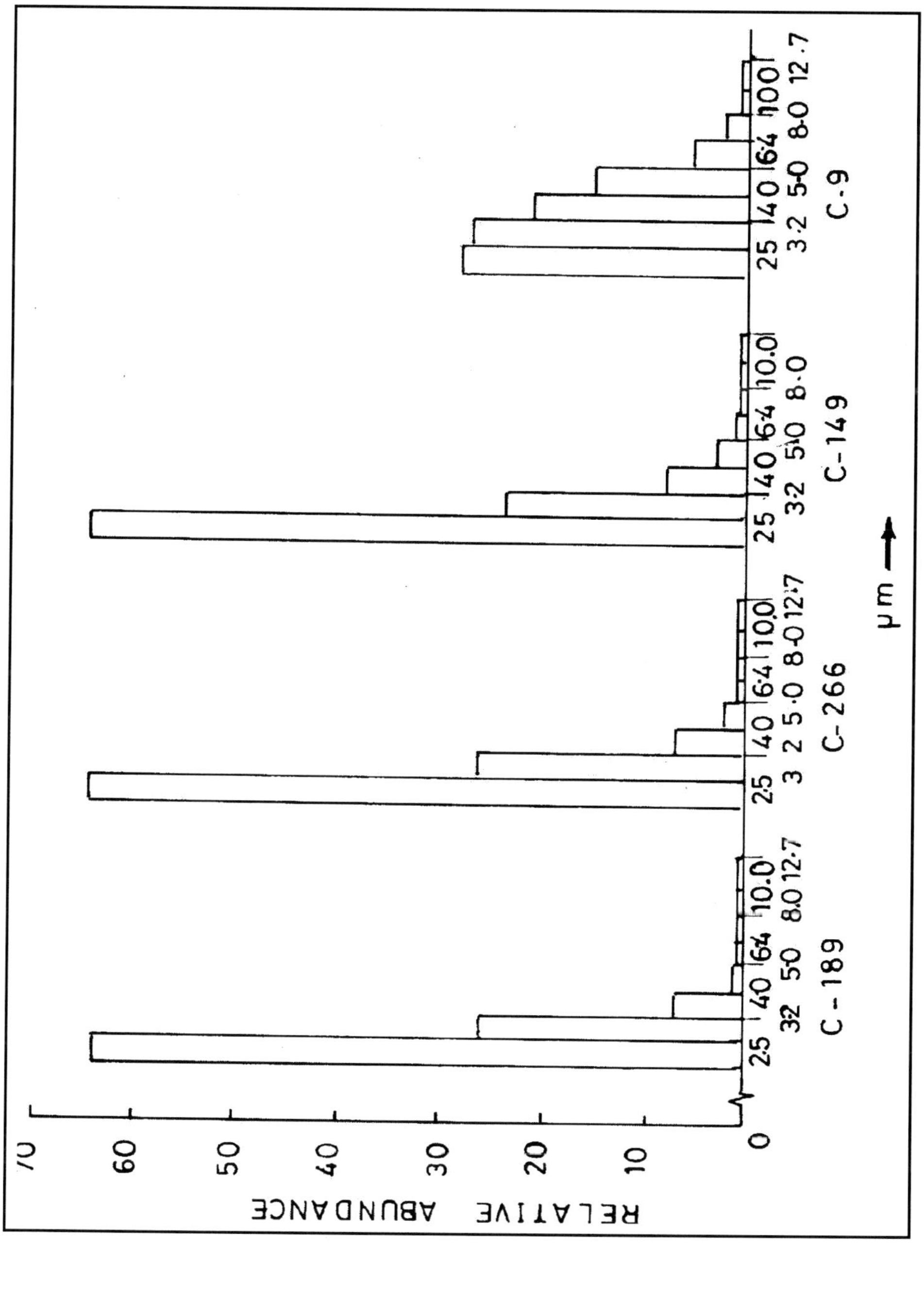

Fig. 1. Granule size Distribution in Colocasia Varities

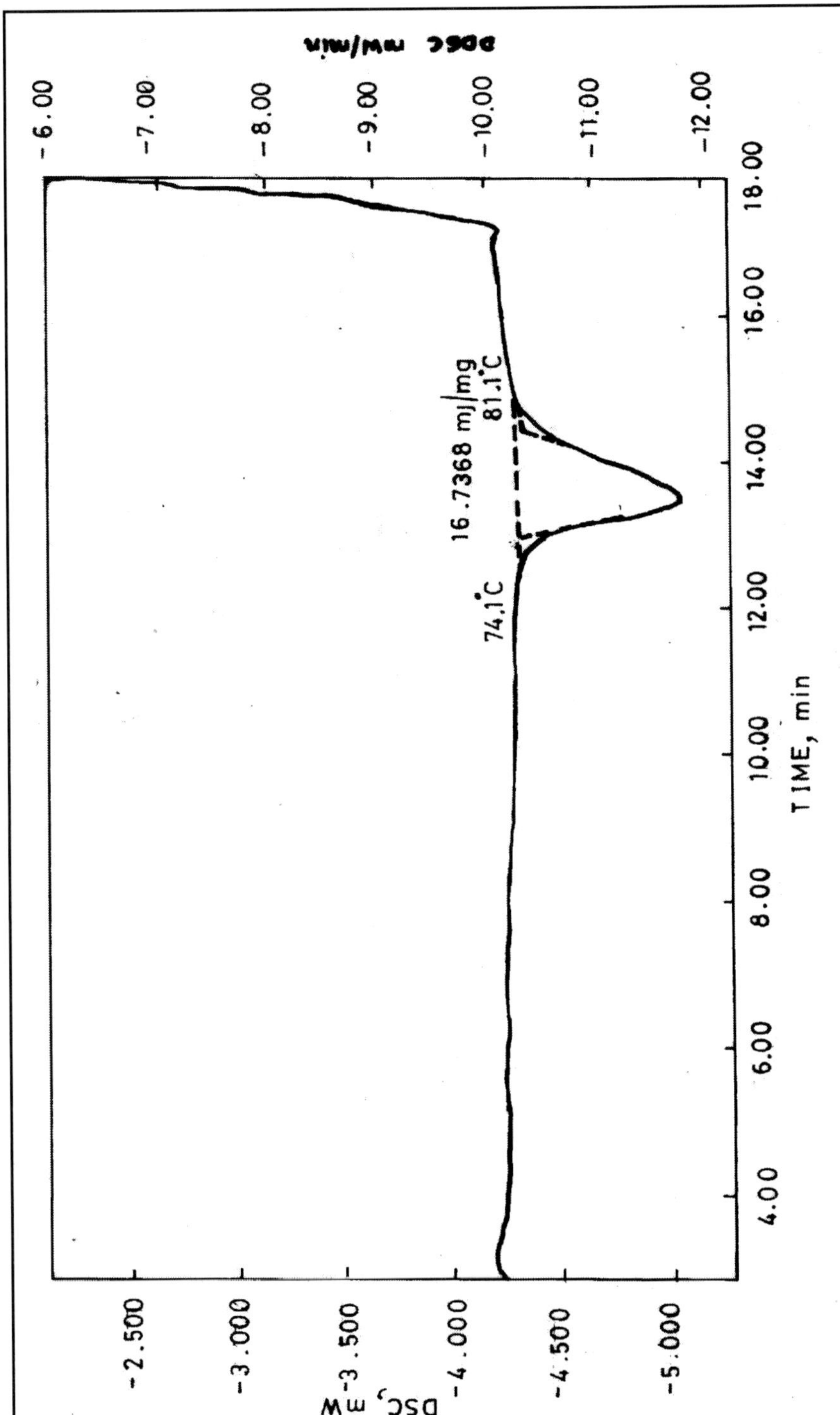

Fig. 2. A Typical DSC Curve of starch (cassava)

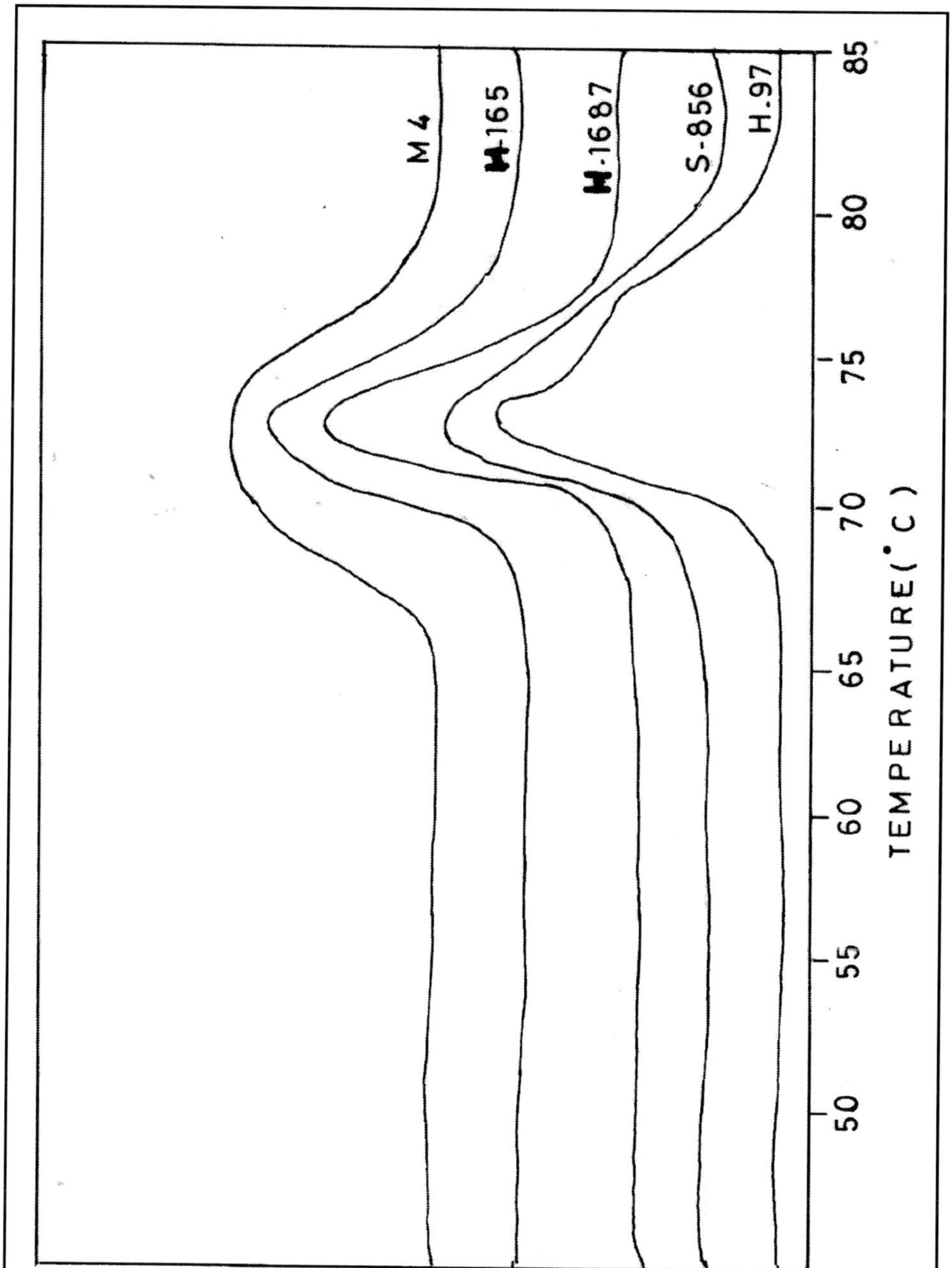

Fig. 3. DSC Patterns of starch from some cassava varieties

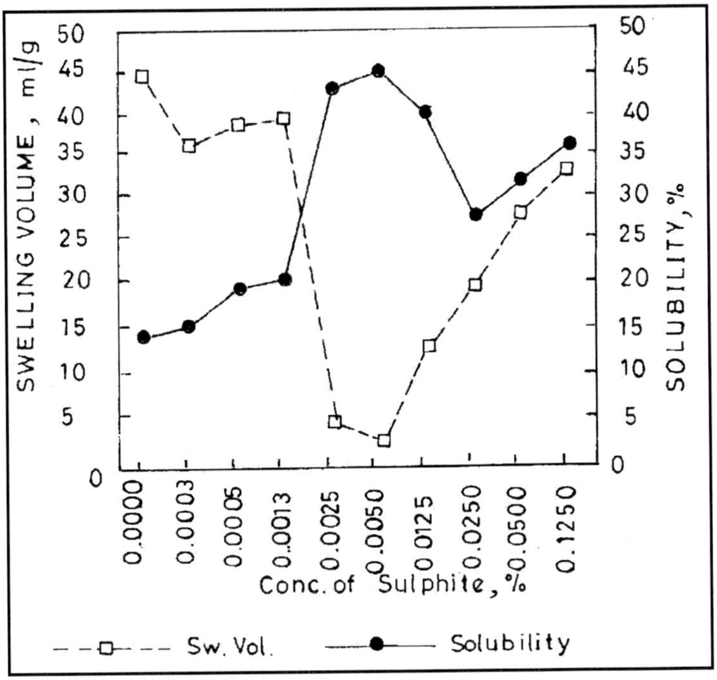

Fig. 4. Effect of sodium sulphate on swelling volume and solubility of cassava starch

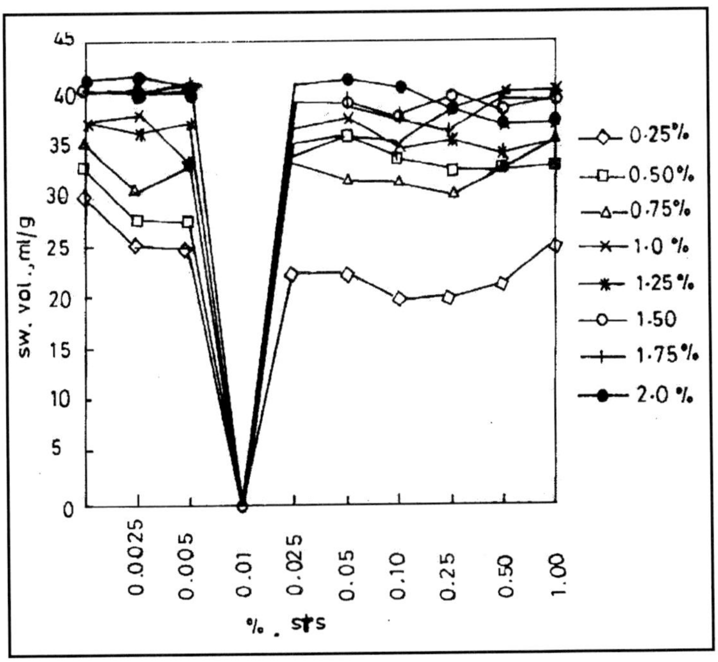

Fig. 5. Sw. vol. of starch with STS (sodium thiosulphate)

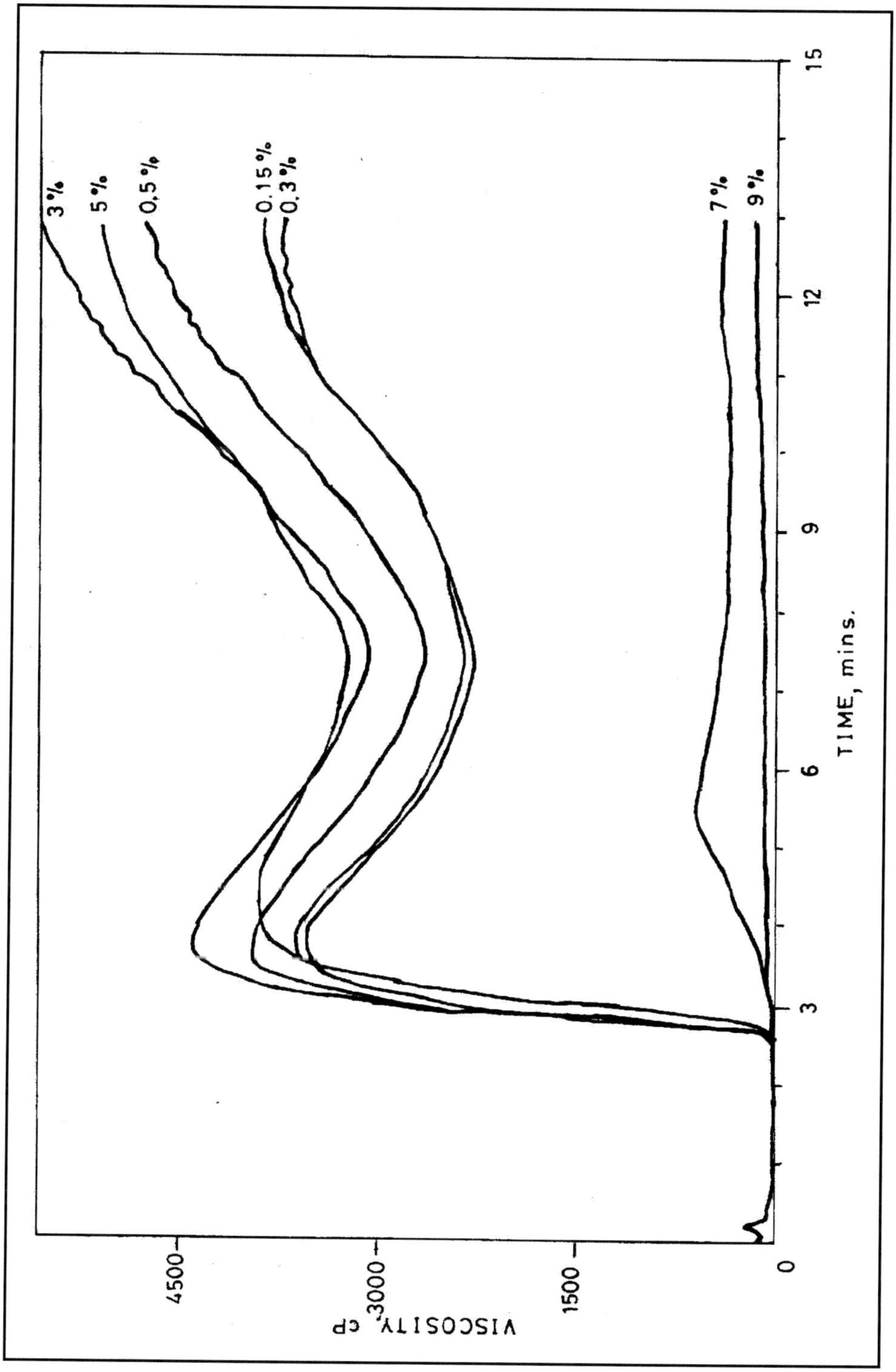

Fig. 6. Viscosity patterns of cassava starch treated with different levels of formaldehyde

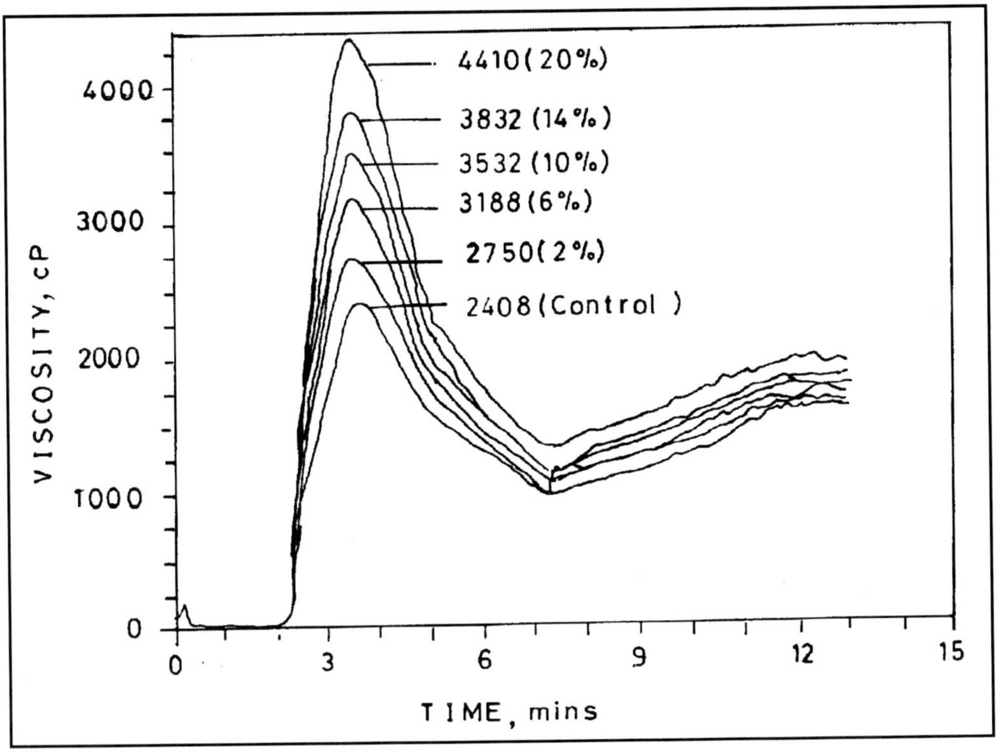

Fig. 7. Effect of different concentrations of CMS on starch viscosity

□□□

12

Post Harvest Technologies and Utilization

Tuber crops are rated as one of the richest sources of energy, besides being a treasure house of vitamins and minerals. The highest rate of dry matter production per day per unit area makes tuber crops important in the food economy, especially in meeting rural feed security. The tuber crop produce can also be processed into various food products to suit the taste and need of the urban, semi-urban and rural folk as well. Better post harvest management and diversification for production of value added products is one of the dependable methods to make tuber crops lucrative to both farmers and entrepreneurs.

1. Storage and Packaging

The root and tuber crops, including cassava, sweet potato, yams and aroids, enjoy considerable importance as a vegetable, staple food, or raw material for small-scale industries at a global level, particularly in the less developed tropical countries. The perishability and post harvest losses of root and tuber crops are the major constraints in the utilization of these crops. Simple, low-cost traditional methods in different parts of the world and modern techniques to store different tuber crops in fresh state were extensively reviewed by Ravi *et al* (1996).

1.1 Storage of Cassava tubers

A major constraint in post harvest marketing and utilization of cassava is the rapid perishability of tubers. Normally cassava tubers cannot be stored without spoilage for more than 3-5 days after harvest. High cost of preservation techniques seems to have limited practical value since cost of cassava tuber itself is low. Attempts were made to develop a low cost preservation technology to enhance shelf life of cassava tubers. Fresh undamaged tubers along with a portion of main stem could be stored in moist sand and soil in pits under shade. Moisture in sand/soil is maintained in the range of 10-15%. Observations made at weekly intervals up to 2 months showed the recovery of undamaged stored tubers to be in the range of 80-85% (Balagopalan *et al*, 1998).

1.2 Packaging and storage of dehydrated Cassava products

Cassava chips are infested by a large number of insect pests. Parboiled chips could be stored for 9 months with 3 % loss in weight. In plain dried chips the loss is quite high. Cassava flour and starch are prone to pick up moisture quickly due to their powdery nature, which render open storage conditions unfavourable for dried cassava products. Drying the starch, chips or flour to low moisture content at production level itself is an essential factor in minimising attack by fungi and insects. Subsequently hygienic storage and effective packaging facilities have to be provided to offer continued protection against biochemical deterioration, microbial growth and insect infestation.

Various modes of conventional packaging and storing of cassava products (starch, chips and flour) were evaluated to assess their shelf life to minimise the qualitative and quantitative losses during storage. HDPE bag and polythene impregnated bags were the most suitable packaging system for storing cassava starch against microbes up to 180 days (Nanda and Potty, 1985). Metal container and polythene lined jute bags were the most effective packaging materials for safe storage of cassava chips against insects up to 90 days (Nanda and Pillai, 1996). Storing chips in metal bins and fumigating with aluminium phosphide tablets were found effective for bulk storage of dehydrated chips. Cassava chips dipped in sodium hypochlorite (0.5%) for 2-3 minutes and washed with fresh water and dried at 70°C have more shelf life (90 days) as compared to untreated chips (25 days). The bacterial load is also reduced in treated chips. Only polythene-lined jute bag was effective for short-term storage of cassava flour up to 90 days (Nanda and Pillai, 1996). Both steamed and roasted sago pearls and grains are generally insect resistant and store well in containers for 12 months (McFarlene, 1982).

1.3 Post Harvest curing of Sweet potato

Sweet potato tubers are subjected to several types of post harvest losses; physical damage, weight loss, pathological decay, sprouting and weevil infestation. Sweet Potato tubers under storage are most severely damaged by sweet potato weevil (*Cylas formicarius*). The process of curing results in healing of wounds during post harvest handling. Curing is carried out at 30-33°C and 85-95% relative humidity for 5-7 days. This process provides a barrier to further moisture loss and stops microbial invasion of the tissue. The final stages of wound healing is formation of a wound periderm beneath the suberized cells. Cured roots can be stored as long as one year under optimal conditions.

1.4 Curing for storage of Yams

Use of yam tubers as planting materials and as valuable food necessitate proper post harvest storage of yams. Storage is normally preceeded by curing, which helps to seal the cuts and bruises inflicted in the tuber at the time of harvest. This protects the tuber from entry of pathogens during storage.

Curing of yam tubers before storage enhances the storage life. It involves exposure of freshly harvested tubers to 29-40°C temperature and 90-95% relative humidity for 5-7 days. Wound healing occurs during curing due to deposition of suberin at wound sites followed by development of a corky layer of dead cells which inhibit excessive water loss and infection by wound pathogens.

The best traditional storage method for yam is the shaded well-ventilated yam barn.

1.5 Storage of Aroids

The three major aroids used as human food or animal feed are taro or Colocasia, tannia or Xanthosoma and elephant foot yam or Amorphophallus. Though crop duration for the aroids is usually 10-12 months, harvest can be delayed for a few weeks after maturity indicated by the senescence of leaves. The tubers can be stored in cool, well ventilated dry rooms. The tubers are also stored in open platforms in well-aerated surroundings.

Storing *Amorphophallus* tubers in fine sand medium is effective against pests and loss of weight in comparison to tubers kept exposed. Amorphophallus tubers could be successfully stored in heaps of white sand. Dipping tubers in cow dung slurry followed by surface drying is practised in rural areas for storing amorphophallus tubers.

2. Starch from Tuber crops

Tropical tuber crops contain starch as the major component and thus act as important source of starch. Except cassava and to a smaller extent sweet potato, starch from other tuber crops has not been exploited for industrial applications partly because of difficulty in extraction of pure starches and partly because of non-availability of information about properties of these lesser known starches. An excellent review by Moorthy (2002) collated available data on physicochemical and functional characteristics of tropical tuber starches, highlighting their unique properties and potential fields of application. The starch granules of *Colocasia esculenta* and *Dioscorea esculenta* tubers are very small whereas those of *Canna edulis* are very large. Viscosity is high for cassava and *C. edulis* starches, but low for most aroid starches. Clarity is good for cassava and yam starches compared to others. Digestibility also varies among starches. The diversity available in tuber starches shows that tuber starches can be used in place of chemically modified starches available in the market. Realization of their importance can help in value addition of these neglected crops besides providing starch with special properties for specific applications.

Brabet *et al* (1998) analysed starch content and properties of 106 sweet potato clones from world germplasm collection held at International Potato Centre (CIP). Research investigations on minor tuber crops are scanty. Perez *et al* (2005) and Sefa-Dedeh and Sackey (2002) characterized the structure and properties of starch from taro/ tannia / cocoyam.

2.1 Utilization of starch

Availability of starch in large quantity and at low cost resulted in various industrial applications of starch and its derivatives, besides its principal utilization as food. In food industries, starch is used to impart functional properties to processed foods such as thickening, binding, filling, and taste. Starch is used as a component in puddings, pies, jellies, biscuits, bakery products, ice-cream, canned soups, instant desserts, processed meats, sauces, and in various infant and invalid food mixtures. Starch is also used for manufacturing a number of sweeteners and syrups and to feed enzymes for manufacture of MSG, a popular taste enhancer.

Almost all major industries have found some use for starch. Major users of starch include the textile, paper, plywood, and adhesive industries. Starch is mainly used by textiles industries in sizing, finishing and printing operations. Starch and its derivatives are used in paper and paper-board manufacture in wet end application, size press application, calendering, regular and colour paper coating and also in carbonless paper.

Starch is very versatile and has a wide variety of uses in the manufacture of myriad nonfood products such as in foundry core binding, drilling mud, in the laundry industry as a stiffener, and as filler and/or binder in pharmaceutical and cosmetic applications.

2.2 Starch from Cassava

Cassava is a commercial source of food and industrial grade starch in India and several other countries in south-east Asia, Africa and Latin America. Most of the starch production from cassava is in small scale sector. In India, it is a thriving agro-industry with about 900 starch and sago factories in north-central Tamil Nadu and about 35 in East Godavari district of Andhra Pradesh. For extracting food grade starch, the tubers are first washed and peeled to remove both the outer skin and the rind as well. The peeled roots are again washed and then rasped into a fine pulp releasing most of the starch granules. The pulped tuber is pumped on to a series of vibrating screens of diminishing mesh size (usually of 80, 150, 260 and 300 mesh) which retain the fibre. Residual fibrous pulp from the screening, called *thippi*, is considered as a by-product of cassava starch industry and finds use as animal feed. It is about 10% by weight of cassava roots and on the basis of dry matter, consists of about 56.0% starch. The starch milk is dewatered by gravitational sedimentation in settling tanks or by means of hydroclones and centrifuges. After removal of free water from the starch by sedimentation, a cake is obtained containing 35 to 40 % moisture. The starch cake is dried and pulverised before packing into bags.

2.3 Sago from Cassava

For manufacturing sago, the partially dried starch cake is globulated into spherical beads on gyratory shakers and the globules are graded by size, partially gelatinised by roasting on heated metal pans and then dried. The dried lumps are separated by passing through a spike beater and finally polished before packing in gunny bags. The yield of sago is around 25% of weight of fresh tubers. Sago is used mainly as infant and invalid food, and in preparation of puddings and sago wafer.

3. Value added Food products from Tuber crops

Poor shelf life and bulkiness of tubers pose a great problem in their storage and transportation. To overcome this difficulty in the marketing and utilization of cassava and to avoid heavy post-harvest losses, roots need to be processed into some form of dried product with longer storage life. The simplest and the most common mode of processing cassava is conversion of tubers into chips. The hydrogen cyanide is also reduced from root tissues during slicing and drying operations. Cassava chips are used for edible purposes and preparation of flour. Dried chips/flour are also used in animal feed formulations. In Industry, it serves as a raw material for manufacturing starch, dextrin, glucose and ethyl alcohol. Colocasia chips are popular in Hawaii (Onwueme, 1999),

Bangladesh (Hussain and Chowdhury, 1986) and in some parts of India (Nanda, 1984).

3.1 Food products from Cassava

Cassava Rava and Porridge

Rava or semolina is normally a wheat based convenience food used for preparation of various breakfast recipes like *uppuma* and *kesari halwa*. Attempts were made at CTCRI to develop a simple economic process for production of cassava based rava as a substitute for wheat rava. The method of preparation requires controlled gelatinisation of starch. The process for producing cassava rava (Balagopalan *et al.*, 1998) consists of following steps.

1. Partial gelatinization of peeled and sliced tubers by parboiling for 10 minutes.
2. Drying to about 15% moisture content.
3. Pulverization of dried chips.

The fraction which passes through 20 mesh but does not pass through 80 mesh has a size range 0.5 to 3 mm and is the most suitable for wheat semolina substitution. It can be used for preparation of products such as 'uppuma' and 'kesari'. The fraction passing through 80 mesh is too fine and possesses a cohesive texture and is useful in the preparation of sweets, puddings etc., products which require fast miscibility of starch in milk or in water. The fine grade pre-gelatinised cassava starch (porridge) can be utilised to make instant energy drink using hot milk or hot water. Two tea spoon full pregelatinised starch could be added to a cup of hot milk or water after adding sugar to taste and served to infants and invalids as an energy drink. Addition of cardamom powder to yuca porridge adds flavour to the product .

Pappads

Cassava pappad is an important snack food item prepared from flour. The preparation involves gelatinisation of flour with minimum quantity of hot water, mixing of salt and pepper, spreading out the paste on a mat, cloth or polythene sheet and drying in sun. The pappad is consumed by deep-frying in oil. The final product undergoes 2-3 times expansion on frying. It is crisp and can be consumed as a side dish.

Sago Wafers

Sago wafer is an important product made in cottage level scale in many parts of Tamil Nadu. The wafers are deep fried in oil and consumed as an adjunctant. Preparation involves packing the sago in round aluminium trays. The trays are then introduced into steam boilers and allowed to be exposed to steam for 20 minutes. The gelatinisation taking place makes the pearls stick to each other and give round shape. The trays are then sun dried and resulting wafers are peeled out. Natural colours and salt are added to taste.

Wafers

Wafers are made from cassava starch similar to sago wafer. In this case, the starch cake containing about 35-40% moisture is used instead of sago. Wafers can be made in different shapes and sizes like round, square, flowery patterns etc. The product on frying undergoes 3-4 fold increase in size. It is also used as an adjunctant.

Crisps

Fried chips or crisps are made by deep fat frying very thin chips of cassava. The tubers are washed thoroughly, the peel and rind removed and then sliced as thin as possible. Tubers of correct maturity from varieties with relatively lower dry matter should be used for quality chips. The slices may be dipped into sodium chloride, or sodium bisulphite solution for 5-10 minutes, and then taken out, washed with water, and surface dried on a filter paper or cloth for improving colour. Compared to potato chips, cassava chips are slightly hardened, but the major advantage lies in the fact that the chips do not become soft like potato chips within a few minutes of exposure and maintain the crispness.

3.2 Food products from Sweet potato

Sweet Potato Noodles

Sweet potato starch is commonly used in noodle production in China. Unlike wheat flour noodles, cassava or sweet potato starch noodles are gluten-free. Therefore, a small amount of partially gelatinized starch is added before mixing with the ungelatinized starch, in order to function as a binder and to facilitate extrusion or sheeting of the starch mixture to produce noodles. Around 28% of sweetpotato is processed to starch noodles (Collado and Corke 1997).

The traditional method of processing sweet potato noodles in China consists of taking a part of sweet potato starch and boiling water to it to make a gelatinous soup. The rest of the starch is mixed with it and shifted through holes in a manual or mechanical press. Resulting noodles are hanged, arranged and cooled down in a pool of water. The noodles are then treated in sulphur room and dried before packing (Jiang *et al.*, 2004).

Sweet Potato Jam

In the process developed at CTCRI to prepare sweet potato jam, fresh tubers are washed, peeled and sliced into 0.3 cm thick slices and cooked to extract the pulp (Balagopalan *et al.*, 2000). The pulp from fruits like banana or apple is also taken and mixed with sweet potato pulp and cooked together. The mixture is stirred well after adding sugar. Food flavour and food colours are added to pulp and cooking continued till the mass is fairly uniform and attains a heavy body form. A spoon is dipped into the mass and checked whether the product runs off the sides of the spoon. On cooling, if the product fall off in the form of a sheet instead of flowing readily in a single stream, the end point has reached and the product is ready for filling into the containers. The hot jam is filled into clean dry jars and stored in a cool dry place.

Sweet Potato Sauce

To prepare sweet potato sauce, the pulp from sweet potato is extracted as in sweet potato jam. This pulp is then filtered to remove the fibre. Tomato juice is mixed with sweet potato pulp and then cooked together. Spices placed in a muslin or cloth bag are kept immersed in the pulp. The cooking is continued till the pulp content is reduced to about half the original volume. The spice bag is then removed and squeezed into the pulp. Vinegar, salt and sugar are added and the content mixed well. The mass is heated for a few minutes. The finished sauce is then filled into clean, sterilised plain can, sealed and immersed for 45 minutes in boiling water before storing for subsequent use.

Pre-Gelatinised Starch Powder (Batata Shake)

Pre-gelatinised sweet potato starch powder, called Batata shake can be made by the following process developed at CTCRI. Sweet potato tubers are washed, peeled, sliced and then cooked in boiling water for 5 to10 minutes. After draining water, the parboiled chips are sun dried for 72 hrs and pulverized. The fine fraction sieved through 80 mesh forms an excellent energy drink made with hot milk in 1:5 ratio. Cardamom for flavour and sugar to taste may also be added.

Sweet Potato Curd

Recently, anthocyanin and β-carotene rich sweet potatoes were utilized for preparing curd (Ray *et al.*, 2005). Fresh tubers are boiled, peeled and macerated. The resulting puree (12-16 %) is incorporated into boiled milk and cane sugar (10%). About 2 % of curd as starter culture is inoculated into milk- sweet potato mixture and allowed to ferment for 18-24 hr at room temperature to obtain sweet potato curd ready for consumption.

Pickles from Tuber Crops

Fresh sweet potato, *Colocasia* and *Amorphophallus* corms can be used for preparation of pickles. Peeled and washed tubers are partially dried after or without boiling, as desired. Roasted and powdered spicy ingredients such as mustard, cumin (*jeera*), asafoetida and fenugreek are fried in oil along with chilli powder, turmeric powder, green chillies and garlic. Tuber pieces and salt are added and stirred well while cooking for a few minutes. Vinegar is then stirred in and after cooling the pickles can be filled in sterilized airtight bottles.

3.3 Food products from Aroids

The use of taro in both traditional and non traditional foods was reviewed based on scientific literature by Maga (1992). The versatility of this plant is reflected by the fact that not only the rhizome but also the stem and leaves are used as vegetable (Aggarwal *et al.,* 1999). Poi, a purplish paste made of cooked taro (*Colocasia esculenta*), hardens rapidly to a rubbery gel at a refrigeration temperature of 4°C. This gelling phenomenon was averted by an addition of α-amylase in poi prior to refrigeration at usage levels of 200 units and 2,000 units. Both the viscosity readings and textural analyses indicated that the enzymatic treated poi maintained a fresh-poi like softness at 4°C for up to seven days. The sensory evaluation indicated that the textural attributes were in agreement with the instrumental measurements (Huang *et al,* 2002).

Taro Flour

Textural characteristics of taro (*Colocasia esculenta* L. Schott) cormels are measured during storage at tropical ambient conditions, evaporatively cooled room (EC room) and refrigerator and also after cooking. The textural parameters (hardness, toughness, cutting force, and cutting energy) significantly increased with storage period. Most of the rheological characteristics were lowered for flours prepared from stored samples (Sajeev *et al.,* 2004).

A study was carried out to determine dehydration kinetics of pregelled 0.5 cm thick taro slices (Njintang and Mbofung 2003) in an air convection oven at 50, 60, 70 or 80° C, before milling into flour. Drying at high temperatures affected colour and the susceptibility to gelatinisation of the flour especially that obtained from gelatinised slices.

Soup Thickener

Cocoyam tuber (*Colocasia esculenta*) is mostly used as conventional soup thickener in Southern Nigeria. A study was carried out to know the effect of heat-treatment and defatting on its proximate composition (Onveike et al, 1995). There was no statistically significant difference in proximate composition between raw and heat-processed samples.

Fermented Products from Taro

Sapal is a traditional fermented food made by mixing cooked, grated taro [*Colocasia esculenta* (L.) Schott] corm with coconut cream and allowing it to ferment at ambient temperature. The fermentation is primarily due to lactic acid bacteria (Gubag *et al.,* 1996). The initial microbial flora were derived from coconut cream. Yeasts grew on the surface of sapal in later stages of fermentation. Overnight storage of grated taro corm resulted in glucose concentration increasing from 1.1 to about 5 g l^{-1}. During fermentation , the glucose concentration decreased to undetectable levels while the pH value fell from an initial value of 6.1 to 4.1 after 24 hr.

Nnam (2001) evaluated nutrient composition of blends of malted maize, African yam bean and fermented cocoyam. Maize and African yam bean were separately germinated for 48 hr. Cocoyam was fermented for the same period. The samples were separately dried and milled to fine flours. Ratios of 70:28:2, 60:38:2 and 50:48:2 of maize, African yam bean and cocoyam were used to formulate blends. Porridges prepared from blends were tested for viscosity. The blend containing 50:48:2 mix had more protein, lipid and minerals but lower carbohydrate and energy. The carbohydrate levels of blends were higher than that of Nutrend, a commercial weaning food. The porridge from 70:28:2 mixture had lower viscosity than the porridges from Nutrend and the other blends. The blends have promising nutrient potential as base ingredients for formulation of complementary food in Nigeria.

Extruded Taro Products

Taro flour is unique because of its extremely small particle size (1-5 micron) and high mucilage or gum content, making it a possible replacement for corn or wheat starch in weaning foods. Taro flour, derived from corm of *Colocasia esculenta*, was extruded with whey protein concentrate (WPC), whey protein isolate (WPI) or lactalbumin (LAC), to derive blends containing 20% protein, simulating the protein content of some weaning foods (Onwulata and

Konstance, 2002). Extrusion processing temperatures were from 100 to 130° C and moisture of the feed blends was held constant at 18%. The extrudates were pulverized, made into powders, and rehydrated to make a paste. Viscosities of feed blends before extrusion and the pastes made from the extrudates were determined using a Rapid Visco Analyzer (RVA) to find ouy peak, final, and breakdown viscosities. Water solubility and absorption indices were also determined. Extrudates made from taro containing whey products expanded more than taro alone; were easier to grind into powders; and rehydrated readily in water to form pastes. Taro extrudates without protein absorbed most of water, and were more soluble than products containing whey proteins or LAC. The addition of whey proteins reduced peak viscosities, but WPI and taro pastes were characteristics of weaning foods. Both extrusion cooking and addition of whey proteins, significantly reduced the gummy properties of taro mucilage.

Taro, containing 15% moisture, was extruded at 80, 100, 120 and 140 °C (Maga *et al.*, 1993). Resulting extrudate expansion ratio, breaking strength and water absorption properties were determined. Optimum properties occured at an extrusion temperature of 120 °C.

Taro Noodles

Blends of hard red wheat flour and taro were prepared for production of noodles (Arnaud-Vinas and Lorenz, 1999). Mixing properties and pasting characteristics of blends were determined. Noodles were evaluated for cooked weight, cooking losses, color, compressibility and sensory properties. Increasing taro levels increased mixograph peak times. Rapid Visco-Analyzer peak viscosities and final set-back values increased with increasing levels of taro in the blend. As the percent of taro in the blend increased, cooked weight of noodles increased. Cooking losses for all samples were below 6%. Statistically significant differences in compressibility were found between raw and cooked noodles from the control and those samples made with 20% or more of taro. Colour of the noodles was affected by addition of taro and produced a brown-gray tint. Taro addition negatively affected aroma and flavor descriptors and, consequently, overall acceptability of noodles.

Colocasia Cookie

Application of taro flour in cookie (biscuit) formulations was tested (Nip *et al.*, 1994). Snap-type cookie formulations with taro flour as one of the major ingredients were developed by first modifying taro flour and wheat flour ratios, followed by modification of taro flour, wheat flour, sugar and shortening ratios. A formulation with same proportion of taro flour, shortening and sugar, a lesser amount of wheat flour and good dough handling was selected for sensory evaluation. A drop-type chocolate chip cookie formulation was developed by replacing 51% of the wheat flour with taro flour in a commercial formulation. Both cookie formulations developed were subjected to taste evaluation and were highly acceptable.

Biscuits were prepared by varying the amounts of basic ingredients to arrive at the most acceptable recipe. The final recipe contained 40 g of mix, 40 g of jaggery and 10 g of ghee. Sensory evaluation of biscuits containing 5 or 10% colocasia leaf powder by composite scoring test and hedonic scale showed that the former type of biscuit was preferred over the latter. The acceptability trial conducted on 2 to 6 yeas old children showed that biscuits from either malted mix with or without 7.5% *Colocasia* leaf powder and raw mix were equally acceptable as shown by the analysis of variance (Goyle and Gujral, 1992).

Colocasia Sevian

Sevian (Khara sev-like product), prepared from a blend containing 30 parts of rice flour and 70 parts of cooked *colocasia*, yielded a crisp and crunchy product upon deep-fat-frying (Kulkarni *et al.*, 1994). Sorption-isotherm studies revealed 18.4% equilibrium relative humidity with 1.2% initial moisture content. Critical moisture content of the product was 2.8%. Product was acceptable upto 60 days storage at room temperature (22-35 ^{0}C), when packed in 250 gauge low density polyethylene and 120 gauge polypropylene bags.

4. Processing techniques to eliminate toxins and irritants

4.1 Detoxification of Cassava Through Processing

Cassava is often classified as 'bitter' or 'sweet' according to the amount of hydrogen cyanide (HCN) present in it in the form of cyanogenic glucoside (CNG). According to a tentative classification, tubers with less than 50 mg HCN / kg are considered innocuous while those having over 100mg HCN/ kg of tuber should be regarded as dangerously poisonous. Traditional processing methods, such as drying and boiling, reduce the cyanide content of cassava to an appreciably low level making cassava relatively safe for consumption (Padmaja, 1995).

4.2 Processing techniques to control acridity in Aroids

Colocasia esculenta and *Amorphophallus paeoniifolius* are two of the major aroids with tubers rich in starch and a good source of certain minerals and vitamins. However, when the tubers are consumed, presence of needle-shaped calcium oxalate crystals causes an intense irritation in the mouth and throat followed by a burning sensation that may last for a few minutes to hours depending on cultivar. The intensity of irritation or acridity is very high in raw tubers, but is reduced on cooking (Sundaresan, 2001).

The following methods were effective in reducing acridity in highly acrid *Colocasia* and *Amorphophallus* cultivars.

1. Tuber size used for cooking appears to play a key role in reduction of acridity. Large pieces of tubers retain higher degree of acridity. When tubers are cut into smaller pieces, acridity is reduced considerably. Maximum loss of calcium oxalate was observed in 3 mm size.
2. Traditionally, tamarind is used in cooking medium to reduce acridity. Addition of citric acid (1%) or tartaric acid (1%) in cooking medium was more effective in reducing acridity. These organic acids facilitate solubilization of calcium oxalate, leading to marked reduction in calcium oxalate levels.
3. Sensory evaluation shows complete elimination of acridity in samples cooked in microwave oven (Sundaresan, 2005).

5. Industrial Products from Tuber Crops

5.1 Cassava alcohol

Ethanol from cassava can be used for drinking purpose, for industrial use and for blending with petrol as biofuel. Fresh cassava roots, or flour of dried cassava chips can be used for ethanol production. More recently thippi, a fibrous waste from cassava starch and sago industries, has also been utilized for this purpose. The first step is hydrolysis of gelatinized starch to glucose by a process called saccharification, accomplished with the help of mild acids or amylase enzymes. The saccharified starch is fed into fermentation vessels and incubated with the yeast *Saccharomyces cerevisiae*. The optimum concentration of sugars for ethanol fermentation is 12 to 18% at 4 to 4.5 pH and 28 to 32°C temperature. Alcohol is recovered from the fermented mash after 48 to 72 hr and is distilled to desired purity (Balagopalan *et al.*, 2000).

The yield of alcohol ranges from 130-150 litres/tonne of fresh cassava tubers and 420-440 litres/tonne of dry cassava starch. Economics of production depends on the initial raw material cost and is slightly higher compared to molasses ethanol.

Ethanol can be blended with petrol to run automobiles and internal combustion engines, popularly referred to as gasohol and/or petrol. Presently, the need and interest to blend ethanol with petrol is revived, particularly in Latin American and south-east Asian countries, including China. The technology is likely to get a second lease in the immediate future.

Productivity and cost of Ethanol from cassava and sugarcane sources

Source	Yield of Raw material (t/ha)	Fermentables (%)	Cost of Raw Material Rs./tonne	Yield of Alcohol (l/tonne)	Alcohol Productivity (l/ha)	*Cost of Power Alcohol (Rs/l)
Cassava tubers	18	30	2000	180	3240	11.10
Cassava chips/flour	7.2	80	8000	442	3182	18.10
Cassava starch	5.4	94.6	11000	520	2808	21.15
Cassava thippi	0.9	55	2000	197	177	10.15
Sugarcane	70	9.6	620	48	3920	12.90
Molasses	3.5	50	2500	220	770	11.40

Ability of cassava to compete with sugar crops for alcohol production largely depends on total production cost. For ethanol production from cassava, approximately 65% of the final production cost is accountable to the cost of raw materials alone. The CTCRI technology for manufacturing ethyl alcohol using cassava chips, flour or starch has been transferred to M/s Superstar Distilleries, Palakkadu, Kerala and M/s Vairam Agro Fuels, Chennai, TN.

5.2 Sweeteners

Liquid Glucose And Dextrose

Starch is a polymer of glucose and therefore the raw material for glucose. The glucose syrup (43% db dextrose) is obtained by acid hydrolysis or enzyme hydrolysis may be further purified to obtain dextrose crystals. Glucose is used for various confectionery and pharmaceutical purposes.

High Fructose Syrups

Fructose is 1.7 x sweeter than sucrose and 4 x sweeter than glucose. Glucose can be isomerised to fructose by alkali or by an enzyme. Fructose syrup has gained importance in view of the fluctuating prices of sugar and harmful effects of synthetic sweeteners.

Maltose

Maltose is a disaccharide that can be obtained by enzyme treatment of saccharified starch. The maltose syrups find use in brewing, baking, soft drink, canning and confectionery industries.

Maltodextrins

Maltodextrins are partially hydrolysed starch with dextrose equivalent (DE) less than 20. Typically malto-dextrins contain 0.3 to 1.6% glucose, 0.9 to 5.8% maltose and the rest other saccharides. Manufactured by action of α-amylase on starch, they are approved as food ingredient and find extensive use in ice creams, hard candies and as fat and oil substitutes because of their low calorific value.

5.3 Adhesives

Dextrins

Production of dextrin involves predrying of starch to less than 5% moisture content, acidification (pH of 4.5 to 5.5) by spraying a dilute solution of HCl and conversion by heating up to 95 - 120°C for white dextrin, 150 - 180°C for canary dextrins and 170 - 195°C for British gum. The adhesive industry is the major consumer of dextrins.

Liquid Adhesive

Liquid adhesives and gum pastes can be made from cassava starch using simple low cost technologies. Starch is cooked with water, cooled and preservatives (viz. formaldehyde or copper sulphate) are added. Shelf life of gums can be improved by adding borax, urea, glycerol, carboxy methyl cellulose etc. These chemicals help in improving and stabilising the paste viscosity. Liquid adhesives possess the advantage of good flow characteristics and are 'ready-for-use' by the consumer.

A pilot plant was developed at CTCRI having a cooking digester of 100 litres capacity, ideal for small scale entrepreneurs. The digester consists of three concentric cylinders. The innermost stainless steel cylinder serves as digester while the intermediate and the outermost cylinders serve for containing heating medium and insulation. Other components of pilot plant are the motive power transmission mechanism, heater controls, the agitating mechanism and the inlet and outlet provisions.

Liquid adhesives find application in carton sealing, laminated board, metal foil-to-paper laminating, corrugated board, bottle and container labeling, bill posting, cigarette seaming and paper bag making.

5.4 Cold water soluble starch

Cold water miscible cassava starch is one of the new innovations of CTCRI. The technology was developed and process optimised through a series of trials. The basic process consists of treating starch with some inorganic phosphates and urea in minimum quantity of water, followed by removal of moisture, heat treatment and purification. The resulting powder is miscible in cold water and gives higher viscosity than native starch.

5.5 Biodegradable plastics from starch

About 32% of our requirements of commodity plastics are met by imports. Agricultural and packaging sectors account for about 50% of plastics consumed in India. Use of plastics now has accelerated to such an extent that the disposal of used products has become increasingly difficult. Starch, a natural polymer as well as a renewable raw material, has captured the interest of academic and industrial research workers across the globe pursuing environmentally degradable polymers for easier disposal. Biodegradation is potentially the most promising approach to meet the increasing menace of environmental pollution by plastics waste. Starch is used in many different ways to induce biodegradability in synthetic polymers. Biodegradable polymers such as polylactic acid (PLA) and polyhydroxy-butyrate (PHB) have mass processing limitations and production costs are likely to remain very high. Cassava (tapioca) or Corn (maize) starch can be blended with polyolefin to make a processible and functional plastic article while inducing biodegradability in appropriate environment.

CTCRI process for production of starch-based plastics (Nanda, 1996) involves blending corn (maize) or cassava (tapioca) starch with suitable synthetic polyolefins (viz. LDPE, HDPE and LLDPE) as stabilizing agents and suitable amounts of appropriate coupling, gelatinizing and plasticizing agents. Compounding of blend prior to extrusion film blowing is adopted to attain proper melt mixing. Successful extrusion film blowing and injection moulding are possible with formulations containing 10 to 40% starch. The process is patented in India and Europe. The technology on starch-based biodegradable plastics is transferred through NRDC to 4 licensees. M/s Shivalik Agro-Poly products Ltd., Parwanoo (Himachal Pradesh) commenced production on a small scale basis. The production cost is about 50% higher than popular plastics which is deterrent for marketing in India to both manufacturers and consumers.

These starch-based plastic films possess adequate mechanical strength and flexibility to make them suitable for various potential agricultural applications. The tensile strength of these plastics films containing 10, 25 and 40% starch are 12.56, 17.34 and 10.67 MPa respectively. The elongation at break values for these films varied from 210.8 to 122% as the starch content varied from 10 to 40%. In comparison, the tensile strength and elongation values of the LDPE control films were 10.97 MPa and 384.1%. The storage stability of these films, with regard to changes in tensile strength and elongation, was almost equivalent to that of the ordinary polyethylene films, the granule form of the material being more stable than the film form (Nanda, 1996).

Suitability of these films for potential areas of application in agriculture and single-use disposable packaging was assessed through outdoor weathering and soil burial which showed drastic reduction in mechanical strength and elongation values resulting in brittleness disintegration with the biodegradation time varying from 6 months to 5 years depending on its composition and soil conditions. Deterioration of strength and of flexibility were progressively greater with increase in starch content of the film and the duration of environmental exposure. Scisson of macromolecular chain was evident from morphological studies. The product is also suitable for injection moulding to manufacture toys and household articles. Rapid biodegradation (6 months) of these films could also be achieved by incorporating a suitable catalytic agent into the film composition. Films of latter type would be much more suited for making nursery bags and single-use disposable packaging. Relatively easier dispersal and absorption of starch-based biodegradable plastics into the soil after a specific time interval would make it an ecologically satisfactory mode for disposal of plastics waste.

5.6 Industrial utilization of sweet potato

Sweet potato starch is mostly used by the food industry as an ingredient in cakes, breads, biscuits, cookies and noodles. Starch is also processed into glucose syrups and various chemicals through enzymatic, microbial and chemical processes. Asian noodles are generally starch noodles produced from purified starch. Demand for starch noodles rose as quality improved in recent years not only in China but also in other parts of Asia such as Taiwan, Korea and Vietnam (Ilangantileke, 2005). In China, sweet potato starch utilization by noodle industry grew sharply over the past several years. The rapid expansion of starch utilization in Asia has enhanced market demand for tropical root and tuber crops, especially cassava and sweet potato. Starch processing from sweet potato has also created new economic and employment activities for farm and rural households and added value to this commodity.

Sweet Potato Wine

A biotechnological process was developed for preparing wine from sweet potato (Chowdhury, *et al.*, 2005). Sweet potato wine is good in taste, aroma and has pH of 2.9-3.3 and titrable acidity of 3.3-12.9%. If residual sugar is present in wine samples, it is subjected for further conversion into lactic acid or Malolactic fermentation (MLF). Malolactic fermentation (MLF) in wine is a secondary fermentation by lactic acid bacteria that usually occurs at end of wine fermentation by yeast although at some times, occurs earlier. In tropical climate, MLF is considered as spoilage (because of increasing sourness) if the lactic acid content increased above 15 mg/100 ml wine.

Sweet Potato Alcohol

The starch content in sweet potato flour and fresh tubers was liquefied by treatment with 0.08% commercial thermostable α-amylase (Termamyl^R, Novozyme) and 0.33% amyloglucosidase (AMG^R, Novozyme). The hydrolysate was subsequently fermented with alcohol strain of *Saccharomyces cerevisiae* into ethanol. The result showed that the final ethanol yield of 225 and 202 g kg^{-1} (flour) and 73 and 66 g kg^{-1} (tuber), was little influenced by liquefaction conditions i.e. 80°C/2 hrs and 90°C/1 hr, respectively. Similarly it was observed that the treatment of AMG for 48 hrs/45°C and 24 hrs/45°C had little significance in the ultimate ethanol yield i.e 263 and 260 g kg^{-1} (flour) and 96 and 97 g kg^{-1} (tuber), respectively after liquefaction (90°C/hr) treatment. Another experiment was designed to study simultaneous saccharification and fermentation (SSF) by saccharifying liquefied (90°C/1hr) hydrolysate and co-fermenting with *S. cerevisiae* at 35°C for 48 hrs. The ethanol yield was 322 g kg^{-1} for flour and 128 g kg^{-1} for tuber, respectively (Sahoo *et al.*, 2005).

5.7 Industrial utilization of aroids

Vanillin Flavour

Spherical aggregates ranging in size from 7.5 to 45 mm obtained from four small sized starch granules, isolated in the laboratory from colocasia (*Colocasia esculenta* L.) in the presence of polysaccharide bonding agents (such as gum Arabic; carboxymethyl cellulose and carrageenan) at 0.1-1.0% were obtained by spray drying a 20% (w/w) starch dispersion at 120 °C. These were used for entrapment of vanillin, a model flavouring compound, at 5% based on starch (Tari and Singhal, 2002).

The study indicated a negative correlation of amylose on the extent of entrapment of vanillin held within the spherical aggregates. With respect to bonding agent the stability of entrapped vanillin decreased in the order, gum Arabic > CMC > carrageenan (Tari *et al.*, 2003).

Citric Acid

Tubers from crops belonging to family Araceae namely *Colocasia antiquorum* and Aponogetonnatans were suitably treated and used as efficient substrates tor citric acid production (Angumeenal *et al.*, 2003) by fermentation using *Aspergillus niger*. The quantities of citric acid produced using these materials as substrates for bioconversion using Aspergillus niger are compared with those produced in synthetic medium comprising glucose as substrate. Transition metal ions such as, chromium, molybdenum, cadmium and lead are added at optimum concentration as nutritional supplements for their effect on the biosynthetic route of citric acid cycle. Experimentally observed growth stages were also used for mathematical modeling to evaluate the kinetic parameters and the values obtained by calculation agree well with observed ones.

Cocoyam Beer

A widely cultivated cocoyam variety, *Xanthosoma sagittifollium*, was assessed for its suitability for lager beer production, using malted *Sorghum vulgare* (white variety) for saccharification of the substrate (Onwuka and Eneh, 1996). The three-mash decoction method was used. Results showed that the cocoyam was superior to barley and sorghum as a substrate because of its potentially higher carbohydrate content (71-78%) compared to barley (65%) and sorghum (70-73%). The percentage wort extract was 13.3-14.5 degrees plato compared to 7-12 for barley and 13° for cassava. The kiln dried *Xanthosoma sagittifollium* gave a dark lager beer with good aroma (68%) and very good flavour (73%) when compared with a commercial lager beer (Monarch) as standard (100%). The alcohol content (4.16% w/v) and the specific gravity (1013) were within the range (4.0-6.4%) and (1011-1019), respectively, specified by Standard Organization of Nigeria.

5.8 Utilization of *Amorphophallus*

Significant analgesic activity of *Amorphophallus campanulatus* tubers was demonstrated (Shilpi *et al.*, 2005) in mice by giving its methanol extract orally at the doses of 250 and 500 mg kg^{-1}. Glucomannan as the fiber supplement was beneficial in the treatment of constipation with and without encopresis in children and recommended in the diet of constipated children with and without encopresis (Loening-Baucke *et al.*, 2004). Glucomannan is a fibre gel polysaccharide from the tubers of Japanese Konjac (*Amorphophallus*) plant. The study recruited 31 children (16 boys and 15 girls) of 4.5 to 11.7 years in age having functional constipation. No significant side effects were reported during treatment period.

The bactericidal effects of Konjac (*Amorphophallus*) as an alkaline food on food-poisoning bacteria was evaluated. Konjac fluids (the pH ranging from 11.42 to 12.53) completely inactivated *Escherichia coli*, enterohemorrhagic *E. coli* O157:H7 and *E. coli* O26:H9, *Salmonella enteritidis*, *Vibrio parahemolyticus* and *Staphylococcus aureus*. No viable gram-negative bacterium cells could be detected within 1 to 2 days and no viable *S. aureus* cells could be detected within 3 to 5 days. Treatment with konjac fluid was also effective in reducing levels of spore-forming bacteria (*Bacillus subtilis, Bacillus cereus, Clostridium perfringens,* and *Clostridium botulinum* type E and type A). At least, a 4-log reduction of spore-forming bacteria was obtained in konjac fluid within 7 to 14 days. Our studies indicate that konjac fluid, which has a long history of use in food, will control food-poisoning, bacterial contamination during production or preservation of konjac and other foods and has a preventive effect on bacteria that can cause severe disease at uniquely low levels (Kitamoto *et al.*, 2003).

6. Use of Tuber crops for Animal feed

Tuber crops being rich in carbohydrate, either alone or incorporated with other protein rich ingredients can effectively play an important role in the compound feed sector. Use of cassava as animal feed can partly replace use of cereals in animal feed formulations which then may be spared for human consumption.

6.1 Ensiling for Animal feed

Cassava Silage for Cattle Feed

Considering the poor shelf life of fresh roots and susceptibility of dried products to insect attack, ensiling emerges as one of the ideal techniques for preserving the nutritive value of cassava, enhancing shelf life and increasing the palatability through lactic acid enrichment.

The cassava ensiling technology developed at CTCRI was scaled up to plastic silos of 100 litres and 225 litres capacity. The process consists of four steps:

1) Fresh cassava roots are washed free of dirt and chopped along with the rind into small chips. This is essential to achieve sufficient consolidation during ensiling.
2) The chopped whole root cassava chips are then spread in the sun for 4 hr to enable partial loss of moisture and also to reduce the initial cyanogen load entering the silo.
3) Dehydrated rice straw is cut into small pieces and then thoroughly mixed with the exposed whole root chips in the ratio 10:90 (10 kg rice straw for 90 kg chopped cassava).
4) The mix is packed tightly into plastic silos and the lid is closed airtight.

The ensiling process is completed in about a week and the stabilised silage is preserved in the silo till it is opened for utilisation. Up to 7 days of utilisation phase, there is no aerobic decay in the silage. In the event of aerobic damage afterwards, the top layer (1-2 cm) may be discarded before being fed to animals.

Proximate analysis of the cassava+rice straw silage after 72 days of ensiling indicated that most of the nutrients were still conserved. Exposing the chopped cassava roots to sunlight for 4 hr before it is mixed with rice straw helped to reduce initial cyanogen load entering the silo. Rice straw at a rate of 10% could serve as good absorbent of silage effluent.

Sweet Potato Silage for Pig Feed

Machine-grated sweet potato roots ensiled with various special additives have also been assessed for nutritional value and feeding efficiency of sweet potato silage as pig feed at 0, 2, 4, 6, 9, 12 and 25 weeks of ensiling (Zuohua et al, 2004). Nutritional value of sweet potato silage did not differ significantly over time except for fat content and pH value. Pig-feeding test was done on weaner and finishing pigs with varying proportions of sweet potato silage to total feed such as 0, 20, 40 and 60 % sweet potato silage for a period of 115 days of weaner- finishing pigs. The growth rates could be increased with sweet potato silage. Also, it could be stored for up to six months without spoilage.

6.2 Feed uses of other Tuber crops

Studies on use of *Colocasia esculenta* (taro cocoyam) cormels as replacements for maize in the diets of weaned pigs (Angwunobi *et al.,* 2002) in five groups of 12 pigs/group showed that there were no significant differences in feed intake, weight gain or feed efficiency between diets containing boiled taro cocoyam cormels. However, for unboiled, sun-dried taro cocoyam cormels, there were significant differences ($p<0.05$) in weight gain and feed efficiency, these being depressed at more than 50% replacement of maize. This may be due to relatively high amounts of antinutritional factors in the unboiled, sun-dried taro cormels. Boiled taro cocoyam cormels were comparable to maize as an energy source in the diets of weanling pigs.

Raw and processed *Colocasia* corn meal was evaluated as an energy-source for broilers by Samarasinghe and Rajaguru (1992). Colocasia corms processed by three methods (dehydrated raw, boiled and soaked, boiled) were used in the preparation of corm meals. Processed *Colocasia* corm meal had gross energy 14.9-15.0 MJ kg-1 dry matter and was comparable with maize meal as an energy source. However, unprocessed *Colocasia*, when compared with control, reduced feed intake by 70%, growth by 83% and dressing percentage by 12%.

Pancreatic hypertrophy and growth inhibition were observed in the raw *Colocasia* group. Boiling removed all these limitations. It was concluded that the feeding value of *Colocasia* corms could be easily improved by boiling in water for 30-45 min and processed corm meal could be used to replace maize in broiler diets at levels up to 100 g kg^{-1} without any adverse effects.

Ravindran *et al.,* (1996) investigated the nutritive value of raw and processed colocasia (*Colocasia esculenta*) corm meal for poultry. Raw, unpeeled *Colocasia* corm meal contained (dry matter basis) 90.7 g kg^{-1} crude protein, 796 g kg^{-1} nitrogen-free extracts, 1234 mg kg^{-1} total oxalates and 8.66 MJ kg^{-1} nitrogen-corrected apparent metabolizable energy. About a third of the oxalates were removed by peeling outer skin of corms, and when peeling was combined with boiling, oxalate contents were lowered to 272 mg kg^{-1}. Results suggested oxalate to be a major factor contributing to anti-palatability and anti-nutritive effects of raw *Colocasia* corms.

Inference

Importance of tuber crops is well established through research as a high energy food, animal feed mix and as an industrial raw material for adhesives, sweeteners, ethanol and other commodity chemicals. However, the industrial potential of tuber crops have not been still commercially tapped except for cassava to some extent. Plantation production on a cash crop footing and assured bulk supply of tuber crops, though major constraints at present, are essential requisites for their processing into value-added food products and industrial commodities at competitive cost.

Traditionally tuber crops supplemented food grain deficit and used by the poor people to supplement their staple diet during periods of scarcity. These have gradually become a subsidiary food in normal years of prosperity. Increased utilization of the tropical tubers as food by urban and affluent section of society can be achieved by processing tubers into various convenience and fast food products, followed by effective marketing, to keep pace with changing consumption pattern and preferred tastes of people, particularly younger generation.

References

Angumeenal, A. R., Kamalakannan, P., Prabhu. H. J. and Venkappayya, D. 2003. Bioconversion of Colocasia antiquorum and Aponogetonnatans to citric acid by Aspergillus niger - Effect of metal ions and kinetics. *J. Scientific and Industrial Research.* 62(5):447-452.

Aggarwal, R.A.K., Arya, P.S. and Mahajan, N.C. 1999. Nutritional evaluation of vegetable *Colocasia* grown in Himachal Pradesh. *J. Food Science and Technology*, 36(5):457-458

Angwunobi, L.N., Angwukam, P.O., Cora, O.O. and Isika, M.A. 2002). Studies on the use of *Colocasia esculenta* (taro cocoyam) in the diets of weaned pigs. *Tropical Animal Health and Production*. 34(3):241-247

Arnaud-Vinas, M.D. and Lorenz, K. 1999. Pasta products containing taro (*Colocasia esculenta* L-Schott) and chaya (*Cnidoscolus chayamansa* L-McVaugh). *J. Food Processing and Preservation*. 23(1):1-20

Balagopalan, C., Padmaja, G., Nanda, S.K. and Moorthy. S.N. 1998. Cassava in Food, Feed and Industry. *C.R.C. Press Inc., Boca Raton, Florida, USA*. 205 pp.

Balagopalan, C., Padmaja, C., Nanda S.K. and Moorthy. S.N. 2000. Integrated Technologies for Value Addition and Post Harvest Management in Tropical Tuber Crops. CTCRI, Trivandrum, 147 pp.

Brabet, C., Reynoso, D., Dufour. D., Mestres, C., Arredondo, J. and Scott, G. 1998. Starch content and properties of 106 sweet potato clones from the world germplasm collection held at CIP. CIP Program Report 1997-98, Lima. pp. 279-286.

Chowdhury,P., Swain. M.R. and Ray, R.C. 2005. Malolactic fermentation of sweet potato wine. Second National Seminar on Root and Tuber Crops. CTCRI, Trivandrum.

CTCRI, 2005. Patented Technologies from CTCRI. Central Tuber Crops research Institute, Trivandrum. 6 pp.

Collado, L.S. and Corke, H. 1997. Properties of starch noodles as affected by sweet potato genotype. *Cereal Chemistry*. 74 (2): 182-187.

Goyle, A. and Gujral, S. 1992. Sensory evaluation of and acceptability trials on biscuits prepared from raw and malted wheat (*Triticum aestivum*)-bengal gram (*Cicer arietinum*) mixes with or without a green leafy vegetable. *Plant Foods for Human Nutrition*. 42 (4): 291-296.

Gubag, R. Omoloso, D.A. and Owens, J.D. 1996. Sapal: A traditional fermented taro [*Colocasia esculenta* (L) Schott] corm and coconut cream mixture from Papua New Guinea. *International J. Food Microbiology*. 28(3):361-367

Huang, A.S., Komarasamy, K. and Liu, X.X. 2002. Textural and sensory properties of alpha-amylase treated poi stored at 4°C. *J. Food Processing and Preservation*. 26(1):1-10.

Hussain, M. and Chowdhury, B.L. 1986. Preparation of dehydrated mukhi (*Colocasia esculenta)* chips. *J. Food Science and Technology (India)*. 23 (3): 167-170.

Ilangantileke, S. and Attaluri, S. 2005. Sweet potato utilization for starch and starch-based products in Asia. Second National Seminar on Root and Tuber Crops. CTCRI, Trivandrum.

Jiang, X., Jianjun, H. and Yi, W. 2004. Sweet potato Processing and Product Research and Development at the Sichuan Academy of Agricultural Sciences. *(In:)* Fuglie, Keith

O. and Hermann, Michael (Eds.). Sweet potato Post-Harvest Research and Development in China. Proceedings of an International Workshop held in Chengdu, Sichuan, PR China, Nov. 7-8, 2001. Bogor, Indonesia: International Potato Center (CIP). pp.114-126

Kitamoto, N., Kato, Y., Ohnaka, T., Yokota, M., Tanaka, T. and Tsuji, K. 2003. Bactericidal effects of konjac fluid on several food-poisoning bacteria. *J Food Prot.*, 66(10):1822-1831.

Kulkarni, S.G., Manan, J.K. and Shukla, I.C. 1994. Studies on deep-fat-fried sevian made from rice flour and *Colocasia. J. Food Science and Technology (India).* 31 (3): 207-210.

Loening-Baucke. V., Miele, E. and Staiano, A. 2004. Fiber (glucomannan) is beneficial in the treatment of childhood constipation. *Pediatrics.* 113(3 Pt 1): 259-264.

Maga, J.A. 1992. Taro - composition and food uses. *Food Reviews International.* 8 (3): 443-473.

Maga J.A., Liu, M.B. and Rey, T. 1993. Taro (*Colocasia-esculenta*) extrusion. *Carbohydrate Polymers.* 21 (2-3): 177-178.

McFarlene, J.A. 1982. Cassava storage II. Storage of dried cassava products. *Tropical Science.* 24:205.

Moorthy, S.N. 2002. Physicochemical and functional properties of tropical tuber starches: A review. *Starch-Starke.* 54(12):559-592.

Nanda, S.K. 1984. Post-harvest practices of tuber crops. *Indian Farming.* 33(12):64-65.

Nanda, S.K. 1996. Biodegradability and starch-based plastics. XI Carbohydrate Conference, 21-22 Nov. 1996, Calcutta, India.

Nanda, S.K. and Pillai, K.S. 1996. Short-term storage of cassava chips and flour : Effect of packaging materials, *(In:)* Tropical Tuber Crops Problems, Prospects and Future Strategies, G.T. Kurup *et al.,* Eds., Oxford & IBH Publishing Co. Pvt. Ltd., New Delhi, India. pp 452 - 457.

Nanda, S.K. and Potty, V.P. 1985. Evaluation of packaging materials for storage of cassava starch. *J. Root Crops.* 11 (1 & 2) : 45-50.

Njintang, Y.N. and Mbofung, C.M.F. 2003. Development of taro (*Colocasia esculenta* (L.) Schott) flour as an ingredient for food processing: effect of gelatinisation and drying temperature on the dehydration kinetics and colour of flour. *J. Food Engineering.* 58(3):259-265

Nip, W.K., Whitaker, C.S. and Vargo, D. 1994. Application of taro flour in cookie formulations. *International Journal of Food Science and Technology.* 29(4): 463-468.

Nnam, N.M. 2001. Nutrient potential of blends of malted maize (Zea mays. L.), African yam bean (*Sphenostylis stenocarpa*) and fermented cocoyam (*Colocasia esculenta*) flours. *Ecology of Food and Nutrition.* 40(4):311-319.

Onveike, E.N., Olungwe, T. and Uwakwe, A.A. 1995. Effect of heat-treatment and defatting on the proximate composition of some Nigerian local soup thickeners. *Food Chemistry*. 53(2):173-175.

Onwueme, I. 1999. Taro Cultivation in Asia and the Pacific. Food and Agriculture Organization of the United Nations, Regional Office for Asia and the Pacific, Bangkok.

Onwuka, N.D. and Eneh, C.O. 1996. The cocoyam, *Xanthosoma sagittifollium*, as a potential raw material source for beer brewing. *Plant Foods Hum Nutr*. 49(4):283-293.

Onwulata, C.I. and Konstance, R.P. 2002. Viscous properties of taro flour extruded with whey proteins to simulate weaning foods. *J. Food Processing and Preservation*. 26(3): 179-194.

Padmaja, G. 1995. Cyanide detoxification in cassava for food and feed uses. *Crit. Rev. Food Sci. Nutr*. 35(4):299-339.

Perez, E., Schultz, F.S. and de Delahaye, E.P. 2005. Characterization of some properties of starches isolated from *Xanthosoma sagittifolium* (tannia) and *Colocassia esculenta* (taro). *Carbohydrate Polymers*. 60(2):139-145.

Ravi, V. Aked, J. and Balagopalan, C. 1996. Review on tropical root and tuber crops .1. Storage methods and quality changes. *Critical Reviews in Food Science and Nutrition*. 36(7):661-709.

Ravindran, V.R., Sivakanesan, A. and Cyril, H.W. 1996. Nutritive value of raw and processed *Colocasia* (*Colocasia esculenta*) corm meal for poultry. *Animal Feed Science and Technology*. 57(4):335-345.

Ray, R.C., Naskar, S.K. and Sivakumar, P.S. 2005. Sweet Potato Curd. Technical Bulletin 39. Central Tuber Crops Research Institute, Trivandrum. 24 pp.

Sahoo, A.K., Sharma, P., Pattnaik, M.S. and Ray, R.C. 2005. Ethanol production from sweet potato (*Ipomoea batatas*) by enzymatic liquefaction and simultaneous saccharification and fermentation (SSF) process. Second National Seminar on Root and Tuber Crops. CTCRI, Trivandrum.

Sajeev, M.S., Manikantan, M.R., Kingsly, A.R.P., Moorty, S.N. and Sreekumar, J. 2004. Texture analysis of Taro (*Colocasia esculenta* L. Schott) cormels during storage and cooking. *J. Food Science*. 69(7):E315-E321.

Samarasinghe, K. and Rajaguru, A.S.B. 1992. Raw and processed wild *Colocasia* corn meal (*Colocasia esculenta* (L) Schott, var *esculenta*) as an energy-source for broilers. *Animal Feed Science and Technology*. 36 (1-2): 143-151.

Sefa-Dedeh, S. and Sackey, E.K.A. 2002. Starch structure and some properties of cocoyam (*Xanthosoma sagittifolium* and *Colocasia esculenta*) starch and raphides. *Food Chemistry*. 79(4):435-444

Shilpi, J. A, Ray, P. K., Sarder, M. M. and Uddin, S. J. 2005. Analgesic activity of *Amorphophallus campanulatus* tuber. *Fitoterapia*. 76(3-4):367-369.

Sundaresan, S. 2001. Post harvest management of tropical tuber crops. *(In:)* Handbook of Horticulture (K.L. Chadha Ed.), ICAR, New Delhi. pp. 957-963.

Sundaresan, S. 2005. Acridity of Edible Aroids. Technical Bulletin 38. Central Tuber Crops Research Institute, Trivandrum. 21 pp.

Tari, T.A. and Singhal, R.S. 2002. Starch-based spherical aggregates: stability of a model flavouring compound, vanillin entrapped therein. *Carbohydrate Polymers.* 50(4): 417-421

Tari, T.A., Annapure, U.S., Singhal, R.S. and Kulkarni, P.R. 2003. Starch-based spherical aggregates: screening of small granule sized starches for entrapment of a model flavouring compound, vanillin. *Carbohydrate Polymers.* 53(1):45-51

Zuohua, L., Zonghui, L., Jian, H., Feiyun, Y., Zhengze, Z. and Wen, L. 2004. Sweet potato roots silage for efficient feeding of weaner and finishing pigs in China. (In:) Fuglie, Keith O. and Hermann, Michael (Eds.). Sweet potato Post-Harvest Research and Development in China. Proceedings of an International Workshop held in Chengdu, Sichuan, PR China, Nov. 7-8, 2001. Bogor, Indonesia: International Potato Center (CIP). pp. 88-99.

❑❑❑

13

Tuber Crops Biotechnology

Tuber crops grown in India include Cassava, Sweet potato, *Dioscorea, Colocasia, Amorphophallus* and minor tuber crops like Chinese potato, Arrowroot, Canna, Yam bean etc. Biotechnology can contribute solution to the major problems and realize greater benefits for tuber crops. Traditional breeding in tuber crops is difficult due to their heterozygous nature, low fertility, out crossing coupled with strong inbreeding depression. Many of the agronomic traits are polygenic and/or recessive which make breeding for such characteristics complicated. Biotechnology is a powerful tool to complement traditional breeding. Developments in the field of tuber crops biotechnology are discussed below.

1. Cassava

In cassava, great advances are made in the field of in vitro conservation, characterization of genome, gene isolation, development of molecular maps, map based cloning and genetic engineering.

1.1 Meristem culture

In vitro propagation and storage techniques are well established in cassava making it possible to multiply and store a large number of germplasm accessions. *In vitro* techniques are developed for collection of wild and cultivated genotypes. Meristem culture was reported to be the most effective method in tissue culture technique, to develop virus free plants *in vitro.* In cassava, meristem culture is used widely to clean plants infected by cassava mosaic disease (CMD) caused by Gemini group of virus and spread by white fly, *Bemisia tabaci*. Elimination of viruses through shoot apical meristem culture in cassava was first reported by Kartha *et al.* (1974). Meristem culture for elimination of Cassava Mosaic Disease was done in MS medium in the presence of BA (5×10^{-7}M) and NAA (10^{-6} M) and GA_3 (10^{-7}M) and elimination of virus was reported (Kartha *et al.*, 1974). Meristem culture combined with thermotherapy is being used for large-scale virus-free conservation of germplasm at Centro International de Agricultura Tropical (CIAT), Cali, Columbia. At CTCRI, Indian cassava mosaic disease (ICMD) free plantlets were obtained by culturing of meristems tips of about 0.4 mm size on Murashige and Skoog (MS) (Murashige and Skoog, 1962) medium supplemented with 0.02 mg l^{-1} NAA, 0.1 mg l^{-1} BAP and 0.03 mg l^{-1} GA_3 (Unnikrishnan and Sheela, 2000). Nodal explants and apical buds also could be used for initiation of cultures. BA was the most effective cytokinin in cassava. Multiple shoot cultures could be produced from meristems by increasing concentration of BA from 2.5 to 5.0 μM (Roca *et al.*, 1984, 2001). Response variations among accessions were documented.

1.2 Slow growth culture

Cassava cultures are amenable to *in vitro* conservation under normal conditions as well as reduced growth rates. *In vitro* genebanks are set up using slow-growth cultures at various cassava genetic resource centers in Africa, South America and Asia. Various conservation methods viz. low temperature incubation, use of low light intensity, use of osmotically active agents or growth retardants, manipulations in concentration of growth hormones, macro and micro-nutrients and sucrose are used individually or in combination for medium term storage of cassava germplasm. Experiments with minimal and mannitol media indicated that combination of both media was effective to induce slow growth. Osmotic retardants like mannitol or sorbitol (3%) along with sucrose (3%) in the medium was effective in extending subculture period from 10 to 12 months. Sorbitol was better in imparting slow and healthy growth than mannitol (Unnikrishnan *et al.*, 2002). Silver nitrate (1mM) and activated charcoal (1g l^{-1}) in MS medium with sucrose (3%) and growth regulators (0.05 μM NAA, 0.1μM BA and 0.3μM GA3) helped in maintaining cultures upto 18 months. Addenda like silver nitrate and activated charcoal was effective in reducing leaf chlorosis and preventing leaf shedding, thereby contributing to culture life and regeneration capacity. Throughout storage period, cultures continued to produce axillary buds, indicating viability of cultures.

1.3 Cryopreservation

Cryopreservation technique for long-term conservation of cassava using shoot tips was reported. Isolated meristems precultured on a medium containing DMSO and treated with a mixture of DMSO, sucrose and glycerol, could be frozen in liquid nitrogen using rapid freezing as well as slow-freezing techniques. By following method of preculture followed by slow freezing using programmable freezer, protocol for cryopreservation, using shoot-tips was developed and tested for 15 genotypes of cassava at CIAT. In some cases, regeneration up to 70% was achieved.

1.4 Somatic embryogenesis in Cassava

Somatic embryogenesis forms the basis of all current transformation methods of cassava. Cyclic embryogenesis is the most routinely used method of *de novo* plant regeneration of cassava *in vitro* and provides constant source of material for transformation. Plant regeneration from somatic embryos can be established via either direct germination or shoot organogenesis. Currently, shoot organogenesis from cotyledons of somatic embryos and embryogenic suspensions derived from friable embryogenic callus are the two most frequently used systems for transformed cassava plants.

To date, all reported cases of somatic embryogenesis in cassava have relied on inducing organized embryogenic structures from immature leaf lobes or cotyledons (Stamp and Henshaw, 1982; Szabados *et al.*, 1987). These tissues are exposed to Murashige and Skoog (1962) media with various types and concentrations of auxin. Similar media are then used to promote subsequent multiplication of these tissues *via* a secondary embryogenesis process (Raemakers *et al.*,1993, 2001; Stamp and Henshaw, 1987).

In organogenesis system, development of shoot primordia is induced directly on cytokinin containing media from explants of germinating somatic embryos. A cyclic system in which secondary somatic embryos are induced on cotyledon explants from maturing somatic embryos is used to provide a constant source of regeneration competent explant material. In some cultivars, a fraction of cyclic somatic embryos maintained on MS medium supplemented with picloram (12mg/l) produces a new tissue type which consists mainly of small globular embryo like structures. When isolated, it produces a highly friable embryogenic callus (Taylor *et al.*, 1996). Frequency of friable embryogenic callus (FEC) production can be increased by substituting the MS medium with GD medium (Gresshoff and Doy, 1974). Once pure FEC is obtained, it can be transferred easily to liquid medium to establish a rapidly proliferating embryogenic suspension in SH medium (Schenk and

Hildebrandt,1972) medium supplemented with 6% sucrose and picloram (12mgl^{-1}). Conversion of globular embryos is still the limiting step of regeneration protocol , as only a small fraction of embryos is able to develop into torpedo stage embryos, and genotypic variation is high. At CTCRI, somatic embryogenesis occured in five out of eleven Indian varieties viz., Sree Vijaya, Sree Harsha, Sree Sahya, Sree Jaya and H165 showed embryogenesis in liquid medium with 2,4-D (8mg l^{-1}) (Sree Rekha *et al.*,2005).

1.5 Molecular markers in Cassava

Diversity analysis

A number of works are reported on molecular finger printing of cassava. Starch and polyacrilamide gel electrophoretic analysis was carried out for determining isozyme patterns of 16 enzymes of cassava varieties as potential genotype markers. Varieties could be distinguished by sequential classification on basis of isozyme patterns of acid phosphatase, esterase, glutamate oxalaoacetate, transaminase and phosphoglucoisomerase. Of these, esterase was the most stable isozyme system for cassava (Ramirez *et al.*, 1987).

Later DNA fingerprinting using RAPD markers was reported by Yu and Nyugen 1994. Variable number of Tandom Report loci was used by Bertram (1993) to examine individual variations within accession. Report on cassava germplasm characterization is also available using RAPD and SSR primed PCR markers for assessing genetic diversity (Carvalho *et al.*, 2000). RFLP technique was successfully employed in genetic diversity studies of an *in vitro* germplasm collection of African cassava clones (Beeching *et al.*, 1995). DNA markers are also employed in cassava for multilocus finger printing by Bertram (1993).

Phylogenetic studies

Recently genetic variation at microsatellite loci was used for investigation on evolutionary and geographical origin of cassava from its wild relatives (Olsen and Schaal, 1999). RFLP analysis of chloroplast DNA and nuclear ribosomal (rDNA) sequences using heterologous probes were also applied to assess phylogenetical relationship among *Manihot* species. Evidence from molecular analysis points towards possible origin of cassava from Brazilian wild species *viz. M.tristis* and *M.esculenta* sub sp. *flabellifolia*. Conclusive evidence of origin of cassava from *M.esculenta* sub sp *flabellifolia* came from phylogenetic study based on sequencing single copy nuclear gene glyceraldehydes 3 phosphate dehydrogenase, *G3pdh*. DNA fingerprinting was carried out using M 13 (cla 1-B sml fragment) and cassava genomic probes, improving discriminatory resolution among suspected duplicates (Ocampo *et al.* 1994). Amplified Length Fragment Polymorphism (AFLP) was used to study phylogeny of cassava and its taxonomic relationship with other species of genus *Manihot*. AFLP evaluation was done with six representative samples of crop's diversity ('core collection') and six wild taxa to estimate genetic relatedness within genus. Studies helped in uncovering unique genetic diversity in *M. esculenta* subsp. *flabellifolia* and *M. esculenta* subsp. *peruviana* and their genetic similarity to cassava supporting the hypothesis that these two species might be ancestors of cassava.

Marker assisted breeding

Over 500 microsatellite (SSR) markers are developed in cassava so far. Many of these are placed on framework map. Markers are used to obtain quantitative assessment of genetic similarity in cassava (Beeching *et al.*, 1995, Second *et al.*, 1997, Elias *et al.*, 2000) and to study genetic nature of germplasm resistant to disease (Fregene *et al.*, 1997, 2000, 2001, Mba *et al.*, 2001).

Resistance breeding using molecular marker

Predominant source of resistance to the cassava mosaic disease is polygenic requiring evaluation in multiple environments to characterize resistant genotypes, which make detection of genes for resistance using segregation analysis inefficient. Bulked segregant analysis (BSA) was used to quickly identify a simple sequence repeat (SSR) marker linked to CMD resistance gene. The marker, SSRY28, identified is located on linkage group R of male parent derived molecular genetic map. Gene designated as *CMD2*, is flanked by SSR, RFLP marker GY1 at 9 and 8 cM, respectively. This is the first report of qualitative virus resistance in cassava and of molecular markers that tag CMD resistance in cassava (Akano *et al.*, 2002). Recently, some Nigerian landraces are identified which exhibit high levels of resistance to CMD. Based on work of Akano *et al.*, (2002), SSR markers SSRY 28 and NS158 are the closest markers to gene *CMD2*, which confers high level of resistance to CMD. Fine-mapping was carried out using a population of 1690 individuals from a cross between TME3 and TMS30572. Population was evaluated with above mentioned SSR markers along with other AFLP and RAPD markers. Two polymorphic RAPD markers were identified. One, RME-1, is 1cM from the gene, it was cloned and converted into a sequence characterized amplified region (SCAR) marker. A bacterial artificial chromosome (BAC) library with more than 10x coverage for cassava genome was constructed from cassava variety TME3, donor parent of *CMD2*. Successive hybridizations of BAC end markers to BAC library will be carried out until markers that flank the gene on both sides are identified (Lopez *et al.*, 2003).

Lokko *et al.* (2005) studied molecular markers associated with resistance to CMD in a resistant Nigerian landrace, using F_1 progenies derived from a cross between TME7 x TMS30555. Quantitative Trait Loci (QTLs) associated with resistance to CMD were identified using F_1 progeny derived from a cross between CMD resistant (TMS30572) and susceptible landrace (TME117). Seven highly significant marker-associated (SSRY6, 7, 21^a, 21^b, 42, 77, 324) with QTL effects, explaining between 7.31 and 12.15 %of total phenotypic variation, were detected by regression and Kruskal Wallis analysis. Significant marker trait associations were due to markers donated by both parents, which confirms the polygenic and recessive nature of this source of resistance (Lokko *et al.*, 2004). Analysis further showed that another SSR marker, SSRY106 and an AFLP marker, E-ACC/M-CTC-225, accounted for 35.59 and 22.5 % of the total phenotypic variation for resistance respectively.

Productivity and plant architecture

Okogbenin and Fregene (2003) attempted to identify quantitative trait loci for several productivity and plant architecture traits in a full sib progeny of 144 individuals from two non inbred parents in cassava. A molecular genetic map of this cross constructed previously with over 250 markers was the source of molecular markers. A total of 30 primary QTLs and 84 secondary QTLs were detected. All QTLs identified for fresh shoot yield were associated with either component traits of productivity or architecture traits. Hence QTLs for plant architecture can be used to improve productivity . An exhaustive search and analysis of QTLs controlling architecture is required for initiating marker assisted selection for productivity in cassava.

Early bulking is another important trait evaluated in F_1 mapping progeny. QTL analysis was done for traits significantly linked to early bulking. Three QTLs were found for dry foliage weight that explains 25-33% of phenotypic variance while five QTLs were found for harvest index that explains 18-27% of phenotypic variance. Based on marker analysis of early bulking lines, a selection criteria involving molecular markers for fresh root yield and harvest will be useful.

Apomixis

Microsatellites (simple sequence repeats) are powerful molecular markers used to understand detailed patterns of parentage composition and for individual discrimination in clone identification in cassava. Using microsatellite markers, genetic facts with respect of apomixis inheritance in cassava was studied by Nassar and Collevatti (2005). Results revealed that apomixis is determined by recessive alleles and also it is controlled by more than one pair of genes, probably located in different chromosomes. One of the most important consequences is possibility of mapping and characterising genes of apomixis by molecular markers and isolating it in future.

1.6 Transgenics in Cassava

Transgenic technology offers an excellent tool for transferring a single or even quantitative traits without problems of linkage encountered in traditional breeding programmes. Transformation technologies are utilized to produce genetically transformed cassava expressing traits of agronomical importance. Important areas for cassava genetic engineering includes yield and root quality improvement, pest and insect resistance and production of noble compounds for value added products in cassava (NWeke *et al.,* 2002).

A range of selectable and visual markers were tested and developed for use in cassava transformation systems. Several laboratories use *npt II* gene that imparts resistance to amyloglucoside antibiotics, routinely with selection for successful transformation events possible by culturing tissues on medium containing kanamycin, paramomycin or geneticin. Efficacy of hygromycin resistance gene(*hpt*) was also studied, but is reported to suppress frequency of plant recovery compared to transgenic tissues selected by expression of *nptII*. Use of *bar* gene to impart resistance to herbicide phosphinothricin acetyltransferase is utilized both as a gene of agronomic interest and as a selection system. Use of the *E.coli phosphomannose isomerase (pmi)* gene was used to transform FEC and cotyledon explants and procedures for effective recovery of transgenic tissues and plants using positive selection were developed. In addition, different visual marker systems like *GUS, luciferase* and *GFP* (Green fluorescent protein) were used routinely for production of transgenic cassava (Taylor *et al.,* 2001, 2004). Summary of transgenic research in cassava is given in Table 1.

Cassava mosaic resistance

The most important disease in cassava is cassava mosaic disease (CMD) caused by Gemini group of virus and spread by white fly, *Bemisia tabaci.* The CMD is a potential threat to cassava production causing yield loss ranging from 20 to 80%. The virus has two covalently closed circular single stranded DNA (ssDNA) molecules known as DNA A and DNA B (Stanley, 1983). DNA A is required for replication of both strands and proliferation and together with DNA B for vector transmission and spread. Both components are needed for systemic infection of susceptible plants. DNA B is involved in cell to cell and long distance viral spread. For control of Cassava mosaic disease,

Table 1. Stable integration of foreign genes in transformed cassava tissues and plants

Target tissue	Transfer method	Gene(s) inserted	Transgenic product	Reference
Leaf, stem, somatic embryos	A	*npt*II, *bar*, *uid*A	Callus	Calderon 1988
Somatic embryos	A	*npt*II, *uid*A	Chimeric embryos	Chavarriaga *et al.* 1993
Leaves	A	*npt*II, *bar*, *uid*A, ACMV-CP, CsMV-CP	Callus	Schopke *et al.* 1993
Embryogenic suspensions	P	*npt*II, *uid*A	Transgenic Plants	Schopke *et al.* 1996
Embryogenic suspensions	P	*bar*, *luc*, *npt*II,	Transgenic Plants	Raemakers *et al.* 1996
Cotyledons from somatic embryos	A	*npt*II, *uid*A, *hpt*	Transgenic Plants	Li *et al.* 1996
Embryogenic suspensions	A	*npt*II, *uid*A	Transgenic Plants	Gonzalez *et al.* 1998
Embryogenic suspensions	P	*Luc, par, cassava AGPase B antisense gene*	Transgenic Plants	Munyikwa *et al.* 1998
Cotyledons from somatic embryos	A	*npt*II, *bar*, *uid*A	Transgenic Plants	Sarria *et al.* 2000
Embryogenic suspensions	P	*npt*II, CsMV-CP	Transgenic Plants	Schopke *et al.* 2000
Embryogenic suspensions protoplasts	E	*Pat, luc*	Transgenic Plants	Raemakers *et al.* 2000
Embryogenic suspensions	A	*ASP1,hpt, uidA*	Transgenic Plants	Zhang *et al.*, 2000, 2003
Germinated somatic embryos	A	*CYP79D1/ CYP79D2, nptII*	Transgenic Plants	Siritunga and Sayre, 2003

A : *Agrobacterium*-mediated transformation, E: electroporation, P: Particle bombardment, *ACMV-CP* : African cassava mosaic virus – coat protein, *AGPase B* : ADP glucose phosphorylase B, *bar*, *pat*: Phosphinothricin acetyl transferase, *CsMV-CP* : Cassava common vein mosaic virus coat protein, *hpt* : *hygromycin phosphotransferase*, *luc* : luciferase, *nptII* : neomycin phospho transferase, *uid A*: β-glucuronidase.

integration of a defective interfering (DL) sequence derived from a deleted B component of a Kenyan strain of African cassava mosaic virus (ACMV) could impart elevated resistance to ACMV (Frischumuth and Stanley, 1991). Transgenic plants were recovered from West Africa cv. 60444 containing integrated copies of a complete DI dimer from ACMV – Kenya (Taylor *et al.*,2003). Efficacy of resistance induced by integration of AC1 gene from ACMV-Kenya was significantly more effective (Verdageuur *et al.*, 1996). Increased ACMV resistance in transgenic cassava was developed using antisense RNA technology at ETH, Zurich. Schopke *et al*(2000) used different constructs carrying AC1 gene for genetic modification of cassava. Three constructs were designed for expression of viral antisense RNA's as 3'untranslated regions (3'UTR) of a selectable hygromycin phosphotransferase (*hpt*) gene driven by *CaMV 35S* promoter. In another approach, a hypersensitive reaction is mimicked by using bacterial barnase and barstar genes from *Bacillus amyloliquefaciens* under control of ACMV DNA, a bi-directional promoter. Results show that it is possible to reduce viral replication by 37-99% through transgenic approach and hence it offers genetic modification of popular varieties into resistant ones (Schopke *et al.,* 1998; Schreuder *et al.,* 2001).

A programme for transgenic control of stem borer and horn worm was initiated at CIAT and transgenics were produced in which *cry 1Ab* genes under control of CaMV 35S promoter were integrated into economically important Columbian cassava cultivars.

Reduced cyanogen content

Two approaches were taken to produce cassava with reduced cyanogen content. In the first approach, genes encoding a small (CYPDI and CYP79D2) family of cytochrome P450s that catalyze first dedicated step in linamarin and lotaustralin synthesis were expressed in an antisense orientation in transgenic cassava lines of cultivar Col215 that showed reduced cyanogen content (34-94%). Second approach is by over expressing HNL (Hydroxy nitrile lyase) that helps in detoxification (Siritunga and Sayre,2003).

Herbicide resistance

Transgenic plants of cassava cultivar Per183, expressing *bar* gene that encodes phosphinothricin (*ppt*) and provides resistance to herbicide basta were regenerated at CIAT(Sarria *et al.*,2000).

Starch quality

Work is also carried out to produce transgenic cassava with altered starch quantity and quality. cDNAs encoding ADP glucose pyrophosphorylase, branching enzymes (BEI,BEII) and granule bound starch synthase (*GBSS*) involved in starch biosynthetic pathway were cloned from cassava. Researchers at Wageningen University, Netherlands cloned *gbss* gene from cassava in antisense orientation under control of *CaMV35S* promoter and *gbss* promoter from potato. A total of 560 transgenic plants with *gbss* antisense gene were recovered, two of which subsequently produced and stored amylase free starch that enabled production of transparent starch with stable viscosity for industrial application (Raemakers *et al.,* 2003). Same work is being carried out successfully at CIAT also. Researchers at Ohio University were successful in upregulating overall starch biosynthesis by expressing a modified version of *E.coli glgC* gene which encodes *AGPase* which is the rate limiting step in the cassava starch biosynthesis, under the control of the tuber specific patatin gene from potato(Uzoma *et al.*, 2003). Sixty five % increase in *AGPase* activity was observed in transgenics as compared to control.

Protein content

Researchers at ETH, Switzerland produced transgenic plants of cv.60444 with an artificial storage protein *ASP1* gene designed to be rich in essential amino acids(Zhang *et al.*, 2003). Research is also being undertaken to produce transgenics having high leaf retention by expressing cytokinin regulation ipt gene under control of antisense controlling promoter. Work has been initiated on the production of transgenic cassava resistant to post harvest deterioration. Long term goals of transgenic improvement of cassava include induction of apomixis, bioplastic production , modified plant architecture and rapid bulking of storage roots.

Sweet Potato

Because of biological complexities of sweet potato, sexual hybridization strategies have not been very effective in developing improved cultivars. Therefore, biotechnological tools, such as gene transfer, are very attractive in sweet potato improvement, as they enable direct introduction of desirable genes from other sources into preadapted cultivars (Prakash, 1994).

In vitro conservation

In vitro methods in sweet potato were used earlier at several international centers like Centro International de la Papa, (CIP), Peru, the International Institute of Tropical Agriculture (IITA), Nigeria and AVRDC, Taiwan for conservation of sweet potato collections. Besides, several national programmes in USA, Costa Rica and India are maintaining a large number of *in vitro* collections of sweet potato for last over 10 years. Meristem cultures in conjunction with thermotherapy and use of antiviral drugs are effective in virus elimination in sweet potato (Ashmore, 1997). Meristem culture technique coupled with indexing by ELISA and grafting on indicator plants was used to establish healthy cultures, at NBPGR (Mandal and Chandel, 1996). *In vitro* propagation technique is refined at NBPGR, to develop a single medium capable of eliciting desirable response from all varieties and related species. MS medium supplemented with 4% sucrose, 0.2 mg l^{-1} kinetin and 0.1 mg l^{-1} IAA is used to establish cultures of 260 sweet potato accessions maintained at NBPGR. For medium term storage of sweet potato, various methods are used successfully. These include incubation at low temperature (16-20°C) (Ng and Hahn, 1985), addition of mannitol to medium coupled with reduction in incubation temperature and use of sucrose (3%) with 3% mannitol (Dodds and Roberts, 1985). Though low temperature incubation (18°C) is effective in reducing growth rate, maintenance of large collections at this temperature during summers when ambient temperature exceeds 40°C is expensive and risky in view of frequent power breakdowns or equipment failure Among several modifications tested, supplementation of medium with 1-2% mannitol was the most effective (Mandal and Chandel, 1991). Using this method, 260 accessions are maintained at culture room temperature of 25±2°C for ten years with an annual subculture. While genotypic variation in response to mannitol does exist, concentration of 1-2% was suitable for all accessions.

Application of cryopreservation technology for sweet potato conservation has not been satisfactory. Use of somatic embryos or embryogenic tissues seems to hold more promise since recent studies (Bhatti *et al.*, 1997; Blackesley *et al.*, 1997) demonstrated recovery of complete plantlets from Liquid Nitrogen exposed embryogenic tissues.

Molecular fingerprinting

Sweet potato cultivars were identified using starch gel electrograms of nine enzymes (Kennedy and Thompson, 1991).Genetic polymorphism was identified using DNA amplification fingerprinting (DAF) technique. Seven, out of twenty-eight octamer, screened, were suitable and detected a high degree of variability in sweet potato genotypes. DAF profiles of 70 accessions collected from around the world showed very high variability (Prakash *et al.* 1996). Lakhanpaul *et al.* (1990) also studied polymorphism using Random Amplified Polymorphic DNA (RAPD) analysis.

Dhillon and Ishiki (1999) used random amplified polymorphic DNA (RAPD) markers for fingerprinting and analyzing genetic relationship among 29 *Ipomoea* accessions from different geographical locations around world including unique wild species. Reproducible profiles were obtained for all accessions using random decamer primer. Primers generated 46 polymorphic markers , one primer alone having 10 products enabling discrimination of all 29 accessions studied. Based on RAPD profile, *Ipomoea trifida* was identified as probable progenitor of *Ipomoea batatus*. RAPDs proved to be useful for sweet potato systematics and should be valuable for germplasm management, gene tagging, and for identifying diverse parents for breeding programmes.

Molecular characterization and expression of starch granule bound starch synthase in the sink and a source tissue of sweet potato was studied by Shu-

Jen Wang *et al.*,1999. A cDNA clone for starch granule bound starch synthase (*GBSSI*) which encoded a 67 kDa protein was isolated from a cDNA library prepared from tuberous roots of sweet potato.

Transgenics

Biotechnology is not well developed for sweet potato because of difficulty of establishing an efficient and reproducible system for plant regeneration from cultured tissues or cells. In 1996, Otani and Shimada developed an efficient and variety independent method for production of embryogenic callus from meristem tissues using a medium containing altered plant growth regulators, picloram, dicamba, or 4-flourophenoxyacetic acid (4FA). Otani *et al.*(1998) reported an efficient method for production of transgenic sweet potato plants via *Agrobacterium tumefaciens* mediated transformation using embryogenic calli.Summary of reports on transformation is given in Table 2.

Table 2. Summary of published studies on transformation of Sweet potato

Target tissue	Transfer method	Gene(s) inserted	Transgenic product	Reference
In vitro whole plant	*A.rhizogenus*	Synthetic genes	Transgenic plants	Dodds *et al.* 1991
Leaves and petioles	Particle bombardment	*Gus* A, *npt* II	Transformed calli	Prakash and Varadarajan 1992
Leaves	*A.rhizogenus15384*	*Gus* A, *npt* II	Transgenic plants	Otani *et al.* 1993
Storage roots	*A.tumefaciens* LBA 4404	*Gus* A, *npt* II, cowpea trypsin inhibitor, snowdrop lectin	Transgenic plants	Newell *et al.* 1995
Embryogenic callus	*A.tumefaciens* EHA 101	*Gus* A, *npt* II	Transgenic plants	Gama *et al.* 1996
Embryogenic callus	*A.tumefaciens* EHA 101	*Gus* A, *hpt*	Transgenic plants	Otani *et al.* 1998
Leaves	*A.tumefaciens* C58C1	*cry*IIIA, *npt* II	Transgenic plants	Moran *et al.* 1998
Embryogenic callus	*A.tumefaciens* EHA 101	*GBSS, hpt*	Transgenic plants	Kimura *et al.* 1999
Mesophyll protoplasts	Electroporation	*hpt*, SPFMV-S coat protein gene	Transgenic plants	Okado *et al.* 2001

One of the most serious diseases of sweet potato is russet crack disease caused by sweet potato feathery mottle virus (SPFMV). Okado *et al.* 2001 reported development of transgenic plant having resistance to this dreadful disease. They developed an expression vector carrying coat protein (CP) and hygromycin phosphotransferase (*hpt*) genes driven by cauliflower mosaic virus

35S promoters. Transgenics developed through electroporation had not only primary but also secondary resistance to SPFMV strain (Murata *et al.*, 1997).

Among pests, sweet potato weevil (*Cylas formicarius* ssp) is the major pest worldwide (Palaniswami, 2000). Expression of foreign genes in plants to confer resistance is an appropriate alternative to reduce crop losses. Moran *et al.*(1998) transformed sweet potato plants of variety Jewel cultivar with *cry III A* delta -endotoxin gene from *Bacillus thuringiensis* var. *tenebrionis* using *A.tumfaciens* mediated transformation.

Development of new cultivars that have a novel physico-chemical property of starch is an important objective of sweet potato breeding. Amylase -free starch with very low amylase content will have a major role in new industrial developments. Full length cDNA and gene for sweet potato granule bound starch synthase I (GBSSI) were isolated and characterized (Shu-Jen Wang *et al.*,1999; Kimura *et al.*, 1999). This full length cDNA for sweet potato *GBSS I* driven by the CaMV 35S promoter was introduced into sweet potato by *Agrobacterium tumefaciens* mediated transformation . Transgenic plant obtained showed absence of amylase in tuberous root as determined by iodine calorific method.

Yams

Yam (*Dioscorea* spp.) is an important crop for sustainable food production in tropical and subtropical regions. *Dioscorea* species probably originated from tropical areas of three separate continents: Africa (*D. cayenensis–D. rotundata* complex), South-East Asia and South Pacific (*D. alata*) and South America (*D. trifida*) (Coursey 1976). Of 600 known species of yams, only six are edible or food yams. *Dioscorea rotundata* and *D. cayenensis* (both known as Guinea yam) are the most popular and economically important yams in West and Central Africa where they are indigenous, while *D. alata* (referred to as water or greater yam) is the most widely distributed species globally. Greater Yam is the main *Dioscorea* species grown in India. Yam production increased steadily in the last decade, from 18 million metric tonnes in 1990 to recent estimates of over 39 million (FAO, 2002).

Yams are characterized by vegetative propagation, dioecious nature, rare and erratic flowering leading to non synchrony of sexes and prevalence of polyploidy. Early testing of progenies from hybridization was complicated by yam's slow multiplication rate and juvenile phase during which it is unreliable to use conventional screens for some economic traits. Progress was made in recent years in germplasm characterization and development of molecular markers for genome analysis. Tissue culture protocols are being employed for medium and long term conservation of germplasm and for elimination of pathogen from planting material.

In vitro conservation

Being a vegetatively propagated crop, yam is seriously affected by an accumulation of pathogens. To overcome problems of field maintenance, IITA, and IPGRI recommended use of *in vitro* techniques for yam conservation (Hanson, 1986). In yams, two kinds of *in vitro* germplasm preservation were considered : slow growth condition for medium term preservation and cryopreservation using encapsulation/dehydration technique for long term conservation.

Many species of yams are successfully established under *in vitro* conditions. Axillary buds and nodal cultures are used most frequently. In India, work related to micropropagation of yams was carried out mainly at NBPGR and CTCRI (Mandal and Chandel 1996; Nair and Chandra Babu 1996).

Using nodal cuttings as explants, *in vitro* tuber development was recorded in several species (Ng and Ng, 1997; Mandal, 1999). Nodal segments with one or two axillary buds were cultured on MS medium alone or supplemented with kinetin or NAA @ 0.25 mg l^{-1} each. For routine sub culturing, a single medium *i.e.* MS supplemented with 0.15 mg l^{-1} NAA, is applicable to all accessions maintained in repository.

Incubation temperatures play a significant role in growth rate of yam plantlets. The storage period could be extended to 1-2 years by incubating at 18-20°C. Use of mannitol and high level of sucrose either alone or in combination, has significant effect on plantlet growth rate at culture-room temperature. Addition of mannitol to medium can prolong storage at normal culture room temperature of 25-30°C for a period up to 12 months with high recovery rates (80%). Use of 2.0-mg l^{-1} BAP or kinetin maintained cultures of *D.alata* for two years. Induction of micro tubers increased shelf life by an additional 3-6 months. At CTCRI, *D. alata, D. rotundata* and *D. esculenta* nodal cultures could be conserved up to 12 months without subculture on MS basal medium containing 3% mannitol and no growth regulators (Unnikrishnan *et al.,* 2002).

Cryopreservation of yam shoot tips of two cultivated edible species (*Dioscorea alata, D. bulbifera*), a medicinal yam (*D. floribunda*) and one wild edible type (*D.wallichii*) were successfully carried out using technique of encapsulation-dehydration (Mandal and Chandel, 1996). Though, survival and growth initiation from cryopreserved shoot-tips was recorded at a frequency of 28-71 %, regenerations of complete plantlets were recorded at a frequency of 21-37 % in *D. alata* and *D. wallichii* (Mandal *et al.*, 1996). Malaurie *et al.* (1998) reported successful cryopreservation of shoot tips *of D. alata* and *D. bulbifera* using technique of encapsulation - dehydration. Success of cryopreservation of shoot-tips/somatic embryos of several edible and medicinal yams using both vitrification and encapsulation-dehydration was registered. These include shoot-tips of *D. rotundata* by vitrification (Ahuja *et al.*, 2000), *D. floribunda* (Mandal, 2000), and *D. pentaphylla* (Ahuja *et al.* 2000), *D. deltoidea* (Mandal and Dixit, 2000) and somatic embryos of *D.bulbifera* (Dixit *et al.* 2000). Regenerated plants obtained from cryopreserved shoot-tips/somatic embryos were transferred to field with 90-100 % establishment. Field grown plants obtained from cryopreserved explants were morphologically similar to those of their mother stocks maintained *in vitro* conditions and transferred to field.

Molecular Breeding

Advent of molecular markers, genome studies, and plant genetic transformation opened avenues for circumventing breeding obstacles in yams. A better understanding of available genetic diversity and breeding potential of specific accessions is important for choice of parents for use in breeding programs. Molecular markers that are linked to genes controlling these economic traits would be useful in selection at an early stage of plant's growth, thereby enhancing speed and efficiency of selection .

Polymorphism was studied in different species of yams using isoelectric focussing of soluble peroxidase and acid phosphatase isozymes (Twyford, 1990). Isozyme pattern studies were useful in characterization of varieties of *D. cayenesis- rotundata* complex (Hamon and Toure, 1990). Zoundjihekpon *et al.* (1994) used isozyme markers *viz.* esterase, isocitrate dehydrogenase, malate dehydrogenase, 6-phosphogluconate dehydrogenase, shikimate dehydrogenase, and phosphoglucoisomerase to check progenies and detect outbreeding in yams. While isozymes are valuable tools for genetic discrimination of yam cultivars (Dansi *et al.*, 2000), polymorphic isozyme systems are often inadequate for unequivocal identification of individual cultivars within broad cultivar groups or morphotypes in yams. Co-dominant molecular markers were used (i) to study relationships between African cultivated species *D. rotundata* and its assumed ancestors *Dioscorea praehensilis* and *Dioscorea abyssinica,* (ii) to study relationship between cultivated species *D. alata* and related species *Dioscorea nummularia* from South–East Asia and Oceania, and (iii) to genetically identify cultivars in three diversification areas.

Several microsatellite markers were identified and characterized from wild yams also. Terauchi and Konuma (1994) characterized six microsatellite loci in *Dioscorea tokoro,* a wild yam species in East Asia.

Hochu *et al.*(2006) developed microsatellite markers in American yam (*Dioscorea trifida*). A microsatellite sequence-enriched genomic library was screened, and after sequencing, primers were designed for 20 microsatellites. Among these, eight primer pairs yielded amplification products that were both interpretable and polymorphic in 24 yam cultivars. Alleles / locus ranged from two to 13 and overall expected heterozygosity was around 0.5. Six of eight *Dioscorea trifida* microsatellite loci gave amplification products in other *Dioscorea* species. Large cross-species applicability indicates that primers will be useful for other studies within *Dioscorea* genus in future.

Tostain *et al.* (2006) developed new simple sequence repeat (SSR) markers in different species of yam (*Dioscorea* sp.). A microsatellite-enriched bank was created from *Dioscorea alata, Dioscorea abyssinica* and *Dioscorea praehensilis.* Sixteen polymorphic loci were characterized. Several of these markers are transferable to species of other *Dioscorea* sections.

Molecular marker Assessment of Genetic diversity

Molecular markers are now used to elucidate genetic architecture of yams. Earlier cultivated yam species white yam, *D. rotundata* (2n=40) was considered by most authors as tetraploid species with a basic chromosome number of ten. Using two isozyme loci and six microsatellite markers Scarcelli *et al.* (2005) conducted genetic analysis of progeny derived from a self fertilizing monoecious plant of *D. rotundata*. They found that white yam is a diploid species with 2n=20.

Similarly based on Microsatellite segregation analysis and cytogenetic evidence Mustapha *et al.* (2006) reported tetrasomic inheritance in American yam *Dioscorea trifida* and a new basic chromosome number 10 or 9 in *Dioscoreae*.

Impact of traditional farmers' management on genetic diversity of vegetatively propagated crops was also studied using microsatellites (Scarcelli *et al.*, 2006). They used 11 microsatellite markers on yam tubers and demonstrated that wild × cultivated hybrids are spontaneously formed.

Relative efficiency of AFLP, RAPD and SSR markers for assessment of genetic relationships, as well as for cultivar identification and discrimination was studied by Mignouna *et al.*,(2003)among 45 West and Central African white yam cultivars belonging to 22 morphotype/cultivar groups. Results showed that each of three marker-systems could unequivocally identify each cultivar, but that marker-systems differed in mean number of profiles generated / primer (or primer combination) / cultivar, referred to as genotype index (GI). Order of merit based on this criterion was AFLP (GI =1.0), SSR (GI=

0.39) and RAPD (GI =0.35). Thus AFLP showed the highest efficiency in detecting polymorphism and revealed genetic relationships that most closely reflected morphological classification. Mignouna *et al.* (1998) previously showed that AFLP is a sensitive and robust technique for yam genome fingerprinting. In spite of their robustness, technical and financial burdens associated with AFLP and SSRs limit their availability to laboratories in developing countries where most of research on yam is conducted. Thus cost considerations may make RAPD the method of choice for laboratories with rather limited resources.

Genome Mapping in Yams

Ability to use molecular marker biotechnology to exploit information at gene level, based on genome mapping and marker-assisted genetic analysis and breeding, now offers a practical solution to many of intractable problems in yam breeding. Molecular genetic maps and marker-aided analysis of complex traits can be used to elucidate genetic control of yield potential and tuber quality, and locate genes of pest and disease resistance, nutrient use efficiency, and flowering (Mignouna *et al.,* 2003).Yam genome mapping was initiated in mid-1990s. Lack of mapping populations and adequate polymorphic molecular markers hampered rapid progress. By end of 2001, considerable progress was made in developing appropriate molecular tools and mapping populations for developing linkage maps of two most important species, *D. alata* and *D. rotundata*

Mignouna *et al.* (2002a) also constructed a genetic linkage map of white yam (*D. rotundata* Poir.), based on 341 co-dominantly scored amplified fragment length polymorphism (AFLP) markers found segregating in an intraspecific F1 cross. Mapping population was produced by crossing a landrace TDr 93-1 as female parent to a breeding line TDr 87/00211 as male parent. Marker segregation data were spilt into maternal and paternal data sets and separate genetic linkage maps were constructed since mapping population was an F1 cross between two presumed heterozygous parents. Markers segregate like a diploid cross pollinated population suggesting that D.rotundata genome is an allotetraploid (2n=4x= 40). They also constructed a genetic linkage map of greater yam (D. *alata* L), based on 469 co-dominantly scored amplified fragment length polymorphism (AFLP) markers segregating in an intraspecific F1 cross. Mapping population was produced by crossing two improved breeding lines TDa 95/00328 as female parent to a breeding line TDa 87/01091 as male parent. The 469 markers were mapped on 20 linkage groups with a total map length of 1.233 cM and a mean marker spacing of 2.62cM. In greater yam also, markers segregated like a diploid cross pollinated population suggesting that *D.alata* genome is an allotetraploid (2n=4x= 40) . This map covered 65% of yam genome and was the first linkage map reported in *D.alata*. Map provided a tool for further genetic analysis of traits of agronomic importance and for using marker assisted selection in *D.alata* breeding programmes. QTL mapping opened new avenues for accumulating anthracnose resistance genes in preferred *D.alata* cultivars.

Markers for agronomic traits

Selective genotyping coupled with DNA sample pooling proved very efficient for initial identification of disease resistance-linked markers. Approach was successfully applied in yams for identification of Yam Mosaic Virus (YMV) and anthracnose resistance genes (Mignouna *et al.*, 2002 a, b).

Yam Mosaic Virus is a major limiting factor for stable production of yams, and D. rotundata is particularly susceptible to the virus Genetic resistance is the only economically viable method of control. A study of genetic control of YMV resistance in three *D. rotundata* cultivars to a Nigerian isolate of YMV showed that resistance is manifested differentially as the action of a single dominant gene in simplex condition or a major recessive gene in duplex condition (Mignouna *et al.*, 2001b). Dominant locus that contributes to YMV resistance was tentatively named Ymv-1, until tests of allelism are conducted. Two RAPD markers OPW18850 and OPX15850 closely linked in coupling phase with dominant YMV-resistance locus Ymv-1 were identified. These markers successfully identified resistance gene in resistant genotypes among a sample of 12 *D. rotundata* varieties. Based on map constructed in white yam, Mignouna *et al.*(2001b) also identified three QTL's on maternal linkage map and one on paternal linkage map were linked to yam mosaic virus resistance

Anthracnose disease, caused by *Colletotrichum gloeosporioides* is a major constraint to production of yam worldwide with *D. alata*, the most widely distributed species, being particularly susceptible to the disease. Resistant yam cultivars could form basis of sustainable management strategies for anthracnose. Initial genetic inheritance studies showed that resistance to yam anthracnose in *D. alata* is dominantly but quantitatively inherited and controlled by gene designated Dcg-1, (Mignouna *et al.*, 2001a). Two RAPD markers, OPI171700 and OPE6950, closely linked in coupling phase with Dcg-1 were identified. Both markers successfully identified *Dcg1* in resistant D. alata genotypes among 34 breeding lines, indicating their potential use in marker-assisted selection. Recently QTL mapping revealed one AFLP marker E-14/M52-307 located on linkage group 2 was associated with anthracnose disease resistance explaining 100 % of phenotypic variance (Mignouna *et al.*,

2003). Further AFLP mapping is needed to identify additional QTLs and strengthen existing marker-QTL linkages. Candidate gene analyses are yet to be employed to investigate a variety of traits. Initial QTL studies indicated that pyramiding of major resistance genes may be a valuable strategy for YMV and anthracnose resistance breeding.

Aroids

Aroids like taro (*Colocasia esculenta*) and elephant foot yam (*Amorphophallus paeoniifolius*) are used as vegetables in India and south east Asian countries. Taro starch granules are very small in size and is ideal for preparation of baby food, and cosmetics. *Amorphophallus* spp are traditionally used in ayurvedic preparations against piles and dysentery.

In vitro technology in aroids is well developed and is being used routinely in national and international centers for conservation of these crops (Staritsky *et al.* 1986; Bessembinder *et al.* 1993). There are several reports on production of virus-free taro. Shoot tips (meristems dome plus one or two-leaf primordial) 0.2 to 0.3 mm in size were cultured on MS medium with 0.1 mg l^{-1} NAA at 25°C under 12 h day length (Morishita and Yamada, 1978). These meristems regenerated to shoots and produced roots after four months of culture. Successful cormel tip culture of *Colocasia esculenta* leading to elimination of Dasheen mosaic virus was done at CTCRI (Nair and Chandara Babu, 1996). In *Xanthosoma sagittifolium*, addition of spermine, arginine and ornithine enhanced plant regeneration from shoot tip cultures (Sabapathy and Nair, 1995). In India work on medium term *in vitro* conservation was carried out at CTCRI and NBPGR. Taro cultures could be stored upto ten months under light (3000 lux) on MS basal medium containing sucrose and mannitol (3 %) without vitamins or growth regulators. Half strength basal medium also induced slow growth (Unnikrishnan and Sheela, 2000). *Colocasia* and *Alocasia* cultures could be conserved at normal culture room temperature of 25°C±2°C upto 10 months without subculturing (Mandal, 1999). For long-term conservation, slow freezing of embryogenic callus prior to cryopreservation resulted in 70 % survival, with no loss of regenerability (Takagi *et al.*, 1997). Vitrification of axillary buds, which initially met with only partial success, was modified for cryopreservatioon of *in vitro* shoot tips resulting in almost 80 % shoot recovery (Takagi *et al.*, 1997; Thinh *et al.*, 2000). Shoot tips were precultured for 16 h on 0.3 M sucrose-containing medium loaded with mixture of 2M glycerol and 0.4 M sucrose for 20 min. at 25°C and treated with PVS_2 (Plant Vitrification solution- 2) for 20 min at 25°C prior to immersion in LN.

Limitations in flowering, lack of synchrony, pollen sterility and protogyny are main constraints in utilization of taro in breeding program. Molecular techniques are used routinely for fingerprinting of taro germplasm . Also efforts are made to characterize genes controlling important traits especially protein.

Electrophoretic patterns of peroxidase and esterase were studied for identifying genetic diversity in different varieties of *Colocasia esculenta* and *C. gigantia*. Cluster analysis, based on 17 morphological characters and isoenzyme patterns, was done besides studying chromosome number. Genetic diversity was analysed in taro germplasm using isozyme and RAPD markers. Moderate to high polymorphism was observed in the isozyme patterns of enzymes acid phosphatase, peroxidase and SOD, while monomorphism with shikimate dehydrogenase and MDH. Amplification of genomic DNA using random decamer primers revealed high polymorphism as only one of the 22 primers analyzed, gave monomorphic patterns. Cluster analysis of allozymes and RAPD data also revealed prevalence of high genetic diversity among the morphotypes studied (Lakhanpaul *et al.*, 1990, Noyer *et al.*, 2003).).

Limited studies were made to unravel expression genes associated with protein and corm development in Taro. Hirai *et al.* (1993) isolatd cDNAs encoding for storage proteins in tubers of taro (*Colocasia esculenta* Schott). Two major protein groups of taro (*Colocasia esculenta*) tuber were purified, and their antisera were used for screening of cDNA library constructed from poly(A)+ RNA of taro tuber. Amino acid sequence deduced from nucleotide sequence of this clone indicated that 25 kD proteins were homologous to trypsin inhibitors of soybean and winged bean as well as storage proteins, sporamins of sweet potato.

Bezerra *et al.* (1995) isolated a gene encoding a globulin from a major taro corm protein family, tarin (G1, ca. 28 kDa) was isolated from a lambda Charon 35 library, using a cDNA derived from a highly abundant corm-specific mRNA, as probe. The gene, named tar1, and corresponding cDNA were characterized and it has an open reading frame (ORF) of 765 bp, and deduced amino acid sequence indicated a precursor polypeptide of 255 residues that is post-translationally processed into two subunits of about 12.5 kDa each. Significant similarity was also found at nucleic acid sequence level with genes encoding lectins from plant species of Amaryllidaceae and Lilliaceae families. Luiz *et al.* (1992) studied gene expression pattern that occurs during taro corm development. They found that corm development is associated with differentiation of specialized cells and tissues and that these differentiation events are coupled with temporal and spatial expression of corm specific genes.

Minor Tuber Crops

There are many minor tubers cultivated as vegetable or medicinal crop. Arrow root is grown for extraction of starch which has medicinal value. Research on improvement of minor tubers using biotechnological interventions is meagre and restricted mainly to standardization of *in vitro* regeneration protocols. In Chinese potato (*Solenostemon rotundiifolius*) a widely grown vegetable crop, standardized protocols for *in vitro* conservation and micropropagation are available. Direct regeneration was obtained from lamina explants and internodal cuttings at higher levels of BAP (1 to 10mM). Higher levels of NAA (1 to 10mM) produced callus and indirect regeneration (Nair, 1992). Nodal cultures grew on hormone free MS medium. Nodal explants developed shoot and root directly in MS medium.

Progress achieved so far in fields of transgenic technology and molecular breeding with limited resources will help to increase productivity and utility of tropical tuber crops. There is need to develop/consider approaches such as use of microarrays technology to study expression of genes, development of a large number of ESTs ,isolation and cloning of novel genes, etc. Also crops like sweet potato can be used for molecular farming.

References

Ahuja Sangeeta, Mandal, B. B. and Srivastava, P. S. 2000. Vitrification: A potential cryopreservation technique for long-term conservation of yam germplasm. Abst. *National symposium on Prospects and Potentials of Plant Biotechnology in India in the 21st Century. Jodhpur*, Oct. 18th -21st, 2000. 159 pp.

Akano, A.O., Dixon, A.G.O., Mba, C., Barrera, E. and Fregene, M. 2002. Molecular genetic mapping of resistance to the African cassava mosaic disease. *Theoret. Applied Genetics*, 105 :521-525.

Ashmore, S .E.1997. *Status report on the development and application of in vitro techniques for the conservation and use of plant genetic resources.* IPGRI, Rome.

Bhatti, M. H., Percival, T, Davey, C. D. M, Henshaw, G.G. and Blakesley, D. 1997. Cryopreservation of embryogenic tissue of a range of genotypes of sweet potato (*Ipomoea batatas* (L.) Lam.) using an encapsulation protocol. *Pl. Cell Rep.* 16: 802-806.

Beeching, J. R., Marmey, P., Hughes .M. A. and Charier, A. 1995 . Evaluation of Molecular Approaches for determining Genetic diversity in cassava Germplasm, *Proceeding of the second International Scientific meeting of the cassava Biotechnology Network*, Boyor, Indonesia (working document 150). 2: 80-89.

Bertram, R. B.1993. Application of Molecular Technique to genetic resources of cassava. *(Manihot esculenta* Crantz) (*Euphorbiaceae*) inter specific evolutionary relationship and intra specific characterization Ph.D dissertation University of Meryland, USA 465.

Bessembinder, J. J. E., Stritsky, G. and Zandvoort, E .A.1993. Long term *in vitro* storage of *Colocasia esculenta* (L.) Schott under minimal growth conditions. *Pl. Cell Tiss. Org. Cult.* 33: 121-127.

Bezerra, I.C., Castro, L.A., Neshich,G., de Almeida, E.R, de Sa, M. F, Mello, L. V, Monte-Neshich, D.C. 1995. A corm-specific gene encodes tarin, a major globulin of taro (*Colocasia esculenta* L. Schott). *Plant Mol Biol.* 28(1):137-44.

Blackesley, D., Percival, T., Bhatti, M. H. and Henshaw, G. G. 1997. A simplified protocol for cryopreservation of embryogenic tissue of sweet potato (*Ipomoea batatas* (L.) Lam.) utilizing sucrose preculture only. *Cryo-lett.* 18: 77-80.

Calderon, A. 1988. Transformation of *Manihot esculenta* Crantz (cassava) using *Agrobacterium tumefaciens* and expression of the introduced genes in transformed cell lines. M. Sc. Thesis Vrije Universiteit Brussel, Belgium, 37 pp.

Carvalho, L.J.C.B., Schaal B.A., and Fukuda,W.M.G. 2000. Agronomic DNA (RAPD) marker were used to assess the genetic diversity of cassava *(Manihot esculenta* Crantz*)* in Carlvalho J.C.B, Thro . Thro A.M. and IV Intl sci Meet – CBN. EMBR. APA – Genetic Res. Biotech. Brasilia.- DF Brazil PP – 51- 64.

Chaavariaga P., Scopke, C., Sangare A., Fauquet, C. M., Beachy, R.N. 1993. Transformation of transgenic cassava (*Manihot esculenta* Crantz) embryogenic tissues using *Agrobacterium tumefaciens* (*In:*) Roca, W.M., Thro, A.M. (eds), *Proc. 1st Int. Sci. Meet Cassava Biotechno. Network*, Cartagena, Colombia, 25-28 August 1992. *Centro Internacional de Agricultura Tropical*, Cali, Colombia, pp : 222-228.

Coursey, D.G.1976. The origins and domestication of yams in Africa. (*In:*) *Origin of African Plant Domestication* (eds Harlan JR, De Wet JMJ, Stemler Mouton ABL), pp. 383–408. Mouton, Hague, the Netherlands.

Dansi, A., Mignouna, H. D., Zoundjihékpon, J., Sangaré, A., Asiedu, R., Ahoussou, N . 2000. Using isozyme polymorphism to assess genetic variation within cultivated yams (*Dioscorea cayenensis*/ *Dioscorea rotundata* complex) of the Benin Republic. *Genet. Resourc. Crop Evol.* 47:371-383.

Dhillon, N. P. S. and Ishiki, K. 1999. Genomic variation and genetic relationships in *Ipomoea* spp. *Plant Breeding* 118 : 161-165.

Dixit Sonali, Mandal B B, and Srivastava P S. 2000. Cryopreservation of somatic embryos/ embryogenic tissues: Application in Conservation and Biotechnology. Abst. *(In:) National symposium on Prospects and Potentials of Plant Biotechnology in India in the 21 st Century*. Jodhpur, Oct. 18th -21st 2000. 161 pp.

Dodds, J. H. and Roberts, L.W. 1985. *Experiments in Plant Tissue Culture*, 2nd edn., Cambridge, NY: Cambridge University Press.

Dodds, J.H., Merzdorf, C., Zambrano, V., Siguenas, C . and Jaynes, J. 1991. Potential use of *Agrobacterium* mediated gene transfer to confer insect resistance in sweet potato. In: R K Jansson , K V Raman eds. *Sweet potato Pest Management: A Global Perspective.* Boulder, CO: Westview, 1991, pp 203-219.

Elias, M., Panaud, O. and Robert, T. 2000. Assessment of genetic variability in a traditional cassava (*Manihot esculenta* Crantz) farming system using AFLP markers. *Heredity,* 85 : 219-230.

FAO 2002. FAOSTAT Agriculture data. Food and Agriculture Organisation of the United Nations. http://apps.fao.org/collections.

Frischmuth, T. and Stanley, J. 1991. African cassava mosaic virus DI DNA interferes with the replication of both genomic components. *Virology,* 183:539-534.

Fregene, M., Angel, F., Gomez,R., Rodríguez, F., Chavaariaga, P., Roca,W., Tohme,J., Bonierbale, M. 1997. A molecular genetic map of cassava (*Manihot esculenta* Crantz.). *Theor. Appl. Genet.,* 95: 431-441.

Fregene,M., Bernal,A., Duque,M., Dixon,A. and Tohme,J., Bonierbale, M. 2000. AFLP analysis of African cassava (*Manihot esculenta* Crantz) . germplasm resistant to the cassava mosaic disease (CMD). *Theor. Appl. Genet.,* 100: 678-685.

Fregene, M., Okogbenin,E., Mba, C., Angel, F., Suarez, M. B.,Guitierez, J., Chavaariaga, P., Roca, W., Bonierbale, M. and Tohme,J. 2001. Genome mapping in cassava improvement: challenges, achievements and opportunities. *Euphytica,* 120 : 159-165.

Gama, M.I.C.S., Leite Jr, R.P., Cordeiro, A.R. Cantliffe, D.J.1996. Transgenic sweet potato plant obtained by *Agrobacterium tumefaciens* – mediated transformation. *Plant Cell Tissue Org .Cult.,* 46:237-244.

Gonzalez,A.E., C.Schopke, Taylor,N.J. Beachy R.N and Fauquet CM. 1998. Regeneration of transgenic cassava plants (*Manihot esculenta* Crantz) through *Agrobacterium* – mediated transformation of embryogenic suspension cultures. *Plant Cell Reports,* 17: 827-831.

Gresshoff, P. and Doy, C.1974. Derivation of a haploid cell line from Vitis vinifera and the importance of the stage of meiotic development of the anthers for haploid culture of this and other genera. *Zeitschrift fir Pflanzenphysiologie* 73, 132-141.

Hamon P and Toure B.1990.Charaterisation of traditional yam varieties belonging to the *Dioscorea cayenensis-rotundata* complex by their isozymic pattern. *Euphytica,* 46: 101-107.

Hanson, J.1986. Methods of storing tropical root crop germplasm with special reference to yam. *FAO/IBPGR Pl. Genet. Res. Newsl.,* 64: 24-32.

Hirai, M., Nakamura, K., Imai, T, Sato, T., 1993. cDNAs encoding for storage proteins in the tubers of taro (*Colocasia esculenta* Schott). *Jpn.J. Genet.,* 68 (3) : 229-36.

Hochu, I., Santoni, S. and Bousalem, M. 2006.Isolation, characterization and cross-species amplification of microsatellite DNA loci in the tropical American yam *Dioscorea trifida* . *Molecular Ecology Notes.*, 6 (1) : 137 - 140.

Kartha, K.K., Gamborg, O.L., Constenbel, F., and Shyluk, J.D. 1974 Regeneration of cassava plant from Apical meristem plant. *Sci.Lett.*, 2:107-113.

Kennedy, L. S. and Thompson, P.G.1991.Identification of sweet potato cultivars using isozyme analysis. *Hort. Sci.*, 26: 300-302.

Kimura,T., Otani,M., Noda.T., Ideta, O., Shimada, T., Saito, A. 1999. Decrease of amylose content in transgenic sweet potato. *Breed.Sci.*, 1: 142.

Lakhanpaul, S., Mandal,B. B. and Chandel, K. P. S. 1990. Isozyme studies in the *in vitro* regenerated plants of *Ipomoea batatas* (L.) (Lamarck). *J.Root Crops,* ISRC Nat. Sym. 1990, Special: 305-310.

Li,H.Q., Sautter,C., Potrykus, I. And Pounti Kaerlas J.1996. Genetic transformation of cassava (*Manihot esculenta* Crantz) . *Nature Biotech.*, 14: 736-740.

Lokko, Y., Gedil, M. and Dixon, A. G. O. 2004. QTLs associated with resistance to the cassava mosaic disease: new directions for a diverse plant. In: Proceedings of the 4th International Crop Science Congress, 26th September-1stOctober, 2004, Brisbane, Australia. http://www.cropscience.org.au/icsc2004/poster/3/7/4/1154_lokko.htm.

Lokko, Y., Danquah, E. Y., Offei, S. K., Dixon A. G. O. and Gedil, M. A. 2005. Molecular markers associated with a new source of resistance to the cassava mosaic disease. *African J. Biotech.*, 4: 873-881.

Lopez, C.E., Zulugua, A.P., Cooke, R., Delseny,M., Tohme,J. and Verdier,V.2003. Isolation of resistance gene candidates (RGCs) and characterization of an RGC cluster in cassava. *Mol. Gen. Genomics,* 268 : 658-671.

Luiz,A.B. de Castro, Mauro Carneiro, Damares de C.M. Neshich and Genaro R. de Paiva. 1992. Spatial *and Temporal Gene Expression Pattern Occur during Corm development. The Plant Cell,* 4 : 1549-1559.

Malaurie, B., Trouslot, J., Berthaud, J., Bousalem, M., Pinel, A. and Dubern, J.. Medium term and Long term *in vitro* conservation and safe international exchange of yam *Dioscorea* germplasm. *Electronic Journal of Biotechnology,* 1(3) , ISSN: 0717-3458.

Mandal,B.B. 1999. Conservation biotechnology of endemic and other economically important plant species of India. *(In:)* E.E. Benson (ed.) *Plant Conservation Biotechnology,* Taylor and Francis, London. pp. 211-225.

Mandal, . B. 2000. Cryopreservation of yam apices: comparative study with three different techniques. *In*: F. Engelmann and H. Takagi (*eds.*) *Cryopreservation of tropical plant germplasm: Current research progress and application, JIRCAS proceedings of International workshop* 20th -23rd oct, Japan/ IPGRI, Rome Italy. pp. 233-237.

Mandal,B.B. and Chandel, K.P.S.1991.Utilization of tissue culture technique in preservation of sweet potato germplasm. *J. Root Crops,* 17:291-295.

Mandal, B.B. and Chandel,K.P.S. 1996. Conservation of genetic diversity in sweet potato and yams using *in vitro* strategies. In G.T.Kurup, M.S. Palaniswami, V.P. Potty, G.Padmaja, S. Kabeerathumma and Santha V. Pillai, (*Eds.*) *Tropical Tuber Crops: Problems, Prospects and Future Strategies.* Oxford and IBH, New Delhi p.49-59.

Mandal, B.B .and Sonali Dixit. 2000. Cryopreservation of shoot-tips of *Dioscorea deltoidea* Wall. - An endangered medicinal yam, for long-term conservation. *IPGRI News Letter for Asia, the Pacific and Oceania.* No. 33. Sept.-Dec.2000. pp. 23.

Mandal B B, Chandel K P S and Dwivedi S. 1996. Cryopreservation of yams (*Dioscorea spp.*) shoot apices by encapsulation - dehydration. *Cryo-Letters,* 17: 165-174.

Mba,R.E.C., Stephenson, P., Edwards, K., Melzer, S., Kumbira,J. N., Gulberg, U., Apel, K., Gale, M., Tohme, J. and Fregene, M. 2001. Simple sequence Repeat (SSR) marker suevey of the cassava genome towards an SSR based molecular genetic map of cassava. *Theoret. Applied Genetics,* 102 : 21-31.

Mignouna, H.D., Abang,M. M., Green, K. R., Asiedu, R. 2001a. Inheritance of resistance in water yam (*Dioscorea alata*) to anthracnose (*Colletotrichum gloeosporioides*). *Theor. Appl. Genet.,* 103:52-55.

Mignouna, H.D., Abang, M. M., Fagbemi, S.A. 2003. A comparative assessment of molecular marker assays (AFLP, RAPD and SSR) for white yam (*Dioscorea rotundata* Poir.) germplasm characterisation. *Ann. Appl. Biol.,* 142:269-276.

Mignouna, H.D., Abang, M.M., Onasanya, A., Agindotan, B., Asiedu, R . 2002b. Identification and potential use of RAPD markers linked to *Yam mosaic virus* resistance in white yam (*Dioscorea rotundata* Poir.). *Ann. Appl. Biol.,* 140:163-169.

Mignouna, H.D., Ellis, T.H.N., Knox, M., Asiedu, R. and Ng, S.Y.C. 1998. Analysis of genetic diversity in Guinea yams (*Dioscorea* sp) using AFLP fingerprinting. *Tropical Agriculture,* 75 : 224 – 229.

Mignouna, H.D., Mank, R.A. , Ellis, T.H.N., van den Bosch, Asiedu, R., Ng, S.Y.C. , Peleman, J. 2002a. A genetic linkage map of Guinea yam (*Dioscorea rotundata* Poir.) based on AFLP markers. *Theor. Appl. Genet.,* 105 (3): 716-725.

Mignouna, H.D., Njukeng,P., Abang, M. M, Asiedu, R. 2001b. Inheritance of resistance to Yam Mosaic Virus, genus Potyvirus, in white yam (*Dioscorea rotundata*). *Theor. Appl. Genet.,* 103:1196-1200.

Mignouna,H.D., Abang ,M.A. and Asiedu,R. 2003.Harnessing modern biotechnology for tropical tuber crop improvement: Yam (*Dioscorea* spp.) molecular breeding . *African Journal of Biotechnology,* 2(12) : 478-485.

Moran ,R., Garcia, R., Lopez, A., Zasldua, Z., Mena,J., Garcia, M., Armas ,R., Somonte,D., Rodriguez, J.,Gomez, M. and Primentel, E. 1998. Transgenic sweet potato plant carrying the delta endotoxin gene from *Bacillus thuringiensis var. tenebrionis. Plant Sci.*,139:175-184.

Morishita, M. and Yamada, K.1978. Studies on tissue culture of taro (*Colocasia esculenta* Schott) (1) Inducing plantlets from apical meristem. *Bull. Osaka Agri. Res. Cent.*, 15:9-12.

Munyikwa, T.R.I., Raemakers, C. J. J.M., Schreuder, M., Kok, R., Rozeboom, M., Jacobsen, E., Visser, R.G.F. 1998. Pinpointing towards improved transformation and regeneration of cassava. *Plant Sci.*, 135:87-101.

Murashige, T. and Skoog, F. 1962. A revised medium for rapid growth and bio-assays with tobacco tissue cultures. *Physiologia Plantarum*, 15: 473–97.

Murata,T., Okada, Y., Fukuoka, H., Saito, A., Kumura, T., Mori, M., Nishiguchi, M., Handa, K., Sakai, J. and Fukuoka.H.1997. Transformation by directly gene transfer in sweet potato [*Ipomoea batatus* (L.) Lam.]. Proceedings of International workshop on sweet potato production system towards the 21 st century. Kyusyu National Agricultural Experimental Station. Miyakonojo, Japan, pp 159-180.

Mustapha Bousalem, Gemma Arnau, Isabelle Hochu, Richard Arnolin, VÃ©ronique Viader, Sylvain Santoni and Jacques David 2006. Microsatellite segregation analysis and cytogenetic evidence for tetrasomic inheritance in the American yam *Dioscorea trifida* and a new basic chromosome number in the *Dioscoreae. Theor. Appl. Genet.*, 113 (3): 439-451.

Nagib Nassar A. and Rosane G. Collevatti. 2005. Microsatellite markers confirm high apomixis level in cassava bred clones. *Hereditas*, 142: 33-37.

Nair, N.G. and Chandra Babu, S.1996. *In vitro* production and micropropagation of three species of edible yams. *(In:)* Kurup, G.T. Palaniswami, M.S. Potty, V.P. Padamaja, G. Kabeerathumma, S. and Pillai, S.V. (Eds.). *Tropical Tuber Crops: Problems, prospects and future strategies*, pp.55-60, Oxford and IBH Publishing Co. Pvt. Ltd., New Delhi.

Nair, N.G.1992. Growth and morphogenesis of Chinese potato(*Coleus parviflorus* Benth) in culture. *(In:)* Subha Rao, N.S., Balagopalan, C., Ramakrishna,S.V. (Eds). New trends in Biotechnology, pp 57-62. Oxford & IBF publishing Co. Pvt.Ltd., New Delhi., India.

Newell, C.A., Lowe, J., Merryweather, M.A., Rooke, L.M. and Hamilton, W.D.O. 1995. Transformation of sweet potato [*Ipomoea batatus* (L.) Lam.] with *Agrobacterium tumefaciens* and regeneration of plants expressing cowpea trypsin inhibitor and snowdrop lectin. *Plant Sci.*, 107:215-227.

Ng, S.Y.C and Hahn, S. K.1985. Application of tissue culture to tuber crops at IITA. *(In:) Biotechnology in International Agricultural Research*, pp. 27-40, International Rice Research Institute, Manila, Phillipines.

Ng, S.Y.C. and Ng, N. Q. 1997. Germplasm conservation in food yams (*Dioscorea* spp.): constraints, applications and future prospects. *(In:)* Razdan,M. K and Cocking,

E.C.(Eds.) *Conservation of Plant Genetic Resources in vitro. Vol. 1: General Aspects,* pp. 257-286, Scientific Publishers Inc.,USA.

Noyer, J. L., Billot, C., Weber, A., Brottier, P., Guero-Garcia and Lebot, V. 2003. Genetic Diversity of Taro, assessed by SSR markers. *Proc. of Taro Genetic Symposium,* Nadi, Fiji, May 22-24, 2003.

Nweke, F. I., Spencer, D. D. C. and Lynam, J. K., 2002. The Cassava transformation. pp.101-114. Michigan State University Press, East Lansing.

Ocampo, C., Angel, F., Jim Crez, A., Jaramillo, G., Hershey, C., Granados, E. and Iglesias,C. 1994. DNA fingerprinting to confirm possible genetic duplicates in cassava germplasm. *Proc.* 2nd Intern. Sci. Meeting CBN II. 22-26, August 1994, Bogor, Indonesia, Abst. 44 pp.

Okado,Y., Saito, M., Nishiguchi, M., Kimura, T., Mori, M., Hanada, K., Sakai, J., Miyazaki, Matsudo,Y and Murata,T. 2001. Virus resistance in transgenic sweet potato (*Ipomoea batatus* (L.) Lam) expressing the coat protein gene of sweet potato feathery mottle virus. *Theor. Appl. Genet.,* 103:743-451.

Okogbenin, E. and Fregene, M. 2003. Genetic mapping of QTLs affecting productivity and plant architecture in a full sib cross from non-inbred parents in cassava (*Manihot esculenta* Crantz). *Theor. Appl. Genet.,* 107:1452-1462.

Olsen, K. M., and Schaal, B. A. 1999. Evidence on the origin of cassava : phylogeography of *Manihot esculenta. Proceedings of the National academy of Sci.,*96: 5586-5591.

Otani, M. and Shimada, T.1996. Efficient embryogenic callus formation in sweet potato [*Ipomoea batatus* (L.) Lam.]. *Breed. Sci.* , 46 : 257-260.

Otani, M., Shimada, T., Kimura, T., and Saito, A. 1998. Transgenic plant production from embryogenic callus of sweet potato (*Ipomoea batatus* (L.) Lam.) using *Agrobacterium tumefaciens. Plant Biotechnol.,* 15:11-16, 1998.

Otani,M., Mii, M., Handa, T., Kamada, H. and Shimada T. 1993. Transformation of sweet potato [*Ipomoea batatus* (L.) Lam.] by *Agrobacterium rhizogenes. Plant Sci.,* 94:151-159.

Palaniswami, M.S. 2000. Biological and parabiological approaches for sweet potato weevil management in India. *CIP (SWA) News Letter,* 3(2): 6-9.

Prakash, C.S. 1994. Sweet Potato Biotechnology: Progress and potential. *Biotechnology and Development Monitor,* No. 18 : 1819-22.

Prakash, C. S, Desai, A. P, Ramana Murthy, G., Korsi Dumenyo, C, He G, Zheng, Q., Egnin, M. and Kanyand, M.1996. Biotechnological approaches for improving sweet potato. *(In:)* Kurup G T, Palaniswami M S, Potty V P, Padmaja G, Kabeerathumma S and Pillai S V (Eds.). *Tropical tuber crops: Problems, Prospects and Future Strategies,* pp. 27-36. Oxford and IBH publishing Co. Pvt. Ltd. New Delhi.

Prakash, C. S. and Varadarajan, U. 1992. Genetic transformation of sweet potato by particle bombardment. *Plant Cell Rep.,* 11:53-57.

Raemakers, C. J. J. M., Amati, M., Staritsky, G., Jacobsen, E. and Visser, R.G. F. 1993. Cyclic somatic embryogenesis and plant regeneration in Cassava. *Annals of Botany*, 71: 289–294.

Raemakers, C.J.J.M., Schreuder, M., Munyikwa, T., Jacobsen, E. and Visser, R.G.F. 2000. Towards a routine transformation procedure for cassava. *(In:)* Carvalho, L., Thro, A.M., Vilarinhos, A.D. (eds) Proc. 4th Int. Sci. Meet Cassava Biotechno. Network, Cartagena, Colombia, 3-7 November 1998, Salvador, de Bahia, Brazil, Brazilian Agricultural Research Corporation – EMBRAPA, Brazil, pp 236-243. August 1992. Centro Internacional de Agricultura Tropical, Cali, Colombia. pp. 250-266.

Raemakers, C. J. J. M., Schreuder, M., Pereira, I., Munyikwa, T., Jacobsen, E. and Visser, R.G. F. 2001. Progress made in FEC transformation of cassava. *Euphytica.* 120:15-24.

Raemakers, C.J.J.M., Schreuder,M., Pereira,I., Munyikwa,T., Jacobsen, E. and Visser,R.G.F. 1996. Production of transgenic cassava plants by particle bombardment using luciferase activity as the selection marker. *Mol. Breeding*, 2: 339-349.

Raemakers, C.J.J.M., Sofiari,E., Jacobson, E., and Visser, R.G.F. 1997. Regeneration and transformation of cassava. *Euphytica*, 96: 153-161.

Ramirez, H., Hussain, A., Roca, W. and Bushuk, W. 1987. Isozyme electrophoregrms of sixteen enzymes in five tissues of cassava (*Manihot esculenta* Crantz.) varieties. *Euphytica*, 36: 39-48.

Roca, W. M. 1984. Cassava *(In:)* Sharp ,W.R ., Evans , D. A., Ammirato, P. V., Yumada, Y.(eds).*Handbook of plant cell culture* .vol 2 .crop species Mac Millan New York. 269-301.

Roca, A. G., Maya, M. M., Daque,M. C., Tohme, J., Allem, A.C., and Bonier bala, H. W. 2000. AFLP analysis of relationship among cassava and other Manihot species. *Theor Applic.Genetic.*, 95.741-750.

Sabapathy, S. and Nair, H. 1995. *In vitro* propagation of taro, with spermine, arginine and ornthine: Plantlet regeneration via callus. *Pl. Cell Rep.*, 14: 520-524.

Sarria, R., Torres . E., Angel, F., Chavarriaga, P. and Roca, W.M. 2000. Transgenic plants of cassava (*Manihot esculenta*) with resistance to Basta obtained by *Agrobacterium* mediated transformation. *Plant Cell Rep.*, 19: 339-344.

Scarcelli, N., Daïnou, O., Agbangla, C., Tostain, S. and Pham., J.L. 2005. Segregation patterns of isozyme loci and micro satellite markers show the diploidy of African Yam (Dioscorea rotundata (2n=40). *Theor. Appl. Genet.*, 111 : 226-232.

Scarcelli, N., Tostain, S., Vigouroux, Y., Agbangla, C. , Daïnou, O. and Pham., J.-L. 2006. Farmers' use of wild relative and sexual reproduction in a vegetatively propagated crop. The case of yam in Benin. *Molecular Ecology*, 15 (9) 2421-2431.

Schenk, R.U., and Hildebrandt, A.C. 1972. Medium and techniques for induction and growth of monocotyledonous and dicotyledonous plant cell cultures. *Canada J. Bot.*, 50 : 199-204.

Schopke, C., Chavarriaga, Fauquet, C.M. and Beachy, R.N., 1993. Cassava tissue culture and transformation : improvement of culture media and the effect of different antibiotics on cassava. *(In:)* Roca, W.M., Thro, A.M.(eds), *Proc. Ist Int. Sci. Meet Cassava Biotechno. Network*, Cartagena, Colombia, 25-28 August 1992. *Centro Internacional de Agricultura Tropical*, Cali, Colombia, pp 140-145.

Schopke, C.,T Masona M.V., Taylor, N.J., Carcamo, R., Beachy, R. N., and Fauquet, C. M. 2000. Production and characterization of transgenic cassava plants expressing the coat protein gene of cassava common mosaic virus *(In:)* Carvalho, L., Thro, A.M., Vilarinhos, A.D. (eds) *Proc. 4th Int. Sci. Meet Cassava Biotechno.* Network, Cartagena, Colombia, 3-7 November 1998, Salvador, de Bahia, Brazil, Brazilian Agricultural Research Corporation - EMBRAPA, Brazil, pp 236-243. August 1992. *Centro Internacional de Agricultura Tropical*, Cali, Colombia. pp 140-145.

Schopke, C., Taylor, N. J., Carcamo, R., Konan, N. K., Marmey, P., Henshaw, G.G., Beachy, R.N., and Fauquet, C.M. 1998. Regeneration of transgenic cassava plants (*Manihot esculenta*) from micro bombarded embryogenic suspension cultures. *Nature Biotech.*, 14: 731-735.

Schreuder, M.M., Raemakers, C.J.J.M., Jacobsen, E. and Visser, R.G.F. 2001. Efficient production of transgenic cassava plants by Agrobacterium mediated transformation of cassava (*Manihot esculenta* Crantz). *Euphytica.*, 120: 35-42.

Second, G., Allem, A., Emperaire, L., Ingram, C., Colombo, C., Mandez, R., and Carvalho, L., 1997. AFLP based *Manihot* and cassava numerical taxonomy and genetic structure analysis in progress: implications for dynamic conservation genetic mapping. *African J. Root and Tuber Crops.*, 2 : 196-200.

Shu-Jen Wang, Kai-Wun Yeh and Chia-Yin Tsai. 1999. Molecular characterization and expression of starch granule bound starch synthase in the sink and source tissues of sweet potato. *Physiologia Plantarum* , 106 : 253-261.

Siritunga, D. and Richard T. Sayre. 2003. Generation of cyanogen free transgenic cassava. *Planta.*, 217:367-373.

Sree Rekha, V. R., Asha, V., Unnikrishnan, M., Sheela, M. N., Subhra Chakraborty., Niranjan Chakraborty and Asis Datta, 2005. Response to callusing and somatic embryogenesis in some indigenous cassava varieties in 2,4 D Medium. *J. Root Crops.*, 31: 111-117.

Stamp, A. A. and Henshaw, G.G.1982. Somatic embryogenesis and plant regeneration in cassava. *ZPflanzenphysiol.*, 105: 183-187.

Stamp, J. A., and Henshaw, G. G. 1987. Somatic embryogenesis from clonal leaf tissues of cassava. *Anna. Bot.*, 59: 445-450.

Stanley, J. 1983. Infectivity of cloned Geminivirus genome requires sequences from both DNA s. *Nature*, 305 :643-645.

Staritsky, G., Dekkers, A. J., Lowwaars, N .P. and Zandvoort, E. A.1986. *In vitro* conservation of aroid germplasm at reduced temperature and under osmotic stress. *(In:)* Withers

L A and Anderson P G (Eds.) *Plant Tissue Culture and its Agriculture applications.* Butterworth, London. pp. 277-283.

Szabados, L., Hoyos, R., and Roca, W.M. 1987. *In vitro* somatic embryogenesis and plant regeneration of cassava. Plant Cell Rep., 6: 248-251.

Takagi, H., Tien Tninh, N., Islam, O. M., Senboku, T .and Sakai, A.1997.Cryopreservation of *in vitro*- grown shoot tips of taro (*Colocasia esculenta* (L.)Schott) by vitrification. 1. Investigation on basic conditions of the vitrification procedure. *Pl. Cell Rep.,* 16:594-599.

Taylor, N.J., Edwards, M., Kieman, R.J., Davey, C.D.M., Blakesley, D., and Henshaw, G.G. 1996. Development of friable embryogenic callus and embryogenic suspension systems in cassava (*Manihot esculenta* Crantz.). *Nature Biotechnol.,* 14 : 726-730.

Taylor, N.J., Masona, M.V., Carcamo, R., Ho, T., Schopke, C. and Fauquet, C.M. 2001. Production of embryogenic tissues and regeneration of transgenic plants in cassava (*Manihot esculenta* Crantz) . *Euphytica,* 120: 25-34.

Taylor, N.J., Masona, M.V., Ogbe, F. and Fauquet, C.M. 2003. *(In:)* C.M. Fauquet and N.J. Taylor., (Eds).Cassava: An ancient crop for modern times (Compact disc 3). *Proceedings of the 5th International Meeting of the Cassava Biotechnology Network* (4-9 November, 2001, St.Louis, MO.

Taylor, N.J., Chaavariaga P., Raemakers, K., Siritunga, D., and Zhang, P. 2004. Development and application of transgenic technologies in cassava. *Plant molecular Biology,* 56:671-688.

Terauchi, R. and Konuma, A. 1994.Microsatellite polymorphism in *Dioscorea tokoro,* a wild yam species. *Genome,* 37(5) : 794-801.

Thinh, T. N., Takagi, H. and Sakai, A. 2000. Cryopreservation of *in vitro* grown shoot tips of five vegetatively propagated monocots by vitrification. *In*: Cryopreservation of tropical plant germplasm. *Current Research progress and application* (F. Engelmann and H. Takagi (*Eds.*) JIRCAS/IPGRI, pp.227-232.

Tostain, S., Scarcelli, N., Brottier, P., Marchand, J.-L ., Pham, J.-L. and Noyer, J.-L. 2006. Development of DNA microsatellite markers in tropical yam (*Dioscorea* sp.) .*Molecular Ecology Notes,* 6(1) : 173-175.

Twyford C T. 1990.Charaterization of species and vegetative clones of *Dioscorea* using isoelectric focussing of peroxidase and acid phosphatese isoenzymes. *Trop. Agri.,* 67: 337-341.

Unnikrishnan, M. and Sheela, M. N. 2000. Biotechnology in conservation and improvement of tuber crops. *(In:)* Biotechnology in Horticultural and Plantation Crops : Chadha K L, P N Ravindran and Leela Sahijram(Eds.), Malhotra Publishing House, New Delhi, India.

Unnikrishnan, M., Sheela, M. N., Gayathri, V. and Nair, J.R. 2002. Genotype response variation to micro propagation and effect of retardants on *in vitro* growth in cassava.

CBN-V. *Fifth International Scientific meeting of the cassava Biotechnology Network,* Nov.4-9, 2001. Donald Danforth Plant Science centre, St.Louis - Missouri - USA. Abstract Book eds. N.J.Taylor, F.Ogbe and C.M.Fauquet. S7-33.

Uzoma, E.I., Arias-Garzon, D.I. and Sayre, R.T. 2003. Genetic modification of cassava for enhanced starch production . Harnessing Crop Technologies to Alleviate hunger in Africa. Proceedings of the 6th Biennial Conference of the African Crop Science society, 12-16th October, 2003, Nairobi, Kenya. 65 pp.

Verdaguer, B., de kochko, A., Fux, C.I., Beachy, R.N. and Fauquet, C.M. 1998. Functional organization of the cassava vein mosaic virus (CsMV) promoter. *Plant Mol. Biol.,* 37: 1055-1067.

Zhang,P., Jaynes, J., Potrykus, I., Gruissem, W. and Pounti Kaerlas, J. 2003. Transfer and expression of an artificial storage protein (ASP1) gene in cassava (*Manihot esculenta* Crantz). *Transgenic Res.,* 12: 243-250.

Zhang, P., Phansiri, S. and Pounti Kaerlas, J. 2000. PIG mediated cassava transformation using positive and negative selection. *Plant Cell Rep.,* 19: 1041-1048.

Zoundjihekpon J., Hamon, S. , Tio-Touré B. and Hamon, P. 1994. First controlled progenies checked by isozyme markers in cultivated yams *Dioscorea cayenensis-rotundata. Theor. Appl. Genet.,* 88 (8): 1011-1016.

□□□

14

Bioprocessing for Value Addition

Tropical root and tuber crops are important staples for food security for about a fifth of the world population. Tropical root and tuber crops like cassava, sweet potato, yams and aroids [taro and elephant foot yam] come under tropical root and tuber crops. Sweet potato, yams and colocasia are processed mainly as fresh vegetable, whereas 75% of the cassava in Africa, 75-80% in Asia and 65% in Latin America are processed either as fermented foods and feeds or as food additives such as starch, starch based sweeteners (glucose and fructose syrups), organic acids (acetic, lactic, glutamic and gluconic), mono-sodium glutamate and microbial polysaccharides (xanthan, pullulan, scleroglucan, etc). The fermented foods from cassava are *gari, fufu, lafun, chickwanghe, agbelima, attieke, kivunde* and *peujeum* in Africa, *gaplek* and *putto* in Asia and *'cheese" bread*, and *'coated peanut'* in Latin America. Similarly, sweet potatoes are either consumed as fresh vegetables or processed into starch, starch derived foods (noodles, macaroni, etc) and fermented products such as soy sauce, vinegar, lacto-juices, lacto-pickles and *sochu* (an alcoholic drink produced in China and Japan). The bagasse (residues after starch extraction) of these root crops serves as suitable substrate for production of microbial enzymes, organic acids, growth regulators, vitamins and low- calorific sweeteners in submerged and solid substrate fermentations. In this chapter, bioprocessing of root and tuber crops by microorganisms into fermented food, feed, food additives, and biofuels has been discussed.

Bioprocessing: Foods and Beverages

1. Cassava

Cassava roots are usually processed before consumption and fermented cassava foods constitute a major part of the daily diets of many homes in Nigeria, most parts of West Africa and Latin America (Moorthy and George, 1998; Ray and Ward, 2006). Food fermentation process can be broadly categorized into solid state (without soaking, i.e. *gari*) and submerged (soaking in water, *i.e. fufu*). Fermentation improves palatability, textural quality and upgrades nutritive value of this crop by enrichment with proteins (Ray *et al.*, 2007a) and reduces toxic factors such as cyanogenic glucosides, linamarin and lotaustralin (Lei *et al.*, 1999; Sanni *et al.*, 2002; Mc Feeters, 2004; Kostinek *et al.*, 2005).

1.1 Fermented foods in Africa

Gari

Cassava roots are harvested, peeled, washed and grated into coarsely knit bags. A heavy object is placed on top of the bag to express some of the juice. The bag contents are then left to undergo natural (solid state) fermentation for several days. On *garifying* (the process to make *gari*), the grated cassava is dried to about 10% moisture content and the starch is probably partially dextrinized (Akingbala *et al.*, 1993; Sokari and Karibo, 1996; Osho and Dashiell, 2002). At this stage, a little palm oil is added to give it colour; the final dry granular product is *gari* (Akingbala *et al.*, 2005). Cassava fermentation to *gari* is associated with a community of microorganisms including yeasts (*Saccharomyces cerevisiae* and *Candida* sp.) and bacteria (*Lactobacillus, Leuconostoc, Alcaligenes* and *Corynebacterium*) (Almazan, 1992; Giraud *et al.*, 1993; Sanni, 1994; Sokari and Karibo, 1996; Sanni *et al.*, 2002) (Table 1).

Fufu

Fufu is a fermented wet paste product processed from cassava. For production of *fufu*, the preliminary preparation for cassava fermentation is identical with the method for *gari* production (Sanni *et al.*, 1998) except that it is processed in submerged state. Species of *Lactobacillus, Leuconostoc* and *Streptococcus* are the predominant microorganisms in *fufu* along with *Bacillus subtilis, Klebsiella* and *Candida krusei* (Okolie *et al.*, 1997; Oyewole, 2001). Production of *gari* and *fufu* has been commercialized in Nigeria (Ezedinma, 2006).

Lafun

Lafun is a fine powdery cassava product that is prepared by fermentation and is commonly consumed in the western states of Nigeria (Uzogara *et al.*, 1990). The whole or peeled roots are immersed in a stream, in stationary water, or in an earthenware vessel for 3-4 days and fermented until they become soft. The fermented roots are then taken out and the pulp broken into small crumbs and sun-dried on mats, racks and on house roofs. The dried crumbs are milled into flour. The flour is added into boiling water with constant stirring until a smooth thick paste is formed. The paste is cooled to about 35°C and is then served with soups. The fermented and dried cassava pulp, *lafun*, is similar to *cossettes* in Zaire and Rwanda, *kanyanga* and *mapanga* in Malawi and *makopa* in Tanzania. Microorganisms involved in *lafun* preparartion include species of *Bacillus, Klebsiella, Leuconostoc, Corynebacterium, Candida* and *Lactobacillus* (Treche and Massamba, 1995).

Chickwanghe (kwanga)

Chickwanghe is the most popular processed food form of cassava in Zaire. *Myondo* and *bobolo* in Cameroon, *mboung* in Gabon and *mangbele* in Central African Republic also belong to this group. Cassava roots are peeled, steeped in water and left for 3 to 5 days to ferment until they become soft. Fibres are removed from the pulp, which are heaped on a rack for further fermentation or covered with leaves and pressed using heavy objects to drain off excess liquid. The pulp is then ground on a stone, or pounded in a mortar. The fine pulp is steamed in pots. *Chickwanghe* is a very viscous paste, much thicker than *fufu*.

Fifteen microbial species, belonging to 10 microbial genera, have been isolated from the cassava dough to be used for chickwanghe preparation. When two of these species, *Streptomyces faecalis* and *Bacillus licheniformis*, were used as a starter culture to aseptically inoculate a submerged cassava fermentation, a product having all the characteristics of chickwanghe was obtained (Nwankwo *et al.*, 1989).

Agbelima

Agbelima is a fermented product popular in the Ivory Coast. The roots are peeled, steeped in water for initial fermentation and then ground to paste like *fufu* and the paste is again left to ferment for two days in jute sacks and then pressed. Finally, the paste is removed from the sacks, crumbled or granulated, and then steamed. The difference from *fufu* preparation is the steaming step in the final stages. Microflora for fermenting cassava dough into *agbelima* showed a dominance of lactic acid bacteria consisting of species of *Lactobacillus*, i.e. *Lb. brevis*, *Lb. plantarum* and *Leuconostoc mesenteroides* (Sefa-Dedeh, 1995; Amoa-Awua *et al.*, 1996; Kofi *et al.*, 1996).

Attieke, placali and kivunde

Attieke and *placali* are two other cassava-fermented products consumed in Ivory Coast. *Attieke* is similar to *agbelima* but has a slightly sour taste and is eaten with milk or meat or vegetables (Aboua, 1992, 1993). *Kivunde* is another traditional fermented food from cassava that is popular in Tanzania (Kimaryo *et al.*, 2000).

1.2 Fermented foods in Asia

Peujeum

Peujeum is a traditional food prepared from cassava in Java, Indonesia. The roots are peeled and steamed until tender, then allowed to cool and dusted with finely powdered *ragi* (a mixture of flour and spices in which yeast and fungi have been active). The cassava mash, mixed with *ragi*, is wrapped in banana leaves and left for 1 to 2 days to ferment in earthenware containers. The *peujeum* has a refreshing and slightly alcoholic flavour and is eaten either as it is or after baking (Ray and Ward, 2006).

Gaplek and putto

The simplest and most common form of processing, used widely in Indonesia, is to peel the roots, wash and slice and then sun-dry for 2-3 days to produce dry cassava chips or chunks, in Indonesia known as *gaplek*. When needed, the dry root pieces are pounded into flour, which is shaken on a bamboo screen with some water to produce granules, called *tiwal*. *Putto* is the most popular fermented rice and sweet potato preparation in Philippines (Yoshi and Garcia, 1984), similar to *iddli* in India (Steinkraus, 1996).

Fermented sweet and sour flour

A technology has been developed for extraction of fermented sweet and sour flour from cassava. In this process, a mixed starter culture of *Lactobacillus cellobiosus, Streptococus lactis, and Corynebacterium* sp. and *Pichia membranaefaciens* was used to ferment cassava roots. It improved the extraction of starchy flour without liberating foul smells. This approach beneficially modifies the functional properties of the starch while leaving the granular structure intact (Moorthy *et al.*, 1993; Nanda and George, 1996). It also lowers the cyanide content and improves the shelf-life, raising property and cooking quality (Padmaja *et al.*, 1994). The technology for production of sweet and sour flour from cassava has been patented and commercialized in baking industries in South India.

1.3 Fermented foods in Latin America

Fermented cassava starch in Brazil

The total consumption of fermented cassava starch in Brazil is around 50,000 tons per year (Cereda and Vilpoux, 2006). Some uses of the fermented starch in Latin America are described below.

Fermented starch allows biscuits to become light and voluminous during baking due to mass expansion. In some Brazilian rural regions these biscuits are substitute for bread. With suitable packaging these biscuits can maintain good texture and flavour for several months. "Cheese" bread is another product made with fermented starch or with a mixture of 50: 50 fermented and non- fermented starches supplemented with milk, fat, eggs, meat and cured cheese of the Parmesan. The total market for cheese bread in Brazil was estimated to be around 80,000 tons in 2002, with a total use of fermented starch of around 16,000 tons (Cereda and Vilpoux, 2006).

Fermented Cassava Starch in Colombia and Bolivia

In Colombia fermented starch is called *almidon agrio* and is produced in very similar manner to the way it is made in Brazil (Cereda and Vilpoux, 2006). The Colombian fermented product is mainly used in bakery or confectionery goods. Some of those products (*Pan de Bono, Buñuelo* and *Pan de Queso*) are similar to Brazilian cheese bread and fermented cassava biscuits (*Besitos*).

2. Sweet potato

2.1 Sour starch and flour

Starch is the prime component of interest for food and industrial uses of sweet potatoes. The efficiency of starch extraction from sweet potato roots was improved by lactic acid fermentation using a mixed culture (*Lactobacillus cellobiosus, Streptococcus lactis* and *Corynebacterium* sp.) inoculum. Study of the properties of the starchy flour showed that there was a significant reduction in the starch content and consequently the soluble ands apparent amylase contents of fermented samples from all six varieties of sweet potato used (Timmins *et al.*,1991; Jyothi *et al.*, 2005a,b).

2.2 Gari

Gari from sweet potato is not a common fermented food item. In Ghana, Oduro et al. (2000) produced sweet potato *gari*. The method was quite similar to that of cassava *gari* production. The product had a moisture content (3.83%), protein (0.60%), crude fibers (3.5%), ash (1.42%), pH (3.97), TTA (0.50%) and swelling capacity (3.05), which were within the recommended levels.

2.3 Lacto- pickles

Lactic acid bacteria influence the flavour of fermented foods in a variety of ways. In many cases, the most obvious change in lactic acid fermentation is the production of acid and lowering of pH those results in an increase in sourness. It not only produced lactic acid which imparted taste and flavour to lacto- pickles, but also preserved ascorbic acid, phenols, and coloured pigments (β-carotene and anthocyanin contents of sweet potato); all these are considered as anti-oxidants (Kaur and Kapoor, 2001; Shivashankara *et al.*, 2004). Sweet potato roots were pickled by lactic fermentation by brining the cut and blanched roots in common salt (NaCl, 2-10%) solution (*sauerkraut* process) and subsequently inoculated with a strain of *Lactobacillus plantarum* (MTCC 1407) for 28 days. Treatment with 8-10 % brine solution was found to be organoleptically most acceptable. The final product with 8 and 10 % brine solutions had a pH (2.9-3.0), titratable acidity (2.9 - 3.7 g kg^{-1}), lactic acid (2.6-3.2 g kg^{-1}) and starch (58-68 g kg^{-1}) on fresh weight basis. Sensory evaluation rated the sweet potato lacto-pickle acceptable based on texture, taste, aroma, flavor and after taste. Principal component analyses reduced the six original analytical variables to two independent components (factors), which accounted for 92 % of the total variations (Panda *et al.*, 2007).

2.4 Lacto-juice

Lacto- juices processed by lactic acid fermentation bring about a change in the beverage assortment for their high nutritive value, vitamins and minerals which are beneficial to human health when consumed. Sweet potato roots (non-boiled/ fully-boiled) were fermented with *Lactobacillus plantarum* MTCC 1407 at 28± 2^{0}C for 48 h to make lacto- juice. During fermentation both analytical [pH, titratable acidity, lactic acid, starch, total sugar, reducing sugar (g/kg roots), total phenol and β-Carotene (mg kg^{-1} roots) and sensory (texture, taste, aroma, flavour and after taste) analyses of sweet potato lacto-juice were evaluated. The fermented juice was subjected to panelist evaluation for acceptability. There were no significant variations in biochemical constituents (pH, 2.2-3.3; lactic acid, 1.19-1.27 g kg^{-1} root; titratable acidity, 1.23-1.46 g kg^{-1} root, etc) of lacto- juices prepared from non-boiled and fully-boiled sweet potato roots except β-Carotene concentration [130 ± 7.5 mg kg^{-1} (fully- boiled roots) and 165± 8.1 mg kg^{-1} (non-boiled roots)]. The panelist evaluation scores ranged from 3 - 4.8 (in a hedonic scale of 1-5) from moderate liking to very much liking of sweet potato lacto-juice. Principal component analyses reduced the eight original analytical variables to three independent components (factors), which accounted for 99.9% of the total variations. Similarly, five original sensory variables were reduced to two independent components, which accounted for 83.1% of the total variations (Panda and Ray, 2007).

2.5 Sweet potato curd and yoghurt

Yoghurt and curd are consumed by lactase-deficient individuals because much of the lactose in milk is converted to digestive lactic acid by curd- or yoghurt- producing bacteria during fermentation. While the starter culture for curd is a mixture of undefined cocktail of lactic acid producing microorganisms, i.e. *Lactobacillus bulgaricus, Streptococcus clemoris , St. thermophilus*, etc, the starter culture for yoghurt is the use of specific symbiotic or mixed culture of *Lb. bulgaricus* and *Streptococcus thermophilus*. In a recent study, a curd like product was prepared by co-fermenting boiled sweet potato pulp (8-16%) from β-carotene and anthocyanin-rich variety, sugar and curd inoculum (Panda *et al.*, 2005, 2006; Mohapatra *et al.*, 2007). Curd with 12-16% sweet potato pulp was most preferred by consumers' panelists (Ray *et al.*, 2005).

2.6 Shochu

The production of *shochu*, a Japanese distilled liquor, which may also be made from rice or barley as well as sweet potato, has a long tradition (Yamakawa, 2000). During fermentation of sweet potato mash, simultaneous starch conversion to sugar by the *koji* enzymes and fermentation of sugar to alcohol by the yeast *Saccharomyces cerevisiae* takes place. The final alcohol concentration is 13-15%.

2.7 Wine and beer

Yellow, red and black coloured beverages like beer (sparkling liquor) and wine are being sold in the Kyushu Province in Japan prepared from anthocyanin-rich sweet potato (Yamakawa, 2000). A process of enzymatically saccharifying the sweet potato pulp (starch) and fermenting the mash into wine using wine yeast *S. cerevisiae* has been developed by Ray *et al.* (2006). The wine prepared from an anthocyanin sweet potato cultivar is bright red in colour and fulfills all the quality attributes of fruit wines. Similarly, wine with medicinal properties against common ailments like cold, cough, muscle pain, stomach ache, can be prepared by co-fermenting sweet potato and desired medicinal herbs, fruits or bark. A rice beverage was prepared by adding barley sprouts, sweet potato, or a mixture of barley sprouts and sweet potato (1:1). The use of sweet potato resulted in an increase of sweetness, flavour and improved preference in rice beverage (Suh *et al.*, 2003).

Bioprocessing: Protein Enriched Feed

1. Silage

Ensiling or silage making is an age-old technique for the conservation of nutrients in green fodder grasses through fermentation and storing them for round-the-year feeding of animals. Successful ensiling technology has been developed for cassava roots and leaves (Padmaja *et al.*, 1995). Sweet potato tops, especially leaves, are very rich in nutrients such as proteins, vitamins and minerals. Besides, they have amazing amounts of polyphenols, especially chlorogenic, isochlorogenic and caffeic acids (Takenaka *et al.*, 2006), which are anti-oxidants (Kaur and Kapoor, 2001). However, one of the constraints to the use of the large quantities of sweet potato foliage for forage is that its availability is concentrated into one or two seasons of the year after harvest, which in general does not coincide with pasture scarcity. Preservation as hay (dehydration) or silage may therefore be needed (Otieno *et al.*, 2006). The latter method is more practical in the humid tropics where quick drying in the field is not possible. Simple small silos can be made with discarded gunny bags, polyethylene sugar bags and related materials, all of which can be afforded by rural farmers in the tropics (CIP, 2000). Tinh *et al.* (2000, 2001) reported that sweet potato vines ensiled with chicken manure resulted in the highest quantity feed, increasing diet crude protein content and dry matter conservation rates. Giang *et al.* (2004) standardized the various proportions of sweet potato roots and vines to be mixed for silage preparation without any additives, which could be successfully stored for three months.

2 Single - cell protein (SCP)

The attractive characteristics in the use of microorganisms for single cell protein (SCP) include (1) their fast growth rate in semi-solid and solid media; (2) their high level of protein; (3) their comparable good nutritional values and (4) their easy genetically modification to growth under specific conditions on particular substrates (Ray *et al.*, 2006; Ward *et al.*, 2007). Cassava roots with 25-30% carbohydrate and 0.17% protein content is a good source for production of SCP. Besides, cassava starch processing wastes (bagasse), constituting about 20-30% by weight of the cassava chips and roots processed (Ray, 2004), are good substrates for protein enrichment by cultivating microorganisms through solid substrate fermentation (Belewu and Musa, 2003). Fermented sweet potato feed stuffs are reported to be produced and utilized for livestock in China (Jiang *et al.*, 1993) and Korea (Ahn *et al.*, 1997). In recent studies (Onifade *et al.*, 2004; Aziz and Mohsen, 2002), sweet potato roots and flour were fermented using pure strains of four species, *Aspergillus niger*, *Rhizopus stolonifer*, *Neurospora sitophila* and *Saccharomyces cerevisiae* using solid substrate fermentation at 30^0 C for 72h. The protein content of the samples increased significantly from 2.3% in the non-fermented roots to 11.5 - 12. 6% in the fermented roots. Colocasia leaves could be included up to 10% as nitrogen source in SCP enrichment of taro roots for feed. No toxicity was observed during feeding the mixtures to weaned calves for 8 weeks as components of their crop diets (Anigbogu and Anigbogu, 1999). Co-fermentation of cassava roots with cowpea flour inoculated with the fungus, *Rhizopus stolonifer* CNF6, resulted in rapid softening of cassava roots and an increase in protein content to 7.43 % 9w/w, dry weight) of the flour (Tivana and Bvochora, 2006; Tivana *et al.,* 2006).

Bioprocessing: Bagasse as Substrate Food and Feed Additives

The various products which can be made from root crops such as cassava and sweet potato starch and dried chips, as well as the peels and bagasse are projected in Fig. 1. Recent studies focus more on production of microbial enzymes, organic acids, amino acids, sugar syrups and microbial polysaccharides using root crop bagasse as solid substrate.

1. Microbial enzymes

Microbial enzymes find a wide range of applications in the feed, food industry, including the hydrolysis of starch, alcoholic fermentation, clarification of beers and fruit juices, in the manufacture of glucose and fructose syrups, de-hiding of leather, etc. Root crop starch, flour or bagasse in particular could be used as cheap fermentation substrates for the production of microbial enzymes (Table 2) (Pandey *et al.,* 2000; Ray *et al.,* 2006). Cassava peels served as the substrate for production of cellulolytic enzymes (Ofuya and Nwajiuba, 1990, 1996). Similarly, *Aspergillus flavus* and *A. niger* produced extracellular amylase into the culture medium when grown on basal medium containing 2% (w/v) cassava peel (Sani *et al.,* 1992) or raw cassava starch (1%) (Okolo *et al.,* 1995) as the sole carbon source. *Rhizopus oryzae* and *B. theobromae* produced á- amylases in a submerged bioreactor when grown on a medium containing 2% (w/v) cassava fibrous residue as carbon source (Nishise *et al.,* 1996; Ray, 2004; Ray *et al.,* 2007b). Soccol *et al.* (1994) reported á-amylase, glucomylase and protein enrichment of raw and cooked cassava roots by *Rhizopus* strains in submerged and solid state conditions.

2. Organic acids

Organic acid made from cassava and sweet potato starch and flour or bagasse include citric acid, acetic acid, fumaric acid, lactic acid and itaconic acid (Ray *et al.,* 2006), which are used in the food industry as well as for the production of plastics, synthetic resins, rubber products, etc. Lactic acid could be produced by *Rhizopus oryzae* from raw cassava starch (Wang and Rakshit, 2000; Nor Anong *et al.,* 2001) and by *Lactobacillus* sp. from potato and sweet potato flour (Ray *et al.,* 1991). Several countries, notably China, Japan and

Vietnam, are now manufacturing citric acid from sweet potato starch or from its by-products. The process necessitates the initial breakdown of starch to sugars before these sugars are fermented by moulds, i.e, *Aspergillus niger* to citric acid (Ray and Ward, 2006).

3. Monosodium glutamate (MSG)

MSG is a well- known flavour enhancing agent used in many Asian kitchens. China is the largest producer and consumer of MSG in the world. It is made through the microbial fermentation of starch or sugar (molasses) in the presence of ammonium salts. The organism commonly used is *Brevibacterium glutamicum* which converts sugar to glutamic acid. Jyothi *et al.* (2005a) optimized glutamic acid production from cassava starch factory residue hydrolysate using *Brevibacterium divaricatum.*

4. Soya sauce

Soya sauce, a popular condiment used every day with Asian dishes is traditionally prepared from a mixture of Soya beans and wheat, fermented by moulds, especially *Aspergillus oryzae* or *A.sojae*, to give a dark brown salty liquid used as a flavouring agent. Wheat flour could be replaced by sweet potato flour in making soya sauce (Data *et al.,* 1986).

5. Vinegar

Vinegar has been processed from high â-carotene sweet potato cultivars (Ray *et al.,* 2006). The appearance and taste is similar to that of while wine vinegar. Recently, a new red vinegar has been developed in Japan via fermentation with the storage root of purple fleshed sweet potato of Japanese variety (*Ipomoea batatas* L. cv. *Ayamurasaki*). The red vinegar had a higher antioxidative activity than white or black vinegars (Matsui *et al.,* 2003).

6. Sugar syrups

Root crop starch, particularly from cassava and sweet potato can be used for the production of many types of sweeteners after hydrolyzation by either acids or enzymes, or both. These sweeteners include maltose, glucose syrups, glucose and fructose, which can be further processed into various oligo-saccharides (Jin Shuren, 2000, 2001).

7. Hydrogenated sweeteners

These include sorbitol, mannitol and maltol. They are produced by treating starch with hydrogen gas in high pr essure tanks, using special catalyst and ion-exchange resins. Sorbitol is used mainly for the production of vitamin C and as a moisture conditioner in toothpastes (Jin Shuren, 2000, 2001).

8. Microbial polysaccharides

Many microorganisms elaborate polysaccharides extracellularly. These microbial polysaccharides can be produced industrially by using low-value and low- cost raw materials such as cassava or sweet potato bagasse by either solid substrate or submerged fermentation (Ray *et al.,* 2006). Some examples in which cassava bagasse had been used for producing microbial polysaccharides are pullulan (by *Aureobasidium pullulans*) (Ray and Moorthy, 2007) and xanthan gum (by *Xanthomonas campestris*) (Woiciechowski *et al.,* 2004), which have a number of uses in pharmaceutical, cosmetic, brewing and electronics industries.

Baker's yeast and other products

Baker's yeast (*S. cerevisiae*) was cultured on hydrolyzed waste cassava starch (Ejiofor *et al.,* 1996). Volatile compounds are produced by the fungus *Geotrichum fragrans* using cassava waste water as the substrate (Damasceno *et al.,* 2003). Cassava bagasse was utilized for production of fruity aroma compounds in solid culture using the fungus *Ceraocystis fimbriata* (Christen *et al.,* 1997; Bramorski *et al.,* 1998).

Mushrooms Production

Cassava bagasse has been used for bioconversion mainly in solid substrate fermentation into mushrooms (Pandey *et al.,* 2000; Ray *et al.,* 2007c). Beux *et al.* (1995) compared the cultivation of *Lentinus edodes* on cassava bagasse and sugarcane bagasse, individually or in their mixture. Both the substrates were found suitable for mushroom production, but the best results were obtained when a mixture of cassava bagasse (80%) and sugarcane bagasse (20%) was used.

Various types of starches are being used for the production of biodegradable plastics. Starch will be converted to lactic acid by microorganisms and polymerization of lactic acid by thermo-chemical reaction lead to the production of polylactic acid, which is then blended with various other polymers to yield biodegradable plastics (Sriroth *et al.,* 2001). Big American company (Cargill) produces polylactic acid from corn starch; however, cassava and sweet potato starch can be of equal potential for production of polylactic acid. Biodegradable plastics based on cassava starch have been developed and the technology has been patented and commercialized (Nanda, 1996).

Bioprocessing: Biofuel

Ethanol

Cane and beet molasses are the first choice for ethanol production all over the world; however, the gap between the demand and supply of molasses has been widened recently world over. Hence, demand for alternative sources of energy for ethanol production from diverse raw materials (starchy crops like cassava and sweet potato) (Ward *et al.*, 2006) and sugar-rich crops like mahua (*Madhuca latifolia* L.) (Swain *et al.*, 2007) have increased. However, cassava has to compete with sugarcane in tropical countries and with maize in the temperate zone. Ethanol from cassava or sweet potato is mainly used as 'biofuel' to mix with gasoline to produce 'gasohol'. The basic processes in the production of ethanol from cassava or sweet potatoes are somewhat cumbersome as compared to molasses, because of the requirement to hydrolyse starch in roots and bagasse and include the following processes: (1) milling cassava and sweet potato chips or flour through sieve of 0.4 mm, (2) liquefaction (using dilute inorganic acid or thermostable α-amylase), (3) saccharification (using amyloglucosidase), (4) fermentation, and (5) distillation (Ray *et al.*, 2007a).

In some countries cassava is used for the production of ethanol. In the late 1970s several alcohol distilling factories were set up in Brazil using fresh cassava roots as raw material. The commercial production of ethanol from cassava is obviously not new in some parts of Asia like India and China. In China several factories in Guangxi Province are now using the solid waste (bagasse) of the cassava starch industry for the production of ethanol (Gu Bi and Ye Gurzhen, 2000; Cheng *et al.*, 2006). Other alcohol factories in China are switching from the use of molasses to that of cassava or sweet potato chips for alcohol production, because of strict pollution control requirements that makes the use of molasses uneconomical. In addition, it is expected that the Chinese government will support the use of "gasohol" (about 10-20% ethanol mixed with gasoline) as automotive fuel by 2008; this is expected to markedly increase demand for cassava and sweet potato chips. In, Thailand "gasohol', containing 10% ethanol, is presently available in some gas stations in Bangkok (Howeler, 2006; Watananonta and Howeler, 2006). In the Philippines, the Department of Energy has focused that by 5% blend of ethanol to gasoline the country would need 166,000,000 litres of ethanol converted into 63,700 ha of cassava production. The biofuel programme will envision to have 105 blend nationwide in year 2010 (Mariscal, 2006).

Bioethanol has been recognized as an alternative fuel and this has been adopted in Brazil. The production alcohol from cassava or sweet potato has basically three steps *viz.*, liquefaction, saccharification and fermentation.

Liquefaction is achieved using á amylase enzyme and then it is hydrolyzed to glucose with glucomylase enzymes and finally it is fermented with yeast to alcohol. The end produce alcohol is distilled and redistilled to get 98% pure ethanol. From one tonne of fresh tuber with 25% starch content we are to get 150 litres of ethanol 98% pure.

In China, ethanol from sweet potato is the most important product after starch and noodles; in Korea, however, ethanol production has been steadily rising, and in the 1990s represented more than 30% of total production, compared to only about 8% in the case of starch (Scott *et al.,* 2000). Japan uses only 6-8% of its sweet potato production for ethanol (Yamakawa, 1998) and in India, technology has been standardized for ethanol production from sweet potato flour and roots (Ray *et al.,* 2006).

Manilal *et al.*, (1990) studied anaerobic digestion of cassava starch factory effluents for methane production. The maximum methane yield was recorded at the C: N ratio of 20:1.

Conclusion

Root crops such as cassava or sweet potato for human consumption do not have major growth potential as rice and wheat remains the preferred food in most of the world. Total food demand may increase globally due to increase in population, but as more and more people in major continents become affluent, they are likely to reduce their consumption of high-energy staples like rice, cassava and sweet potato in preference for meat products or convenience foods (Howeler, 2004; Panda *et al.,* 2005). However, in Africa and Latin America, cassava is still the preferred food, and fermentation will probably remain the most important mode of processing of this root crop into edible foods and feeds, because of their low cost and the desirable flavour they impart. Similarly, sweet potato remains as the secondary staple crop in China. The development of starter cultures would facilitate process standardization and quality control, and would set the stage for strain-improvement programmes through conventional microbiology and genetic engineering techniques, for better quality fermented products from root crops. Further, the study of fermentation microorganisms could lead to the identification of microbial strains particularly suited for the over-secretion of valuable biological products such as enzymes, amino acids, vitamins, antibiotics and flavour enhancers like monosodium glutamate and tocopherol (Howeler, 2004; Ray and Isobe, 2007). Among the bioproducts, biofuel particularly ethanol from root crops will have a huge demand in future particularly in Asian countries like India, China and Thailand, and Latin America.

Table 1. Microorganisms associated with fermented foods from root and tuber crops

Fermented Foods	Microorganisms
Gari	*Lactobacillus plantarum, Leuconostoc, Corynebacterium*
Fufu	*Alcaligenes, Candida,Lactobacillus* spp., *Leuconostoc* spp.
Lafun	*Bacillus subtilis, Lactobacillus, Leuconostoc, Streptococcus*
Chickwanghe	Streptococcus faecalis, Bacillus lichenoformis
Agbelima	*Lactobacillus* spp. (*Lb. plantarum, Lb. brevis*); *Leuconostoc mesenteroides, Bacillus* spp. (*B. subtilis, B. lichenoformis, B. cereus, B. mycoides* and *B. polymixa*)
Sweet potato sour starch	*Lactobacillus* spp.; *Leuconostoc* spp.
Sweet potato curd & yoghurt	*Lactobacillus bulgaricus, Streptococcus thermophilus*

Table 2. Fermentation of root crops as substrate for production of bio-based products

Bio- based Products	Substrate	Microorganisms
Enzymes		
Amylase	Cassava peel, Raw cassava starch Cassava waste	*Aspergillus flavus, A. niger Rhizopus oryzae, Botryodiplodia theobromae*
Cellulase	Cassava peel	*Trichoderma* sp.
Linamarase	Cassava roots	*Rhizopus* spp. , *A. niger*
Organic acids		
Lactic acid	Raw cassava starch	*R. oryzae, Lactobacillus plantarum*
Citric acid	Sweet potato flour	*A. niger*
Glutamic acid	Cassava fibrous waste	*Brevibacterium divaricatum*
Food Additives		
Monosodium glutamate	Sweet potato starch	*Brevibacterium glutamicum*
Pullulan	Cassava fibrous waste	*Aureobasidium pullulans*
Xanthan	Cassava fibrous waste	*Xanthomonas campestris*
Falvouring compound	Cassava fibrous waste	*Geotrichum fragrans*
Baker's yeast	Cassava starch	*Saccharomyces cerevisiae*
Other Products		
Ethanol	Cassava and sweet potato starch, flour and fibrous wastes	*Zymomonas mobilis,* ***S. cerevisiae***

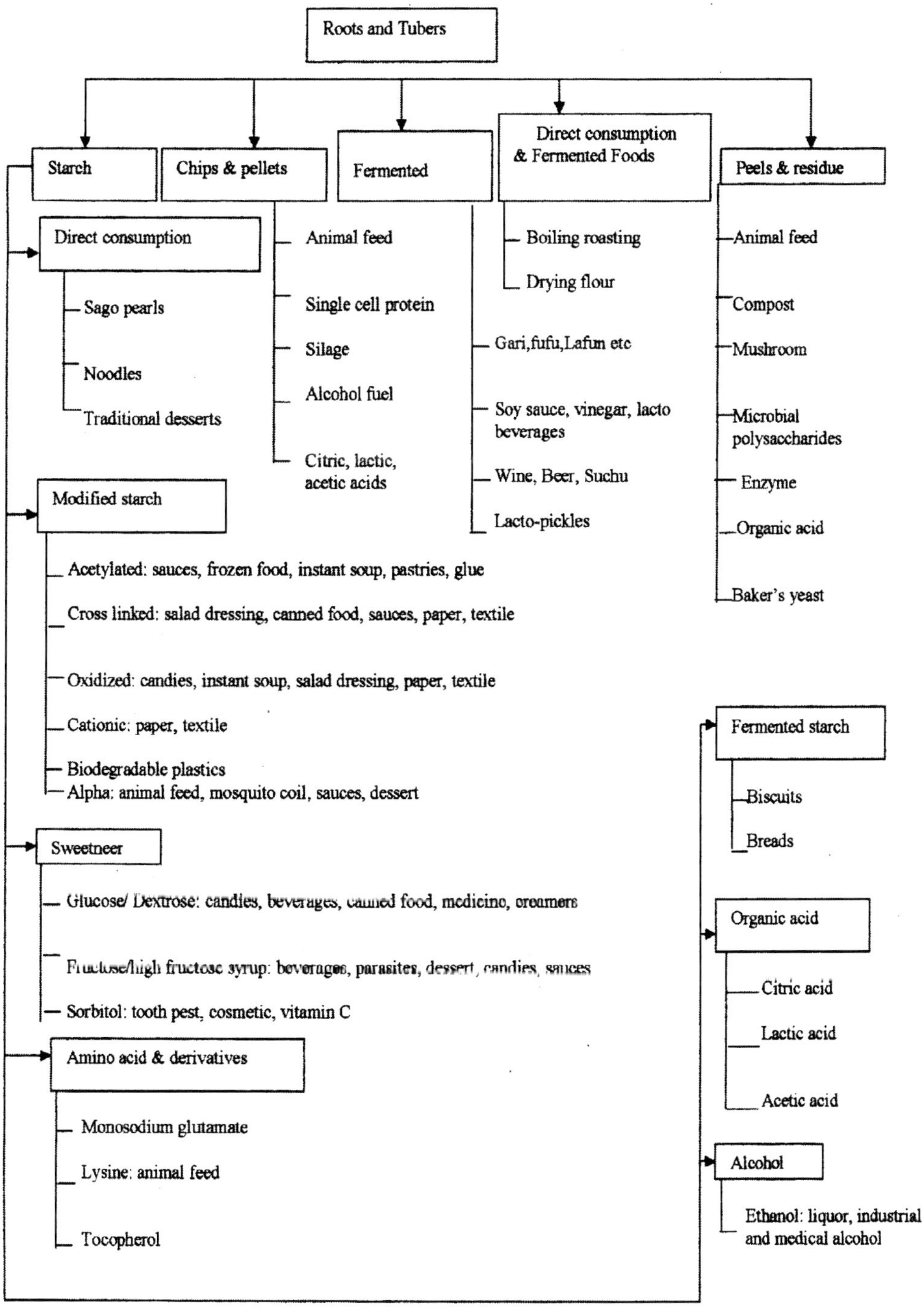

Fig. 1. Processing of root tuber crops into value added products

References

Aboua, F. 1992. Chemical changes in Cassava tubers (*Manihot esculenta* Crantz) during production of placali. *Trop. Sci.*, 32 : 353-360.

Aboua, F. 1993. Some biochemical and microbiological changes during dehydrated attieke storage. *Food Chem.*, 47: 345-347.

Ahn, Y.S., Jeong, B.C. and Oh, Y.B. 1997. Production and utilization of sweet potato in Korea. (In:) Proceedings of the International Workshop on Sweet potato Production System toward the 21st Century, Dec 9-10, 1997, Miyakonojo, Miyazaki, Japan. pp. 137- 147.

Akingbala, J.O., Oguntimein, G.B. and Bolade, M.K. 1993. Effect of unit operations of production on the cyanide content and the acceptability of gari. *J. Food Processing Preserv.*, 17: 337-350.

Akingbala, J.O., Oyewole, O.B., Uzo-Peters, P.I., Karim, R.O. and Baccus- Taylor, G.S.H. 2005. Evaluating stored cassava quality in gari production. *J. Food Agric. Environ.* 3: 75- 80.

Almazan, A. M. 1992. Influence of cassava variety and storage on gari quality. *Trop. Agric.*, 69: 386-390.

Amoa-Awua, W.K.A. Appoch, F. and Jakobsen. 1996. Lactic acid fermentation of cassava. into agbelima. *Int. J. Food Microbiol.*, 31: 87-98.

Anigbogu, N.M. and Anigbogu, C.C. 1999. Single -cell protein enriched taro (*Coloasia escuenta* (L) Schott) and the role of natural nitrogenous source and its value in crop diet for weaned calves. *J. Root Crops, India* 25: 147-152.

Aziz, N. H. and Mohsen, G. I. 2002. Bioconversion of acid- and gamma ray- treated sweet potato residue to microbial protein by mixed cultures. *J. Ind. Microbiol. Biotechnol.*, 29: 264 - 267.

Belewu, M.A. and Musa, A.K. 2003. Effect of selected *Azotobacter* bacterial strains on the enrichment of cassava waste during solid state fermentation. *Trop. J. Anim. Sci.*, 6: 1-7.

Beux, M.R., Soccol, C.R., Martin, B., Tonial, T. and Roussos, S. 1995. Cultivation of *Lentinus edodes* on the mixture of cassava bagasse and sugarcane bagasse. (In:) Advances in Solid State Fermentation (eds.) S. Roussos, B.K.Lonsane, M. Raimbault, and G. Viniegra- Gonzalez.), Kluwer Academic Publishers, Dordecht, Netherlands. pp. 499- 511.

Bramorski, A., Soccol, C.R., Christen, P. and Revah, S. 1998. Fruity aroma production by *Ceratocystis fimbriata* in solid cultures from agro-industrial wastes. *Rev. Microbiol.*, 29: 208- 212.

Cereda, M.P. and Vilpoux, O.E. 2006. Cassava fermentation in Latin America: Fermented starch: an overview. (In:) Microbial Biotechnology in Horticulture, Volume 1 (ed. Ray, R.C and Ward, O.P.), Science Publishers, Enfield, New Hampshire, USA. pp. 397- 420.

Cheng, L., Haiyan, W., Wenli, Z., Kaimian, L. and Wenquan, W. 2006. Production and utilization of cassava in China: A developing bio-energy industry. (In:) 14th Triennial Symp. Int. Soc. Trop. Root Crops, 20-26 November, 2006, Central Tuber Crops Research Institute, Thiruvanathapuram, India. 133 pp.

Christen, P., Meza, J.C. and Revah, S. 1997. Fruity aroma production in solid state fermentation by *Ceratocystis fimbriata*: Influence of the substrate type and the presence of precursors. *Mycol. Res.*, 101: 911- 919.

CIP (International Potato Center). 2000. Fermented sweet potato feed for pig: a boon to farmers. Scientist and Farmers: Partners in research for 21st century. CIP Program Report 1999. 312 pp.

Damasceno, S., Cereda, M.P., Pastone, G.M. and Oliveira, J.G. 2003. Production of volatile compounds by *Geotrichum fragrans* using cassava waste water as substrate. *Process Biochem.*, 39: 411-415.

Data, E.S., Diamante, J.C. and Forio, E.E. 1986. Soy sauce production utilizing root crops flour as substitute for wheat flour (100% substitution). *Ann. Trop. Res., (Phillip)* 8: 42-50.

Ejiofor, A.O., Chisti, Y. and Moo Young, M. 1996. Culture of *Saccharomyces cerevisiae* on hydrolyzed waste cassava starch for production of baking quality yeast. *Enz. Microb. Technol.*, 18: 519- 525.

Ezedinma, C. 2006. Domestic and cross border trade in traditional cassava food products in Nigeria. *(In:) 14th Triennial Symp. Int. Soc. Trop. Root Crops,* 20-26 November, 2006, Central Tuber Crops Research Institute, Thiruvanathapuram, India. pp.254- 255.

Giang, H.H., Ly, L.V. and Ogle, B. 2004. Evaluation of ensiling methods to preserve sweet potato roots and vines as pig feed. *Livestock Res. Rural Dev.*, 16: 12-21.

Giraud E., Gosselin, L. and Raimbault, M. 1993. Production of a *Lactobacillus plantarum* starter with linamarase and amylase activities for cassava fermentation. *J. Sci. Food Agric.*, 62 : 77-82.

Gu Bi and Ye Guozhen. 2000. Commercial scale production of ethanol from cassava pulp. *(In:) Cassava Starch and Starch Derivatives*, Proc. Int. Symp. (eds. R.H. Howeler, C.G. Oates and G.M. O'Brien), held in Nannking, Guangxi, China, November 11-15, 2000. pp. 11 -15.

Howeler, R. H. 2004. Cassava in Asia- present situation and its future potential in agro-industry. *J. Root Crops, India.* 30: 81- 92.

Howeler, R.H. 2006. Production and utilization of cassava in Asia and its potential as a bio-fuel. (In:) 14th Triennial Symp. Int. Soc. Trop. Root Crops, 20-26 November, 2006, Central Tuber Crops Research Institute, Thiruvanathapuram, India. 116 pp.

Jiang, L., Rozelle, S. and Huang, J. 1993. Production, technology and post harvest processing of sweet potatoes in Sichuan Province, Unpublished Report of CIP Sub-project 6211. 63 pp.

Jin Shuren. 2001. Production and use of modified starch and starch derivatives in China. (In:) Cassava's Potential in Asia in the 21st Century, Present Situation and Future Research and development Needs. (eds. R.H. Howeler and S.L. Tan), Proc. 6th Regional Workshop, held in Ho Chi Minh City, Vietnam, February 21- 25, 2001. pp. 553- 563.

Jin Shuren. 2000. The current situation and prospects for further development of China's sorbitol industry. (In:) Cassava Starch and Starch Derivatives, Proc. Int. Symp. (eds) R.H. Howeler, C.G. Oates, C.G. and G.M. O'Brien), held in Nannking, Guangxi, China, November 11- 15. pp. 32- 36.

Jyothi, A.N., Sasikiran, K., Nambisan ,B. and Balagopalan, C. 2005 a. Optimisation of glutamic acid production from cassava starch factory residues using *Brevibacterium divaricatum*. *Process Biochem.*, 40: 3576- 3579.

Jyothi, A.N., Wilson, B., Moorthy, S.N. and George, M. 2005 b. Physiochemical properties of the starchy flour extracted from sweet potato tubers through lactic acid fermentation. *J. Sci. Food Agric.*, 85: 1558-1563.

Kaur, C. and Kapoor, H.C. 2001. Antioxidants in fruits and vegetables- the millennium's health. *Int. J. Food Sci. Technol.*, 36: 703- 725.

Kimaryo, V.M., Massawe, G.A., Olasupo, N.A. and Holzapfel, W.H. 2000. The use of starter culture in the fermentation of cassava for the production of 'kivunde' a traditional Tanzanian food product. *Int. J. Food Microbiol.*, 56: 179- 180.

Kofi, W., Amoa-Awua, A., Appok, F.E. and Jacobsen, M. 1996. Lactic acid fermentation of cassava dough into agbelima. *Int. J. Food Microbiol.*, 96 : 87-98.

Kostinek, M., Specht, I., Edward, V.A., Schillinger, U., Hertel, C., Holzapfel, W.H. and Franz, C. M. A. P. 2005. Diversity and technological properties of predominant lactic acid bacteria from fermented cassava used for the preparation of Gari, a traditional African food. *Syst. Appl. Microbiol.*, 28 (6): 527-540.

Lei, V., Amoa-Awua, W.K.A. and Brimer, L. 1999. Degradation of cyanogenic glycosides by *Lactobacillus plantarum* strains from spontaneous cassava fermentation and other microorganisms. *Int. J. Food Microbiol.*, 53 : 169-184.

Manilal, V.B., Narayanan, C.S. and Balagopalan, C. 1990. Anaerobic digestion of cassava starch factory effluents. *World J. Microbiol. Biotechnol.*, 6: 149-154.

Mariscal, A. 2006. Present status of root crops production, utilization and marketing in the Philippines. (In:) 14th Triennial Symp. Int. Soc. Trop. Root Crops, 20-26 November, 2006, Central Tuber Crops Research Institute, Thiruvanathapuram, India. pp. 11-12.

Matsui, T.N., Fukui, T., Sugita, K. and Matsumoto, K. 2003. Caffeoylsophorose in a red vinegar produced through fermentation with purple sweet potato. *J. Agric. Food Chem.*, 51: 2539- 2543.

McFeeters, R.F. 2004. Fermentation microorganisms and flavor changes in fermented food. *J. Food Sci.*, 69 (1): 35-37.

Mohapatra, S., Panda, S.H., Sahoo, A.K., Sivakumar, P.S. and Ray, R.C. 2007. β-carotene rich sweet potato curd: Production, nutritional and proximate composition. *Int. J. Food Sci. Technol.*, 42: in press.

Moorthy, S.N. and George, M. 1998. Cassava fermentation and associated changes in physicochemical and functional properties. *Crit. Rev. Food Sci. Nutr.*, 18 (2): 73-121.

Moorthy, S.N., George, M. and Padmaja, G. 1993. Functional properties of the starchy flour extracted from cassava as fermentation with a mixed culture inoculum. *J. Sci. Food Agric.*, 61 : 443-447.

Nanda, S.K. 1996. Biodegradibility and starch based plastics. XI Carbohydrate Conference. 21-22 November, 1996. Kolkata.

Nanda, S. K. and George, M. 1996. Physical aspects of the softening of cassava tubers upon fermentation with a mixed culture inoculum. *J. Food Eng.*, 29: 129-137.

Nar Anong, N., Poocharoean, D., Nuanphan, N. and Duangduan, P. 2001. Production of L- lactic acid from raw cassava starch by *Rhizopus oryzae* NRRL 395. *Kasetsart J. Natural Sci.*, 19: 1111- 1122.

Nishise, H., Fuji, A., Ueno, M.J. and Tani, Y. 1996. Production of raw cassava starch digestive glucoamylase by *Rhizopus* sp. in liquid culture. *J. Ferment. Technol.*, 66 : 397-402.

Nwankwo, D., Anadu, E. and Usoro, R. 1989. Cassava fermenting organisms. *Mircen J. Appl. Microbiol. Biotechnol.*, 5: 169- 179.

Oduro, I., Ellis, W.O., Otoo, J.A., Akuffo, A.O. and Moses, E. 2000. Production of gari from sweet potato tubers (In:) Potential of Root Crops for Food and Industrial Resources (eds.) M. Nakatani and K. Komaki, Twelfth Symposium of International Society of Tropical Root Crops (ISTRC), 10-16 Sept., 2000, Tsukuba, Japan. pp. 292-294.

Ofuya, C.O. and Nwajiuba, C.J. 1990. Microbial degradation of cassava peel. *World J. Microbiol. Biotechnol.*, 6 : 144-148.

Ofuya, C.O. and Nwajiuba, C.J. 1996. Fermentation of cassava peels for the production of cellulolytic enzymes. *J. Appl. Bacteriol.*, 68 : 171-177.

Okolie, N.P., Lbeh, I.N. and Ugochukwu, E.N. 1997. Production of improved cassava fufu, 'akpu' through controlled fermentation. *Food Chem.*, 44(2) : 137-139.

Okolo, B.N., Ezeogu, L.I and Mba, C.N. 1995. Production of raw starch digesting amylase by *Aspergillus niger* grown on native starch sources. *J . Sci. Food Agric.*, 69 : 109-115.

Onifade, A.K., Atum, H.N. and Adebolu, T.T. 2004. Nutrient enrichment of sweet potato (*Ipomoea batatas* L.) by solid substrate fermentation using four fungal species. *Global J. Pure Appl. Sci.*, 10: 31- 36.

Osho, S.M. and Dashiell, K.E. 2002. The processing and acceptability of fortified cassava based product (gari) with soybean. *Discovery and Innovation* 14(3/4): 186- 191.

Otieno, K., Leon- Velarde, C., Agili, S., Kimathi, M., Quiroz, R. and Potts, M.J. 2006. The use of sweet potato feed in East Africa: an overview. *(In:) 14th Triennial Symp. Int. Soc. Trop. Root Crops*, 20-26 November, 2006, Central Tuber Crops Research Institute, Thiruvanathapuram, India. pp. 121- 122.

Oyewole, O.B. 2001. Characteristics and significance of yeasts involved in cassava fermentation for fufu production. *Int. J. Food Microbiol.*, 65: 213- 222.

Padmaja, G. George, M. and Balagopalan, C. 1995. Ensiling as an innovative biotechnological approach for conservation of high cyanide tubers for feed use. *(In:) The Cassava Biotechnology Network*: Proc. of the Second International Scientific Meeting, Bogor, Indonesia, 22-26 August, 1994, Cali, Colombia, CIAT, Working document No.150. pp. 784-794.

Padmaja, G., George, M. and Moorthy, S.N. 1994. Nutritional evaluation of the starchy flour obtained from cassava tubers on fermentation with a mixed culture inoculum. *J. Agric. Food Chem.*, 42: 766-770.

Panda, S. H., Naskar, S.K. and Ray, R.C. 2006. Production, proximate and nutritional evaluation of sweet potato curd. *J. Food Agric. Environ.*, 4: 124 -127.

Panda, S. H., Paramanick, M. and Ray, R.C. 2007. Lactic acid fermentation of sweet potato (*Ipomoea batatas* L.) into pickles. *J. Food Process. Preserv.*, 31 (1): 83- 101.

Panda, S.H. and Ray, R.C. 2007. Lactic Acid Fermentation of β-Carotene rich Sweet potato (*Ipomoea Batatas* L.) into Lacto- juice. *Plant Food Human Nutr.*, 62: In press.

Panda, S.H., Parmanik, M, Sharma, P., Panda, S. and Ray, R. C. 2005. Microorganisms in food biotechnology. (In:) Microbes in Our Lives (eds) R.C.. Mohanty, R.C. and P.K. Chand, Department of Botany and Microbiology, Utkal University, Bhubaneswar, India, 2005. pp. 47- 54.

Pandey, A., Soccol, C.R., Nigam, P., Soccol, V.T., Vandenberge, L.P.S. and Moham, R. 2000. Biotechnological potential of agro-industrial residues. II. Cassava bagassae. *Bioresource Technol.*, 74: 81- 87.

Ray, L.G., Mukherjee, G. and Majumdar, S.K. 1991. Production of lactic acid from potato fermentation. *Indian J. Exptl. Biol.*, 29: 681-682.

Ray, R.C. 2004. Extracellular amylase(s) production by fungi *Botryodiplodia theobromae* and *Rhizopus oryzae* grown on cassava starch residue. *J. Environ. Biol.*, 25(4): 489-495.

Ray, R.C., Naskar, S.K. and Sivakumar, P.S. 2005. *Sweet potato Curd*. Technical Bulletin Series: 39, Central Tuber Crops research Institute, Thiruvanathapuram, India. 24 pp.

Ray, R.C., Sahoo, A., Asano, K. and Tomita, F. 2006. Bio-processing of agricultural residues into food and food additives. (In:) Microbial Biotechnology in Agriculture and Aquaculture, Volume II (ed.) R.C. Ray.), Science Publishers, New Hampshire, New Hampshire, USA, pp. 511- 552.

Ray, R.C. and Isobe, S. 2007. Agro –industrial bioprocessing of tropical root and tuber crops: Current research and future prospects. (In:) Biotechnology of Microorganisms- an Agro-industrial Approach (eds.) D.K. Maheshwari and R.C.Dubey, I.K. International Pvt. Ltd., New Delhi, in press.

Ray, R.C. and Moorthy, S.N. 2007. Microbial exopolysaccharide production by *Aureobasidium pullulans* from cassava fibrous residue. *J. Sci. Ind. Res.* in press.

Ray, R.C., Mohapatra, S., Panda, S. and Kar, S. 2007a. Solid substrate fermentation of cassava fibrous residue for production of alpha amylase, lactic acid and ethanol. *J. Environ. Biol.,* 28: in press.

Ray, R.C., Shetty, K. and Ward, O. P. 2007b. Solid state fermentation and value-added utilization of horticultural processing wastes, (In:) Microbial Biotechnology in Horticulture, Volume 2 (ed.) R.C. Ray. and O.P.Ward), Science Publishers, New Hampshire, in press.

Ray, R. C. and Ward, O. P. 2006, Post harvest microbial biotechnology of tropical root and tuber crops. (In:) Microbial Biotechnology in Horticulture, Volume 1, (ed. Ray, R.C. and Ward, O.P.), Science Publishers, Enfield, New Hampshire, USA, pp. 345- 396.

Ray, R.C., Ward, O.P., Singh, A. and Isobe, S. 2007c. Commercialization of microbial biotechnology in horticulture: Summary outlook of achievements, constraints and prospects. (In:) Microbial Biotechnology in Horticulture, Volume 2, (ed. Ray, R.C. and Ward, O.P.), Science Publishers, New Hampshire.

Sani, A., Awe, F.A and Akinyanju, J.A. 1992. Amylase synthesis in *Aspergillus flavus* and *Aspergillus niger* grown on cassava peel. *J. Ind. Microbiol.,* 10: 55-59.

Sanni, M.O. 1994. Gari processing in Ibadan metropolis -factor controlling at the small scale level. (In:) Root Crop and food security in Africa (ed.) M.O. Akoroda, International Society for Tropical Root Crops, African Branch, Ibadan, Nigeria.

Sanni, A.I., Morlon- Guyot, J. and Guyot, J.P. 2002. New efficient amylase-producing strains of *Lactobacillus plantarum* and *L. fermentum* isolated from different Nigerian traditional fermented food. *Int. J. Food Micobiol.,* 72 : 53-62.

Sanni, L.O., Akingbala, J.O. Oguntunde, Z.A., Bainbridge, Graffham, A.J. and Westby, A. 1998. Processing of fufu from cassava in Nigeria: Problems and Prospects for development. *Sci. Technol. Develop.* 16: 58-71.

Scott, G.H., Best, R., Rosegrant, M. and Bokanga, M. 2000. Roots and tubers in the global food system: a vision statement to the year 2020. A Co-Publication of the CIP/ CIAT/IFPRI/IITA/IPGRI, Printed in Lima, Peru, International Potato Centre.

Sefa-Dedeh, S. 1995. Process and product characteristics of agbelima, a fermented cassava product. *Trop. Sci.* 35: 359-364.

Shivashankara, K.S., Isobe, S., Al-Haq, M.J., Takenaka, M. and Shiina, T. 2004. Fruit-antioxidant activity, ascorbic acid, total phenol, quercetin, and carotene of Irwin mango fruits stored at low temperature after high electric field pretreatment. *J. Agric. Food Chem.*, 52: 1281- 1286.

Soccol , C.R , Iloki, I., Martin, B. and Raimbault, M. 1994. Comparative production of alpha-amylase, glucoamylase and protein enrichment of raw and cooked cassava by *Rhizopus* strain in submerged and solid state fermentations. *J.Food Sci.Technol.*, 31(4): 320-323.

Sokari, T.G. and Karibo, P.S. 1996. Short duration processing of cassava for high quality gari production. *J. Sci. Food Agric.* 70(4): 427-432.

Sriroth, K., Chollakup, R., Choineeranat, S., Piyachomkwan, K. and Oates, C.G. 2001. Biodegradable plastics from cassava starch in Thailand. *(In:)* Cassava's Potential in Asia in the 21st Century: Present Situation and Future Research and development Needs (eds.) R.H. Howeler and S.L. Tan, Proc. 6th Regional Workshop, held in Ho Chi Minh City, Vietnam, February 21- 25, 2000. pp. 538- 552.

Steinkraus, S.H. 1996. *Handbook of Indigenous Fermented Foods*, 2nd edition, Marcel Dekker, New York. 776 pp.

Suh, H.J., Kim, J.M. and Choi, Y.M. 2003. The incorporation of sweet potato application in the preparation of a rice beverage. *Int. J. Food Sci. Technol.* 38: 145- 151.

Swain, M.R., Kar, S., Sahoo, A.K. and Ray, R. C. 2007. Ethanol fermentation of mahula (*Madhuca latifolia* L.) flowers using free and immobilised yeast *Saccharomyces cerevisiae. Microbiol. Res.* in press.

Takenaka, M., Nanayama, K., Isobe, S. and Murata, M. 2006. Changes in caffeic acid derivatives in sweet potato (*Ipomoea batatas* L.) during cooking and processing. *Biosci. Biotechnol. Biochem.*, 70: 172- 177.

Timmins, W.H., Marter, A.D., Westby, A. and Rickard, J.E. 1991. Aspects of sweet potato processing in Siachuan Province, People's Republic's of China. (In:) Product Development for Root and Tuber Crops Volume I - Asia, (eds Scott, G.G., Wiersema, S. and Ferguson, P.I.), International Potato Centre, Lima, Peru. pp. 217-227.

Tinh, N.T., Thuy, T.P.T., Thach, P.N.T. and Peters, D. 2000. Ensiling sweet potato vines as feed to fattening pigs. Information of Science and Technologies on Animal Production. No 4/2000. (In Vietnamese language). Hanoi Agricultural Publishing House, Hanoi. pp. 18-30.

Tinh, N.T., Thuy, T.P.T., Thach, P.N.T. and Peters, D. 2001. Processing, storing sweet potato root by fermentation and using the fermented feed to fatten pigs. Information of Science and Technologies on Animal Production. No 2/2001. (In Vietnamese language). Hanoi Agricultural Publishing House, Hanoi. pp. 18- 30.

Tivana, L.D. and Bvochora, J. 2006. Co-fermentation of cassava roots with cowpea using isolated moulds for protein enrichment of cassava flour. (In:) 14th Triennial Symp. Int. Soc. Trop. Root Crops, 20-26 November, 2006, Central Tuber Crops Research Institute, Thiruvanathapuram, India. 40 pp.

Tivana, L.D., Bvochora, J. and Owens, J.D. 2006. Effect of slicing of cassava roots on growth rate of moulds furing solid state fermentation. (In:) 14th Triennial Symp. Int. Soc. Trop. Root Crops, 20-26 November, 2006, Central Tuber Crops Research Institute, Thiruvanathapuram, India. 124 pp.

Treche, S. and Massamba, J. 1995. Les modes de transformation traditionnels du manioc du congo. (In:) Cassava Food Processing (eds.) E. Agbor, A. Brauman, D. Griffon and S. Treche, ORSTOM, Paris. pp. 133- 150.

Uzogara, S.G., Agu, L.N. and Uzogara, E.O. 1990. A review of traditional fermented foods, condiments and beverages in Nigeria: their benefits and possible problems. *Ecol. Food Nutr.*, 24:267-277.

Wang, X.G.X. and Rakshit, S.K. 2000. Study of lactic acid fermentation with *Lactobacillus amylovorus* using cassava starch. *(In:) Cassava's Potential in Asia in the 21st Century*, Present Situation and Future Research and development Needs (eds Howeler, R.H. and Tan, S.L.), Proc. 6th Regional Workshop, held in Ho Chi Minh City, Vietnam, February 21- 25, 2000. pp.198- 201.

Ward, O.P., Singh, A. and Ray, R.C. 2006. Production of renewable energy from agricultural and horticultural substrates and wastes. (In:) Microbial Biotechnology in Horticulture, Volume 1 (ed. Ray, R.C. and Ward, O.P.), Science Publishers, Enfield, New Hampshire, USA. pp. 517- 557.

Ward, O.P., Singh, A. and Ray, R.C. 2007. Single cell protein from horticultural and related food processing wastes. (In:) Microbial Biotechnology in Horticulture, Volume 2 (ed. Ray, R.C. and Ward, O.P.), Science Publishers, Enfield, New Hampshire, USA, 2006. in press.

Watananonta, W. and Howeler, R.H. 2006. Present situation and future potential of cassava production and utilization in Thailand. (In:) 14th Triennial Symp. Int. Soc. Trop. Root Crops, 20-26 November, 2006, Central Tuber Crops Research Institute, Thiruvanathapuram, India. pp. 16-17.

Woiciechowski, A.L., Soccol, C.R., Rocha, S.N. and Pandey, A. 2004. Xanthan gum production from cassava bagasse hydrolysate with *Xanthomonas campestris* using alternative sources of nitrogen. *Appl. Biochem. Biotechnol.*, 118: 305- 312.

Yamakawa, O. 1998. Development of new cultivation and utilization system for sweet potato towards 21st century. *(In:)* Proceedings of the International Workshop on Sweet potato Production System toward the 21st Century, Dec 9-10, 1997, Miyakonojo, Miyazaki, Japan, pp. 1-8.

Yamakawa, O. 2000. New cultivation and utilization system for sweet potato toward the 21st century. (In:) Potential of Root Crops for Food and Industrial Resources (eds

Nakatani, M. and Komaki, K.). Twelfth Symposium of International Society of Tropical Root Crops (ISTRC), 10-16 Sept., Tsukuba, Japan. pp. 8-13.

Yoshi, H. and Garcia, V.V. 1984. Processing of cassava into fermented foods. (*In:) Tropical Root Crops: Post harvest Physiology and Processing* (eds Uritani, I. and Reyes, D.), Japan Scientific Society Press, Tokyo. 205 pp.

□□□

15

Indigenous Traditional Knowledge on Root and Tuber Crops in India

After attaining a rich experience over a period of time, our forefathers have developed very pertinent primitive knowledge in a wide variety of fields (Rohilla *et al.*, 1999). These location specific wisdom are known as indigenous traditional knowleges (ITKs). Indigenous traditional knowledge are the upshots of the pragmatic approach by our sons of the soil. As such, there are many ITKs emerged due to the incremental and cumulative learning among societies living in close connection with nature. The traditional communities are often called as indigenous or local or ethinic people. They have accumulated lots of empirical knowledge gained from the day-to-day life. The variations in availability of raw materials and environmental conditions clubbed with time tested traditional knowledge and wisdom, have made people of different regions of the North Eastern hill states to formulate, develop and perpetuate consumption of a wide range of traditional foods and beverages unique to its places since ages (Savitri and Bhalla, 2007).

The root and tuber crops are cultivated in several pockets of India. The cultivation of cassava is very popular in Southern States like Tamil Nadu, Kerala and Andhra Pradesh. The crops like sweet potato, elephant foot yam, *Colocasia* and *Xanthosoma* are grown commercially in eastern and northeastern states. About 70% of the area under sweet potato is distributed in Orissa, West Bengal, Uttar Pradesh and Bihar. Since time immemorial root and tuber crops are being cultivated in India. Learning about such edible plants, processing of foods and medicine using location specific wisdom and conservation of food related resources has been in the large part due to incremental and cumulative learning of the tribal community.

Predominantly, these ITKs are rarely in documented form and generally transmitted to the next generation. The systematic documentation of ITKs will not only help the researchers to understand the scientific base behind them, but will also pave way for the researchers to focus on the location specific problems (Prakash *et al.*, 1999). Hence, an effort was made to compile available ITKs on tuber crops looking into the importance that they clutch to their credit.

ITKs on tuber crops could be sorted into following categories *viz.*,

- As food
- In Medicine
- In feed and fodder
- In propagation and cultivation
- In post harvest management

1. Food and Medicine

Since many years of hunter-gatherers, our knowledge base has evolved with the ecosystems within which it existed and has further developed as a result of historical continuity of local resource dependence. Knowing wild animals and plants which are palatable and which have nutritious content, has long been a survival strategy for the rural population, indigenous people and tribal communities, particularly those living in harsh adverse environmental conditions. Basic information is essential for supplementing diets when harvests fail due to unfore seen natural calamities. Hence wild nutritional resources are often termed as the *hidden harvest* (Pilgrim *et al.*, 2007) especially when it comes to roots and tubers which are really the hidden treasures for the needy.

Traditional foods are important and inseparable constituents of diets of tribals/tribes ad play a vital role in their indigenous traditional life style. The traditional and fermented foods of Monpa tribe in West Kameng district (Singh *et al.*, 2007), *Adis* tribe, *Nyishis* tribe, *Apatanis* tribe of Arunachal Pradesh (Tiwari and Mahanta, 2007), Pnar tribe of Jaintia hills of Meghalaya, Naga tribes of Nagaland (Mao and Odyuo, 2007), tribal women of North East (Singh *et al.*, 2007), Kandha tribe in Kalahandi district of Orissa (Panda and Padhy, 2007), *Kadars, Pulaiyars, Malasars, Malaimalasars* and *Mudhuvars* of Nilgiris district of Tamil Nadu (Ramachandran, 2007) and Muthuvan tribes of Idukki district of Kerala (Manithottam and Francis, 2007) involve varieties of tuber crops. Chiefly, they encompass yams *(Dioscorea* spp.), taro, bunda, swamp taro, elephant foot yam, sweet potato, cassava, giant taro, *etc.*

Tribal women of North Eastern region have a wide range of variability in the ethnic food made of tubers, leaf and other edible plant parts. In each state, the processing method of these food is somewhat different based on the culture, variability in the materials used in the food, climate and overall knowledge of the processing and preparation. Food materials used in the dietary system were found to be nutritionally rich and culturally important in various festivals and ceremonies. Ethnic food prepared and consumed by women cannot be seen in the isolated mode, instead it is a complex dynamics in which nutrition, health, food security, culture, ethics, subsistence economy and ecological sustainability are integral components (Singh *et al.,* 2007).

There are about 110 wild plants growing in Meghalaya. The plant parts, *viz.,* roots, tubers, stems, leaves, flowers, fruits and seeds are used in raw or cooked forms which are eaten whole or in part by the local people, *Khasi, Jaintia and Garo* tribes of Meghalaya (Kayand, 2007).

An ethnobotanical survey of food practices of an aboriginal, hill-dwelling *Kandha* tribe of Kalahandi district, revealed that in addition to their conventional foods, rice, finger millet and a few popular pulses, they use many types of naturally occurring unusual additional food items such as several types of tubers of the genus *Dioscorea,* Kernels of mango, wild bean *Mucuna utilis* Wall. ex Wight, *Madhuca indica* J.F. Gmel. flowers, *Caryota urens* pith, *Tamarindus indica* seeds, young seedlings of bamboo *(Dendrocalamus strictus)* and wild mushrooms but non poisonous (Panda and Padhy, 2007).

Traditional medical system of Indian Ayurveda indicates variety of alcoholic beverages named generally as *Madya* in which roots and tubers form an integral part (Sekar, 2007). Brahmaputra valley of Assam is very unique in its plant bioresources and is inhabited by different communities. Besides the cultivated crops, they are also dependent on various wild plants, which are used for consumption as well as in ethnomedicine. There are 38 wild plants of Majuli island and Darrang districts (Barua *et al.,* 2007). The wild edible food plants of *Palliyars* in the Southeastern slopes of the Western Ghats, Tamil Nadu covers about 171 species belonging to 67 families which include many of the root and tuber crops (Arinathan *et al.,* 2007). Traditional wild food plants used among tribal people of Andhra Pradesh include 156 species. Among those species, 56 species are herbs followed by 55 trees, 27 shrubs and 18 climbers. Mostly, leafy vegetables, tubers, fruits etc., are used (Reddy *et al.,* 2007).

Antioxidant properties are noted in roots and tubers which are associated with slowing the ageing process and improving overall health. Phyto oestrogens are present in *Dioscorea* spp (especially in wild yams) and in

Asparagus spp. Many tribals are aware of this properties and make use of such roots and tubers.

In Manipur, giant taro (*Alocasia indica* Schoot.) leaves are used for wrapping the bamboo basket internally, which is used mainly for fermentation of alcoholic beverages before distillation (Singh and Singh, 2006).

Young leaves of taro are cut in to small pieces and mixed with onion, garlic, green chilli , turmeric powder, besan and salt and deep fried. This preparation called *Colocasia* pakoda is very popular among Bengali communities in Tripura and Assam.

Elephant foot yam chutney with or without dry fish is common dish among the tribes of Tripura and Meghalaya.

Xanthosoma petioles and leaves are eaten as salad (petiole and leaves are cut in to small pieces and it is taken along with tamarind, green chilli and salt).

Colocasia powder is used in treatment of joint pains. *Colocasia* corms with high acridity (high calcium oxalate) are used in the preparation *Colocasia* powder, which is very much useful in joint pains and stomach ache. The procedure is as follows

Selection of high acrid corms	Raw rice
↓	↓
Peeling	Soak it in water for 20 minute
↓	↓
Cut in to small piece	Decant and dry it in sun
↓	↓
Drying	Grinding
↓	↓
Grinding	Rice flour

Flour
↓
Mixing 1: 1 ratio of colocasia and rice flour
↓
Add little red chilli powder and salt and roasting in frying pan
↓
Bottle it after cooling
↓
Powder can be stored for 30-45 days.

Elephant foot yam dry powder balls

Elephant foot yam is locally called as Mayungkapher (Mayung means elephant and kapher means the foot impression in Kokbrok language). The tribal people collect the corms from forest. The corms are peeled, sun dried and powdered.The powder is again made into small ball size with the addition of water. This ball is covered with the pulp of the seeded banana available locally. This encapsulated ball is taken for controlling stomach disorders and piles. Use of elephant foot yam corms has also been highlighted in Ayurveda literature for piles and gastrointestinal problems.

It is also effective in curing dyspepsia, splenic disorders etc. It is more useful if used along with diet. It is excellent laxative and stimulates liver secretion. It also possesses rejuvenative and aphrodisiac properties. It regulates menstural cycle. In Indian Alchemy called 'Rasasasthra' elephant foot yam adorns significant role. Many formulations described in Alchemy Consisting of 'Choorna' is indicated for inscreased sexual potency (Raghu *et al.,* 1999).

Colocasia Leaf Biscuit (Aniishe)

Aniishe is an ethnic food item of Ao Naga community under Mokochung district of Nagaland.It is used as tastemaker also for use during lean period when the availability of fresh leaf is less. For preparation, large quantities of *Colocasia* leaves are required. The leaves are collected and kept for some period till they become yellowish. Ribs of the leaves are excluded with preparation. Then the leaves are pound nicely till they are made into a paste. The pound leaves are then wrapped with banana leaves in which small holes are made in the lower part. Then it is placed on a elevated structure over which a heavy object like stone is kept for about 3-4hrs to drain out juice. Again, it is transferred to the metallic plate, which is placed over the fire. Ash is spread over the metallic plate before placing the wrapped *Colocasia* leaves over it. Heavy objects are again placed over it and time-to-time it is turned over. After few hours, it is ground again with addition of dry chilli, salt and local spices. This paste is made in to biscuit form, having a diameter of about 2 inches, for drying over the fire. The taste is better if dried quickly. Hence, people go for over night drying (Pongen, 2003).

Dried *Aniishe* is heated in pan with continuous stirring and then pound in large mortar into powder. Cooked smoked meat, dry chilli, salt and other local spices are added to it and ground again. This can be taken as such with rice. This ethnic food has very good market value, fetching Rs.160-180 kg^{-1} and has been in practice for long time among the Ao Naga community (Pongen,2003).

Cooking the elephant foot yam in bamboo shoot ash water and after decanting water, the cooked tubers are made in to paste and dried in the form of cake. After drying, the cake is again cooked in bamboo shoot ash water and dried in sun after decanting ash water. This dried cake can be stored for 30-45 days with out any quality deterioration. This preparation helps to remove the acridity and enhance the nutrient availability in terms P and K. This preparation is popular among the tribes of Tripura.

Xanthosoma petiole and tender leaves are used for making curries and fry.

Swamp taro stolon are used as vegetables by both Bengali and tribal communities.

Dioscorea spp. are well known in traditional medicine of Africa, China and Asia. It is widely used among tribals. Tuber of *Dioscorea deltoidea* is used in rheumatics and ophthalmic illness. *D. floribunda* are used as contraceptive by the tribals in the North East Hill region as well as in Andaman and Nicobar Islands. Many tribal farmers of Andaman and Nicobar Islands grow many species of *Dioscorea* especially *D. alata, D. esculenta, D. bulbifera, D. pentaphylla, D. hamiltonii, D. opposipolila, D. deltoidea, D. floribunda* and *D. sylvatica* in their jhum fields for its medicinal and food values.

The crushed pod of *P. tuberosus* mixed with largl is used to cure itch (Kay, 1987).

2. Reducing Bitterness in Sweet Potato

Flowers of Mahua *(Madhuca indica)* have characteristic feature of minimizing the bitterness which is caused by feeding of sweet potato weevil. *Cylas formicarius* F. is an ubiquitous pest of sweet potato. Adult and grubs of *C. formicarius* produce pentacyclic triterpenoids in tubers while feeding. This is bitter in taste and unsuitable for consumption. At the time of planting of sweet potato vines, flowers of mahua *(Madhuca indica)* are placed in the furrow of soil at the interval of about 30 cm. The practice is followed in villages of Ranchi district of Jharkhand and has potentially for its further spread where mahua plants are available. This helps in reducing the weevil damage and there by reducing bitterness in tubers to a great extent (Kumar, 2003). These tubers with reduced bitterness are used mainly as feed material for livestock besides human consumption.

3. Propagation and Cultivation

(i) Indigenous practice for raising planting material

In Maharashtra, the farmers follow the practice of storing various kinds of yams and aroids for seed tuber purpose in a pit of 30x30x30 cm at the bottom of the which white sand is spread in a layer of 5 cm thickness. The rhizomes(or) seed yams are placed in the rows inside the pit again covered with a thin layer of white sand. Finally, the pit is covered with soil. This process preserves rhizomes (or) seed yams in good and viable condition (Tawade, 2003).

(ii) Planting material production

Colocasia is cultivated in kharif season, for planting purpose using small cormel/corm cuttings. To produce rooted planting, a pit of 1x1x1M is dug and filled with mango leaf and dry cowdung and burned into ash. The small cuttings are planted in the ash and subsequently, the pit is filled with dry twigs and branches of mango, guava etc. It is irrigated at an interval of about 3-4 days. When the plant reaches the height of 30 cm above the ground level, planting materials are ready for planting (small rooted cuttings) (Marandi,2003).

(iii) Establishment of setts

In Kerala, some farmers used cones of jack fruit leaves and fixed them over the planted setts of cassava in summer. This practice of putting jack leaf cone over the cassava setts provides partial shade, thus preventing the moisture loss from the cut ends of the setts in summer, leading to better establishment of the setts by 12-15% and saves 30% irrigation. This is in practice for more than 25 years (Velayudhan, 2003). In south India, the application of fresh cowdung on the cut surface of planted setts is common practice. This practice prevents the moisture loss from the cut ends and fresh cowdung also supplies the auxin like compound to the setts, which results in early rooting and shooting.

In Tripura and Meghalaya, the practice of treating ash in the cut ends of elephant foot yam and *Dioscorea* prevent the tuber rot and increase the shellife of tubers. Ash acts as a deterrant and disinfectant for the chewing and sucking type of insect pest (Lal and Verma, 2006). Alder based farming practice is very common among the indigenous communities of Meghalaya in which sweet potato, elephant foot yam and *Colocasia* are grown.

Pest management

In elephant foot yam corm borer attacks at later stage of the crop when tuberization is at its peak. It enters into tuber and eats away the tuber resulting in reduction of yield. The larva is about one inch long with stripes of brownish green colour. This larvae is locally called as 'laskari iyal' (Prodenia/Spodoptera) in Valsad district of Gujarat.To control this pest, some farmers use the residual waste obtained during preparation of country wine. Approximately one litre of this alcoholic liquid is applied to the soil near the base of the plant (Patel, 1996).

It has been reported that the cultivation of elephant foot yam as intercrop in arecanut garden would control the root grub. The exudates from elephant foot yam inhibit the growth of root grub (Anonymous, 2002). Besides, the incidence of Koleroga in arecanut is also kept under check by growing elephant foot yam as intercrop.

Firewood or dung cake burnt ash is dusted on tubers to avoid pest infestation. It is being practiced in Huzurabad and surrounding villages of Karimnagar district in Andhra Pradesh under tuber crops based farming system (Narayanareddy, 2003).

Wood ash applied over or around the tuber crops or pasted on tubers for controlling insect pest damage. Ash suffocates the eggs and larvae of insects and prevents the contact of adult insect pests on the leaves. These enhances the productivity of the crops. In most of the villages, in hilly area, farmers use dry cow dung and wood for fuel purpose. The ash obtained is spread over the soil and around the tuber crops. This ITK, which is cheap and eco-friendly, is practiced in most of the villages in hilly mountain terrain of Himachal Pradesh since time immemorial (Verma, 2003).

4. Post Harvest Management

In order to keep the harvested cormels of taro fresh till next sowing season and for table use, it is stored in pits about three feet deep. After two layer of *Colocasia*, one foot layer of soil is put on the top. By this method, *Colocasia* cormels can be stored as fresh for longer time. This method has been practiced by the farmers over several decades in the Himalayan region (Verma, 1998)

In Uttaranchal, the farmers store *Colocasia* cormels with the green leaves of *Malu*'. During the storage, the cormels are covered with the green leaves of *Malu* from all sides. After covering it properly, it is kept inside a 2-3 inch layer of cowdung. The tubers could be safely stored for the next season without losing moisture and vigour. The leaves of *Malu are* locally available and hence the cost is nil (Mehta and Bisht, 2003).

Freshly harvested cassava tubers can be stored by burying them in loose soil to avoid damage. Freshly harvested cassava can also be stored for at least one month without damage if it is buried intact with out removing from the stem in a cool and humid place with out water logging.

In Kumaun hills, tubers buried deep into the pit and covered with heep of hay for its preservation. This technique preserved tubers for long time for future use (Chander and Mukherjee, 1994).

Other ITK in tuber crops

- Planting giant taro in uplands gives lesser acridity corms.
- Use of small cormels for production of seed corms in elephant foot yam is popular in Gujarat and Tripura.
- Planting of *Dioscorea* in the cement bags kept in side a pit method is used for easy harvest of deeply grown tuber of *D.alata* varieties. The pit made for planting *Dioscorea spp* is filled with banana pseudostem and planting is done above the pseudostem. This method is used for production of long tuber and higher yield .
- Cassava stems are used in ponds for protection and periphyton development. The cassava stem has got dual advantage in using ponds. It will give give protection like that of bamboo in ponds. The semi decayed peels from the stems would be ideal food for omnivorous carps. Unlike bamboo, the decayed and damaged stems will float on the water and can be removed easily at the time of harvesting and it will not damage the nets.
- Cassava and sweet potato leafy foliage is used for feeding the grass carp.
- *Colocasia* flowers are popularly used for preparation of a kind of beverage, which is common among the tribes of Tripura

Conclusion

Indigenous traditional knowledges developed over years and passed on from generation to generation are the nature's gift for the welfare and livelihood of humankind. Traditional foods processed and prepared through region specific ITK are intimately connected to socio-cultural, ecological, spiritual life and health of local community. Processing and preparation of ethnic foods not only demonstrate the creativity and treasure of food heritage of tribals but also their incremental learning to sustain the life and ecosystem as a whole. ITKs on tuber crops are the tools that help to raise the socio-economic status of the rural people, in general and tribals, in particular. Indigenous traditional knowledges are the essence of tribal life style and have indispensable utility in all the walks of life in one or other arena.

References

Anonymous, 2002. Control of root grub in arecanut *(In:)* Inventory of Indigenous Technical Knowledge in Agriculture, Division of Agrl.Extension, ICAR, New Delhi. Vol:1, 174 pp.

Arinathan, Y., Mohan, V.R., Britto, A.J.D. and Murugan, C. 2007. Wild edibles used by *Palliyars* of the Western Ghats, Tamil Nadu, *Indian J. Traditional Knowledge,* 6(1):163-168.

Barua, U., Hore, D.K. and Sarma, R. 2007. Wild edible plants of Majuli island and Darrang districts of Assam. *Indian J. Traditional Knowledge,* 6(1):191-194.

Chandler, M. and Mukherjee, R. 1994. Traditional agriculture and animal husbandry practices for sustainable agriculture in Kumuan Hills of Uttar Pradesh. *J. Rural Development,* NIRD, 13 (3):443-449.

Kayand, H. 2007. Tribal knowledge on wild edible plants of Meghalaya, Northest India. *Indian J. Traditional Knowledge,* 6(1):177-181.

Kay, D.E. 1987. Crop digest and Product No. 2. Root crops (Revised by Gooding, E.G.B.). Tropical Development and Research Institute, London. 380 pp.

Kumar, V. 2003. Pest control in sweet potato *(In:)* Inventory of Indigenous Technical Knowledge in Agriculture, Division of Agrl.Extension, ICAR, New Delhi. Vol:2. 282 pp.

Lal, C and Verma, L. R. 2006. Use of certain bio products for insect and pest control. *Indian J. Trad.Knowledge*. 5(1), 29-82.

Manithottam, J. and Francis, M.S. 2007. *Arenga wightii* Griff. - A unique source of starch and beverage for Muthuvan tribe of Idukki district, Kerala. *Indian J. Traditional Knowledge,* 6(1):195-198.

Mao, A. A. and Odyuo, N. 2007. Traditional fermented foods if Naga tribes of Northeastern India. *Indian J. Traditional Knowledge,* 6(1):37-41.

Marandi, M. 2003. Planting material production *(In:)* Inventory of Indigenous Technical Knowledge in Agriculture, Division of Agrl.Extension, ICAR, New Delhi. Vol:2, 283 pp.

Mehta, P.S. and Bisht, I. S. 2003. Storage technique for *Colocasia*, ginger and turmeric *(In:)* Inventory of Indigenous Technical Knowledge in Agriculture, Division of Agrl.Extension, ICAR, New Delhi. Vol:2, 245 pp.

Narayanareddy, P. 2003. Dusting cow dung ash on vegetable plant *(In:)* Inventory of Indigenous Technical Knowledge in Agriculture, Division of Agrl. Extension, ICAR, New Delhi, Vol:2, 287 pp.

Panda, T. and Padhy, R.N. 2007. Sustainable food habits of the hill-dwelling Kandha tribe in Kalahandi district of Orissa, *Indian J. Traditional Knowledge,* 6(1):103-105.

Patel, D. D. 1996. Control of larvae in tuber. *Honey bee* 7(3): 16.

Pilgrim, S., Cullen, L., Smith, D. and Pretty, J. 2007. *Hidden harvest* or hidden revenue – A local resource use in a remote region of Southeast Sulawesi, Indonesia. *Indian J. Traditional Knowledge,* 6(1):150-159.

Pongen, T. 2003. *(In:)* Inventory of Indigenous Technical Knowledge in Agriculture, Division of Agrl.Extension, ICAR, New Delhi. Vol:2, 613 pp.

Prakash, N., Pal, P.P., Sundarambal, P., Kumar, R. and Bihari, B. 1999. Indigenous technological knowledge followed by the tribal farmers of north eastern hill region in Agriculture. *Indian J. Hill Farmg* 12 (1&2): 134-135.

Raghu, V., Deepa, V.C. and Sundaran, K. 1999. A study on soorna (*Annorphophallus paeoniifolius*) the king of tubers. In : Tropical Tuber crops. Oxford & IBH, New Delhi. pp. 10-14.

Ramachandran, V.S. 2007. Wild edible plants of the Anamalais, Coimbatore district, Western Ghats, Tamil Nadu. *Indian J. Traditional Knowledge,* 5(2):173-176.

Reddy, K.N., Pattanaik, C., Reddy, C.S. and Raju, V.S. 2007. Traditional knowledge on wild food plants in Andhra Pradesh. *Indian J. Traditional Knowledge,* 6(1):223-229.

Rohilla, P.P., Bujarbaruah, K.M., Singh, G. and Kumar, M. 1999. Indigenous technical knowledge used for animals treatment in foothills of Nagaland. *Indian J. Hill Farmg.* 12 (1&2):121-123.

Savitri and Bhalla, T.C. 2007. Traditional foods and beverages of Himachal Pradesh. *Indian J. Traditional Knowledge,* 6(1):17-24.

Sekar, S. 2007. Traditional alcoholic beverages from Ayurveda and their role on human health. *Indian J. Traditional Knowledge,* 6(1):144-149.

Singh, P. K. and Singh, K. I. 2006. Traditional alcoholic beverage, yu of meitei communities of Manipur. *Indian J. Traditional Knowledge* 5(2) 184-190.

Singh, R.K., Singh, A. and Sureja, A.K. 2007. Traditional Foods of *Monpa* tribe of West Kameng, Arunachal Pradesh. *Indian J. Traditional Knowledge,* 6(1):79-94.

Tawade, N. D. 2003. Indigenous practice for raising planting material *(In:)* Inventory of Indigenous Technical Knowledge in Agriculture, Division of Agrl.Extension, ICAR, New Delhi . Vol:2 308 pp.

Tiwari, S.C. and Mahanta, D. 2007. Ethnological observations on fermented food products of certain tribes of Arunachal Pradesh. *Indian J. Traditional Knowledge,* 6(1):106-110.

Velayudhan, S. 2003. Establishment of setts *(In:)* Inventory of Indigenous Technical Knowledge in Agriculture, Division of Agrl.Extension, ICAR, New Delhi. Vol:2, 281 pp.

Verma, L. R. 1998. A Glimse of Indigenous Technical Knowledge for watershed management in upper north west Himalayas of India (Aparna Negi and Prem N

Sharma Eds.) WATMATEC, YSP University of Horticulture and Forestry, Solan, HP and PWMTA, Netharlands/ FAO (UN), Kathmandu, Nepal, pp 1-97.

Verma, L. R. 2003. Wood ash as pesticide for tuber crops. *(In:)* Inventory of Indigenous Technical Knowledge in Agriculture, Division of Agrl.Extension, ICAR, New Delhi. Vol:2. 314 pp.

□□□

16

Potentials of Tropical Tuber Crops

In tropics, many crops are grown for their edible roots or tubers. Broad group of "tuber crops" encompasses a wide range of starchy food crops, some of which are true botanical roots and some others are stem tubers or corms. In global scenario, they are in a neglected state mainly due to their negative image as poor people's food. They are highly perishable and are normally consumed locally in fresh or home processed state. It is time to bring out the positive attributes of these important foods and the increased contribution they can make to the nutritional welfare, food security and economic wellbeing of people, especially in developing countries.

Food Security

Tuber crops make an important contribution to the diet of many impoverished people in tropical countries assuring their food security. In general, tuber crops are consumed as a cheap source of dietary energy either as a basic source or as a supplement to cereals. For example, cost of calories from cassava is about 25 to 50 % to that of the locally produced traditional grains and pulses (Goering, 1979). This energy is about one-third of that of an equivalent weight of grains such as rice or wheat. Cassava and yams have been traditionally associated with poverty and was a major factor contributing to development of *Kwashiorkor*, a disease due to malnutrition. Although tuber crops are deficient in protein, regions where tuber crops are consumed do not necessarily coincide with a high incidence of malnutrition. The nutritive deficiencies of tuber crops need not be a cause for concern when it is consumed with other supplementary foods. The *Kwashiorkor* disease problem from cassava is mainly due to consumption of excessive quantities because it supplies sufficient calories and gives a feeling of satiety in their diet (Grace, 1977). Therefore, many suffer from deficiency of protein and vitamins.

The case of Kerala is a classic example in defense of cassava. Kerala state has a population of 32 million, whose staple food is rice. Production of rice in Kerala is hardly sufficient to meet one-fifth of its requirement. As the land suitable for cultivation of rice is in short supply, and it can not be grown in marginal uplands, such uplands areas of low fertility are planted mainly with cassava. Thus, along with rice, cassava also attained the status of a staple food. This might have been expected to adversely affect nutritional status of people. However, the infant mortality rate, an index of nutritional status, remained relatively low in Kerala. Although cassava supplied very little protein, some of the deficit in the diet was made up by the intake of rice and fish. Fish and fish products are very common in the diet of people of Kerala. Therefore, it can be concluded that as tuber crops are inferior in protein and lipids, some cereals and animal protein or pulses can be used to balance the diet. This has been confirmed by the data discussed by Goering (1979) indicating that serious protein deficiency is not necessarily common in countries where tuber crops are one of the sources of calories.

Daily protein requirement for optimal health status is estimated at about 0.8 g kg^{-1} of body weight (WHO, 1985) indicating need for substantial amounts of high-quality dietary protein. Unfortunately, in most developing countries, high quality proteins are scarce and of limited availability. In certain developing countries, where there is substantial consumption of cassava, protein malnutrition does exist. Animal protein is particularly limited in areas of drought and in areas of poor soil quality or limited land. In urban areas, meats and animal products such as milk and eggs are mostly imported or even when produced locally, are generally expensive. As such protein malnutrition is prevalent, particularly among dependent groups such as children and prenatal women who have the highest needs for high quality protein. Genetic modification can be applied for improving availability and quality of protein in food crops (Niba, 2003), and in future, protein rich tuber crops may be released especially if the global biotechnological initiatives become fruitful.

Tuber crops can be used for preparation of baby foods too, although their energy content on a wet basis is low compared to cereals. The energy density can be increased by drying and thus be used in baby foods (Oke, 1990). Cassava is used in a number of commercial baby foods in industrialized countries. Composite flours prepared from tuber crops and cereals could be used in baby food formulae, if appropriately supplemented. Addition of malted cereals to cassava flour increases energy density of gruels prepared from it, by reducing their viscosity through action of amylolytic enzymes. The use of fresh cassava products as infant weaning foods should be discouraged, because of probable toxicity, low protein content and energy density.

As tubers are very low in lipids, they can not be rich sources of lipid soluble vitamins. Apart from yellow variety of sweet potato, which contains a high amount of beta-carotenes (up to 30 mg retinol equivalent %) most of the other root crops contain only negligible amounts (Oke, 1990). Their leaves contain a substantial amount of beta-carotenes that could contribute significantly to daily requirement of vitamin A, especially for children, thereby helping to eradicate ocular diseases affecting about six to eight million children from Asia, Africa and Latin America. Dietary retinol obtained from consumption of animal foods is relatively expensive and contributes about 14 and 20 % to the vitamin A intake of people in Asia and Africa respectively. Beta-carotenes from leaves such as sweet potato or cassava contain about 800 mg 100 g^{-1}, which are about the same as beef liver, contribute as much as 86 % in Asia and 80 % in Africa. Certain yams are also promising in this regard. A few tuber crops contain appreciable amounts of Vitamin C. Tuber crops, if correctly prepared, can make a significant contribution to the vitamin C content of the diet (Augustin *et al.*, 1975). Most of the tuber crops contain small amounts of vitamin B group, sufficient to supplement normal dietary sources. The B-group of vitamins acts as a co-factor in enzyme systems involved in oxidation of food and production of energy. These vitamins are found mainly in cereals, milk and milk products, meat and green vegetables, including leaves of roots and tubers.

Leaves of tuber crops are also used as food or feed. As a vegetable, leaves of cassava, sweet potato, and aroids are used in many countries. They are also utilized in a big way as animal food either fresh or in processed form. In many African and tropical Asian countries, tuber crops are usually subsistence crops grown mainly as food, therefore, farmers keep sufficient quantity to feed their families and sell only the surplus. In Kerala also the situation is almost same. With the changes in standard of living and urbanization, people shift from a tuber crop predominant diet to a cereal predominant diet. In urban areas, direct consumption of tuber crops is negligible. However, there may be a growing commercial market for value added products from them.

Starch and its products

From a global perspective, cassava and sweet potato will be increasingly used in processed form for food, feed and starch derived products such as high fructose syrup and monosodium glutamate (Scott *et al*, 2000a). Non-food, non-feed uses will also increase. Yams will be used largely as food in fresh from. A better understanding on quality characteristics of yam starch may identify additional market segments for yams.

According to Moorthy (2001), tuber crops starches have a bright future. The high viscosity of cassava and canna starches make them very useful in many food and industrial applications especially where high thickening powder is desired. The low viscosity of aroids can be exploited in paper industries which prefer low viscosity and good film forming capacity. The smaller granular structure of taro starches makes them ideal as filler in bio degradable plastics and in aerosols, and talcum powders. The good gel strength of these starches is another merit. Easy gelatinization of cassava and sweet potato starches can make them suitable in manufacture of hydrolytic product derived from starch. It is also possible to modify starch properties by simple physical methods like hydrothermal or steam pressure treatments. Fermentation of starch by use of selective organisms or enzyme modifications can also bring about specific substitutions.

Fermented foods constitute an important food ingredient in many developing countries. These include fermented grains and tuber crops and milk products. Fermented cassava and yam foods are staples in many parts of Africa. Fermentation accomplished by bacteria and yeasts facilitate preservation and storage. It also serves as a means to reduce detrimental components such as cyanogenic glycosides often present in tuber crops such as cassava. Levels of anti-nutritional factors such as haemagglutinins and tannins are depleted by fermentation (Mbugua, 1992). In addition, fermentation may modify starches to form resistant starches which are associated with protection against colon cancer and gastrointestinal disease (Ahmed *et al.*, 2000).

There is great potential for modification and improvement of starch and carbohydrate quality. Tuber crop starches are typically high in amylopectin with lower ratios of amylose to amylopectin. This yields highly digestible, high-glycemic index foods. These high-glycemic index starches are associated with diet-related conditions such as type 2 diabetes and insulin resistance. Modification of starch to increase amylose to amylopectin ratios would be greatly useful in improving glycemic index and physiological responses to starchy foods that are prevalent in diet in many developing countries (Niba, 2003). Modifications in starch synthesis pathway were effected to modify ratios of amylose to amylopectin in potato and cassava starch (Visser *et al.*, 1991; 1997). Improving the glycemic indices of staple starchy food crops such as tuber crops will improve their nutritional quality, and subsequently have an effect on prevalence of attendant disease conditions.

Tuber crops also contain non-starch polysaccharides, including celluloses, pectins and hemicelluloses, and other associated structural proteins and lignins, which are collectively referred to as dietary fibre (Oke,1990). Role of dietary fibre in nutrition aroused a lot of interest in recent years. Some epidemiological evidence suggests that increased fibre consumption contributes to a reduction in incidence of certain diseases, including diabetes, coronary heart disease, colon cancer, and various digestive disorders. Fibre acts as a molecular sieve, trapping carcinogens which would otherwise have been recirculated into the body. It also absorbs water thus producing soft and bulky stools. Sweet potato is a significant source of dietary fibre as its pectin content can be as high as 5 % of fresh weight or 20 % of dry matter at harvest (Collins and Walter, 1982). There is, therefore, a need to pay more attention to the significance of fibrous components in these tuber crops, especially sweet potato, and to determine their composition and dietary function. Other tuber crops, particularly yam, contain mucilages, which have a considerable influence on their cooking qualities.

Another major area of use is production of ethanol from tuber crops. Already, several ethanol producing factories based on cassava and sweet potato have been set up in many Southeastern countries. The alcohol can be used as an automotive fuel either mixed with petrol or as anhydrous ethanol.

Animal feed

Non-availability of low cost feed is a major hindrance restricting development of animal production in many developing countries. Traditionally, maize is a major ingredient of animal feed. When feed costs increase, animal products become very expensive. If part of the feed could be substituted with less expensive tuber root crops like cassava and sweet potato, then feed cost can be reduced substantially. In addition, a part of the maize ration could be freed for human consumption. In cassava, apart from tubers, cassava tops, stems and leaves are also available as animal feed and are comparatively high in utilizable protein. Sweet potato tubers, vines and leaves can also be used as animal feed.

Production of single cell protein for livestock feed is another opportunity for tuber crops (Oke, 1990). These tuber crops could serve as suitable substrates for a range of micro-organisms, and additional feed can be produced. Under optimal conditions, from 100 kg of sweet potato containing 6.9 kg protein, certain fungi could produce 8.12 kg of dried mycelia, and a residue of unutilized sweet potato tissue for livestock feed, contain 31.6 kg protein i.e. protein concentration can be increased fourfold. Fungal mycelia contain more lysine, histidine, tryptophan, methionine and tyrosine than casein (Gray and Abou-el-Seoud, 1966).

Having considered the potential of tuber crops in general, now, a detailed account of potential of major tuber crops is enunciated.

1. Cassava

Cassava plays a vital role in many areas influencing livelihoods of millions of people in developing world. It tolerates drought, soils with low nutrient availability and poor husbandry (Burrel, 2003). Cassava tubers and young leaves have multiple end uses. This includes direct human consumption of fresh tubers and leaves after cooking, on-farm animal feeding, commercial production of animal feed, and production of starch and starch derivatives. Demand for fresh cassava depends on improved storage methods, but markets for cassava as a substitute for cereal flours in bakery products and as energy source in animal feed rations are likely to expand. Cassava leaves, which contain appreciable quantities of protein and vitamins, are also consumed as a vegetable. Cassava starch is one of the most important plant products to humans. Cassava tubers and leaves are also used as livestock feeds either directly or after processing into pellets or silage.

1.1 Fresh tubers for human consumption

Fresh peeled cassava tubers are consumed directly after boiling or roasting. This type of consumption is predominant in African countries, India and certain parts of China and Vietnam. Market also includes pre-processed cassava, where tubers are peeled and in some cases pre-cooked vacuum packed and frozen. As Howeler (2003) remarked, fresh cassava for human consumption does not have much growth potential in Asian countries as rice remains the preferred food in the region. Although total food demand may increase due to population growth, most Asian countries may become more affluent and are likely to reduce their consumption of high energy staple foods like rice and cassava in preference for meat products or convenient foods. Probably in African countries, demand may increase due to population growth. It is projected that by 2020, more than 50 % of the Asian population will be urban than rural. Urbanization changes food consumption pattern as urban population have to buy all their food and they prefer clean attractively packaged as convenient foods. In urban areas, there is greater potential for processed foods and snack foods. Thus cassava based processed products may find a niche in markets.

1.2 Parboiled and dried chips for human consumption

This type of consumption is prevalent in India. Tubers after removing outer leathery skin is cut into slices by a specially made thin broad knife and put into boiled water for a few minutes. It is then taken out and dried under sun. Product locally called *Vattukappa* can be stored for about a year or so without any problems from storage pests. This is an ingenuous method developed by local people to tide over the problem of non-availability of tubers in lean season. Demand for *Vattukappa* declines in future as more and more people change to cereal based diet instead of preserved cassava.

1.3 Flour for human consumption

Tubers are peeled, washed, and cut into slices or big chunks and dried. Dry cassava chips or chunks can be stored and traded. It is mainly used for producing cassava flour. In Indonesia it is known as *Gaplek*, and in Kerala it is known as *Vellukappa*. Cassava flour is used to prepare many food items. In Brazil, which ranks first in production of Cassava, cassava flour is a common food item (Norman *et al.*, 1995). In future, demand for flour will increase.

1.4 Starch for food and industry

In future, main use of cassava will be to produce starch for industrial use and other products used in processed food. From cassava, both native starch and modified starch can be produced. One tonne of fresh roots, normally produces 450 kg of dried chips (85%DM) and about 250-300 kg of native starch (Howeler , 2003). Starch is produced from both fresh tubers and dried chips. Starch is a big business worldwide, and it is finding application in several industries. Cassava starch performs most of the functions where maize, rice and wheat starch are currently used. Cassava starch has special characteristics for food processors. It is readily gelatinized by cooking with water and solution after cooling remains comparatively fluid. Solutions are relatively stable and do not separate again into an insoluble form (retrogradation) as is the case with maize and potato starch. High viscosity of cassava starches make them very useful in many food and industrial applications especially where high thickening power is desired. Starch is utilized in sizing and dyeing in textiles industries to increase brightness and weight of cloth. In pharmaceutical industries, starch serves as a filler material and bonding agent for making tablets.

Pearl sago is manufactured from cassava starch and this almost displaced sago from sago palm in India . Sago is mainly used as a convenience food, especially for invalids and infants. Wet starch containing 40 % moisture is rolled into small globules and these are roasted dried and finished. Cassava starch also has several other uses such as an additive in cement to improve setting time, and it is used to improve viscosity of drilling muds in oil wells. It is also used to seal walls of bore holes and prevent fluid loss. Starch is also main raw material in glue and adhesive industries. In paper production, cassava starch is currently used as glue to achieve brightness and strength. Starch is also an important raw material for powder in cosmetics industries. In detergent soap manufacture, starch is used to get better recovery and to improve shelf life of detergents. While in rubber and foam industries, starch is employed for getting better foaming and color. Cassava starch is also used for production of biodegradable plastics.

Cassava starch can be used for production of many types of sweeteners after hydrolyzation by either acids or enzymes or both. Cassava starch can be converted to maltotriose, maltose, glucose and other modified sugars and organic acids such as citric acid, acetic acid, and lactic acid (Tan *et al.*, 1984). Lactic acid is produced by fermentation of starch with *Lactocbacillus amylovorus* (Wang Xiaodong *et al.*, 2000). Roble *et al.*(2003) demonstrated production of L-Lactic acid from raw cassava starch in a bioreactor using the fungus, *Aspergillus awamori* and the bacterium, *Lactococcus lactis lactis.*

Starch from cassava can be used to make fructose syrups (Vuilleumier, 1993) and formulate gelatin capsules (Nduele *et al.* 1993). Hydrogenated sweeteners, which include sorbitol, mannitol and maltol, can be produced by treating starch with hydrogen gas in high-pressure tanks, using a special catalyst and ion exchange resins. Sorbitol is used mainly for production of vitamin C and as a moisture conditioner in tooth pastes (Jin Shuren, 2000). Monosodium glutamate (MSG), a flavour enhancing agent used in many Asian countries, and lysine, an important amino acid used as a supplement in animal feed, can also be produced from cassava. MSG, is made through microbial fermentation of starch in the presence of ammonium salt (Maneepan, 1996). In Thailand, MSG production is the main consumer of native cassava starch.

The world starch market is showing fast growth. According to estimates of Maneepan (1996), world starch market grew from 15.1 million tonnes in 1980 to 33.7 million tonnes in 1993. Although cassava starch represents only 6.0 % marketed annually (Ostertag *et al.*, 1996), it holds much promise for propelling cassava into global market as starch and derivatives.

Cassava starch has both regional and global significance. An outstanding example is its increased production in Southeast Asia. Once sub-Saharan Africa and South America begin to produce and export sizeable quantities of this income-generating product, its prospects for capturing a major proportion of the international market are high. In India, cassava starch is produced and is widely used. Several comparative advantages in production and use are encouraging many industries to adopt cassava starch as an alternative to traditional sources. Cassava starch is unique for its very low production costs, neutral taste, high purity, resistance to acid media and sheer stress, resistance to structural change under freezing, high viscosity and clear paste nature, and production of transparent gels. These qualities give cassava starch potential to replace most modified starches used in food industry. It is hoped that through genetic transformation cassava starch's functional properties can be modified in near future. This will definitely open up new potential avenues for the crop.

1.5 Ethanol from cassava

Use of cassava as a source of ethanol for fuel is already being exploited and is very promising. Mixing of ethanol upto 20 % with petrol not only does not cause any difficulty in running automobiles but results in less atmospheric pollution than use of 100 % petrol (Nair, 1978). The starch-to-alcohol conversion ratio is 1.76 kg starch for one litre of alcohol (Oke, 1990). Although sugar cane is the most energy efficient crop for alcohol production, use of cassava starch is increasing because it can be produced under conditions unsuitable for sugar cane. Howeler (2003) reports that in China, alcohol factories are switching from use of molasses to that of cassava chips for alcohol production, because of ever stricter pollution control requirements that makes use of molasses uneconomical. With unabated increase in prizes of petroleum products, this is an attractive avenue. Southeast Asian countries like China, Philippines, and Thailand are already promoting "gasohol" (about 10-20% ethanol mixed with gasoline). This trend is expected to markedly increase demand for cassava chips in future, as this is the cheapest raw material.

1.6 Cassava in animal feed

Both fresh and dried cassava tubers, in different forms such as sliced, chopped, ground, or pelleted, are used as animal feed. In the European countries large quantities of cassava chips, pellets and meal are being imported for animal feeding. Inclusion levels of up to 65 %, preferably peletted, do not seem to affect health, carcass quality or overall performance when the diets are carefully balanced. Feed formulae wherein cassava chips form ingredients from 24 to 43% has been recommended in Kerala (KAU, 2001). As cassava tubers contains about 1.0% protein only, its use as animal feeds for monogastric animals is limited. However, it is shown that cassava meal could be included in poultry rations up to 40% without detrimental effect, provided nutrient content of ration meet the poultry requirements (Kompiang *et al.*, 1994).

The quality of cassava tubers can be increased through a microbial process known as solid state fermentation. Crude protein content (dry basis) of cassava tubers can be increased by fermentation with *Aspergillus niger* from 3 to 18-42 %. The product is named as protein enriched cassava or "Cassapro", a brand name which is quite popular in Indonesia (Kompiang *et al.*, 1994). The dried fermented cassava tubers may be fed directly to ruminants or may be grounded to a 1.0-mm mesh before feeding to monogastrics.

Cassava fibre or pomace is the residue obtained after extraction of starch from cassava tubers. Cassava tubers yield approximately equal amounts of starch and fibre, which have less feed value than cassava root meal, but can be included in rations for cattle, as it contains about 24 % crude fibre and 55 to 70 % nitrogen free extract (Devendra, 1977). The inclusion of this by-product in broiler and pig diets is reported to be as much as 10 % (Aritonang and Silalahi, 1992), 30 % in cattle diets (Sumangkut *et al.*, 1976) and 20 % in sheep diets (Pulungan and Bhakti, 1984). A feed formula with 15 % cassava starch waste is recommended in Kerala (KAU, 2001) The nutritive quality of this by-product may also be improved by fermentation using *A. niger*, its crude protein content was increased from 1.55 to 18.5 % after fermentation (Kompiang *et al.*, 1995; Nur, 1995).

Fresh cassava tubers can be directly fed to pigs. In Nigeria, cassava is the main carbohydrate base for swine nutrition. Modebe (1963) formulated cassava-based pig rations for exotic breeds in Nigeria, and whole life cycle swine feeding systems using cassava roots are standardized (Gomez, 1977). Inclusion of cassava in feed rations of poultry and other classes of livestock are recent phenomena. Earlier, they were raised exclusively on maize or grain carbohydrate-based feeds. Scarcity and high cost of maize have brought into sharp focus value of utilizing cassava in livestock production industry because of special inherent attributes such as high energy yields and continuous availability.

1.7 Cassava leaves

Cassava leaves which are high in protein and in important vitamins especially Vitamin A are consumed as fresh vegetable in several African and south American countries. In Africa, leaves are cooked and prepared in different ways. As leaves contain considerable quantities of protein, there is vast scope for its utilization in different dishes. Cassava leaves are rich in protein and contain 20.6 to 36.4% protein by dry weight (Nair, 1978). As reported by Okeke (1978), crude protein content of cassava leaf blades ranges from 24-26% in Nigerian varieties. Müller *et al.* (1974) reported a crude protein content of 23.3%. In Kerala, tender cassava fruits are consumed as a vegetable.

Entire top growth, stem, petiole and leaf blades contain up to 17% crude protein (Montaldo, 1977). A Japanese factory in Thailand exploits this factor by pelletizing the top growth and exporting it to Japan as a protein source for use in animal feeds (Booth and Wholey, 1978). Annual yield of foliage is as high as 20 tonnes/ha (Maini, 1978). Although hydrocyanic acid is high in leaves, it can be brought down to safer levels by steeping in water and boiling and then drying. According to Montaldo (1977), cassava plants can withstand defoliation for several months, if they receive adequate fertilization and irrigation. Under good management, cassava has potential to produce up to 4.0 tonnes of protein per hectare, annually. Ensiling of cassava leaves is an appropriate way to conserve them (Limon, 1992).

Nutritional value of cassava leaves may be increased by fermentation with *A. niger* using a similar process as for cassava tubers. Other advantages obtained through this process are that the toxic content of HCN in leaves may be significantly reduced and fermented product may be stored for an extended period as they are in the form of powder and easily transported to other places if required. Bioprocess technology using *A. niger* to cassava leaf meals increased protein content and digestibility of dry matter and protein (Darma *et al.*, 1994, 1995). As cassava leaf meal is an important source of beta-carotene, it has great potential in feed for poultry.

1.8 Other uses

Cassava wastes could be employed for phytase production after addition of a nitrogen source and mineral salts (Hong *et al.* 2001). According to Rajeshwarisivaraj *et al.* (2001), activated carbons prepared from waste cassava peel are efficient as adsorbents for dyes and metal ions.

Cassava stems which are available in plenty during harvest season have many uses in rural areas. Stems are dried after removal of bark by scrapping with a knife. The stem shavings are fed to cattle and the stem after drying is used as fuel. Tender stems are also fused for feeding after crushing into pieces. Stems can also be utilized in preparation of boards, paper and related items.

1.9 Cassava toxicity

Cassava plant carries two cyanogenic glucosides, linamarin (hydroxy isobutyro nitrile-β-D glucose) and its methyl homologue, lotaustralin (L-hydroxy-L-methyl butyronitrile- β-D glucose), in its edible roots and leaves (Coursey, 1973, Maini,1978). Traditional processing and cooking methods for cassava can reduce cyanide content to non-toxic levels. Boiling and drying are sufficient to make low-cyanogen cultivars safe for consumption, but more rigorous procedures like grating, fermenting, and sun-drying, are necessary to effectively remove cyanogens from cultivars of higher toxicity (Bolhuis,

1954) . Cyanogens present in consumed cassava can be efficiently eliminated by the body, if protein intake is adequate.

Many researchers claim that linamarin could lead to self-destruction of cancer cells and may play a great role in the treatment of cancer. Corte *et al.,* (1998) developed a new strategy for cancer gene therapy using a plant gene which encodes the enzyme, linamarase. Retroviral vectors that carry linamarase as a potential killer-suicide gene cause a marked sensitization to the innocuous substrate, linamarin, followed by cell death. This means that linamarin could be harnessed to provide treatment for cancer. Although linamarin had yet to be tested on either animal or human being, researchers anticipate a bright future for linamarin as an alternative treatment for cancer.

1.10 Global initiatives

Tropical tuber crops have untapped potential for improvement in both productivity per unit area and for development as commodity products especially in developing countries using tools of biotechnology. Although cassava faces serious challenges as a vegetatively propagated crop , prone to genetic erosion and pest and virus infestation, there are many plus points in favour of it. It responds well to irrigation or favorable rainfall conditions, and use of fertilizers. Its ability to grow on poor soils and under difficult climatic conditions, and advantage of flexible harvesting of tubers whenever there is a need, make it the crop of last resort for farmer families and their domestic animals in tropics (Hillocks *et al.,* 2001). In addition, as the case of Thailand proved, cassava can become a profitable export crop, if successful international breeding partnerships, effective distribution and selection of improved materials, and marketing are combined. According to Kawano (2003), in Taiwan, yield in terms of fresh tubers improved by 100 % and tuber dry matter content by more than 20 %. In other countries such as Costa Rica and Brazil where commercialization of cassava reached an advanced stage, technologies that make planting, harvesting, and postharvest processing more efficient and less time-consuming are developed. Efforts to make cassava production more cost-effective set a revival of global cassava production especially designed for animal feed over the last decade.

Production of enhanced germplasm through conventional breeding is a problem in vegetatively propagated crops. Cassava is characterized by strong heterozygosity and inbreeding depression. Although important contributions to disease resistance and harvest index were achieved through traditional breeding techniques (Jennings and Iglesias, 2002; Kawano, 2003), as Taylor *et al.,*(2004) remarked transgenic technologies offer key to unlocking full potential of cassava. Insertion of genetic material by direct gene transfer allows beneficial traits to be integrated rapidly into elite germplasm without changing genetic background of mother plant. In this way, desired traits can be introgressed into elite parent material prior to sexual crossing or into products of a breeding programme that lack a specific trait required by farmers. In other cases, existing farmer-preferred landraces and varieties can be modified to perform better in a given environment. An example of latter approach is introduction of transgenic resistance against cassava mosaic viruses into susceptible landraces grown by small farmers in Africa or into cultivars grown for commercial-scale starch production in Southern India. Biotechnology also allows introgression of beneficial traits not accessible through conventional means, such as herbicide tolerance and *Bt*-imparted resistance against stem borers and hornworms (Taylor *et al*, 2004).

Cassava Biotechnology Network(CBN) was established in 1988 as a global initiative to bring together researchers in developing and industrialized countries with a shared vision for using advanced molecular tools to improve the crop (Thro *et al.*, 1999).

Now more globlised activities are noted as indicated below (Aerni, 2006).

The Global Cassava Development Strategy and Implementation Plan (GCDS): It was set up in 1996 by FAO and IFAD (International Fund for Agricultural Development). Its aim is to counteract negative trends of cassava food and cash crop agriculture by using a demand-driven approach to promote and develop cassava-based industries.

The Consorcio Latinamericano y del Caribe de Apoyo a la Investigación y al Desarrollo de la Yuca (CLAYUCA): Established in 1999, it is a self-financing consortium that aims at boosting research for technology transfer and information exchange in cassava cash crop agriculture throughout Latin America. Its main partner in Africa is Southern Africa Root Crops Research Network (SARRNET).

The Global Cassava Partnership (GCP21): It constitutes a worldwide, multi-institutional research and development consortium dedicated to employing new tools of biotechnology to deliver improved germplasm to end users in tropics.

The BioCassavaPlus project: This project was established under "Grand Challenges in Global Health" of Bill and Melissa Gates Foundation as an innovative solution to global nutrition issue. It addresses major nutritional challenges related to cassava subsistence agriculture. Ohio State University is leading this $7.5-million, 10-institution cassava improvement project. Research groups include University of Bath in the United Kingdom, Centro Internacional de Agricultura Tropical in Colombia, and Zentrum-ETH in Switzerland. Project is using biotechnology to increase the plant's levels of zinc, iron, protein and vitamins A and E, plus longer shelf life and resistance to geminivirus.

The Double Haploid Initiative: A programme funded by Rockefeller Foundation addresses the problem of inbreeding depression in cassava, and costs associated with germplasm exchange and transportation of this vegetatively propagated crop. The 'Double Haploid Initiative' aims to solve problem of strong inbreeding depression due to strong heterozygosity through rapid and complete achievement of homozygosity in cassava breeding.

Recovery of first transgenic cassava plants was reported in 1996 only (Li *et al.*, 1996; Raemakers *et al.*, 1996; Schopke *et al.*, 1996). This inordinate delay was mainly due to inability to generate totipotent tissues required to produce transgenic plants. Breakthroughs in ability to generate morphogenic culture systems (Li *et al.*, 1996; Taylor *et al.*, 1996) eventually resulted in production of cassava plants transgenic for marker genes both via microparticle bombardment (Raemakers *et al.*, 1996; Schopke *et al.*, 1996) and *Agrobacterium*-mediated gene integration (Li *et al.*, 1996). Since first reports of transgenic plant recovery in cassava, efforts have focused on integrating genes with potential to impart traits of agronomic interest. Genetically transformed plants are recovered with reduced cyanogenic content (Siritunga and Sayre, 2003), expression of Bt proteins (Ladino *et al.*, 2002), modified starch content (Uzoma *et al.*, 2003), and elevated protein content within storage roots (Zhang *et al.*, 2003). In most cases, technologies successfully developed in other crops were transferred to cassava.

2. Sweet potato

Sweet potato has several nutritional advantages than cassava or other tropical tuber crops. These advantages make it an attractive focus for further increase in its production and consumption both directly as fresh or processed or as animal food. Sweet potato can be used in various ways, as direct food, processed food such as gelatinized cakes and noodles, industrial starch, ethanol and animal feed.

2.1 Sweet potato as human food

Compared to other tuber crops, selected varieties of sweet potato are rich in beta carotene. This is one of the nutritional advantages of sweet potato because sufficient and regular ingestion of sweet potato leaves, together with tubers of high beta-carotene varieties can meet consumer's daily requirement of vitamin A. According to van Jaarsveld *et al.*, (2005), beta carotene-rich orange-fleshed sweet potato (OFSP) is an excellent source of provitamin A compared to white fleshed sweet potato (WFSP). In many developing countries, sweet potato is a secondary staple food and plays a role in controlling vitamin A deficiency. Consumption of OFSP improves vitamin A status and plays a significant role in developing countries as a viable long-term food-based strategy for controlling vitamin A deficiency in children.

Sweet potato is a significant source of dietary fibre as its pectin content can be as high as 5.0 % of fresh weight or 20 % of dry matter at harvest (Collins and Walter, 1982). Significance of fibrous components in sweet potato and their composition and dietary function may receive increased attention in future.

2.2 Sweet potato as processed product and raw material for industry

There are efforts in private sector to capitalize on enormous potential for sweet potato products. Value-added products like French fries, chips, and flakes are developed from sweet potato. In the United States, one patent was granted for producing sweet potato bread and flour composed of 100 % sweet potato (Collins, 1993). These products are marketed as hypoallergenic for people who cannot tolerate cereal breads and flours. Several other products are in pipeline involving sweet potato. Where wheat has to be imported or can be grown only with extensive inputs such as in developing countries, sweet potato is a viable alternative component for breads and flours. There are disadvantages such as reduced loaf volume and decreased storage life , but reduced costs and decreased dependability on imports may outweigh disadvantages. Acceptability of component breads and flours is excellent with as much as 10 to 15 % content of sweet potato flour (Sammy, 1984). Some researchers have reported using up to 25 % sweet potato flour with no adverse consumer reaction. In countries, where wheat is plentiful and relatively inexpensive, this alternative use of sweet potato does not appear to be economically justified except as a specialty product.

Sweet potato when sliced, dried under sun and ground, gives a flour that remains in good condition for a long time. Flour is used as a dough conditioner in bread manufacturing and as a stabilizer in ice-cream industry. Sweet potato is processed into chips in much the same way as potato and product is popular in Asia (Kay, 1985, Lin and Chen, 1985). Cho *et al.*, (2003) demonstrated that anthocyanin prepared from purple sweet potato exhibits memory enhancing effects, which may be associated with its antioxidant properties.

Starch is produced from sweet potato in much the same way as from other starchy roots. Starch shows properties intermediate between potato starch and corn/ cassava starch in terms of viscosity and other characteristics. In Japan, about 90 % of starch produced from sweet potato is used in manufacture of starch syrup, glucose and isomerized glucose syrup, lactic acid beverages, bread and other food manufacturing industries. Sweet potato processing into starch, especially to make noodles, has emerged as a major use in China and to a lesser extent in Vietnam (Scott *et al.*, 2000b).

Sweet potato has immense potential as a biomass species for methane and ethanol production. Ethyl alcohol can be produced from sweet potato and can be used in same way as that of ethanol from cassava. Many Southeast Asian countries are developing a number of commercial alcohol factories using raw material from sweet potato. They give special attention to sweet potato production because it can be harvested twice a year compared to cassava, harvested only once a year (Yang, 1982).

2.3 Sweet potato as animal feed

Sweet potato tubers and vines are used as an animal feed and feed supplement in many developing countries (Gohl 1981; Thomas, 2003). Yeh *et al.* (1978) found that digestible energy and metabolizable energy are 91 % of those of corn, and net energy is about 79 % of that of corn as pig feed. Excess and culled tubers are used as fresh or dehydrated in feed rations. According to Yang (1982), sweet potato is a good feed for horses, mules and pigs. For lactating dairy cows, it is feasible when compounded with corn meal feed; and for poultry feed, at 25 % substitution for corn. Upto about 50% dehydrated sweet potato chips are used as supplementary diet to dairy cows. Fresh sweet potatoes can replace 30-50 % cereals in pig diets. The problem of protein digestibility is cited as an obstacle to increased use of raw sweet potatoes as animal feed. Trypsin inhibitor is implicated in poor protein digestibility. Cooking tubers is a sure way to get rid of trypsin inhibitor present in tubers. Another option is to utilize genetic variability as genotypes with no measurable trypsin inhibitor activity are available. Popping chips improved starch and nitrogen digestibility as well as removing trypsin inhibitor which might have contributed to lowering of feed value, but it also resulted in reducing lysine availability (Yeh *et al.*, 1978).

Animal feed policy makers predict great prospects for sweet potato in China. Sweet potato production for animal feed in China has risen from about 10 to 45 % over the last 30years according to Timmins *et al.*, (1992). It is hoped that higher global prizes for corn and strong demand for animal meat will further escalate sweet potato demand as animal feed. This strategy will go a long way in countries where feed or feed materials are imported, resulting in substantial import savings. Sweet potato foliage as feed will also increase, especially in small holder farms.

3. Yams

Edible yams provide staple carbohydrate food in the yam zone of West Africa where daily consumption is 0.5-1.0kg (Purseglove, 1975). Direct consumption of yams declined sharply among upper income groups, as they can easily procure cereals such as rice and wheat. Although yams are still preferred carbohydrate staple in parts of West Africa, their relative importance may decline in future.

Although most tuber crops are very low in vitamin A, certain yams are promising in terms of provitamin A present as pigment beta-carotene, especially *D. cayenensis*, yellow yam. Yellow colour is because of carotenoids, consisting mainly of beta-carotene in quantities of 0.14-1.4 mg per 100 g (Murtin and Ruberté, 1972). Some Pacific Island varieties of yam contain up to 6.0 mg per 100 g of carotene (Coursey, 1967). Yams also contain appreciable amounts of Vitamin C to the tune of 6-10 mg of vitamin C per 100 g and up to 21 mg in a few cases (Onweueme, 1984).

In West Africa, yams are usually eaten as *Fufu* , prepared by boiling peeled cut-up yams and pounding them until a sticky elastic dough is produced (*Fufu* is also prepared from cassava and cocoyam either alone or mixed with yams). For preparation of food items like *Fufu,* a stiff dough is required; and so the rheological properties of starch paste become significant. Viscosity of starch-water pastes of different yams varies considerably from a relatively low value for *D. dumetorum* through increasing viscosity in *D. esculenta* to the highest value in *D. rotundata.* Hence, *D. rotundata* is traditionally the accepted yam for *Fufu*. Most yams give viscous pastes with a much higher gel strength than that of other crops. Therefore, yams are traditionally preferred for *Fufu* (Rasper, 1969, 1971).

The only processed yam product traditionally made at village level is yam flour (Oke, 1990). Nutritional value of yam flour is same as that of pounded yam. Yams have not been processed to any significant extent commercially. Dehydrated yam flours and yam flakes have been produced by sun drying. Manufacture of fried products from *D. alata* is attempted. Both chips and French fries are manufactured.

In general, fully grown edible yam tubers do not contain any toxic principles. Bitter principles tend to accumulate in immature tuber tissues of certain types like *D. rotundata* and *D. cayenensis*. They may be polyphenols or tannin like compounds (Coursey, 1983). Wild forms of *D. dumetorum* also contain bitter principles, and hence are called bitter yam, but bitter yams are not normally eaten except in times of food scarcity. Toxic principles must be dissipated by boiling or soaking in a vessel of salt water, or hot fresh water. Bitter principle is identified as alkaloid dihydrodioscorine, while that of Malayan species, *D. hispida*, is dioscorine. These are water soluble alkaloids which, on ingestion, produce severe and distressing symptoms (Coursey, 1967).

Certain species of *Dioscorea* grown in Southeast Asia contain toxic saponins. They are made locally into a medicinal shampoo, used to destroy head lice. They are also used in an insecticidal powder, similar in effect to derris dust, used to destroy rice parasites in paddy fields in Malaysia. *D. cirrhosa* contains enough tannin for commercial use. S few cultivars of *D. alata* contain 6 to 38 % of tannin which is used in Southeast Asia for tanning fishing nets and in Taiwan for tanning leather, to which it imparts a red colour (Coursey, 1967).

A few yams have considerable pharmaceutical potential. Most of these yams are non-edible. Yams contain steroidal sapogenins, which are useful starting materials for preparation of cortisone and related drugs (Oke, 1990). In Mexico, a wild yam species, *D. mexicana*, contains useful quantities of sapogenins, which can be converted into progesterone intermediates. Yams like *D. floribunda* are good sources of diosgenin, a starting material for manufacture of corticosteroids. Diosgenin is used in manufacture of oral contraceptives, sex hormones and cortisone (Pursegolve, 1975). Oral contraceptives based on progesterone are now widely used in family planning in several tropical countries (Oke, 1972).

Yams are widely used in traditional medicine in Africa, China and Asia. It is a major part of tribal medicine. The Zulus used an extract of *D. sylvatica* for treatment of uterine and mammary disorders in cattle (Oke, 1990).

4. Aroids

Among group of aroids, three tuber crops are important. They are taro or eddoe, and elephant foot yam. Tubers of these crops are rich in carbohydrates, and all the three can be consumed as a staple food or as a vegetable. Taro, because of its potential to yield heavily and throughout the year, could overcome food shortages. Tubers of taro can be consumed directly after boiling and peeling outer skin. It is also roasted or baked. Young leaves including both lamina and petioles can be used as spinach. In Kerala, tubers and leaves of taro are consumed as vegetables.

Taro flour is also made from corms, which can be used for soups, biscuits, bread, beverages, and puddings. Starch grains of taro are extremely small and are easily digestible. Taro starch can be used in baby foods, hypoallergenic foods, and as a cereal substitute in diets for victims of celiac disease (NAS, 1975). It is also sliced and fried into taro chips. In Hawaii, *Poi,* a national dish is made by pounding boiled, peeled and mashed corm fermented a day or two.

Tannia or cocoyam has its origin in South America and the Caribbean. Spanish and Portuguese introduced it to Europe, and were also responsible for spreading it to Asia. The corms, cormels, leaves and young stems of tannia are utilized in different ways. Normally, only cormels are used for human consumption, and corms are used for animal feed and for propagation purposes. It is capable of producing 30-60 tonnes of cormels per hectare (NAS,1975). Cormels compare with potato in qualities. Fresh cormels contain 2-3 % protein, but deficient in lysine, methionine and cysteine. Tubers are also rich in vitamins and minerals. Quality of tannia is inferior as starch grains are larger than those of taro. Although tannia is relatively new to the Pacific region, it has spread rapidly and widely, becoming quite an important crop in many of the islands. It is also widely cultivated in Puerto Rico, the Dominican Republic and Cuba and is important along coastal mountains of South America, in Amazon basin and in Central America. Starchy cormels are eaten boiled, roasted, parboiled, baked, fried in oil or ground into flour. In West Africa, they are preferred to taro as they are more suitable for making *Fufu* (Purseglove, 1975). Ground fresh tubers are sometimes used to thicken stews and soups. They are easier to grow than taro. Starch may be prepared from grated tubers. Young leave can also be used as spinach just as taro.

Elephant yam or elephant foot yam is an important tuber crop of tropics grown as a vegetable. It is also used in Ayurvedic preparations against piles and dysentery. Petioles are also used as a vegetable. It is a heavy yielder and is capable of producing 80 tonnes of corms. Tubers contain irritant needle shaped crystals of calcium oxalate, which must be removed by prolonged washing or boiling (Purseglove, 1974, 1975).

Conclusion

There are great opportunities for diversification and utilization of tuber crops. From a global perspective, cassava and sweet potato will be increasingly used in processed form for food, feed and starch derived products. Non-food, non-feed uses will also increase. Yams and aroids will be used largely as food in fresh form. In urban areas, there is greater potential for processed foods and snack foods. Tuber crop based processed products may find niche markets.

Policy makers foresee a greater potential for tuber crops in the area of starch and starch based products. From cassava, both native starch and modified starch can be produced. Several comparative advantages in production and use are encouraging many industries to adopt cassava starch as an alternative to traditional sources. This is due to increasing demand for starch in processed food and in paper and textile industry. Potential demand for starch exists for manufacture of biodegradable plastic and ethanol.

There are vast opportunities for tuber crops to be used as animal feed in fresh and processed form. If part of animal feed could be substituted with tuber crops such as cassava and sweet potato, then feed cost can be reduced substantially.

Policy initiatives from administrates are also needed for harnessing great opportunities for tuber crops. Both developed and developing countries should adopt appropriate policies to ensure that production of a diversified range of tuber crops for food, feed, and industry will help eradicate poverty, assure food security, and improve income of people.

References

Aerni, P. 2006. Mobilizing science and technology for development: The case of the Cassava Biotechnology Network (CBN). [On-line]. *AgBioForum*, 9(1), 1-14. Available: http://www.agbioforum.org/. [20-6-2006].

Ahmed, R., Segal. I. and Hassan. H. 2000. Fermentation of dietary starch in humans. *American Journal of Gastroenterology* 95: 1017-1020.

Aritonang, D. and Silalahi, M. 1992. Effects of feeding two different levels of protein in the ration to the performance of fattening pigs. (In:) Proceeding of Animal Agro-industry in the Village. Research Institute for Animal Production, CIAWI, Bogor. pp. 104-114.

Augustin, J., McDole, R.E., McMaster, G.M., Painter, C.G. and Sparks, W.C. 1975. Ascorbic acid content in Russet Burbank potatoes. *J. Food Science* (40): 415-416.

Bolhuis, G.G. 1954. The toxicity of cassava roots. *Netherland Agricultural Science* (2): 176-185.

Booth, R.H. and Wholey, D.W. 1978. Cassava processing in Southeast Asia. *(In:)* Ghoninard, A., Cook, J.H. and Weber, E.J(eds.), Cassava Harvesting and Processing. CIAT/IDRC. pp 7-11.

Burrell, M. M. 2003. Starch: the need for improved quality or quantity—an overview. *Journal of Experimental Botany* 54 (382): 451-456.

Cho, J., Kang, J,S,, Long, P.H., Jing ,J., Back, Y. and Chung, K.S..2003. Antioxidant and memory enhancing effects of purple sweet potato anthocyanin and cordyceps mushroom extract. *Archives of Pharmacal Research*, 26(10):821-825.

Collins, W.W. 1993. Root vegetables: New uses for old crops. *(In:)* J. Janick and J.E. Simon (eds.), New crops. Wiley, New York. pp. 533-537.

Collins, W.W. and Walter, W.M.1982. Potential for increasing nutritional value of sweet potatoes. *(In:)* Villareal, R.L.and Griggs,T.D.(eds), International Symposium on Sweet Potato. 1, Tainan, 1982.. Taiwan, AVRDC. pp. 355-363.

Cortés, M. L., Felipe, P. de, Martín, V., Hughes, M. A. and Izquierdo, M.1998. Successful use of a plant gene in the treatment of cancer *in vivo*. *Nature* 5(11):1499-1507.

Coursey, D. G. 1983. Yams. *(In:)* Chan, H.T.(ed.). Handbook of Tropical Foods. Dekker, New York, pp. 555-601.

Coursey, D.C. 1973. Cassava as food: toxicity and technology. *(In:)* Nestel, B. and MacIntyre, R.(eds.), Chronic Cassava Toxicity. Proceedings of an Interdisciplinary Workshop,. London, 29-30 Jan. 1973, IDRC, Ottawa. pp. 27-36.

Coursey, D.G. 1967. Yams. Longman, London.

Darma, J., Purwadaria, T., Haryati, T., Sinurat, A.P. and Dharsana, R. 1994. Upgrading the Nutritional Value of CassavaLeaves Through Fungal Biotechnology. Research Institute for Animal Production Research Report for FAO/ANBAPH. Ciawi, Bogor.

Darma, J., Supriyati, Purwadaria, T., Haryati, T. and Kompiang I.P. 1995. Effect of ambient temperature and water contents before fermentation on the nutritive quality of Cassapro. (In:) Proceedings of the National Seminar on Science and Technology. Research Institute for Animal Production. CIAWI, Bogor. pp. 109-114.

Devendra, C. 1977. Cassava as a feed source for ruminants. *(In:)* Nestel, B. and Graham, M.(eds.), Cassava as Animal Feed. Proceedings of Workshop. University of Guelph. IDRC-095e, 107, Ottawa.

Fiedler, H. and Wood, J.L. 1956. Specificity studies on the Bmercaptopyruvate-cyanide transsulfuration system. *J. Biochemistry,* (222): 387-397.

Goering, T.J. 1979. Tropical Root Crops and Rural Development. World Bank Staff Working Paper No. 324. World Bank, Washington D.C.

Gohl, B. 1981. Tropical Feeds: Feed Information Summaries and Nutritive Values. FAO Animal Production and Health Series No.12, Food and Agriculture Organization of the United Nations, Rome.

Gomez, G. 1977. Utilization of cassava-based diets in swine feeding. *(In:)* Cassava as Animal Feed. Proceedings, Cassava as Animal Feed Workshop, University of Guelph, 18-20 April 1977, IDRC, Ottawa, Canada.. pp 65-71.

Grace , M.R. 1977. Cassava Processing. FAO Palant production and Protection Series No. 3. Food and Agriculture Organization of the United Nations, Rome.

Gray, W.D. and Abou-el-Seoud, M.O. 1966. Fungal protein for food and feeds II. Whole sweet potato as a substrate. *Economic Botany* (20): 119126.

Hillocks, R. J., Tresh, J.M., and Bellotti, A.C. (eds.). 2002. Cassava: Biology, Production and Utilization. New York: CABI Publishing.

Hong, K., Ma, Y. and Li, M. 2001. Solid-state fermentation of phytase from cassava dregs. *Applied Biochemistry and Biotechnology*, 91-93: 777-785.

Howeler, R. 2003. Cassava in Asia: Present Situation and its Future Potential in Agro-Industry [On-line]. Available: http://ciat-library.ciat.cgiar.org/Articulos_Ciat/CASSAVA_IN_ASIA_PRESENT_S. pdf [20-6-2006]

Jennings D.L., and Iglesias, C. 2002. Breeding for crop improvement. *(In:)* Hillocks, R.J. and Thresh, J. M. (eds.), Cassava: Biology, Production and Utilization. CABI, Wallingford, Oxon, UK pp. 149-166.

Jin Shuren, 2000. The current situation and prospects for further developemny of China's sorbitol industry. *(In:)* Howeler, R.H. Oates,C.G. and O'Brien, G.M.(eds.), Cassava, Starch and Starch Derivatives. Proceedings, International Symposium. Nanning, Guangxi, China. Nov.11-15,1996. pp 32-36.

KAU [Kerala Agricultural University].2001. Package of Practices Recommendations: Veterinary and Animal Husbandry-*2001*. Kerala Agricultural University, Thrissur.

Kawano, K. 2003. Thirty years of cassava breeding for productivity: Biological and social factors for success. *Crop Science*, 43: 1325-1335.

Kay, S.J. 1985. Formulated sweet potato products. *In:* Bouwkamp, J.C. (ed.), Sweet Potato Products: A Natural Resource for the Tropics. CRC Press, Boca Raton .pp. 205-218.

Kompiang, I. P., Sinurat, A.P. and Supriyati 1995. The effect using protein enriched sago and its by-product in comparison to fermented cassava fiber in the rations on the performance of broiler chicken. *(In:)* Research Results on Poultry and Miscellaneous Animals. Research Institute for Animal Production. CIAWI, Bogor. pp. 490-498.

Kompiang, I. P., Sinurat, A.P., Supriyati, Purwadaria, T. and Darma, J. 1994. Nutritive value of protein enriched cassava: Cassapro. *Ilmu dan Peternakan*. 7 (2): 22-25.

Ladino, J.J., Echeverry, M., Mancilla, L.I., Lopez, D., Chavarriaga, P., Tohme, J., and Roca, W. 2002. Genetic transformation of cassava: Confirmation of transgenesis in clone 60444 and analysis of CRY1Ab protein in transgenic lines. Preliminary data on transformation of farmer-preferred cultivars SM1219-9 and CM3306-4 (annual report). Centro Internacional de Agricultura Tropical. Cali, Colombia.

Li, H.Q., Sautter, C., Potrykus, I., & Pounti-Kaerlas, J. 1996. Genetic transformation of cassava (*Manihot esculenta* Crantz). *Nature Biotechnology*, 14: 736-740.

Limon R L 1992 Ensilage of cassava products and their use as animal feed. *(In:)* Machin, D. and Speedy,A.W. (eds.), *Roots, tubers, plantains and bananas in animal feeding*. FAO Animal Production and Health. Paper No 95:pp.99-110.

Lin, S.S.M. and Chen, D.M. 1985. Sweet potato production and utilization in Asia and the Pacific. In: Bouwkamp, J.C. (ed.), Sweet Potato Products: A Natural Resource for the Tropics, CRC Press, BocaRaton, pp. 139-148.

Maini, S. B.1978. Quality aspects of cassava. *(In:)* Hrishi,N. and Nair, R.G. (eds.), Cassava Production Technology. Central Tuber Crops Research Institute, Thiruvananthapuram. pp. 49-57.

Maneepan, S. 1996. Perspectives and limitations of cassava sector in the world. Poster presented in the 1st Latin American Conhgress for Tropical; Roots and the 9th Brazilian Cassva Congress, San Pedro, Sao Paulo,Brazil. Oct.7-10, 1996.

Mbugua, S.K., Ahrens. R. H., Kigutha. H.N., Subrahmanian. V. 1992. Effect of fermentation, malted flour treatment and drum drying on nutritional quality of uji. *Ecology of Food and Nutrition*, 28:271-277.

Modebe, A.N.A. 1963. Preliminary trial on the value of dried cassava for pig feeding. *J. West African Science Association*, 7: 127- 133.

Montaldo, A. 1977. Whole plant utilization of cassava for animal feed. *(In:)* Nestel, B. and Graham, M.(eds.), *Cassava as Animal Feed*. Proceedings, Cassava as Animal Feed Workshop, University of Guelph, 18-20 April 1977, IDRC, Ottawa, Canada. pp. 95-105.

Moorthy, S. N. 2001. *Tuber Crop Starches*. Technical Bulletin Series 18, Central Tuber Crops Research Institute, Thiruvananthapuram.

Müller, Z., Chou, K.C. and Nah,K.C.1974. Cassava as a total substitute for cereals in livestock and poultry rations. *World Animal Review* 12(1): 19-24.

Murtin, F.W. and Ruberté, R. 1972. Yam for production of chips and french fries. *J. of Agricultural University, Puerto Rico*, 56: 228-234.

Nair, K.G.1978. Utilization of Cassava. *(In:)* Hrishi,N. and Nair, R.G. (eds.), Cassava Production Technology. Central Tuber Crops Research Institute, Thiruvananthapuram. pp. 37-48.

NAS[National Academy of Sciences].1975. *Underexploited Tropical Plants with Promising Economic Value*. National Academy of Sciences, Washingtin, D. C.

Nduele, M.; Ludwig, A. and van Ooteghem, M.1993. The use of cassava starch in the formulation of gelatin capsules. *Journal de Pharmacie de Belgique*, 48 (5): 325-334.

Niba, L.L. 2003. The relevance of biotechnology in the development of functional foods for improved nutritional and health quality in developing countries. *African J. Biotechnology*, 2 (12): 631-635.

Norman, M.J.T., Pearson, C.J. and Searle, P.G.E. 1995. Tropical Food Crops in Their Environment, (2nd ed.). Cambridge University Press.430p.

Nur, Y.S. 1995. The utilisation of different levels of fermented cassava fibre in the diets of broiler chicken. Proceedings of the National Seminar on Science and Technology. Research Institute for Animal Production. Ciawi, Bogor. pp. 244-248.

Oke, O.L. 1972. Yam: a valuable source of food end drugs. *World Review of Nutrition and Dietics* ,15: 156-184.

Oke, O.L. 1990. Roots, Tubers, Plantains and Bananas in Human Nutrition. Food and Agriculture Organization of the United Nations, Rome.

Okeke, J.E. 1978. Concentration of nutrients in plant parts and growth relationships. PhD thesis, University of Ibadan, Nigeria.

Onwueme, I.C. 1984. Yams. *(In:)* Goldsworthy,P. and Fisher,N.M. (eds.), The Physiology of Tropical Field Crops. John Wiley and Sons.

Ostertag , C. F., Alonso, L. Best, R. and Wheatly, C. 1996. The cassava flour project in Columbia: From opportunity identification to market development . *(In:)* Dufou, D. O' Brien, G.M., and Best, R. (eds.), Cassava Flour and Starch: Progress in Research and Development. CIAT (Centro Internacional de Agricultural Tropical), Cali, Columbia.

Pulungan, H. and Bhakti, B.N. 1984. The substitution of concentrate with cassava root meals in the diets of growing sheep. *LPP Bulletin*. 29: 1-13.

Purseglove, J.W. 1974. Tropical Crops: Dicotyledons. Longman, London.

Purseglove, J.W. 1975. Tropical Crops: Monocotyledons. Longman, London.

Raemakers, C.J.J.M., Sofiari, E., Taylor, N. J., Henshaw, G.G., Jacobsen, E., and Visser, R.G.F.1996. Production of transgenic cassava plants by particle bombardment using luciferase activity as the selection marker. *Molecular Breeding*, 2: 339-349.

Rajeshwarisivaraj, Sivakumar, S.; Senthilkumar, P. and Subburam, V.2001. Carbon from Cassava peel, an agricultural waste, as an adsorbent in the removal of dyes and metal ions from aqueous solution. *Bioresource Technology*, 80(3):233-235.

Rasper, V. 1969. Investigations on starches from major starch crops grown in Ghana. II. Swelling and solubility patterns: amyloclastic susceptibility. *J. Science and Food Agriculture,* 20: 642-646.

Rasper, V. 1971. Investigations on starches from major starch crops grown in Ghana. III. Particle size & particle size distribution. *J. Science and Food Agriculture,* 22:572-580.

Roble, N.D.; Ogbonna, J.C. and Tanaka, H. 2003. L-Lactic acid production from raw cassava starch in a circulating loop bioreactor with cells immobilized in loofa (Luffa cylindrica). *Biotechnology Letters*, 25 (13):1093-1098.

Sammy, G.S. 1984. The processing potential or tropical root crops. *(In:)* Proceedings of the Caribbean Regional Workshop on Tropical Root Crops, Jamaica, April 10-16, 1983. p. 199.

Schopke, C., Taylor, N.J., Carcamo, R., Konan, N.K., Marmey, P., Henshaw, G.G., Beachy, R.N. and Fauquet, C.M. 1996. Regeneration of transgenic cassava plants (*Manihot esculenta* Crantz) from microbombarded embryogenic suspension cultures. *Nature Biotechnology*, 14: 731-735.

Scott, G. J., Best, R., Rosegrant, M. and Bokangam, M. 2000a. Roots and Tubers in the Global Food System: A vision Statement to the Year 2020. International Potato Centre, Lima, Peru.111p.

Scott, G.J. Rosegrant, M. W. and Ringler, C. 2000b. Roots and Tubers for the 21st Century: Trends, Projections, and Policy Options [On-line]. Food, Agriculture and the Environment Discussion; International Food Policy Research Institute, Washington, DC. Available: http://www.ifpri.org/2020/ briefs/ brief.66.pdf [20-6-2006].

Siritunga, D. and Sayre, R.T. 2003. Generation of cyanogen-free transgenic cassava. *Planta*, 217: 367-373.

Sumangkut, M.H., Maksum, K. and Siregar, S.B. 1976. The use of cassava root meals vs. cassava fibre in the concentrate containing urea in the diets of young dairy cattle. *LPP Bulletin*, 16: 1-15.

Tan, K.H.; Ferguson, L.B. and Carlton, C.1984. Conversion of cassava starch to biomass, carbohydrates, and acids by *Aspergillus niger*. *J.Applied Biochemistry*, 6 (1-2: 80-90.

Taylor, N., Kent, L., and Fauquet, C. 2004. Progress and challenges for the deployment of transgenic technologies in cassava [On-line]. *AgBioForum*, 7(1&2), 51-56. Available on the World Wide Web: http://www.agbioforum.org/. [20-6-2006].

Taylor, N.J., Edwards, M., Kiernan, R.J., Davey, C., Blakesley, D., and Henshaw, G.G. 1996. Development of friable embryogenic callus and embryogenic suspension cultures in cassava (*Manihot esculenta* Crantz). *Nature Biotechnology*, 14: 726-730.

Thomas, C.G. 2003.Forage Crop Production in the Tropics. Kalyani Publishers, Ludhiana. 259 pp.

Thro, A.M., Taylor, N.J., Raemakers, C.J.J.M., Pounti-Kaerlas, J., Schöpke, C., Visser, R.G.F., Inglesias, C., Sampaio, M.J., Fauquet, C.M., Roca, W., and Potrykus, I. 1999. Maintaining the cassava biotechnology network. *Nature Biotechnology*, 16: 428-430.

Timmins,W.H.., Marter,A.D., Westby,A., and Rickard,J.E. 1992. Aspects of sweet potato processing in Sichuan province, Poples's Republic of China. *(In:)* Scott, G., Wiersema, and Fergusen, P. (eds.), Product Development for Root and Tuber Crops, Vol I-Asia. Proceedings of the International Workshop, April 22 – May 1, 1991, Leyte, Philippines.

Uzoma, E.I., Arias-Garzon, D.I. and Sayre, R.T. 2003. Genetic modification of cassava for enhanced starch production. *(In:)* Harnessing crop technologies to alleviate hunger in Africa: Proceedings 6th biennial conference of the African Crop Science Society .Nairobi, Kenya. 65 pp.

van Jaarsveld, P.J., Faber, M., Tanumihardjo,S.A., Nestel, P., Lombard, C.J and Spinnler Benadé, A.J..2005. β-Carotene–rich orange-fleshed sweet potato improves the vitamin A status of primary school children assessed with the modified-relative-dose-response test. *American J. Clinical Nutrition* 81(5): 1080-1087.

Visser, R., Somhurst, I., Kuipers, G., Ruys, N., Feenstra, W., Jacobsen, E.1991. Inhibition of the expression of granule-bound starch synthase in potato by anti-sense constructs. *Molecular and General Genetics* 225:289-296.

Visser, R., Suurs, L., Steeneken, P., Jacobsen, E. 1997. Some physicochemical properties of amylose-free potato starch. *Starch/Staerke* 49: 443-448.

Vuilleumier, S.1993.Worldwide production of high-fructose syrup and crystalline fructose. *American J. Clinical Nutrition* 58(5): 733S-736S.

Wang Xiaodong, Guo Xuan and Rakshit, S. K. 2000. Study of lactic acid fermentation with *Lactobacillus amulovarus* using cassava starch. *(In:)* Howeler, R.H. Oates,C.G. and O'Brien, G.M.(eds.), Cassava, Starch and Starch Derivatives. Proceedings, International Symposium. Nanning, Guangxi, China. Nov.11-15,1996. pp198-201.

WHO [World Health Organization].1985.Energy and Protein requirement. Report of a Joint FAO/ WHO/ UNU Expert Consultation. World Health Organization Technical Report Series 724. Geneva, Switzerland.

Yang, T.H. 1982. Sweet potato as a supplemental staple food. In: Villareal, R.L. and Griggs, T.D. (eds.). International Symposium on Sweet Potato. 1. Tainan, 1982, pp. 31-36. Taiwan, AVRDC.

Yeh, T.P., Wung, S.C., Lin, K. & Kuo, C.G. 1978. Studies on different methods of processing some local feed materials to enhance their nutritive value for swine. Anim. Ind. Res. Inst. Taiwan Sugar Corp. Ann. Res. Report: p. 25. (Original Chinese with English summary).

Zhang, P., Jaynes, J., Potrykus, I., Gruissem, W. and Pounti-Kaerlas, J. 2003. Transfer and expression of an artificial storage protein (ASP1) gene in cassava (*Manihot esculenta* Crantz). *Transgenic Research* 12: 243-250.

❑❑❑

17

Future Thrust

The world's population is increasing exponentially. Projected totals for the year 2050 vary but it is acknowledged that access to sufficient quality and quantity of food, water and living space and other natural resources will be limiting factors for the human population. The environmental impact of this huge population and its everyday activities will be significant and possibly irreversible. There is urgency therefore, in addressing issues of sustainable food production, water and land use for the coming decades. Endemic hunger (due to carbohydrate-protein malnutrition), hidden hunger (induced by deficiencies of vitamins and minerals) and transient hunger (caused by natural calamities such as drought, flood, earthquake, tsunami etc.) can be overcome only by integrated strategy with sustainable and equitable use of agricultural biodiversity.

Farmers constitute the basis of food production for both the rural and rapidly growing urban populations. Food security and sustainable economical development will therefore need to be tied to improved productivity and enhanced value of the major food crops. Increased productivity should relieve pressure on the world's diminishing biodiversity, while improved access to quality foods will beneficially impact the health of billions of people.

The tropical root and tuber crops play a central role in food security, culture and economic prosperity in many African, South American and Asian countries. Ranked after rice and maize as the most important source of dietary calories in the tropics, root and tuber crops are eaten daily in one of its many fresh or processed forms, by an estimated 600 million people.

Billions of people will rely on cassava for food security and economic development in the coming decades. Total cassava production grew by 75% over the last 3 decades, but the productivity per unit area increased by less than 0.7% per year. Biotic and agronomic constraints mean that average cassava yields are only 15% of its potential production. Improved production and utilization of cassava would free space and manpower within rural economies and decrease the pressure on fragile ecosystems within the tropics (Scott *et al.,* 2000).

The revolution in molecular biology, recombinant DNA technologies and genetic engineering provide new tools to circumvent many of the problems that previously hindered tuber crops improvement. Inherent heterozygosity and inbreeding depression in cassava can be overcome and improved germplasm delivered to farmers through a combination of conventional breeding and biotechnology. Cassava is highly heterozygous and is suitable for the exploitation of hybrid vigour. Triploid cassava has convincing role in enhancing the productivity which could be largely due to hybrid vigour. For the production of triploids in large scale, production of tetraploids as male parents developed through somatic hybridization incorporating selected elite diploids will prove to be ideal as this will lead to increased heterozygosity.

1. Utilization of Biodiversity

Biodiversity of root and tuber crops and their closely related species needed to be collected, screened and documented for their traits with potential benefits. The disciplines of genomics, molecular biology, gene cloning and genetic engineering will be developed and integrated to provide a platform of tools to enable beneficial traits to be identified, analyzed, isolated and transferred to cultivated root and tuber crops through conventional breeding and/or genetic engineering programmes.

Collection, conservation and documentation of cultivated and wild species act as a germplasm pool for screening for potentially important agronomic traits, such as disease resistance, improved post harvest characteristics and enhanced nutritional qualities as well as apomixes and provide a source of genes for transfer of beneficial traits to farmer-preferred cassava cultivars. Most of the wild relatives of tuber crops are highly medicinal and are used by tribals in indigenous medicine and most of them are endangered or have become extinct due to denudation of forest and over exploitation by traders. Hence, conservation of these endangered species is the need of the hour.

Attitudes have changed or are being changed towards root and tuber crops. A growing number of policy makers, socio-economists and scientists are recognizing their importance in food, feed, industries and economic security within the tropics. Secondly, the rapid emergence of modern DNA technologies offer possibilities to overcome traditional barriers to crop improvement, offering exciting new possibilities to improve these crops and to transform into a major player in 21st century agriculture. Polymorphic DNA markers are invaluable in plant genetic research because of their applications in the development of genetic maps, gene tagging, cloning of useful genes, marker assisted genetic improvement / selection, creation of value added varieties and in studying genetic diversity. The utility of the morphological features, agronomic traits and allozyme etc. is becoming inadequate in the face of growing number of cultivars. This has necessitated the application of molecular genetic markers for characterization of germplasm. Amplified Fragment Length Polymorphism (AFLP) and microsatellite analysis are considered most informative, reproducible, robust and suitable for high resolution genetic comparison. The DNA amplification fingerprinting (DAF) technique can be used for the molecular characterization of sweet potato gene bank.

2. Breeding New Genotypes

Sequencing of the genome of major tuber crops, high density markers map, mapping of desired traits, development of high saturation BAC libraries, development of high saturation ESTs, gene expression profiles and improved genetic transformation technologies capable of producing genetically engineered cassava/sweet potato/taro/elephant foot yam plants across a range of agronomically important cultivars.

Although, breeding had limited impact in the past, breeding will play an essential role in the future due to the availability of modern technologies and the possibility to integrate specific genes into interesting genotypes. Interspecific hybrids will play an important role in modern cassava breeding and the possibility to transfer apomixes from wild to cultivated species will be investigated.

Interspecific transfer of genes for high protein from wild *Dioscorea* spp. to cultivated yams and identification and cloning of dwarfing genes in yams are interesting areas of research.

3. Biotechnological Improvement

3.1 Starch

Cassava is the main source for commercial extractions of starch besides maize in India. The starch production has increased transgenically in potato using the bacterial gene *(E. coli glgC 16 gene)*. Similarly the ADPGPase gene transferred from barley has also been found to enhance starch production in potato. Antisense technology has also been found effective in transgenic potato to improve starch quality. Similar line of transgenic technology can be applied in cassava/sweet potato for commercial exploitation. Transgenic cassava/ sweet potato plant also may be used to produce novel carbohydrates/starches with improved thermoplastic properties and biodegradability.

3.2 Protein/Minerals

Root and tuber crops are an excellent source of carbohydrates but are deficient in protein, vitamins and some minerals. Improving the nutritional quality of these crops would have a significant impact on the health of hundreds of millions of the world's poorest people and help protect post-weaned children from protein-energy deficiency and related conditions. Cassava and sweet potato being ideal materials for bio fortification (for protein, iron etc.) to alleviate this serious problem affecting population in rural and semi urban areas.

Cassava roots contain an average of only 2 to 3% dry weight protein. Enhancing protein content to levels equivalent to that of cereal (7-9%) with balanced amino acid ratios would have significant impact on the health of young children and mothers in many developing countries in Africa, South America and Asia. Investigations are warranted on the feasibility of accumulating protein in cassava roots, either through expression of genes from wild cassava relatives or from other crops such as corn or soybean.

Ferritin gene isolated from soyabean was transferred to rice, tobacco etc and is found to be suitable both for dicots as well as monocots. The production of transgenic cassava with enhanced iron content is a prospective area of research. The popular table varieties having high carotene content can be used as the base material for such gene transfers.

Sweet potato tubers are also low in proteins and vitamins though rich in carbohydrate. Transfer of the synthetic gene *HEAAE* (High essential amino acid encoding gene) to potato has been found to increase the content of essential amino acids. Similarly transfer of *BN2 S* gene from Brazilnut to potato has also resulted in the production of methionine rich protein. High protein sweet potato by recombinant DNA technology with novel storage protein (asp-1) has been achieved by Dr. C. S. Prakash in USA with 2.5 to 5 fold increase of protein content. Collaborative Programme needs to be taken up in cassava and sweet potato by sharing the technology, genetic resources and facilities for development and identification by precise chemical characterization of nutritionally enriched lines of these crops of the third world countries.

3.3 Shelf life

Rapid deterioration of cassava storage roots after harvesting is an important problem for the crop. An effective shelf life of less than three days significantly limits farmers' ability to supply high quality tubers to local markets and processing factories and limits the development of the crop as commercial commodity. Post harvest deterioration means that cassava roots must be consumed or processed within 48 hours of removal from the soil which is a major constraint in cassava and other tubers. Understanding the physiological processes and underlying molecular genetics involved will allow strategies to be designed to extend shelf life of cassava roots and other tubers for increasing marketability and allowing its more effective commercialization. It will be beneficial for the farmer and industry if anti microbial protein genes are incorporated in cassava as has been done in potato to enhance the shelf life.

The diapausing pupae of giant silk worm moth synthesize more than 15 new types of such proteins effective against gram positive and gram negative bacteria. Cercopins, Attacins and Lysozymes are some of the proteins that have been found non-phytotoxic. Work in this line for incorporating such alien genes through transgenic technique will be of great help to the cassava grower and industrialist for fighting post harvest spoilage.

3.4 Biotic factors

Among the various biotic factors, which affect the tuber crops, Indian cassava mosaic disease (ICMD) has emerged as one of the serious limiting factor. It causes yield losses ranging from 50 to 88% depending upon the cultivars. To overcome all these problems only solution could be the use of biotechnological means to develop pathogen derived resistance. In the recent years, availability of information about the putative functions of geminivirus genes have led to the development of virus control strategies based on the expression of virus derived genes like coat protein, replicase, movement protein, the use of defective interfering particles or the expression of dianthin (a RIP) in transgenic plants. Due to the importance of ICMD and lack of resistance source in cassava plant itself, it is worth to attempt for creating pathogen derived resistance in cassava against ICMV by adopting recombinant DNA technology.

One of the interspecific hybrids, *M. caerulescence* X *M. esculentu* has shown stable ICMV resistance for the past ten years and is confirmed through graft studies. The efforts to transfer this ICMV resistance to released varieties through repeated back crossing have not reached full success so far. This could be overcome through somatic hybridization.

The cassava mosaic disease is reported as recessive controlled polygenically by a larde additive component with high narrow (35%) and broad sense (68%) heritability. However the studies on genetic nature of ICMV is hampered by the non availability of segregating progeny for disease resistance in earlier days. Now it is possible to use molecular markers to facilitate marker assisted selection for ICMV resistance. Cloning of genes for cassava mosaic resistance from *M. caerulescence* and production of transgenic CMV resistant plants is a prospective area of research. Similarly cloning of genes for taro leaf blight resistance from wild relatives of taro can be attempted.

Sweet potato feathery mottle virus and mixture of many viruses in sweet potato are cause for serious concern. Higher returns could be obtained only when virus free planting materials are used for planting. Presently virus free plants are identified/selected based on absence of symptoms which often gives heavily infected plants in the field. Hence a sound reliable, sensitive, early virus detection/diagnostic technique is urgently required for identifying virus free planting material and exchange of germplasm among different countries. Resistance to SPFMV disease can be developed in Indian sweet potato cultivars by introducing coat protein and 3' NTR sequences of SPFMV as done in transgenic *Nicotiana benthamiana*. Novel approaches like protoplast technology can also be employed to develop resistance against SPFMV in sweet potato. Introduction of coat protein and antisense RNA genes of SPFMV into sweet potato cultivars has been developed at Tuskegee University, USA.

Lyctic peptide genes - Shiva I and SB-37 have been introduced into elite sweet potato varieties for resistance to bacterial and fungal diseases in USA. Protoplast technology can be utilized in developing resistance for sweet potato virus diseases.

Among the pest and diseases that attack sweet potato, the weevil *Cylas formicarius* is the most serious and ubiquitous one. Resistant lines against SPW are not available anywhere in the world. Inducing resistance in sweet potato either by introducing Bt cry gene selective to weevil or plant derived proteinase inhibitors, peroxidase etc is the need of the hour. IPM technology with 'pheromone trap' in which a septa impregnated with BARC-SPW pheromone is widely tested for its efficacy and accepted by farmers. An effort is needed for commercial production of pheromone and kairomone on pilot plant basis and distribution to sweet potato farmers through Government / NGO agencies. Incorporation of serine protease inhibitor such as soyabean trypsin inhibitor (SBTI) from soyabean and Cowpea trypsin inhibitor (CpTI) from cowpea for creating host resistance in sweet potato to sweet potato weevil.

Arthropod pest management research should concentrate on a more comprehensive understanding of the biodiversity in root and tuber cropping systems. Another important issue is that of plant quarantine measures to prevent the movement of pests, especially into Asia, in light of what occurred in Africa (Frison and Feliu, 1991). There are many exotic diseases and pests found in the neotropics (listed in chapter VIII) that could potentially cause severe crop losses in India and we need stringent plant quarantine especially in the context of liberalization of trade policies. Moreover, what is considered a secondary pest in the neotropics at present could become a major pest outside its center of origin, where native natural enemies or tolerant germplasm are not available.

A successful integrated pest management (IPM) program will depend on having effective, environmentally sound, low-cost pest management technologies available to farmers in developing countries. Biotechnology tools presently available offer the diagnostic kits and potential to develop improved pest-resistant varieties and to enhance the effectiveness of natural control organisms, including parasitoids and entomopathogens. The new generation of genetic pest management technologies presently being integrated with traditional IPM offers alternative technologies for controlling stemborers, leaf-cutting ants, grasshoppers, whitegrubs, and other difficult-to-control pests. Research activities in these areas are needed underway and should be available to farmers in the near future.

3.5 Abiotic factors

Lines with tolerance to salinity in soil can be isolated in sweet potato which can be used in reclamation of waste lands and coastal areas affected with climatic catastrophes like tropical cyclones or tsunami. Sweet potato is the most suited crop for such purposes as it matures within 90 to 120 days. Saline tolerant lines have been isolated in sweet potato using cell suspension cultures. Somatic hybridization of sweet potato with the wild species *Ipomoea perscarpae* and *in vitro* identification of somaclonal variants having salinity tolerance can be effectively used for wider cultivation of this crop in India for better utilization of land and food production.

Sweet potato is a photo induced crop. There are varieties which respond to storage root production either during short days (short day induced) or during long days (long day induced). Basic studies with different wave lengths of light and green house technologies can lead to generation of information on this important aspect that controls sweet potato production.

Growth and tuber production are affected when these tropical tuber crops are cultivated in areas with prolonged and extreme winter conditions due to low soil temperature. Incorporation of genes coding for antifreeze proteins or gene for superoxide dismutase inducing cold/freezing tolerance and other such genes expressing tolerance to abiotic stress like extreme low and high temperatures will help in the spread of these crops in non- traditional areas resulting in higher production.

Round the year cultivation of seasonal tuber crops like coleus and photosensitive crop like sweet potato can be practised in green house condition for higher returns as has been done for many vegetables in India and abroad. Sweet potato can be grown in hydroponics. Aeroponics is still a hi-tech protected system which can be tried for tuber crops and using this technology tuber crops can be cultivated any where in the world. NASA is working on assured production systems for astronauts especially in sweet potato.

4. Site Specific Integrated Nutrient Management

Site specific integrated nutrient management is a site-based approach for managing the nutrient needs of crops. Main idea is feeding the crops with required nutrients as and when needed for maximizing the crop yield and at the same time reducing nutrient wastage and environment pollution and cost of production.

Conventional recommended fertilizers will not hold good for all areas/ region. There is spatial and temporal variability of crop requirement of nutrients and soil supplies of these nutrients. Precision management of nutrient application as per the variation within the fields is a very important strategy for sustainable agriculture.

5. Participatory Technology Development

A promising approach to overcome slow technological diffusion and to increase productivity in small-scale production systems combines grass-roots community development with a holistic approach to crop management and diversification of markets. In some traditional cassava-producing areas, increased productivity has been tied to product diversification in local and regional markets to provide an acceptable return on farmers' investments (CIAT, 1992; Wheatley and Chuzel, 1992). This has been done most successfully by small-scale drying and chipping of cassava for animal feed (Perez-Crespo, 1991). By opening new markets, profitability of cassava production increases, and pest control becomes economically viable through proper agronomic and phytosanitary practices, use of resistant cultivars, and biological and cultural control methods (Bellotti *et al.*, 1992).

Participatory technology development for different crops have emphasized integration of farmers' knowledge and their participation in experimentation (Ashby, 1990, Ashyby *et al.*, 1995; Bentley, 1990; Bentley and Andrews, 1991; Braun *et al.*, 1993). For example, improved cultivars must meet highly localized preferences for characters such as root color, texture, and growth habit. Attempts to incorporate pest resistance or higher yield must consider local preferences as selection criteria through farmer participation in germplasm evaluation (Fukuda *et al.*, 1997).

Key to the success of participatory research are identification of needs by farmers, inclusion of farmers in field research and validation, training of research and extension personnel in participatory methods, and integration of production with expanded markets.

Modern technologies such as genomics and genetic transformation can help to tide over some of the constraints and do 'cut and paste' genes from interesting gene pools, most especially those from wild related species, into modern cultivated cultivars.

Advances have been made in developing some of the core biotechnologies for cassava and sweet potato. However, progress has been slow and piecemeal. New, bold initiatives are required if the great potential of biotechnology is to benefit producers and consumers within the next 10-20 years.

The task at hand is too large and broad based and requires a coordinated and multifaceted approach.

6. Industrial Application

It has been shown that both elastic modulus and ultimate tensile strength of low density polyethylene (LDPE) has been increased on incorporating wild cocoyam *(Caladium bicolor)* and wild sweet yam *(Dioscorea dumetorum)* starches to it. Densities of the composites increase with time in composting environment indicating that the starches can be used as a biodegradable filler. This is a significant advantage judging by the concern being shown in the environmental problems posed by polyolefin films disposal. Therefore both starches have been shown to be dual purpose additive for LDPE (Umoren, 2005).

Cassava flour (even without processing into pure starch) can be a substitute to agar and improve the growth of plant shoots in tissue culture. If cassava flour can provide both the gelling and carbon source requirements in the tissue culture medium then it can substantially reduce the medium cost (Maliro and Lameck, 2004).

Sweet potato can be explored in the preparation of edible vaccine by intriguing technology, not only for human but also for farm animals.

7. Research gaps and Programmes have been identified for strengthening Tuber and Root Crops Research in India (Palaniswami and Shirly, 2006)

I Genetic Resources of Tropical Root and Tuber Crops

- Identifying the unexplored areas in the different regions based on the geographical map of the area and exploring the biodiversity of different species of tropical root and tuber crops (TRTC).
- Characterizing and cataloguing of genetic stock after eliminating duplicates.
- Molecular characterizaiton of important genetic stocks under each species of TRTC for its higher yields and for its biotic and abiotic stress tolerance/resistance.
- Net work researching on *in vitro* conservation of genetic stock of different species of TRTC.
- Investigation on hitherto unexplored minor tuber crops suitable for each region.

II Varietal Development and Improvement

- Sharing promising varieties/ accessions among different centres for research work.
- Identification / development of biofortified accessions/varieties especially on β-carotene, protein, minerals such as Zn, Fe, Cu, Mn etc.
- Developing short duraton varieties with ideal plant type suitable for each region with economic traits.(Suitable varieties of cassava for sago/ starch industires ; suitable sweet potato for noodles, suitable varieties of TRTC for chips making, cassava varieteis with profuse foliage for ericulture, determinate (dwarf) yam etc.).

III Organic Farming

- Organic farming with reduced input of synthetic fertilizers and pesticides.
- Biofertilisers-identification of region specific suitable microbes for enhancing nutritional efficiency and as growth promotors and disease suppressors.

- Effective utilization of indigenious micro organisms suitable for each region for better management of tuber crops.
- Use of liquid biofertilizers in TRTC.
- *In situ* production of organic manure and vermicompost.

Organic and biological farming is very much suitable for tropical root and tuber crops.

IV Tuber crops based cropping system

- Inter cropping in tuber crops and tuber crops as inter crops in plantation/ orchards.
- Developing modalities for micro irrigation and fertigation system for efficient use of water and nutrients for different TRTC.

V Pest and Disease management strategies

- IPM and IDM for major pests and diseases including use of resistant lines, biocontrol agents and chemecology based components (Allemones/ pheromones).
- Survey and surveillance for new pests and diseases.

VI Utilization and Scope for rural employment

- Exploiting the pharmaceutical and nutraceutical values of different TRTC for the benefit of the consumers (like β-carotene is a very good source of vitamin 'A' and β-carotene rich cassava/sweet potato can be encouraged among vitamin 'A' deficient population. Similarly elephant foot yam is very good for piles and rejuvenation etc.). Popularizing orange fleshed varieties of cassava/sweet potato in backward areas of Bihar, West Bengal, Jharkhand, Uttar Pradesh, Madhya Pradesh, Andhra Pradesh, Maharashtra and North-Eastern states.
- Encouraging small scale entrepreneurship through TRTC *vis a vis*- sago, starch, noodle like food products.
- Ericulture using cassava leaves (profusely leaf bearing cassava varieties can be identified for ericulture) for producitn of eri silk.
- Storage of tubers against post harvest deterioration, due to physiological and due to pests and diseases and improving shelf life of tubers.
- Utility of leaves of TRTC for food and feed.
- Biopesticides from yam bean and other TRTC.

VII Extension Education and Awareness programmes

- Area expansion under TRTC in different states especially non-traditional and under /unutilized areas.
- Popularization of different tuber crops through farmers meet, seminar, exhibition etc., and educating the nutritive, nutraceutical and pharmaceutical values of different TRTC for nutrition, health and wellness.
- Dissemination of knowledge on the utilization of different TRTC, small scale/cottage industries of starch, gums, noodles, ericulture, biopesticies etc.
- Developing a network data base of different centres/states with special reference to TRTC.
- Front line demonstration for transfer of technologies developed for each region.
- Human resource development-Training in molecular mapping and molecular marker- based QTL mapping.
- Off campus and on campus training for tuber crops farmers.

VIII Planting Material Production and Distribution

Organizing supply of healthy planting materials of different species of TRTC utilizing minisetts/micro propagation technologes through the AICRPTC centres in liason with Agriculture/Horticulture Departments/ KVKs.

Conclusion

India has wide and varied agroclimatic conditions and regions. Agro-Trade assumes important dimension especially in the context of liberalization of trade policies. More than 65 crores of the population depend on agriculture. The revenue from agriculture has to be enhanced. Sweet potato has been identified as one of the super foods and has found a place even in space travellers' food. Techniques are being perfected for its production in hydroponics and even in aeroponics.

Root and tuber crops also help in alleviating hunger and poverty, especially in endemic hunger caused by malnutrition, hidden hunger due to deficiences of vitamins and minerals and transient hunger caused by drought, floods and other natural calamities. In fact these crops can be termed as life saving crops.

Tropical root and tuber crops offer tremendous scope for entrepreneurial opportunities, employment potentialities and revenue generation from a wide range of value added food and its products, feeds for cattle, pigs, poultry and fishes, biofuels, pharmaceutical and nutraceutical products, industrial starch, biodegradable plastics and many other products.

References

Ashby, J.A. 1990. Small farmers' participation in the design of technologies. In Agroecology and Small Farm Development, ed. MA Altieri, SB Health, 23:245-55. Boca Raton. FL:CRC Press.

Ashby, J.A., Gracia T, Guerrero MDP., Quiros C.A., Roa, J.I. *et al.* 1995. Institutionalising Farmer Participation in Adaptive Technology Testing with the 'CIAL.' London: Overseas Dev. Inst. Agric. Res. Ext. Netw., Pap. 57. 43 pp.

Bellotti, A.C., Arias, B., Guzman, O.L. 1992. Biological control of the cassava hornworm *Erinnyis ello* (Lepidoptera: Sphingidae). *Fla. Entomol.* 75:506-15.

Bentley, J.W. 1990. Conocimiento y experimentos espontaneos de campesinos hondurenos sobre el maiz muerto. *Manejo Integr.* Plagas 17:16-26. (In Spanish).

Bentely, J.W., Andrews, K.L. 1991. Peasants, pests and publications: anthropological and entomological views of an integrated pest management program for small-scale Honduran farmers. *Human Organ.* 50:113-22.

Braun, A.R.. Bellotti, A.C, Lozano, J.C. 1993. Implementation of IPM for small-scale cassava farmers. In: Crop Protection Strategies for Subsistence Farmers, ed. MA Altieri, pp. 103-15. Boulder. CO:Westview.

Centro Internacional de Agricultura Tropical. 1992. *Annual Report Cassava Program, 1987-1991.* Cali, Colomb.: Cent. Int. Agric. Trop. 475 pp.

Frison, E.A., Feliu, E. eds. 1991. FAO/IBPGR Technical Guidelines for the Safe Movement of Cassava Germplasm. Rome: Food Agric. Org. United Nations, Int. Board Plant Genet. Res. 48 pp.

Fukuda, W.M.G., Magalhaes, J.A., Cavalcanti, J., Pina, P.R., Tavares, J.A., *et al.* 1997. Pesquisa Participativa em Melhoramento de Mandioca: Uma experiencia no Semi-adrido do Nordeste do Brasil. Cruz das Almas, Bahia, Brasil: Empr. Bras. Pesqui, Agropec.- Cent. Nac. Pesqui. Mand. Fruticult. 46 pp. (In Portuguese).

Maliro, F.A. M. and Lameck, G. 2004. Potential of cassava flour as a gelling agent in media for plant tissue cultures. *African. J. Biotechnology.* 3(4): 244-247.

Palaniswami, M.S. and Shirly Raichal Anil. 2006. Region specific technologies for tropical root ant tuber crops in India. All India Coordinated Research Project on Tuber Crops, CTCRI, Thiruvananthapuram. 67 pp.

Perez-Crespo, C.A. ed. 1991. Integrated Cassava Projects. Cali, Colomb.: Cent. Int. Agric. Trop. 242 pp.

Scott, G.J., Best, R., Rosegrant, M. and Bokanga, M. 2000. Roots and tubers in the global food system. A vision statement to the year 2020. CIP / CIAT / IFPRI / IITA / IPGRI, Lima, Peru. 111 pp.

Umoren, S. A. 2005. Caladium bicolor and wild *Dioscorea dumetorum* starches as dual purpose polymer additive. *J. Applied Sciences & Environmental Management*, 9:75-78.

Wheatley, C.C., Chuzel, G. 1992. Cassava *(Manihot esculenta)* use as a raw material. (*In:*) Encyclopedia of Food Science, Food Technology and Nutrition, ed. R. Macrae, R.K. Robinson, M.J. Sadler. pp. 734-43. London: Academic.

❑❑❑

Appendix

Descriptors

The IBPGR now uses the following definition in genetic resources documentation

i) Passport data (accession identifiers and information recorded by collectors);

ii) Characterization (consists of recording those characters which are highly heritable, can be easily seen by the eye and are expressed in all environments);

iii) Preliminary evaluation (consists of recording a limited number of additional traits thought desirable by a consensus of users of the particular crop

Characterization and preliminary evaluation will be the responsibility of the curators, while further evaluation should be carried out by the plant breeder. The data from further evaluation should be fed back to the curator who will maintain a data file.

I. Sweet potato

The descriptor list for sweet potato specifies the growth stage in which certain descriptors should be recorded. The growth stages used are listed below:

1. Seedlings (from true seed)
2. Sprouts (from storage roots)
3. Vine cuttings

4. Established in field (beginning to grow)
5. Approximately 50% leaf canopy
6. Approximately 100% leaf canopy
7. One week to normal harvest
8. At harvest
9. Harvested roots

1. Accession Data

1.1 This number serves as a unique identifier for accessions and is assigned by the curator when an accession is entered into his collection. Once assigned this number should never be reassigned to another accession in the collection. Even when an accession is lost, its assigned number is still not available for re-use. Letters occur before the number to identify the gene bank.

1.2 Scientific Name

1.2.1 Genus

1.2.1 Species

1.2.3 Subspecies

1.3 Date of Last Multiplication or Regeneration

Year of last multiplication or regeneration

1.4 Donor Name

1.5 Donor Number

1.6 Any other Names or Numbers Associated with the Accession

E.g. USDA Plant introduction number, common name etc.

(Not collection number, see 2.2)

1.7 Type of Maintenance

Advanced cultivars, primitive cultivars and breeding lines with valuable gene combinations should be maintained vegetatively

1. Vegetative
2. True seed
3. Both

1.8 Reference to Publicaitons

4. Established in field (beginning to grow)
5. Approximately 50% leaf canopy
6. Approximately 100% leaf canopy
7. One week to normal harvest
8. At harvest
9. Harvested roots

1. Accession Data

1.1 This number serves as a unique identifier for accessions and is assigned by the curator when an accession is entered into his collection. Once assigned this number should never be reassigned to another accession in the collection. Even when an accession is lost, its assigned number is still not available for re-use. Letters occur before the number to identify the gene bank.

1.2 Scientific Name

1.2.1 Genus

1.2.1 Species

1.2.3 Subspecies

1.3 Date of Last Multiplication or Regeneration

Year of last multiplication or regeneration

1.4 Donor Name

1.5 Donor Number

1.6 Any other Names or Numbers Associated with the Accession

E.g. USDA Plant introduction number, common name etc.

(Not collection number, see 2.2)

1.7 Type of Maintenance

Advanced cultivars, primitive cultivars and breeding lines with valuable gene combinations should be maintained vegetatively

1. Vegetative
2. True seed
3. Both

1.8 Reference to Publicaitons

2. Collection Data

2.1 Collecting Institute

Institute or person collecting the original sample

2.2 Original Number Assigned by Collector of the Sample

2.3 Date of Collection of Original Sample

Expressed as day/month/year. e.g. 10 February 1981 as 100281

2.4 Country of Collection

Use the three letter abbreviations supported by the Statistical Office of the United Nations. Copies of these abbreviations are available from the IBPGR Secretariat.

2.5 Latitude of Collection Site

Degrees and minutes suffixed by N or S, eg. 1030 S

2.6 Longitude of Collection Site

Degrees and minutes suffixed by E or W, eg. 7625 W

2.7 Location of Collection Site

Number of kilometers and direction from nearest town, village or map reference point.

2.8 Altitude of Collection Site

Elevation above sea level, in metres

2.9 Collection Source

1. Field
2. Market
3. Farm store
4. Agricultural institute

2.10 Type of Sample

1 Vegetative
2 True seed
3 Both

2.11 Status of Sample

1 Wild
2 Weedy
3 Breeders line
4 Primitive cultivar
5 Advanced cultivar

2.12 Other Notes from Collector

Some collectors will record ecological and soil information, whether a cultivated crop was irrigated, season of sowing, topography of land, associated crops or wild plants, etc.

3. General

3.1 State Characterization and Preliminary Evaluation

3.2 Year of Characterization and Preliminary Evaluation

3.3 Evaluator (S) Name and Address

4.4 Planting Date

Expressed as day/month/year

4. Characterization

Characterization Growth Stage

4.1 Twining 5-7

0 Non-twining
+ Twining

4.2 Plant Type 5,6

3 Compact
7 Spreading

4.3 Vine Growth Rate 5

3 Slow
5 Intermediate
7 Fast

4.4 Vine Internode Length 6

3 Short
5 Intermediate
7 Long

4.5 Vine Pigmentation 5-7

3 Green
5 Moderately purple
7 Purple

4.6	Vine Tip Pubescence	5-7
	0 None	
	3 Sparse	
	5 Moderate	
	7 Heavy	
4.7	Mature Leaf Lobing	7
	0 None	
	3 Slight	
	5 Moderate	
	7 Deep	
4.8	Mature Leaf Size	7
	3 Small	
	5 Medium	
	7 Large	
4.9	Mature Leaf Colour	5-7
	1 Yellow	
	2 Yellow-green	
	3 Green	
	4 Purplish-green	
	5 Purple	
4.10	Immature Leaf Colour	5-7
	1 Yellow	
	2 Yellow-green	
	3 Green	
	4 Purplish-green	
	5 Purple	
4.11	Abaxial Leaf Vein Colour	5-7
	1 Yellow	
	2 Green	
	3 Purple spotting	
	4 Half of main rib purple	
	5 Entire main rib purple	
	6 All veins purple	

4.12	Petiole Length	6-7
	3 Short	
	5 Intermediate	
	7 Long	
4.13	Petiole Pigmentaton	6-7
	1 Green	
	2 Moderately purple	
	3 Purple	
4.14	Storage Root Skin Colour	8
	1 White	
	2 Yellow	
	3 Brown	
	4 Red	
	5 Purple	
4.15	Storage Root Skin Colour Intensity	8
	3 Pale	
	5 Intermediate	
	7 Dark	
4.16	Storage Root Flesh Colour	8-9
	1 White	
	2 Yellow	
	3 Orange	
	4 Purple	
4.17	Storage Root Flesh Colour Intensity	8-9
	3 Pale	
	5 Intermediate	
	7 Dark	
4.18	Flowering Habit	5,6
	0 None	
	3 Sparse	
	5 Moderate	
	7 Profuse	

4.19	Flower Colour	5-7
	1 White	
	2 White limb with purple throat	
	3 Purple	
	4 Other (specify)	
4.20	Flower Length	5-7
	Expressed in centimeters	
	Average of ten typical flowers	
4.21	Equality of Sepal Length	5-7
	1 Outer two shorter	
	2 Equal	
4.22	Number of Sepal Veins	5-7
	1 0-1	
	2 2	
	3 3-5	
	4 >5	
4.23	Sepal Shape	5-7
	1 Ovate	
	2 Elliptic	
	3 Obovate	
	4 Oblong	
	5 Lanceolate	
4.24	Sepal Apex	5-7
	1 Acute	
	2 Obtuse	
	3 Acuminte	
	4 Caudate	
4.25	Seed Capsule Set	6,7
	0 None	
	3 Sparse	
	5 Moderate	
	7 Heavy	

5. Prelimary Evaluation

The descriptors in this section should be recorded on commercial size storage roots.

5.1 Storage Root Length — 8,9
Average length of ten storage roots in centimeters

5.2 Storage Root Diameter — 8,9
Average of largest diameter of ten storage roots in centimeters

5.3 Number of Storage Roots Per Plant — 8,9
Average of ten plants (actual number)

5.4 Storage Root Shape Variability — 8,9
1 Uniform
9 Highly variable

6. Storage Root Data

	Storage root data	Growth stage
6.1	Storage Root Dry Matter Percentage	9
6.2	Storage Root Nitrogen Percentage	9
6.3	Storage Root Fibre Percentage American Official Analytical Chemistry on cooked roots of at least 3 cm diameter	9
6.4	Storage Root Starch Percentage	9
6.5	Storage Root Water Soluble Sugar Percentage	9
6.6	Storage Root Carotene Content In milligram per 100 g fresh weight	9

6.7 Keeping Quality of Storage Roots
3 Poor
5 Medium
7 Good

6.8 Sprouting Ability
3 Poor
5 Medium
7 Good

7. Pest Reaction

Coded on a 1-9 scale, where 3 = Resistant and 7 = Susceptible. Each pest needs a separate descriptor. The acronym of the institute supplying the information on a specific pest should be appended to the score of the pest reaction – eg. If the International Institute of Tropical Agriculture Scores an accession as Very Resistant to Tobacco wireworm then the code for descriptor 7.1.3 would be : 1 IITA

7.1 Insect Susceptibility

7.1.1 Sweet potato weevil (*Cylas* spp.)

7.1.2 Southern potato wireworm (*Conoderus falli* Lan.)

7.1.3 Tobacco wirework (*Conoderus vespertinus* Fabricus)

7.1.4 Wireworm (*Melanotus communis* Gyll)

7.1.5 Sweet potato flea beetle (*Chaetocnema confines* Crotch.)

7.1.6 Banded cucumber beetle (*Diabrotica balteata* Le Conte)

7.1.7 Spotted cucumber beetle
(*Diabrotica undecimpunctata howardi* Barber)

7.1.8 Flea beetle (*System frontalis* Fabricus)

7.1.9 Pale-striped flea beetle (*Systena blanda* Melsheimer)

7.1.10 Elongate flea beetle (*Systena elongate* Fabricus)

7.1.11 Weevil (*Euscepes post fasciates* DeGeer).

7.1.12 Grubworm (*Phyllophaga ephilida* Say)

7.1.13 Grubworm (*Plectris aliena* Chapin)

7.1.14 Other (specify)

8. Disease Reaction

Coded on a 1-9 Scale

1 Immune

2 Resistant (Hypersensitive

3 Resistant (High tolerance)

5 Resistant (Moderate tolerance)

7 Susceptible

9 Lethal

Each disease needs a separate descriptor. The acronym of the institute supplying the information on a specific disease should be appended to the score of the disease reaction - *eg.* if the Asian Vegetable Research and Development Center scores an accession as Immune to Black Rot, then the code for descriptor 8.1.3. would be : 1 AVRDC

8.1 Susceptibility to Fungal Diseases

8.1.1 Fusarium Wilt or Stem Rot *(Fusarium oxysporum f. op. batatus* (Wr.)

8.1.2 Sclerotial Blight and Circular Spot *(Sclerotium rolfsii* Sacc.)

8.1.3 Black Rot *(Ceratocystis fimbriata* Ell. & Halst)

8.1.4 Surface Rot (Fusarium) *(Fusarium oxysporium* Schlect.)

8.1.5 Fusarium Root Rot *(Fusarium solani* (Mart.) Appel & Wr.)

8.1.6 Scurf *(Monilochaetes infuscans* EII. & Halst. ex. Harter)

8.1.7 Soft Rot *(Rhizopus stolonifer* (Ehr. ex. Fr.) Lind.), other Rhizopus species

8.1.8 Java Black Rot *(Diplodia tubericola* (Ell. & Ev.) Taub.)

8.1.9 Dipanthe Dry Rot *(Diaporthe batatatis* Harter & Field)

8.1.10 Scab or Spot Anthracnose *(Elsinoe batatas* (Saw.) Viegas & Jenkins)

8.1.11 Phyllosticta Leaf Blight *(Phyllosticta batatasI* (Thuem.) Cbe.)

8.1.12 Cercospora Leaf Spot *(Cercospora batatae* Zimm., other *Cercospora* species)

8.1.13 White Rust *(Albugo ipomoeae - panduratae* (Schw.) Swing.)

8.1.14 Foot Rot *(Plenodomum destruens* Harter)

8.1.15 Charcoal Rot *(Macrophomina phaseoli* (Maubl.) Ashy)

8.1.16 Septoria Leaf Spot *(Septoria bataticola* Taub.)

8.1.17 Other (specify)

8.2 Susceptibility to Bacterial Diseases

8.2.1 Pox or Soil Rot *(Streptomyces ipomoea* (Person & W.J. Martin) Waksm & Henrici

8.2.2 Bacterial Stem and Root Rot (*Erwinia chrysanthemi* Dupes)

8.2.3 Other (specify)

8.3 Susceptibility to Vir A Diseases

8.3.1 Feathery mottle - common strain

8.3.2 Feathery mottle - Russet crack strain

8.3.3 Feathery mottle - Internal cork strain

8.3.4 Mild mottle virus

8.3.5 Vein mottle virus

8.3.6 Sweet potato virus disease (SPVD complex)

8.3.7 Other (specify)

8.4 Susceptibility to Mycoplasma Diseases

8.4.1 Witches broom

8.4.2 Other (specify)

8.5 Susceptibility to Diseases Caused by Nematodes

8.5.1 Southern root-knot nematode (*Meloidogyne incognita* (Kofoid & White Chitwood)

8.5.2 Javanese (Tropical) root-knot nematode (*Meloidogyne javanica* (Treub.) Chitwood)

8.5.3 Northern root-knot nematode (*Meloidogyne hapla* Chitwood)

8.5.4 Reniform nematode (*Rotylenchulus reniformis* Linford and Oliveira)

8.5.5 String nematode (*Belonolaimus longicaudatus* Rau)

8.5.6 String nematode (*B. gracilis* Steiner)

8.5.7 Brown Ring rot (*Ditylenchus dipsaci* (Kuhn) Filipjev)

8.5.8 Root lesion nematode (*Pratylenchus coffeae* (Zimmernann) Goodey)

8.5.9 Other (specify)

II. Cassava

1. *Plant type :* Recorded at the time of harvest at 10th month: EB (Erect Branching); ENB (Erect North-Branching); SB (Spreading and Branching)

2. *Nature of stem :* Recorded from mature plant: St (Straight) Z (Zig-zag)
3. *Leaf :* B (Broad); N (Narrow)
4. *Rind :* Cr (Cream); LPi (Light Pink) ; Pi (Pink)
5. *Flesh :* W (White); VLY (Very Light Yellow); LY (Light Yellow); Y (Yellow)
6. *Taste :* Tested organoleptically at the time of harvest: NB (Non-Bitter); Sl Bi (Slightly Bitter); Bi (Bitter)
7. *Colour of stem :* Individually recorded as per colour chart at 8th month and later broadly classified under 8 groups: (DGy (Dark Grey); Gy (Grey); LGY (Light Grey); LBr (Light Brown); RBr (Reddish Brown); BrY (Brownish Yellow); BlBr (Blackish Brown); Bl (Blackish)
8. *Young stem :* Gr (Green); VLStr (Very Lightly Streaked); LStr (Lightly Streaked); Str (Streaked); HStr (Highly Streaked)
9. *Shoot :* Gr (Green); VLS (Very Light Sepia); LS (Light Sepia); S (Sepia); DS (Dark Sepia)
10. *Stipule :* Gr (Green); Lpur (Light Purple); Pur (Purple)
11. *Leaf :* Gr (Green); LGr (Light Green)
12. *Petiole :* Recorded at 5th month: Gr (Green); VLR (Very Light Random); LR (Light Random); R (Random) ; Pur (Purple)
13. *Dorsal vein :* Gr (Green); VLT (Very Lightly Tinged); LT (Lightly Tinged); T (Tinged)
14. Ventral vein : Gr (Green); VLT (Very Lightly Tinged); LT (Lightly Tinged); T (Tinged
15. *Shape : Tuber :* Cy (Cylindrical); Fu (Fusiform)
16. Yield : Yield per plant (kg)
17. Neck length : Neck length of tuber (cm)
18. Colour of skin : LBr (Light Brown); Br (Brown); DBr (Dark Brown)
19. HCN (μg/g F.W) : Determined by colorimetric method using Sodium picrate impregnated paper: Hydrocyanic acid - microgram per gram (on fresh weight basis)
20. Starch (%) : Determined by hydrolysis to reducing sugars followed by titrimetirc estimation of sugar using alkaline potassium ferricyanide solution and methylene blue as indicator (on fresh weight basis)

21. Male flower perianth : Gr (Green); LGR (Light Green); VLGr (Very Light Green); LT (Lightly Tinged); T (Tinged); HT (Highly Tinged)
22. Male flower disc : Y (Yellow); Or (Orange): Pur (Purple)
23. Pollen fertility (%) : Determined as per stainability with acetocarmine:

    H (High = above 71%)

    M (Medium = 31-70%)

    L (Low = upto 30%)

    MS (Male Sterile)
24. CMD : Recorded at 2nd and 4th months: Cassava Mosaic Disease : P (Present): A (Absent)
25. BLSD : Recorded at 5th month: Brown Leaf Spot Disease: T (Tolerant); S (Susceptible); HS (Highly Susceptible)
26. Spider mite : Recorded during peak period of infestation: L (Low); M (Medium); H (High)

III. *Dioscorea* spp.

Descriptors Used for Characterisation

1. Accession Number
2. Donor number/vernacular name
3. Place of collection
4. Country
5. Vigour

    3 Low

    5 Intermediate

    7 High
6. First leaf emergence
7. Emerging stem colour

    1 Green

    2 Purplish green

    3 Brownish green

    4 Dark brown

5 Purple

8. Young stem colour

   1 Green

   2 Purplish green

   3 Brownish green

   4 Dark brown

   5 Purple

9. Mature stem colour

   1 Green

   2 Purplish green

   3 Brownish green

   4 Dark brown

   5 Purple

   6 Dark purple

10. Internodal length - Mean of five measurements recorded at 20 days after emergence

11. Density of spines

    3 Few

    4 Many

    5 Purple

12. Petiole colour - young leaf

    7 Green

    8 All green with light purple middle portion

    9 All green with dark purple middle portion

13. Petiole colour - mature leaf

    7 Green

    8 All green with light purple middle portion

    9 All green with dark purple middle portion

14. Petiole length

    Mean of five measurements recorded at 5th month

15. Emerging leaf colour

    1 Whitish

    2 Pale green

    3 Dark green

    4 Purplish green

    5 Purple

    6 Light brown

16. Young leaf colour

    1 Whitish

    2 Pale green

    3 Dark green

    4 Purplish green

    5 Purple

    6 Light brown

17. Mature leaf colour

    1 Whitish

    2 Pale green

    3 Dark green

    4 Purplish green

    5 Purple

    6 Light brown

18. Leaf shape

    1 Oval

    2 Cordate

    3 Cordate long

    4 Cordate broad

5 Sagittate long

6 Sagittate broad

19. Leaf base of overlapping - young leaves

    1 Overlapped

    2 Separate

20. Leaf base overlapping - mature leaves

    1 Overlapped

    2 Separate

21. Leaf length
    Mean of five measurements recorded at fifth month

22. Leaf breadth
    Mean of five measurements recorded at fifth month

23. Total number of tubers
    Mean of four plants for three years

24. Number of marketable tubers
    Mean of four plants for three years

25. Tuber weight (kg/hill)
    Mean of four plants for three years

26. Tuber length (cm)
    Mean of five tubers

27. Tuber girth (cm)
    Mean of five tubers recorded at the middle

28. Tuber shape

    1 Round

    2 Oval

    3 Oval oblong

    4 Cylindrical

29. Tuber texture

    1 Smooth

    2 Rough

30. Roots on tuber surface
    3 Few
    7 Many
31. Tuber skin colour
    1 Light brown
    2 Dark brown
    3 Reddish brown
32. Cooked tuber - Softness
    1 Very soft
    2 Soft
    3 Medium hard
    4 Hard
33. Cooked tuber - Stickiness
    () Non sticky
    1 Sticky
34. Cooked tuber - Sweetness
    () Non sweet
    1 Slightly sweet
    2 Sweet
35. Overall acceptance of cooked tuber
    3 Poor
    5 Average
    7 Good
    8 Excellent
36. Dry matter content (%)
    Fresh weight basis
37. Starch content (%)
    Fresh weight basis
38. Sugar content (%)
    Fresh weight basis

39. Nematode attack (dry rot) on tubers at harvest
    1 Absent
    3 Low
    5 Intermediate
    7 High
40. Presence of scale insects
    1 Absent
    3 Few
    7 Many

IV. *Colocasia Esculenta* / Taro

Definition of Descriptors and Explanation for Abbreviations

Accession Number Numbered serially in the order of Collection/ introduction

1. Germplasm Type Indicating the type of germplasm Collected; expressed as i. Cultivated (Cu), ii. Wild (W), iii. Woody (Wc)
2. Growth Habit Plant type *viz.,* Erect (E); Spreading (S)
3. Type of Stem Indicate underground stem Stolen (St); Sucker (Su); Rhizome (Rh)
4. Tillering Low (LT) : 1-3; Medium (MT) : 4-6, High (HT) : > 6
5. Plant Size Dwarf (D) - < 50 cm; Medium (M) – 50-100 cm
   Tall (T) - > 100 cm
6. Leaf Arrangement Clockwise (C), Anticlockwise (A)
7. Lamina Orientation Erect (E1), Droopy (D1), Cup shaped (C1)
8. Leaf Margin Entire (Em), Undulate (Um)
9. Leaf Margin Colour See Note
10. Leaf Colour – UPPER See Note
11. Leaf Colour – LOWER See Note
12. Sinus Colour Line of confluence of the leaf halves posterior to the midrib on the lower side of the leaf (for colour see note).
13. Leaf Variegation Present (V), Absent (Va)

14. Leaf Length : Breadth Ratio Recorded during the peak period of growth. Average of five plants.

15. Leaf Type Narrow (N1) - > 1.3:1

    Medium (M1) - 1.2 - 1.3:1

    Broad (B1) - 1.1:1

16. Petiole Junction
    Pattern and Colour
    For colour See Note
    (Pet.Jun.P & C)

    Pattern Upper i. Spotted Yellow spot (Ysp) and Purple spot (Psp)
    ii. Asterisk (Ask)
    iii. Solid

17. Pet. Jun. P & C
    LOWER i. Semilunar (S1m)
    ii. Colour spread on to Vein (Pv)

18. Vein Colour
    See Note*
    Upper

19. Vein Colour
    Lower

20. Petiole Colour Top ? of the length
    For colour see note

21. Petiole Colour Middle ? of the length

22. Petiole Colour Base ? of the length

23. Petiole : Leaf Length Ratio Recorded during the peak period of growth. Average of five plants.

24. Petiole Type Short (Sp) - < 1.5:1
    Medium (Mp) - 1.5:2.0:1
    Long (Lp) - >2:1

25. Leaf Sheath Pattern Close (Co), Open (O) (See fig.2)

26. Leaf Sheath Colour See note

27. Petiole : Sheath Ratio Recorded during the peak period of growth. Average of five plants.
28. Sheath Type Short (Sh) - <2:1

    Medium (Mh) - 1.5 – 2:1

    Long (Lh) - 1.5:1
29. Flower Formation Flowering (F)
    Non-flowering (NF)
30. Seed Formation Absent (Nil)
    Present (SS)
31. Maturity Counted from the date of planting to harvest.

    Early (EM) - Below 25 weeks

    Normal (NM) - 26-28 weeks

    Late (LM) - 29 weeks and above
32. Corm Shape Conical (Co), Round (Rs)

    Cylindrical (Cys), Elliptical (Es), Dumb bell (Ds).
33. Corm Size Small (Ss), Medium (Ms), Big (Bs)
34. Cormel Shape Conical (Cs), Round (Rs), Cylindrical (Cys), Elliptical (Es), Club shaped (Ls)
35. Cormel Size Small (Ss), Medium (Ms), Big (Bs).
36. Yield Range Low (Ly) - Below 7 t ha^{-1}
    Medium (My) - 8-15 t ha^{-1}
    High (Hy) - Above 15 t ha^{-1}.
37. Edibility of Petiole Edible (EP)
    Non-edible (NP)
38. Edibility of Leaves Edible (EL)
    Non-edible (NL)
39. Edibility of Cooked Corms Edible (EC)
    Non-edible (NC)
40. Edibility of Cooked Cormels Edible (ER)
    Non-edible (NR)

41. Disease Incidence Recorded for the entire cropping season.

    Virus - Mosaic (MD)

    Phytophthora

    Other - Ghost Spot (GD)

    Shot hole (SD)

*** Note on Colour**

Green (G), Light green (LG), Dark Green (DG), Bluish green (BG), Glaucous green (GG), Grey (Gr), Cream (Cr), Purple (P), Purplish green (Green with purple streaks) (PG).

V. Elephant foot yam (*Amorphophallus paeoniifolius* (Dennst.) Nicolson var. *campanulatus*

(Decne.) Sivad.

Family : Araceae

Synonym : *Amorphophallus campanulatus* Denn.

Hindi name : Jimikand

Common name : Elephant foot yam, Telinga-potato

Other species : *Amorphophallus paeoniifolius* (Dennst.) Nicolson var. *paeoniifolius* (Decne.) sivad

Breeding behaviour : Cross-pollinated

Mode of propagation : Vegetatively through corms and cormels

Habit : Tuberous herb

1.	Pseudo stem height (cm) (Recorded at full foliage stage)	Quantitative
2.	Number of primary partitions	Quantitative
3.	Number of secondary partition/primary partition	Quantitative
4.	Number of primary partition/secondary partition	Quantitative
5.	Length of primary partition (cm)	Quantitative
6.	Breadth of primary partition (cm)	Quantitative
7.	Number of leaflets per primary partition	Quantitative
8.	Length of largest leaflet (cm)	Quantitative
9.	Breadth of largest leaflet (cm)	Quantitative

10. Leaflet nature	1 Entire
	2 Not entire
11. Leaflet margin	1 Undulate
	2 Not undulate
12. Surface of pseudostem	1 Smooth
	2 Tubercled
	3 Spiny
13. Thickness of pseudostem base (mm)	Quantitative
14. Pseudostem colour*	1 Green mottled
	2 Dark green mottled
15. Leaf spot	1 Present
	0 Absent
16. Stalk of leaflet	1 Present
	0 Absent
17. Nature of rachis	1 Winged
	2 Not winged
18. Shape of rachis	1 Angled and curved
	2 Cylindrical
19. Date of senescence	Quantitative
20. Fresh weight of corm/plant (kg) (Recorded at harvest)	Quantitative
21. Corm shape	1 Depressed or globose
	2 Flat and globose
	3 Pyramidal and globose
22. Colour* of corm surface top	1 Black
	2 Dark brown
	3 White
23. Colour* of corm surface bottom	1 Black

	2 Dark brown
	3 White
24. Colour* of flesh	1 Light yellow
	2 Dark yellow
	3 Reddish yellow
25. Height of corm (cm)	Quantitative
26. Diameter of corm (cm) Number of cormels/corm	Quantitative
27. Weight of cormel/plant (g)	Quantitative
28. Length of cormel (cm)	Quantitative
29. Thickness of cormel (mm)	Quantitative
30. Reemergence	1 Present
Regeneration after senescence of mother pseudostem	0 Absent
31. Biotic stress susceptibility	1 Very low or no visible sign of susceptibility
	3 Low
	5 Intermediate
	7 High
	9 Very high
32. Biotic notes	
33. Remarks	

*Royal Horticultural Society. 1966, c. 1986. RHS Colour Chart (ed.1,2), Royal Horticultural Society.

Source : Biodiversity International, Rome.

Suggested Readings

Balagopalan, C., Nayar, T.V.R., Sundaresan, S., Premkumar, T and Lakshmi, K.R. 1999. Tropical Tuber Crops in Food Security and Nutrition. Oxford and IBH, New Delhi. 530 pp.

Chadha, K.L. 2001. Hand Book of Horticulture. ICAR, New Delhi. 1031pp.

Chadha, K. L. and Nair, G.G. 1994. Advances in Horticulture Vol. 8. Tuber Crops, M.P.H., New Delhi 714 pp.

Chandra, S. 1984. Edible Aroids. Clarendon Press, Oxford, UK. 252 pp.

Chris Ballard, Paula Brown, R. Michael Bourke and Tracy Harwood. 2005. The Sweet potato in Oceania: a reappraisal. Ethnology monographs 19, Oceania Monograph 56. The University of Sydney, Australia.

Coursey, D.G. 1967. Yams. London: Longman Green and Co. Ltd. London. 230 pp.

Erman, A. M., Mbulamoko, N.M., Delange, F. and Ahluvalia, R. 1980. Role of cassava in the etiology of endemic goitre and cretinism. Ottawa, Ont., IDRC. 182 pp.

FAO, 1992. Roots, tubers, plantains and banans in animal feeding. (D. Machin and S. Nyvold Ed.). Animal Production and Health paper No. 95. FAO, Rome. 289 pp.

FAO, 2001. The global cassava development strategy and implementation plan. Volume I. Proceedings of the validation forum on the global cassava development strategy. CIAT/CIRAD/IITA/NRI/FAD. FAO, Rome. 70 pp.

Ghosh, S.P., Ramanujam, T., Jos, J.S., Moorthy, S.N and Nair, S.G. 1988. Tuber Crops. Oxford and IBH Publishing Co. Pvt. Ltd. New Delhi. Bombay. Calcutta. 403 pp.

Kay, D.E. (revised by Gooding, E.G.B). 1987. Crop and product digest NP.2-Root Crops, Second Edition. London: Tropical Development and Research Institute, XV & 280 pp.

Kurup, G.T., Palaniswami, M.S., Potty, V.P., Padmaja, G., Kabeerathumma, S. and Pillai, S.V. 1996. Tropical Tuber Crops. Oxford and IBH. New Delhi. 597 pp.

Nair, N.G., Palaniswami, M.S., Rajendran, P.G., Abraham, K. and Potty, V.P. 1990. Recent Advances in Production and Utilization of Tropical Tuber Crops. Special Issue. *J. Root Crops.* ISRC, Trivandrum. 343 pp.

Onwueme, I.C. 1978. Tropical Tuber Crops. John Wiley and Sons Ltd. New York. 234 pp.

Purseglove, J.W. 1968. Tropical Crops : Dicotyledons. 2 Vols. London : Longman. 719 pp.

Purseglove, J.W. 1972. Tropical Crops : Monocotyledons 1 and 2. London : Longman. 334 pp and 335-607 pp.

Scott, G.J., Best, R., Rosegrant, M. and Bokanga, M. 2000. Roots and tubers in the global food system. A vision statement to the year 2020. CIP / CIAT / IFPRI / IITA / IPGRI, Lima, Peru. 111 pp.

Sorenson, Marten. 1996. Yam bean (*Pachyrhizus* DC). Promoting the conservation and use of underutilized and neglected crops. 2. Institute of Plant Genetics and Crop Plant Research, Gatersleben/International Plant Genetic Resources Institute, Rome. 141 pp.

Villareal, R.L. and Griggs, T.D. 1982. Proceedings of International Symposium on sweet potato, Taiwan, AVRDC.

Wang, J. K. 1983. Taro, a review of *Colocasia esculenta* and its potentials. University of Hawaii Press, Honolulu. 400 pp.

Woolfe, J. A. 1992. Sweet potato - an untapped food resource. Cambridge University Press, Cambridge. 643 pp.

❑❑❑

Colour Plates Section

CHAPTER : 2
History, Origin and Geographical Distribution

Cassava field

Sweet potato field

Greater yam field

Lesser yam field

White yam field

Taro field

Contd.....

Contd..... Chapter 2

Swamp taro

Amorphophallus field

Yam bean field

Coleus field

Arrowroot field

Canna field

CHAPTER : 3

Botany, Physiology and Biodiversity

Cassava

Yams

Taro

Elephant foot yam

Canna

Yam bean

Inflorescence

CHAPTER : 6 : Improved Varieties in India

Cassava Varieties

CO-2

H-226

CO-3

Sree Vijaya

TCS-Seedling tuber yield of 'Ambakadan' cross

Sree Prakash

Contd.....

Contd..... Chapter 6

H-165

Carotene rich Brazil variety (Cooked tuber)

Sweet Potato Varieties

CO CIP-1

CO-3

Bidhan Jaganath (90/101)

NDSP-10

Contd.....

Contd..... Chapter 6

Indira Madhur

Indira Naveen

Kamalasundari

Sree Kanaka

NDSP-9

Palgar - 1

Contd.....

Contd..... Chapter 6

Gola local (Check)

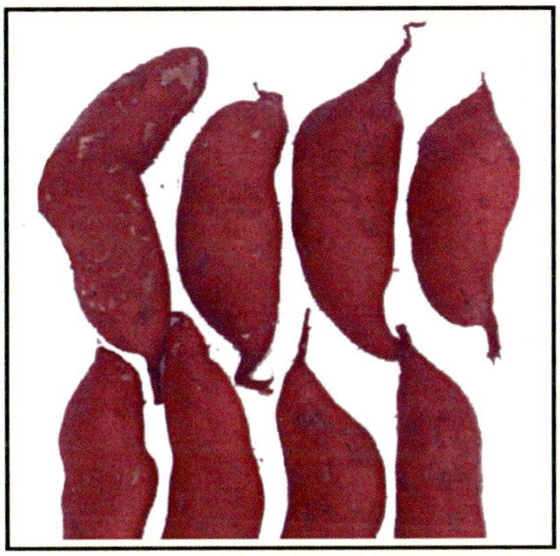

RS - 45

Colocasia taro and bunda varieties

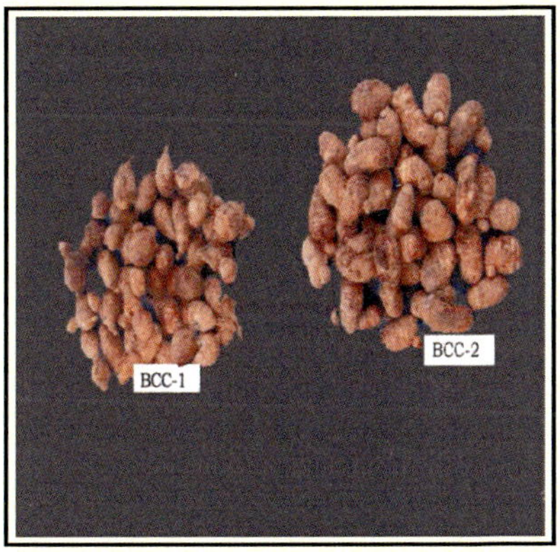

Bidhan Chaithanya Bidhan Joydeb

CO-1

NDB-21 (Bunda)

Jhankuri Sonajhuli

Contd.....

Contd..... Chapter 6

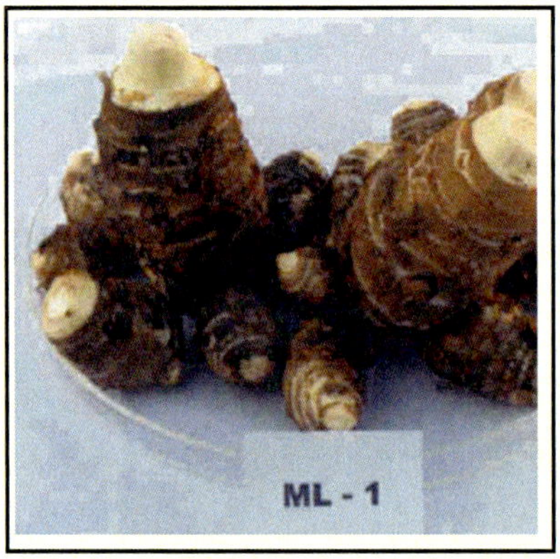

ML-1

Colocasia (Bunda)

Aroids, Country Potato, Yam bean, Yam and Elephant Foot Yam

Colocasia (Swamp taro)

Country Potato

Yam bean (RM-1)

Contd.....

Contd..... Chapter 6

Greater yam (Sree Karthika)

Lesser yam (Konkan kanchan)

White yam

Gajendra

Bidhan Kusum

Contd.....

CHAPTER : 7
Agronomy and Production Technologies

Cassava in coconut gardens

Elephant foot yam in banana plantation

Elephant foot yam in mango orchards

Sweet potato in mango orchard

Cowpea intercrops in elephant foot yam

Green gram in elephant foot yam

Intercropping

CHAPTER : 8

Pests and Diseases : Their Management and Plant Quarantine

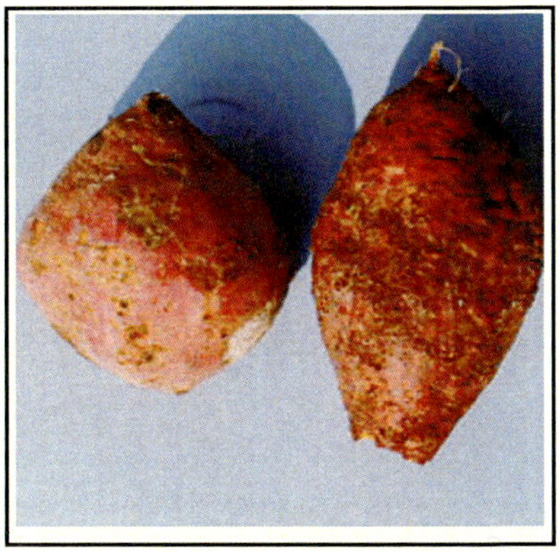

Tuber Damage

Cross-section of the tuber damage

Crown Damage

Pheromone trap

Damage due to sweet potato weevil

Contd..... Chapter 8

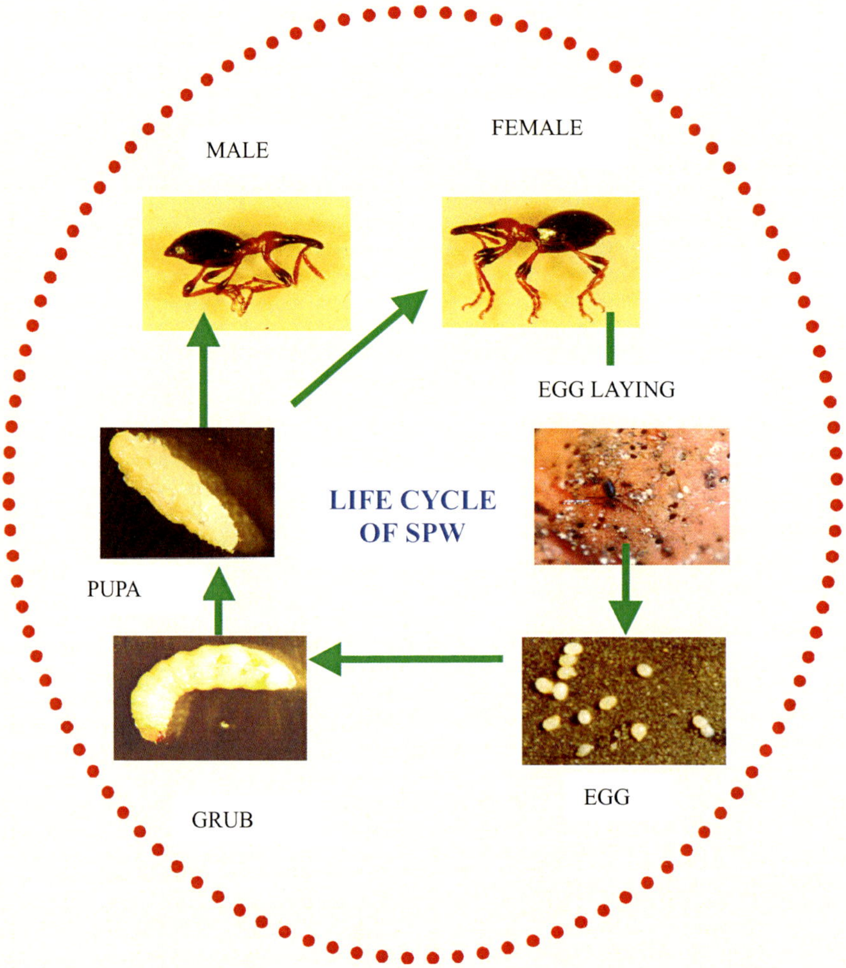

Sweet potato weevil *Cylas formicarius* (Fab.)- Life cycle with different stages

Contd..... Chapter 8

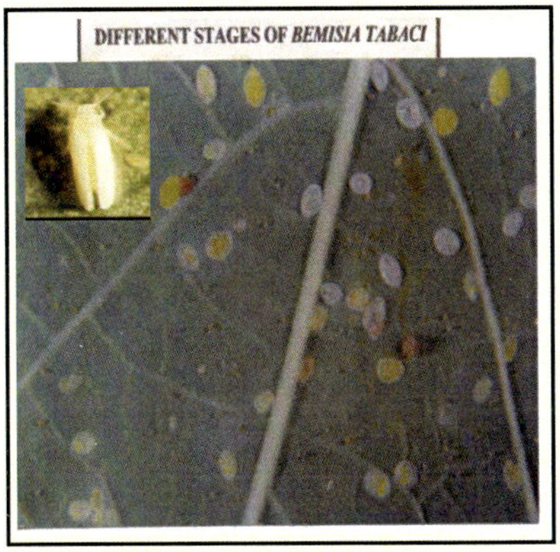

Inset White fly on Cassava

Termite attack on cassava

Spodoptera on taro

Petiole borer in taro

Aphids on taro

Yam scale

Insect Pests

Contd..... Chapter 8

Cassava Mosac Disease (Inset ICMV)

Sweet potato feathery Mottle Virus

Sweet potato witches broom

Sweet potato viral diseases

Dasheen Mosaic

Severe incidence of blight in taro field

Diseases

Contd..... Chapter 8

Elephant foot yam Mosaic

Elephant foot yam Collar Rot

Elephant foot yam Blight

Yam Mosaic

Root Knot Nematode in Coleus

Yam bean Mosaic

Diseases

CHAPTER : 9
Tools and Equipments for Tuber Crops

Thailand cassava chipper

Cassava harvester

CIAT cassava harvester

CHAPTER : 12
Post Harvest Technology and Utilization

Sago Wafers

Cassava Starch Wafers

Biodegradable Plastics